EXS 79

Regulation of Angiogenesis

Edited by I. D. Goldberg
E. M. Rosen

Birkhäuser Verlag
Basel · Boston · Berlin

Editors

Itzhak D. Goldberg, M.D.
Eliot M. Rosen, M.D. Ph.D.
Department of Radiation Oncology
Long Island Jewish Medical Center
270-05 76th Avenue
New Hyde Park, New York, NY 11042
USA

QP
106
.6
R44
1997

A CIP catalogue record for this book is available from the Library of Congress,
Washington, D.C., USA.

Deutsche Bibliothek Cataloging-in-Publication Data
EXS. – Basel ; Boston ; Berlin : Birkhäuser.
   Früher Schriftenreihe
   Fortlaufende Beil. zu: Experientia
Regulation of angiogenesis / ed. by I. D. Goldberg ; E. M. Rosen. – Basel ;
Boston ; Berlin : Birkhäuser, 1997
   (EXS ; 79)
   ISBN 3-7643-5309-0 (Basel ...)
   ISBN 0-8176-5309-0 (Boston)
NE: Goldberg, Itzhak D. [Hrsg.]

The publisher and editors cannot assume any legal responsibility for information on drug dosage
and administration contained in this publication. The respective user must check its accuracy by
consulting other sources of reference in each individual case.

© 1997 Birkhäuser Verlag, P.O. Box 133, CH-4010 Basel, Switzerland
Printed on acid-free paper produced from chlorine-free pulp. TCF ∞
Printed in Germany
ISBN 3-7643-5309-0
ISBN 0-8176-5309-0

9 8 7 6 5 4 3 2 1

Dedicated to the memory of my beloved father
Aryeh Goldberg
a man of uncompromising integrity and valor,
whose love and guidance I will always cherish.

Itzhak D. Goldberg

Contents

Angiogenesis inhibition

*Regulation of angiogenesis by cell-matrix cell-cell and other
interactions*

Regulation of Angiogenesis
ed. by I.D. Goldberg & E.M. Rosen
© 1997 Birkhäuser Verlag Basel/Switzerland

Introduction

Angiogenesis, the formation of new blood vessels from pre-existing micro-vasculature, occurs normally during physiologic and reparative processes. However, lack of proper spatial and temporal regulation leading to excessive angiogenesis may contribute to various pathologic processes, including cancer and chronic inflammatory disorders. Conversely, inadequate angiogenesis may impair processes that require vessel formation, such as healing of wounds and burns or organ repair. Knowledge of the molecular mechanisms by which angiogenesis is controlled is yielding insights which are already being applied in therapeutic trials for diseases characterized by abnormal angiogenesis. This volume is mainly devoted to describing recent insights into the molecular biology of angiogenesis regulation, especially as it pertains to human diseases.

Angiogenesis has been proposed and utilized as both a prognostic indicator in certain types of malignancies (e.g. breast cancer) and as a target for therapeutic intervention. Chapters contributed by Drs. Harris, Johnson and Bruce, and Guerin and Laterra discuss angiogenesis and its prognostic and therapeutic implications in the context of specific types of neoplasms (breast and brain tumors). The macrophage is an interesting cell type that produces both positive and negative regulators of angiogenesis; these cells accumulate in tumors and at sites of chronic inflammation (e.g. joints of patients with rheumatoid arthritis). Dr. Polverini discusses macrophage-mediated angiogenesis and its role in disease in his chapter.

The molecular mechanisms by which specific growth factors induce angiogenesis is a major subject of this volume. Drs. Ferrara and Keyt and Brown and colleagues provide new insights into the mechanisms of action of the increasingly important angiogenic growth factor vascular endothelial growth factor, which is also known as vascular permeability factor. Fibroblast growth factors were the first factors identified as *in vivo* angiogenesis factors. A chapter describing the latest knowledge of the mechanism of action of fibroblast growth factor family members was contributed by Dr. Klein and colleagues. Drs. Rosen and Goldberg describe the angiogenic action of scatter factor (also known as hepatocyte growth factor) a recently recognized invasogenic and angiogenic cytokine related to plasminogen.

We now know that, in addition to factors that stimulate angiogenesis, a growing group of endogenous inhibitors act to prevent angiogenesis or to modulate the process of angiogenesis. Dr. DiPietro discusses thrombospondin-1, a high molecular weight adhesive glycoprotein of the extracellular matrix with extremely potent anti-angiogenic activity. Interestingly, thrombospondin-1 may be encoded by a tumor suppressor gene. Dr. O'Reilly discusses angiostatin, an exciting recently discovered factor derived from the fibrinolytic proenzyme plasminogen that inhibits tumor angiogenesis, primary tumor growth, and formation of metastases.

In addition to the soluble class of angiogenesis-regulatory factors discussed above, interactions of endothelial cells with components of the extracellular matrix and with other cell types are critical for proper formation of vessels. Drs. Grant and Kleinman discuss the role of laminin and other matrix molecules in regulation of capillary formation. Dr. van Hinsbergh and colleagues describe the role of fibrin and the fibrinolytic system in angiogenesis associated with wound repair. Cell surface molecules that interact with the extracellular matrix have been implicated in the regulation of angiogenesis. Dr. Varner discusses some exciting new studies on the roles of specific vascular cell integrins ($\alpha_v\beta_3$ and $\alpha_v\beta_5$) in mediating tumor angiogenesis and angiogenesis associated with wound healing. The pericyte, a vascular smooth muscle-like cell, exerts a powerful regulatory effect during the later stages of angiogenesis in which mature capillaries are formed. These mechanisms are discussed by Drs. Hirschi and D'Amore.

With all the recent progress in the molecular biology of angiogenesis, the contribution of microenvironmental conditions such as hypoxia and pH to angiogenesis is often ignored. Drs. Rockwell and Knisely review this area of investigation and present studies of experimental tumor models. Another area that has received less than adequate attention in discussions of angiogenesis is the formation of new lymphatic vessels (lymphangiogenesis), which may also play a role in tumor growth. Drs. Witte and colleagues discuss the mechanisms and importance of lymphangiogenesis in their chapter.

Given the enormous amount of interest within the research community in the subject of angiogenesis, a volume of this size cannot hope to be all-inclusive. We have attempted to highlight particularly exciting avenues of investigation. Many of the chapters provide reasonably extensive background material, so that a careful reading of the chapters should provide a good general understanding and appreciation of the subject.

We would like to thank the Long Island Jewish Medical Center and its President, Dr. David Dantzker, for its interest in and support of basic science research. We are grateful to Diane Akseizer for her assistance in the coordination of this publication.

I.D. Goldberg
E.M. Rosen

Regulation of Angiogenesis
ed. by I.D. Goldberg & E.M. Rosen
© 1997 Birkhäuser Verlag Basel/Switzerland

Angiogenesis and angiogenesis inhibition: An overview

J. Folkman

Department of Surgery, Children's Hospital, and Departments of Surgery and of Cell Biology, Harvard Medical School, Boston, Massachusetts 02115, USA

The study of angiogenesis attracts investigators from many different fields who find that regulation of blood vessel growth underlies a wide spectrum of biologic processes. These include reproduction, development, repair, angiogenic diseases, and cancer (Folkman, 1995).

The process of angiogenesis appears at first glance to be deceptively simple: capillary blood vessels either grow or they do not. An analogy is blood coagulation. Yet, in both systems the regulation of a "simple" endpoint turns out to be exceedingly complex and subtle.

Physiological vs. pathological angiogenesis

A central theme emerging from the study of angiogenesis is that while vascular endothelial cells seem always poised to proliferate and migrate, an elaborate machinery normally restrains them. Thus, endothelial cells are capable of proliferating as rapidly as any cell in the body, but for most of their life remain out of the cell cycle and are quiescent. Endothelial cells are literally steeped in growth factors, yet these cells rarely respond. Even when endothelial cells do proliferate and migrate during physiological angiogenesis or repair angiogenesis, the resulting neovascularization is short-lived and is turned off abruptly, as for example during ovulation or wound healing. When local administration of an angiogenic protein (such as bFGF or VEGF) is employed to induce neovascularization, discontinuation of this stimulus is followed by rapid involution and disappearance of the new capillary vessels. Thus, neovascularization is not easily sustained. For these reasons, there is increasing interest in uncovering the inhibitory machinery which normally prevents endothelial growth and puts limits on physiological angiogenesis.

This attempt to understand the physiological inhibitors of angiogenesis is also based on the observation that pathological angiogenesis is very prolonged and is not easily reversed therapeutically. The difficulty of inhibiting pathologic angiogenesis suggests that local endogenous inhibitors of blood vessel growth may somehow be downregulated in diseases dominated by abnormal neovascularization such as in diabetic retinopathy, arthritis, psoriasis, atherosclerotic plaques, and cancer.

Normal suppressor mechanisms of angiogenesis

The mechanisms which restrict endothelial growth under normal conditions are being elucidated at the cellular, biochemical and molecular levels.

Cell shape

At the cellular level, the endothelial cell carries its own internal defense against stray growth factors. It is among the most sensitive of all cells to growth control by cell shape (Folkman and Haudenschild, 1980; Wong et al., 1994; Ingber and Folkman, 1989). When the spreading of capillary endothelial cells *in vitro* is limited to < 500 μm^2, basic fibroblast growth factor (bFGF) becomes a differentiation factor and is not mitogenic. In contrast, when an endothelial cell is allowed to spread beyond 500 μm^2, increasing cell area up to 3000 μm^2 correlates with increasing DNA synthesis in the presence of a constant concentration of bFGF (Ingber, 1990). This phenomenon is based on recent findings that mechanical forces on a cell are necessary for growth factors, cytokines, and hormones to function. These soluble molecules will remain inactive unless they are coupled to the mechanical forces generated by specific *insoluble* molecules. The insoluble molecules (i.e. collagen and fibronectin) lie in the *extracellular matrix*, and bind to specific receptors and integrins on the cell surface. This allows a cell to pull against its extracellular matrix and to generate tension over interconnected cytoskeletal linkages. This tension can confer specificity, depending on the receptor or cluster of receptors being pulled (Singhvi et al., 1994). For example, certain signalling pathways can be activated by brief tension on a specific integrin. Magnetic microbeads coated with antibodies which bind to specific integrins and not to others, can be twisted by application of a magnetic field. This local deformation of the cell membrane leads to upregulation of cyclic AMP, or activation of the phosphoinositol-3 kinase system, even before a change in cell shape occurs. Further deformation leads to cell shape changes, a prerequisite for entry into the cell cycle and for gene expression. In fact, for the endothelial cell it is not the area and shape configuration of the *outer* cell membrane which supplies the direct mechanochemical information that permits DNA synthesis (or gene expression), but rather the shape of the nucleus. Nevertheless, nuclear shape *per se* is governed by the shape of the outer cell membrane and by tension forces transmitted to the nucleus over the cytoskeletal network. Thus, in an elegant experiment, Ingber photoetched 1 micron islands of fibronectin attachment points onto a nonadhesive substratum so that endothelial cells could be stretched into any configuration. He showed that any given cell shape which pulled the projected nuclear area beyond 60–70 microns allowed increasing DNA synthesis, which correlated with increasing nuclear area. Thus, there is now a molecular explanation for an older observation that confluent endothelial cells are refractory to mitogenic stimuli (Haudenschild et al., 1976). Based on these studies we can now ask whether the vasodilation of micro- ✲ vessels which invariably *precedes* angiogenesis, is necessary to increase the stretch and to decrease the confluence of endothelial cells in order to make them susceptible to mitogens. This early vasodilation may be mediated by nitric oxide produced by endothelial cells themselves and upregulated in response to stimulation by certain angiogenic factors (Ziche and Gullino, 1982; Polverini and Leibovich, 1984).

Pericytes

Neighboring cells also participate in defending endothelial cells against mitogenic stimuli. Pericytes are normally in close apposition to capillary endothelial cells *in vivo*. Pericytes *in vitro* secrete latent TGF-β (transforming growth factor beta) which becomes activated upon contact with endothelial cells, and this contact also releases active TGF-β from the endothelial cells (Antonelli-Orlidge et al., 1989; Satoh and Rifkin 1989; Satoh et al., 1990; Hirsch and D'Amore, 1996, in press). TGF-β inhibits ✲ endothelial cell proliferation *in vitro*. It is not known if pericytes suppress endothelial proliferation *in vivo*.

Macrophages

Macrophages recruited into tumors elaborate angiogenic activity (Polverini and Leibovich, 1984), but under certain conditions tumor macrophages secrete an elastase capable of cleaving angiostatin from circulating plasminogen (Dong et al., 1996). Angiostatin is a specific inhibitor of endothelial cell proliferation and it is among the most potent of the known angiogenesis inhibitors (O'Reilly et al., 1994).

Extracellular matrix

Extracellular matrix components and in particular certain proteoglycans bind and store endothelial cell growth factors, thus making them inaccessible to endothelial cells (Vlodavsky et al., 1990; Folkman, 1988). Both bFGF and VEGF bind to heparan sulfate proteoglycan. The basement membrane itself may inhibit endothelial cell growth. For example, the laminin B1 chain contains two internal sites which, in the form of synthetic peptides having the sequence RGD and YSGR, inhibit angiogenesis (Grant et al., 1994). Collagen XVIII is localized to the perivascular region of large and small vessels (Oh et al., 1994; Rehn and Pihlajaniemi, 1994).

A 187 amino acid fragment in its carboxy-terminal region, called endostatin, is a potent and specific inhibitor of endothelial proliferation (O'Reilly et al., unpublished 1996).

Angiogenesis suppressor genes

In normal fibroblasts (Dameron et al., 1994) or in cells of the central nervous system (Van Meir et al., 1994), the tumor suppressor gene p53 may also code for inhibitors of angiogenesis. It is not clear whether other tumor suppressor genes will be discovered to regulate angiogenesis, but this is an exciting area of investigation.

Receptors for angiogenic factors

Soluble receptors for the angiogenic proteins VEGF and bFGF may also play a physiological role in the regulation of the biological activity of these mitogens. The soluble Flt-1 receptor, which is encoded in the natural short message of the *fit-1* gene, binds VEGF at a high affinity and could absorb VEGF secreted from solid tumors (Kendall and Thomas, 1993; Shibuya, 1995). It has also been shown that binding proteins for acidic and basic fibroblast growth factors circulate in human plasma. These are each truncated forms of the high-affinity FGF receptor (Hanneken et al., 1994). This is of interest because abnormally elevated levels of biologically active bFGF have been found in the serum and urine of patients with a wide variety of different types of cancer (Nguyen et al., 1994) and also in infants with hemangiomas (Folkman et al., unpublished data). For example, serum levels of bFGF in normal adults range from $5-7$ picograms/ml but in cancer patients are elevated to $100-300$ picograms/ml. These findings suggest that, under normal conditions, circulating levels of potent angiogenic proteins are limited to very low concentrations.

Heterodimers of angiogenic factors

It has recently been reported that natural heterodimers of VEGF and placenta growth factor (PLGF) are secreted into the conditioned media of various human tumor cell lines (Cao et al., 1996). Furthermore, PlGF/VEGF heterodimers display $20-50$-fold less mitogenic activity than $VEGF_{165}$ for endothelial cells *in vitro*. If similar heterodimers are formed by non-neoplastic cells, this may be a subtle mechanism for modulating VEGF-induced angiogenesis.

Regulation of secretion of angiogenic factors

The release of angiogenic factors from tumor cells has been extensively studies (Abraham and Klagsbrun, 1995) although less is known about how normal cells release these proteins. The lack of a signal peptide in bFGF has led to the proposal of alternative non-classical pathways for its secretion (Jackson et al., 1995). It has also been suggested that transient plasma membrane wounds to a cell may release bFGF (McNeill et al., 1989). VEGF contains a signal peptide but its release may be regulated in part by its heparin affinity, i.e. $VEGF_{121}$ and $VEGF_{165}$ have low heparin affinity while $VEGF_{189}$ and $VEGF_{206}$ have high heparin affinity and may be membrane-bound. This is also a fruitful area of research, because the "switch to the angiogenic phenotype", whether it occurs under physiologic conditions (ovulation) or under pathological conditions (psoriasis or neoplasia), may be highly regulated at the level of cellular release of an angiogenic protein.

Endogenous regulators of endothelial proliferation

Several endogenous proteins which are potent inhibitors of endothelial proliferation and which also inhibit angiogenesis have been identified recently (Table 1). In each case the endothelial inhibitor activity is found in a fragment of a larger protein which itself lacks inhibitory activity. Platelet factor 4 is an exception, but even its 17.8 kD subfragment is reported to be up to 50 times more potent as an angiogenesis inhibitor than the parent molecule (Gupta et al., 1995). These findings raise the interesting question of whether there is a general mechanism by which inhibitors of endothelial growth can be produced at a local site by specific proteolytic cleavage of a larger protein. More is known about the mechanism of action of prolactin 16 kD than for the others. This fragment inhibits the mitogenic activity on endothelial cells of both bFGF and VEGF by acting distal to their receptors and proximal to the mitogen-activated protein kinases (MAPK), specifically by inhibition of the activation of Raf-1 (D'Angelo et al., 1995; Weiner and D'Angelo, 1996).

Table 1. Endothelial inhibitors that are fragments of larger proteins

Fibronectin	29 kD	Homandberg et al., 1985
Prolactin	16 kD	Clapp et al., 1993
Thrombospondin	(fragment)	Tolsma et al., 1993
Angiostatin	38 kD	O'Reilly et al., 1994
Platelet Factor-4	7.8 kD	Gupta et al., 1995
Murine EGF	(fragment 33–42)	Nelson et al., 1995

Conclusions

The array of suppressor cells and molecules discussed here, which normally restrict endothelial proliferation and prevent angiogenesis, represent only a partial categorization of the intricate control mechanisms for the growth of capillary blood vessels. Many other mechanisms are likely to be discovered in the future, and it is clear that much further work will be necessary before we can understand the actions of the normal inhibitors of angiogenesis assembled in this overview. However, even an incomplete analysis of the tight controls which guard against proliferation of vascular endothelial cells reveals the powerful inhibitory machinery which must be overcome for the angiogenic phenotype to be expressed and implies that various components of this suppressor machinery itself may need to be downregulated for the angiogenic switch to occur, as previously suggested by Bouck and her colleagues (Bouck, 1990).

References

Abraham, J. and Klagsbrun, M. (1995) The role of the fibroblast growth factor family in would healing. *In:* R.A. Clark (ed): *The Molecular and Cellular Biology of Wound Repair.* Plenum Press, New York, pp 195–248.

Antonelli-Orlidge, A., Saunder, K.B., Smith, S.R. and D'Amore, P.A. (1989) An activated form of transforming growth factor beta is produced by co-cultures of endothelial cells and pericytes. *Proc. Natl. Acad. Sci. USA* 86:4544–4548.

Bouck, N. (1990) Tumor angiogenesis: The role of oncogenes and tumor suppressor genes. *Cancer Cells* 2:180–185.

Cao, Y., Chen, H., Zhou, L., Chiang, M.-K., Anand-Apte, B., Weatherbee, J.A., Wang, Y., Fang, F., Flanagan, J.G. and Tsang, M.L.-S. (1996) Heterodimers of placenta growth factor/vascular endothelial growth factor. *J. Biol. Chem.* 271:3154–3162.

Clapp, C., Martial, J.A., Guzman, R.C., Rentier-Delure, F. and Weiner, R.I. (1993) The 16-kilodalton N-terminal fragment of human prolactin is a potent inhibitor of angiogenesis. *Endocrinology* 133:1292–1299.

Dameron, K.M., Volpert, O.V., Tainsky, M.A. and Bouck, N. (1994) Control of angiogenesis in fibroblasts by p53 regulation of thrombospondin-1. *Science* 265:1582–1584.

D'Angelo, G., Struman, I., Martial, J. and Weiner, R.I. (1995) Activation of mitogen-activated protein kinases by vascular endothelial growth factor and basic fibroblast growth factor in capillary endothelial cells is inhibited by the antiangiogenic factor 16-kDa N-terminal fragment of prolactin. *Proc. Natl. Acad. Sci. USA* 92:6374–6378.

Dong, Z., Kumar, R. and Fidler, I.J. (1996) Generation of the angiogenesis inhibitor, angiostatin, by Lewis lung carcinoma is mediated by macrophage elastase. *Proc. Am. Ass. Cancer Res.* 37:58.

Folkman, J. (1995) Clinical applications of research on angiogenesis. *N. Engl. J. Med.* 333:1757–1763.

Folkman, J. and Haudenschild, C. (1980) Angiogenesis *in vitro. Nature* 288:551–556.

Folkman, J., Klagsbrun, M., Sasse, J., Wadzinski, M., Ingber, D.E. and Vlodavsky, I. (1988) A heparin-binding angiogenic protein – basic fibroblast growth factor – is stored in the basement membrane. *Am. J. Pathol.* 130:393–400.

Grant, D.S., Kibbey, M.C., Kinsella, J.L., Sid, M.C. and Kleinman, H.K. (1994) The role of basement membrane in angiogenesis and tumor growth. *Pathology, Research and Practice* 190:854–863.

Gupta, S.K., Hassel, T. and Singh, J.P. (1995) A potent inhibitor of endothelial cell proliferation is generated by proteolytic cleavage of the chemokine platelet factor 4. *Proc. Natl. Acad. Sci. USA* 92:7799–7803.

Hanneken, A., Ying, W., Ling, N. and Baird, A. (1994) Identification of soluble forms of the fibroblast growth factor receptor in blood. *Proc. Natl. Acad. Sci. USA* 91:9170–9174.

Haudenschild, C.C., Zahniser, D., Folkman, J. and Klagsbrun, M. (1976) Human vascular endothelial cells in culture. Lack of response to serum growth factors. *Exp. Cell Res.* 98:175–183.

Hirsch, K.K. and D'Amore, P.A. (1996) Pericytes in the microvasculature. *Cardiovascular Research, in press.*

Homandberg, G.A., Williams, J.E., Grant, D., Schumacher, B. and Eisenstein, R. (1985) Heparin-binding fragments of fibronectin are potent inhibitors of endothelial cell growth. *Am. J. Pathol.* 120:327–332.

Ingber, D.E. and Folkman, J. (1989) How does extracellular matrix control capillary morphogenesis. *Cell* 58:803–805.

Ingber, D.E. (1990) Fibronectin controls capillary endothelial cell growth by modulating cell shape. *Proc. Natl. Acad. Sci. USA* 87:3579–3583.

Jackson, A., Tarantini, F., Gamble, S., Friedman, S. and Maciag, T. (1995) The release of fibroblast growth factor-1 from NIH 3T3 cells in response to temperature involves the function of cysteine residues. *J. Biol. Chem.* 270:333–36.

Kendall, R.L. and Thomas, K.A. (1993) Inhibition of vascular endothelial cell growth factor activity by an endogenously encoded soluble receptor. *Proc. Natl. Acad. Sci. USA* 90:10705–10709.

McNeil, P.L., Muthukrishnan, L., Warder, E. and D'Amore, P.A. (1989) Growth factors are released by mechanically wounded endothelial cells. *J. Cell Biol.* 109:811–822.

Nelson, J., Allen, W.E., Scott, W.N., Bailie, J.R., Walker, B., McFerran, N.V. and Wilson, D.J. (1995) Murine epidermal growth factor (EGF) fragment (33–42) inhibits both EGF- and laminin-dependent endothelial cell motility and angiogenesis. Cancer Res. 55:3772–3776.

Nguyen, M., Watanabe, H., Budson, A.E., Richie, J.P., Hayes, D.F. and Folkman, J. (1994) Elevated levels of an angiogenic peptide, basic fibroblast growth factor, in the urine of patients with a wide spectrum of cancers. *J. Natl. Cancer Inst.* 86:356–361.

Oh, S.P., Kamagata, Y., Muragaki, Y., Timmons, S., Ooshima, A. and Olsen B.R. (1994) Isolation and sequencing of cDNAs for proteins with multiple domains of Gly-Xaa-Yaa identify a distinct family of collagenous proteins. *Proc. Natl. Acad. Sci. USA* 91:4229–4233.

O'Reilly, M.S., Holmgren, L., Shing, Y., Chen, C., Rosenthal, R.A., Moses, M., Lane, W.S., Cao, Y., Sage, E.H. and Folkman, J. (1994) Angiostatin: A novel angiogenesis inhibitor that mediates the suppression of metastases by a Lewis lung carcinoma. *Cell* 79:315–328.

O'Reilly, M.S., et al. (1996) Unpublished data.

Polverini, P.J. and Leibovich, S.J. (1984) Induction of neovascularization *in vivo* and endothelial proliferation *in vitro* by tumor-associated macrophages. *Lab. Invest.* 51:635–642.

Rehn, M. and Pihlajaniemi, T. (1994) ∂I (XVIII), a collagen chain with frequent interruptions in the collagenous sequence, a distinct tissue distribution, and homology with SV collagen. *Proc. Natl. Acad. Sci. USA* 91:4243–4238.

Satoh, Y. and Rifkin, D.B. (1989) Inhibition of endothelial cell movement by pericytes and smooth muscle cells: Activation of key TGF-β-like molecules by plasmin. *J. Cell Biol.* 109:309–315.

Satoh, Y., Tsuboi, R., Lyons, R., Moses, H. and Rifkin, D.B. (1990) Characterization of the activation of latent TGF-B by co-cultures of endothelial cells and pericytes or smooth muscle cells: A self-regulating system. *J. Cell Biol.* 111:757–763.

Shibuya, M. (1995) Role of VEGF-FLT receptor system in normal and tumor angiogenesis. *Adv. Cancer Res.* 67:281–284.

Singhvi, R., Kuman, A., Lopez, G.P., Stephanopoulos, G.N., Wang, D.I.C., Whitesides, G.M. and Ingber, D.E. (1994) Engineering cell shape and function. *Science* 264:696–698.

Tolsma, S.S., Volpert, O.V., Good, D.J., Frazier, W.A., Polverini, P.H. and Bouck, N. (1993) Peptides derived from two separate domains of the matrix protein thrombospondin-1 have anti-angiogenic activity. *J. Cell Biol.* 122:497–511.

Van Meir, E.G., Polverini, P.J., Chazin, V.R., Huang, H.-J. S., de Tribolet, N. and Cavenee, W.K. (1994) Release of an inhibitor of angiogenesis upon induction of wild type p53 expression in glioblastoma cells. *Nature Genetics* 8:171–176.

Vlodavsky, I., Bashkin, P.K., Korner, G., Bar-Shavit, R. and Fuks, Z. (1990) Extracellular matrix-resident growth factors and enzymes relevance to angiogenesis and metastasis. *Proc. Am. Ass. Cancer Res.* 31:491–493.

Weiner, R.I. and D'Angelo, G. (1996) Signaling for the antiangiogenic action of 16 k prolactin. *Proc. Am. Ass. Cancer Res.* 37:667–668.

Wong, J.Y., Langer, R. and Ingber, D.E. (1994) Electrically conducting polymers can noninvasively control the shape and growth of mammalian cells. *Proc. Natl. Acad. Sci. USA* 91:3201–3204.

Ziche, M. and Gullino, P.M. (1982) Angiogenesis and neoplastic progression *in vitro. J. Natl. Cancer Inst.* 69:483–487.

Significance of angiogenesis in human disease

Regulation of Angiogenesis
ed. by I.D. Goldberg & E.M. Rosen
© 1997 Birkhäuser Verlag Basel/Switzerland

Role of the macrophage in angiogenesis-dependent diseases

P.J. Polverini

Laboratory of Molecular Pathology, Department of Oral Medicine, Pathology, and Surgery, University of Michigan School of Dentistry, Ann Arbor, Michigan 48109-1078, USA

Introduction

It is now well established that the functional domain of the macrophage (Mϕ) extends far beyond its originally recognized role as a scavenger cell. The rich array of secretory products, now numbering in excess of 100 well-characterized molecules, and their widespread anatomic distribution and functional heterogeneity, are unmatched by any other cell type (Nathan, 1987). This remarkable diversity enables the Mϕ to influence virtually every facet of the immune response and inflammation as well as contribute to the etiology and/or pathogenesis of a number of diseases. Angiogenesis, the process which results in the formation of new capillary blood vessels, is an essential component of a number of important physiological processes (Folkman and Cotran, 1976; Auerbach, 1981; Folkman and Klagsbrun, 1987; Klagsbrun and D'Amore, 1991). Furthermore, when angiogenesis occurs in excess or inappropriately, it can contribute to the etiology and/or pathogenesis of several inflammatory, degenerative and developmental diseases. Mϕ are key angiogenesis effector cells that produce a number of growth stimulators and inhibitors, proteolytic enzymes, and cytokines that can influence one or more steps in the angiogenesis cascade. In this review I will summarise the evidence implicating Mϕ as important accessory cells in physiological angiogenic responses, and describe how disruption of the coordinate production of Mϕ-derived stimulators and inhibitors of angiogenesis contributes to tumor progression and the pathogenesis of chronic inflammatory diseases.

Mϕ are anatomically and functionally diverse

Cells belonging to the mononuclear phagocyte system are derived from blood-borne myeloid precursors that originate from bone marrow stem cell populations. Mϕ progenitors, under the influence of colony stimulating factors and differentiating inducing signals, divide, differentiate and migrate into the blood stream as circulating monocytes (Gordon, 1986; Sunderkotter et al., 1991, 1994; Stein and Keshav, 1992). When called upon, as in inflammation or during immune responses, Mϕ undergo further differentiation and specialization where they become highly responsive to environmental signals that further modify their functional and morphological phenotype (Adams and Hamilton, 1984; Johnson, 1988; Van Furth, 1988; Sorg, 1989; Adams, 1989). Mϕ populate virtually every tissue compartment in the body, where they perform several common but also may specialized functions depending upon the tissues or organ in which they reside. Thus alveolar Mϕ, Kuppfer cells of the liver, bone marrow osteoclasts, peritoneal Mϕ, synovial-type A cells, and Langerhans cells and dermal dendritic cells of the skin, while all of similar origin each display unique and specialized functions. Mϕ are capable of rapidly responding to changes in their environment by undergoing further differentiation and specialization. This process, termed activation, is a tightly regulated biological process. When Mϕ are activated they display properties such as increased phagocytic and microbiocidal activity, exhibit enhanced chemotactic activity, and secrete a wide spectrum of biologicaly active molecules including a number of growth factors. Activation thus enables Mϕ to influence numerous biological processes in a precisely defined fashion.

Mϕ are among the most versatile cells in the mammalian organism. Recognized early on for their endocytic and phagocytic activity and their central role in host defense against infection and malignancy (Gordon, 1986). Mϕ have gained considerable attention in recent years as potent secretory cells, having been shown to influence the growth and function of a diverse array of target cells (Nathan, 1987). One of the first studies implicating Mϕ in promoting the growth of other mesenchymal cells was the work of Carrel (1922), who recognized that leukocytes present in inflammatory exudates could stimulate the growth of fibroblasts. In the early seventies a number of studies reported that Mϕ culture fluids contained growth stimulatory and inhibitory activities for lymphoid and bone marrow derived cells (Schrader, 1973; Gordon et al., 1974; Calderon and Unanue, 1975; Calderon et al., 1975; Unanue et al., 1976; Unanue, 1976). The functional importance of the macrophage in wound repair and fibroblast proliferation was first demonstrated when Leibovich and Ross (1975) showed that guinea pigs depleted of monocytes and Mϕ exhibited significantly delayed wound repair. They later showed that macrophage culture fluids contained growth promoting activity(s) for fibroblasts (Leibovich and Ross, 1976). The biochemical identity of this biological activity, termed macrophage-derived growth factor (MDGF), remained uncertain for several years. As the identity of many other growth promoting activities and other secretory products became known, it became increasingly clear that a number of biochemically distinct molecules with a much wider range of cell targets, including vascular endothelium, were responsible for the growth promoting effects of Mφ (Gillespie et al., 1986; Polverini, 1989, for reviews).

Mφ are potent mediators of angiogenesis

Shortly after Mφ were reported to have growth promoting activity for several types of nonlymphoid mesenchymal cells, results from several laboratories implicated Mφ in angiogenesis. Our studies in macrophage-induced angiogenesis stemmed from interests in the mechanisms of vascular proliferation during nonspecific and immune-mediated chronic inflammatory responses (Polverini et al., 1977a, 1977b). At the time there were very little data on the mechanisms responsible for endothelial proliferation induced by nonneoplastic stimuli. While the role of diffusible tumor derived factors in angiogenesis had been firmly established, comparable mechanistic data on endothelial proliferation induced by nonneoplastic stimuli were not available. The occurrence of endothelial cell proliferation during immunologic reactions had been documented in several studies. Light microscopic and ultrastructural studies of hypersensitivity reactions in experimental animals and humans revealed the presence of activated and dividing endothelial cells (Gell, 1959; Dvorak et al., 1976). Graham and Shannon (1972) reported that endothelial mitoses were observed during lymphocyte migration through high endothelial venules in arthritic joints of rabbits. Anderson et al. (1975) demonstrated extensive proliferation of postcapillary venular endothelium in lymph nodes draining skin allographs, and Sidkey and Auerbach (1975) showed that capillary proliferation occurred during local graft vs. host reactions in the skin of mice. It was proposed that the accompanying dermal angiogenesis was induced by immunocompetent donor lymphocytes. Also, a number of *in vitro* studies of Mφ indicated that they produced growth inhibitory and stimulatory factors for several mesenchymal cell types (Calderon et al., 1974, 1975, 1976; Leibovich and Ross, 1975, 1976). It was in this context that we therefore initiated a series of experiments to examine and quantitate the vasoproliferative response in two models of delayed-type hypersensitivity: the tuberculin reaction and contact sensitivity to dinitrochlorobenzene. Using a quantitative autoradiographic approach we demonstrated that incorporation of tritiated thymidine by microvascular endothelial cells in the skin of sensitized guinea pigs coincided with the onset and magnitude of mononuclear infiltration at the reaction site (Polverini et al., 1977a). The mechanisms responsible for endothelial proliferation were at the time conjectural: two explanations were postulated. First, that endothelial replication was a reparative response to nonspecific injury and necrosis induced by humoral or cell-derived toxins. Alternatively, the proliferation might be mediated by growth factors produced by one or more of the mononuclear cell types comprising the infiltrate. We therefore asked whether Mφ might play a similar role in stimulating microvascular endothelial cell growth. At about the time we began these studies a report by Clark et al. (1976) appeared in which they showed that wound Mφ, when introduced into rabbit corneas, stimulated neovascularization in the wake of an acute inflammatory response. Using the corneal bioassay of neovascularization we examined whether Mφ or their conditioned culture media, when introduced into guinea pig corneas, could stimulate ingrowth of capillary blood vessels (Polverini et al., 1977b). Peritoneal Mφ obtained from Balb/c mice and Hartley albino guinea pigs by lavage or following injection of an inflammatory stimulant were processed by standard separation techniques to yield a preparation consisting of 85–90% Mφ. Intracorneal injection of viable Mφ or their conditioned culture media were potently angiogenic in over 75% of corneas tested. In contrast, unactivated resident Mφ were either weakly angiogenic or showed no activity at all. Similarly, preparations of Mφ from inbred strain II guinea pigs yielded essentially similar results, thus precluding the possibility that neovascularization occurred as a result of an immunologically mediated inflammatory response. The requirement that Mφ be activated for expression of angiogenic activity was demonstrated with cultured cells. Brief exposure of nonangiogenic resident peritoneal Mφ to latex induced angiogenic activity. Aliquots of cell free, dialyzed and concentrated cultre media from *in vivo* and *in vitro* activated Mφ also induced angiogenesis when incorporated into Hydron polymer and implanted intracorneally. These responses exhibited the same pattern of capillary growth as was observed with viable Mφ. Collectively these results indicated that activated Mφ were able to induce angiogenesis in the absence of inflammation through an inducible secreted product. These studies were subsequently confirmed by Thakral et al. (1979) and Hunt and coworkers (1984) with wound-derived Mφ and wound fluids rich in Mφ, and by Moore and Sholley (1985) with autologous rabbit peritoneal Mφ.

A variety of environmental stimuli capable of inducing expression of the angiogenic phenotype by Mφ have been identified. Knighton et al. (1983) and Jensen et al. (1986) showed that adherent Mφ can be activated by low oxygen tension and high lactate concentration. Both of these stimuli are present in great abundance in organizing wound and are an important stimulus for activating stimuli in the wound response. Also there is evidence that endothelial cell-derived cytokines and adhesion molecules may function to activate Mφ (Koch et al., 1995). This may represent another mechanism for macrophage activation at sites of inflammation. Before migrating into inflamed tissues Mφ and other leukocytes must adhere to endothelial cells (Carlos and Harlan, 1994). These cytokine-activated post-capillary venular endothelial cells express a high number of adhesion molecules that are able to activate Mφ, which in turn can participate in the generation of soluble adhesion molecules with proangiogenic activity (Szekanecz et al., 1994).

Numerous substances have been reported to induce angiogenesis and many of these mediators are produced by Mφ (Polverini, 1989; Sunderkotter et al., 1991, 1994). These include polypeptide growth factors, cytokines, prostaglandins and proteolytic enzymes (Folkman and Klagsbrun, 1987; Klagsbrun and D'Amore, 1991; Polverini, 1995). As a consequence Mφ can mediate new capillary growth by several different mechanisms. First, the macrophage can produce factors that act directly * to influence the angiogenic cascade. *In vitro* studies have shown that Mφ produce in excess of 20 molecules that stimulate endothelial cell proliferation, migration, and differentiation (Sunderkotter et al., 1991) and many of them have been show to be potentially angiogenic *in vivo*. A second mechanism by which Mφ might promote angiogenesis is by modifying the extracellular matrix (ECM). The composition of the ECM * has been shown to influence endothelial cell shape and morphology dramatically, and may profoundly influence new capillary growth (Ingber and Folkman, 1989; Ingber, 1991). Mφ can influence the composition of the ECM either through the direct production of ECM components, or through the production of proteases which effectively alter the ECM.

Mφ play a central role in wound neovascularization

Wound healing is perhaps the most well studied and best example of a physiological process that is strictly dependent upon the timely ingrowth of new capillary blood vessels. Wound healing is an essential biologic process that is driven by the cooperative interaction of a variety of cell types and mediator systems. Normal tissue repair requires that the cells and mediators of the immune system, the connective tissue, and vascular endothelium function in a coordinated manner in order to effect the repair process in a timely and efficient manner. Although numerous cell types are involved, the monocyte-derived macrophage appears to play a central role in orchestrating the repair process. Mφ migrating into healing* wounds phagocytose wound debris, become activated, and secrete diffusable cytokines that modulate tissue repair (Leibovich and Wiseman, 1988; Riches, 1988). In surgical models of wound healing, activated Mφ have been shown to be the predominant cell type approximately 5 days after injury when the proliferative phase of the repair process is most pronounced (Ross and Odland, 1968; Leibovich and Ross, 1975). Seveal reports suggested that wound Mφ function primarily to augment the repair response. The introduction of activated Mφ into healing wounds results in enhanced wound repair (Danon et al., 1989), and activation of wound Mφ either by systemic or topical administration of the macrophage stimulant, glucan, significantly increased wound breaking strength (Browder et al., 1988).

Mφ are believed to be responsible for coordinating much of the growth and tissue remodeling that occurs during the wound response. A key step in the repair process is the formation of inflammatory granulation tissue: a temporary tissue that is composed largely of new capillary blood vessels. It is now well established that Mφ are key regulators of wound neovascularization. Activated Mφ or their culture supernatants have been shown to induce new capillary growth *in vitro* and angiogenesis *in vivo* (Clark et al., 1976; Thakral et al., 1979; Hunt et al., 1984; Polverini et al., 1977; Koch et al., 1986; Folkman and Klagsbrun, 1987; Sunderkotter et al., 1991, 1994). During the proliferative phase, growing capillaries provide nutrient support for regenerating tissues, while during the resolution phase many of the newly formed capillaries regress as tissue regeneration is complete. Recent evidence suggests that capillary regression may also be mediated by Mφ, as Mφ have been shown to produce substances which are inhibitory to endothelial cell growth, and some which are anti-angiogenic *in vivo* (Jaffe et al., 1985; Polverini, 1989; Vilette et al., 1990; Besner and Klagsbrun, 1991; DiPietro and Polverini; 1993, 1994). Thus, a substantial body of evidence has implicated the macrophage in both the stimulation and suppression of wound angiogenesis.

There has been much speculation regarding which of the known angiogenic substances produced by Mφ might be key positive regulators of wound neovascularization. Employing reverse transcriptase PCR analysis, transcripts from a variety of potential angiogenic factors have been identified (Ford et al., 1989; Grotendorst et al., 1988; Rappolee et al., 1988). However, among these mediators only transforming growth factor-α (TNF-α) and TGF-β have been convincingly demonstrated to mediate angiogenesis *in vivo* (Schreiber et al., 1986, Leibovich et al., 1987, Fajardo et al., 1992). TNF-α represents a likely candidate as an angiogenic regulator as it is 1) found in wound fluid, 2) transcribed in wounds, probably by Mφ, and 3) has been described to be a major angiogenic molecule produced by Mφ (Ford et al., 1989; Fahey et al., 1991; Leibovich et al., 1987). More recently Strieter et al. (1992) and Koch et al. (1992) have demonstrated that interleukin 8 (IL-8), a major macrophage cytokine, is a potent mediator of angiogenesis in rodent corneas. Clearly, the induction and regulation of angiogenesis by Mφ may involve redundancy so that any single factor may not be essential for adequate wound repair. To add to the complexity, wound Mφ are capable of modulating other parameters of wound repair, such as the clearance of debris, and the growth and maturation of tissues (Leibovich and Ross, 1975, 1976). Moreover, it is likely that some macrophage-derived factors may affect more than one aspect of wound repair. Although Mφ influence many aspects of wound repair *in vivo*, the regeneration of the cellular and connective tissue components cannot proceed without the recruitment of new vasculature for nutrient support.

M φ produce inhibitors of neovascularization

One of the most important features that distinguishes physiological from pathological angiogenesis is that in the former setting angiogenesis is strictly delimited in both time and space (Bouck, 1990, 1993; Polverini, 1995). During wound neovascularization the rapid induction of new capillary vessels is tempered by inhibitors of neovascularization the rapid induction of new capillary vessels is tempered by inhibitors of neovascularization. One of the most convincing pieces of evidence linking a well known macrophage-derived product, the multifunction matrix glyco-protein thrombospondin 1 (TSP1), to the inhibition of angiogenesis emerged from investigations into the function of tumor suppressor genes (Rastinejad et al., 1989; Good et al., 1990; Bouck, 1990, 1993). It is well known that the processes of malignant transformation and tumor progression are characterized by the activation of oncogenes and the loss or inactivation of multiple tumor suppressor genes (Weinberg, 1989; Sager, 1989). While both the identity and mechanism of action of many oncogenes are known, by comparison very few tumor suppressor genes have been identified and less is known about their function at the physio-logical level. One of the firststudies that implicated tumor suppressor genes in the control of theproduction of an inhibitor of angiogenesis came from a series of studies by Bouck and colleagues (Bouck et al., 1986; Rastinejad et al., 1989; Good et al., 1990; Dameron et al., 1994). Using the technique of somatic cell hybridization these investigators showed that when normal nonangiogenic fibroblasts were fused to the potently angiogenic trans-formed BHK (baby hamster kidney) fibroblast line that had lost a single tumor suppressor gene, the resultant hybrids behaved like the nontrans-formed fusion partner. One of the transformed phenotypes that was "corrected" in these hybrids was that they were now unable to induce angiogenesis. A more detailed analysis of this phenomenon was undertaken by Rastinejad and colleagues (1989) Good et al. (1990) and Dameron et al. (1994). They showed that nontransformed BHK cells that contained an active tumor suppressor gene secreted an inhibitor of neovascularization whereas cells which had lost this suppressor gene did not. A detailed biochemical, immunologic and functional analysis of this angiogenic inhibitory activity revealed it to be a truncated form of TSP1. More recently Dameron et al. (1994) have directly linked control of the production of TSP1 to the p53 tumor suppressor gene.

TSP1 is one member of a family of five homologous proteins. It is a 450 kDa disulfide-linked trimer that is composed of three identical chains with a monomeric mass of about 140 kDa. Its modular structure in part enables it to interact with a variety of extracellular matrix proteins, cell surface and serum proteins, and cations. TSP1 is present in great abundance in the platelet alpha granules and is secreted by a wide variety of epithelial and mesenchymal cells (Lawler, 1986; Sage and Bornstein,

1991; Frazier, 1987; Frazier, 1991; Bornstein, 1995). It has been shown to participate in cell substrate interactions where many cells have been show to attach, spread and migrate on insoluble TSP1 (Good et al., 1990; Sage and Bornstein, 1991; Bornstein and Sage, 1994). As discussed above, TSP1 was first implicated as an inhibitor of neovascularization when an anti-angiogenic hamster protein whose secretion was controlled by a tumor suppressor gene was found to have an amino acid sequence similar to human platelet TSP1 (Rastinejad et al., 1989). Authentic TSP1 was then purified from platelets and shown to block neovascularization *in vivo* (Good et al., 1990). A role for TSP1 in the inhibition of angiogenesis is supported by several observations. It is present adjacent to mature quiescent vessels and is absent from actively growing sprouts both *in vivo* (O'Shea and Dixit, 1988) and *in vitro* (Iruela-Arispe et al., 1991). Hemangiomas which consists of rapidly proliferating endothelial cells fail to produce detectable TSP1 (Sage and Bornstein, 1982). Antibodies to TSP1 added to endothelial cell cultures enhance sprouting *in vitro* (Iruela-Arispe et al., 1991) and endothelial cells in which TSP1 production has been downregulated by antisense TSP1 exhibit an accelerated rate of growth, enhanced chemotactic activity and an increase in the number of capillary-like cords (DiPietro et al., 1994). More recently, DiPietro et al. (1996) have shown that the addition of antisense TSP1 oligomers to wounds in the skin of mice results in delayed-healing, suggesting that TSP1 is just as important in the initiation of the wound response as it is in the organization phase of wound repair. Also, Polverini et al. (1995) reported that mice with a targeted disruption in TSP1 show delayed wound organization, i.e. prolonged wound neovascularization.

Previous investigations have shown that both resting and activated Mφ produce TSP1. DiPietro et al. (1994) reported an approximately 6-fold increase in the steady state levels of TSP1 mRNA expression in the murine monocyte line WEHI-3 when the cells were treated for 24 hours with the potent activating agent lipopolysaccharide (LPS) with peak secretion of TSP1 protein occurring by 8 hours. The finding that activated Mφ produce the angiogenesis inhibitor TSP1 would seem paradoxical. There are several possible explanation for this apparent functional dichotomy. The angiogenic potential of Mφ *in vivo* may be the result of the balanced production of both positive and negative regulators of angiogenesis. Alternatively, macrophage-derived TSP1 may not exert significant effects on endothelial cells, particularly if TSP1 is rapidly degraded or sequestered into the extracellular matrix. In this instance the influence of activated Mφ may be shifted toward production of proangiogenic mediators of angiogenesis rather than inhibitors such as TSP1. Several other functions of macrophage-derived TSP1 may influence wound repair and angiogenesis. The adhesive capacity of TSP1 may facilitate migration of activated Mφ. Mφ have surface receptors for TSP1 (Silverstein and Nachman, 1987), and thus might lay down TSP1 upon the existing extracellular matrix as a scaffold upon which to migrate.

Mφ and tumor neovascularization

Mononuclear phagocytes are a frequent component of the stroma of neoplastic tissues (Mantovani, 1994). Interest in cells of the mononuclear phagocyte system in relation to the growth of neoplasms stems largely from the observation that these effector cells, when appropriately activated, are able to arrest the growth or kill neoplastic and transformed target cells. Although there is substantial *in vitro* evidence supporting the antitumor activity of activated Mφ, the *in vivo* relevance of these observations has not been unequivocally demonstrated even under conditions in which Mφ are likely to have anti-tumor activity. Mφ express diverse functions essential for tissue remodeling, inflammation and immunity. Analysis of tumor associated Mφ (TAM) functions suggest these multifunctional cells have the capacity to affect diverse aspects of neoplastic development, including vascularization, growth rate and metastasis, stroma formation and destruction. There is evidence that in some neoplasms, including human cancers, the protumor functions of Mφ prevail. These observations emphasize the dual potential of TAM to influence neoplastic growth and progression in opposite directions, with protumor activity often prevailing in the absence of therapeutic interventions in many neoplasms.

The formation of new blood vessels is a crucial step in the growth of neoplastic tissues. Several lines of evidence suggested a relationship between the macrophage content of tumors, the rate of tumor growth, and the extent of vascularization. In studies by Evans (1977a, 1977b, 1978) it was reported that mice depleted of Mφ by whole body X-irradiation showed a delay in the appearance of tumors, while following the administration of azothioprine to tumor bearing mice there was a suppression of the growth of established tumors, and a marked reduction in tumor neovascularization. Mostafa et al. (1980a, 1980b) and Stenzinger et al. (1983) showed that neovascularization of several human tumors grown on the chick chorioallantoic membrane or subcutaneously in nude mice developed at the same time as mononuclear infiltration at the tumor site. This led these workers to speculate that tumor growth was partially dependent on the angiogenic activity of infiltrating macrophages. Polverini and Leibovich (1987) reported that hamsters bearing chemical carcinogen-induced squamous cell carcinomas showed a marked reduction in thymidine incorporation by endothelial cells and neovascularization of tumors when treated with low doses of steroid and antimacrophage serum. A more direct assessment of the role of TAM in tumor neovascularization was conducted by Polverini and Leibovich (1984). They showed that TAM isolated from several transplantable chemical carcinogen-induced rat fibrosarcomas patently stimulated endothelial cell proliferation *in vitro* and neovascularization *in vivo*. Furthermore, they reported that tumors depleted of TAM grew more slowly when introduced into rat corneas. When TAM were added back to TAM-depleted tumors at concentrations equivalent to their original *in situ* concentration, tumor growth rate and the potency of neovascular responses were restored to the levels observed in native tumors. The importance of a sustained influx of activated TAM to tumor neovascularization has recently been investigated by Lingen, Bouck and Polverini (unpublished observations). They showed that human squamous carcinoma treated with the chemopreventive agent retinoic acid failed to activate Mφ to express angiogenic activity and exhibited a diminished capacity to stimulate chemotaxis of Mφ. These results suggest that the ability of retinoic acid to reduce the incidence of secondary tumor growths may be due to inducing the production of an inhibitor of angiogenesis by tumor cells and blocking the sustained infiltration of Mφ into tumors and their subsequent activation for expression of angiogenic activity.

Mφ and chronic inflammatory diseases

The persistent and unrelenting formation of granulation tissue is a feature common to a number of chronic inflammatory diseases. Cirrhosis of the liver, idiopathic pulmonary fibrosis and chronic periodontitis are other examples where the destructive effects of persistent granulation tissue and aberrant angiogenesis contribute significantly to the pathogenesis of the disease. Perhaps the best-studied disorder in which Mφ have been shown to participate in the pathogenesis by inducing persistent angiogenesis is rheumatoid arthritis. Rheumatoid arthritis is a systemic disease that is characterized by the development of a proliferating mass of invasive granulation tissue that degrades cartilage and bone of diarthrodial joints (Harris, 1974). The resulting hyperplastic synovial pannus invades and subsequently degrades cartilage and bone. The formation of this highly vascularized granulation tissues is largely responsible for the irreversible destruction of diarthrodial joints. The rheumatoid synovium contains many different cells types including synovial cells, fibroblasts, lymphocytes, monocytes and Mφ, as well as numerous capillary blood vessels. In a series of studies from the Koch laboratory (1986, 1992, 1995), the role of the Mφ in rheumatoid synovial neovascularization has been extensively studied. These workers were the first to show that subpopulations of Mφ retrieved from the inflamed synovial pannus were patently angiogenic *in vivo* and stimulated endothelial cell chemotaxis *in vitro*. Moreover, synovial Mφ, when fractionated into subpopulations by density gradient sedimentation, were found to be functionally heterogeneous in that only certain subpopulations were angiogenic. It was speculated that by altering either the production of a proangiogenic mediator or by altering the environment that would favor the accumulation of nonangiogenic Mϕ one might be in a position to alter the destructive effect of this disease process. The basis for this functional compartmentalization remains unclear but lends further support to the notion that Mϕ can display diverse and often opposing functions.

Although much of the data on Mϕ and disease-associated angiogenesis support a Mϕ role in the induction of this response, a report by Koch et al. (1992) suggests that their failure to produce angiogenic factors may contribute to the chronic debilitating disease scleroderma. Scleroderma is a poorly understood systemic disorder characterized by Raynaud's phenomenon, proliferative vascular lesions and fibrosis of the skin and organs (Seibold, 1985; Maricq et al., 1976; Maricq, 1981). Patients with scleroderma have a decreased number of capillaries, capillary hemorrhages, and clusters of enlarged distorted capillary loops which can be visualized in the skin folds of the nail beds that correlate with sclerotic changes in multiple organ systems. Since the changes involving large and small vessels are a major feature of this disorder, it has been proposed that endothelial cell dysfunction plays a role in the pathogenesis of this disease. There have only been limited studies, often with conflicting results, on the role of angiogenesis in scleroderma and none have examined the role of Mϕ to the level that has been pursued in other disorders of tissue remodeling. Kahaleh et al. (1986) reported that supernatants from peripheral blood monocytes derived from scleroderma patients inhibited endothelial cell proliferation. Marczak et al. (1986) have shown that mononuclear cell populations that were enriched for monocytes from patients with scleroderma showed enhanced angiogenic activity. Kaminsky et al. (1984) reported that populations of peripheral blood leukocytes enriched for monocytes enhanced angiogenesis. Koch et al. (1992), using highly purified populations of peripheral blood derived monocytes from scleroderma patients, showed that these cells failed to stimulate angiogenesis *in vivo* or endothelial cell proliferation and chemotaxis *in vitro*. This was despite the fact that production of the potent angiogenic mediator TNF-α by monocytes from these patients was not significantly different from that of normal patients. This raised the interesting possibility that the defect in Mϕ angiogenic activity in scleroderma patients may be due to overproduction of an inhibitor of neovascularization. Preliminary studies (Polverini and Koch, unpublished observations) of Mϕ supernatants from scleroderma patients suggest the presence of an angiogenesis inhibitory activity. The identity of this inhibitory activity remain unclear, although the angiogenesis inhibitor TSP1 is a major candidate and thus deserves further attention.

Psoriasis is a common skin disease, linked to both genetic and environmental triggering factors (Farber et al., 1974). It is characterized pathologically by excessive growth of epidermal keratinocytes, inflammation, and microvascular proliferation, which is believed to result from a disruption in the complex and reciprocal molecular cross-talk between activated keratinocytes and dermal cells (Nickoloff, 1991). Folkman (1972a, 1972b) has suggested that hyperkinetic psoriatic keratinocytes produced an angiogenic factor that mediated the neovascularization of the upper dermis, which manifests as vasodilatation and increased tortuosity of capillary loops in the papillary dermis (Braverman and Yen, 1977). Several lines of evidence have implicated both psoriatic keratinocytes and dermal Mϕ and dendritic cells in the persistent vascular proliferation that accompanies this disease. Using fresh human psoriatic lesional tissue which was separated into epidermal and dermal components, the angiogenic potential of the lesion was found by two different groups to reside in both the dermal and in the epidermal compartment (Wolfe and Hubler, 1976; Wolfe, 1989; Malhotra et al., 1989). Psoriatic keratinocytes are known to produce a variety of pro-angiogenic cytokines such as basic fibroblast growth factor, interleukin-1 (IL-1), transforming growth factor-alpha (TGF-α) and IL-8 (Nickoloff, 1991). In addition to expressing several candidate mediators of angiogenesis, keratinocytes are also known to be a source of the angiogenesis inhibitor TSP1 (Nickoloff et al., 1988). In an attempt to define further the molecular mechanism underlying this chronic inflammatory skin disease, Nickoloff et al. (1994) examined the mechanisms responsible for the disregulated vasoproliferation that characterizes psoriasis. These workers showed that psoriatic keratinocytes appeared to have a combined defect in both the overproduction of the pro-angiogenic cytokine, IL-8, and a deficiency in the production of the angiogenesis inhibitor TSP1. Previous studies have described differences between normal and psoriatic keratinocytes with respect to their growth response, (Nickoloff et al. 1989) and immunomodulating capacity (Nickoloff, 1991). It would appear that in addition to these "mutant-like phenotypes", psoriatic keratinocytes also have a defect in which there is an imbalance in the production of positive and negative angiogenic mediators that govern the orderly growth of new capillary endothelial cells. Direct evidence implicating Mϕ and/or dermal dendritic cells in psoriatic angiogenesis is lacking. Polverini and Nickoloff (unpublished data) have found that human monocyte-derived Mϕ exposed to conditioned media from psoriatic keratinocytes as well as from symptomless skin and conditioned media from cultures of dermal dendritic cells obtained from the skin of psoriatic patients, express potent angiogenic activity. The potency of angiogenic response appears to be linked to disease status. Dendritic cells obtained from symptomless skin or the skin of patients with active disease are angiogenic while those from normal skin show no activity. These preliminary observations suggest that Mϕ and dermal dentritic cells can be activated by psoriatic keratinocytes to express angiogenic activity and thus may contribute to the dermal vasoproliferation of this disease.

Future considerations and therapeutic implications

It is now well established that Mφ can influence angiogenesis in several physiological and pathological settings. Mφ have been shown to produce both stimulators and inhibitors of angiogenesis and thus have the potential to modulate angiogenesis in either a positive or negative fashion. The implications of these observations for the development of therapeutic strategies for the treatment of solid tumors and other angiogenesis-dependent disorders are obvious. The use of inhibitors of neovascularization for the treatment of angiogenesis-dependent diseases has long been envisioned as a possible mode of therapy (Folkman, 1972b, Folkman and Klagsbrun, 1987). Several examples already exist where this approach has met with success (Orchard et al., 1989; White et al., 1991). As the tools of genetic engineering move from the laboratory to the bedside, and as the molecular and biochemical basis for the functional diversity of many positive and negative regulators of angiogenesis are defined, it may be possible to up- or downregulate angiogenic responses with exquisite precision. For example, it may be possible to customize therapy by instructing accessory cell populations such as Mφ, that normally enter a focus of inflammation or infiltrate tumors, to produce or even overproduce endogenous or synthetic pro-angiogenic cytokines or inhibitors. Regardless of the diverse settings in which angiogenesis is encountered and the great redundancy of mediator systems that participate in this process, the sorting out of the mechanisms that control the balanced production of positive and negative regulators of angiogenesis by accessory cell populations such as Mφ will undoubtedly prove to be a fruitful area of investigation and ultimately provide improved treatment of patients who suffer from angiogenesis-dependent diseases.

Acknowledgments
Supported in part by NIH grants HL39926 and CA64416.

References

Adams, D.O. (1989) Molecular interactions in macrophage activation. *Immunol. Today* 10:33–35.
Adams, D.O., Hamilton, T.A. (1984) The cell biology of the macrophage activation. *Ann. Rev. Immun.* 2:283–310.
Anderson, N.D., Anderson, A.O., Wyllie, R.G. (1975) Microvascular changes in lymph nodes draining skin allographs. *Am. J. Pathol.* 81:131–153.
Auerbach, R. (1981) Angiogenesis-inducing factors: a review. *Lymphokines* 4:69–88.
Besner, G.E., Klagsbrun, M. (1991) Mφ secrete a heparin-binding inhibitor of endothelial cell growth. *Microvasc. Res.* 42:187–197.
Bornstein, P., Sage, H.E. (1994) Thrombospondins. *Methods in Enzymology* 245:62–85.
Bornstein, P. (1995) Diversity of function is inherent in matrix proteins: An appraisal of thrombospondin 1. *J. Cell Biol.* 130:503–506.
Bouck, N.P., Stoler, A., Polverini, P.J. (1986) Coordinate control of anchorage independence, actin cytoskeleton and angiogenesis by human chromosome 1 in hamster-human hybrids. *Cancer Res.* 46:5101–5105.

Bouck, N. (1990) Tumor angiogenesis: role of oncogenes and tumor suppressor genes. *Cancer Cells* 2:179–185.

Bouck, N. (1993) Angiogenesis: a mechanism by which oncogenes and tumor suppressor genes regulate tumorigenesis. In: *Oncogenes and tumor suppressor genes in human malignancy* (Benz, C.C., Liu, E.T. editors, Boston: Kluwer Academic, pp. 359–371.

Braverman, J.M., Yen, A. (1977) Ultrastructure of the capillary loops in the dermal papillae of psoriasis. *J. Invest. Dermatol.* 68:53–58.

Browder, W., Williams, D., Lucore, P., Pretus, H., Jones, E., McNamee, R. (1988) Effect of enhanced macrophage function on early wound healing. *Surgery* 104:224–230.

Calderon, J., Kiely, J.-M., Lefko, J.L., Unanue, E.R. (1975) The modulation of lymphocyte functions by molecules secreted by Mφ. I. Description and partial biochemical analysis. *J. Exp. Med.* 142:151–164.

Calderon, J., Unanue, E.R. (1975) Two biological activities regulating cell proliferation found in cultures of peritoneal exudate Mφ. *Nature* 253:359–361.

Calderon, J., Williams, R.T., Unanue, E.R. (1974) An inhibitor of cell proliferation released from cultures of Mφ. *Proc. Natl. Acad. Sci. USA.* 71:4273–4277.

Calderon, J., Kiely, J.M., Lefka, J., Ununue, E. (1976) The modulation of lymphocyte functions by molecules secreted by macrophages. II. conditions leading to increased secretion. *J. Exp. Med.* 144:155–160.

Carrell, A. (1922) Growth promoting functions of leukocytes. *J. Exp. Med.* 36:385–391.

Carlos, T.M., Harlan, J.M. (1994) Leukocyte-endothelial adhesion molecules. *Blood* 84:2068–2101.

Clark, R.A. Stone, R.D., Leung, D.Y.K., Silver, I., Hohn, D.D., Hunt, T.K. (1976) Role of Mφ in wound healing. *Surg. Forum* 27:16–18.

Dameron, K.M., Volpert, O.V., Tainsky, M.A., Bouck, N. (1994) Control of angiogenesis in fibroblasts by p53 regulation of thrombospondin-1. *Science* 265:1582–1584.

Danon, D., Kowatch, M.A., Roth, G.S. (1989) Promotion of repair in old mice by local injection of Mφ. *Proc. Natl. Acad. Sci. USA* 86:2018–2020.

DiPietro, L.A., Polverini, P.J. (1993) Role of the macrophage in the positive and negative regulation of wound neovascularization. *Behring Inst. Mitt.* 92:238–247.

DiPietro, L.A., Polverini, P.J. (1994) Angiogenic Mφ produce the angiogenesis inhibitor thrombospondin 1. *Am. J. Pathol.* 143:678–684.

DiPietro, L.A., Nebgen, D.R., Polverini, P.J. (1994) Downregulation of endothelial cell thrombospondin 1 enhances in vitro angiogenesis. *J. Vas. Res.* 31:178–185.

DiPietro, L.A., Nissen, N.N., Gamelli, R.L., Koch, A.E., Pyle, J.M., Polverini, D.T. (1996) Thrombospondin 1 synthesis and function in wound repair. *Am. J. Pathol.* 148:1851–1860.

Dvorak, A.M., Mim, M.C., Dvorak, H.F. (1976) Morphology of delayed-type hypersensitivity reactions in man. II. Ultrastructural alterations affecting the microvasculature and the tissue mast cell. *Lab. Invest.* 34:179–191.

Evans, R. (1977a) Effect of x-irradiation on host cell infiltration and growth of murine fibrosarcoma. *Br. J. Cancer* 35:557–566.

Evans, R. (1977b) The effect of azothioprine on host-cell infiltration and growth of a murine fibrosarcoma. *Int. J. Cancer* 20:120–128.

Evans, R. (1978) Macrophage requirement for growth of murine fibrosarcoma. *Br. J. Cancer* 37:1086–1095.

Fahey, T.J., Sherry, B., Tracey, K.J., van Deventer, S., Jones, W.G., Minei, J.P., Morgello, S., Shires, G.T., Cerami, A. (1991) Cytokine production in a model of wound healing: the appearance of MIP-1, MIP-2, cachetin/TNF, and IL-1. *Cytokine* 2:2–19.

Fajardo, L.F., Kwan, H.H., Kowalski, J., Prionas, S.D., Allison, A.C. (1992) Dual role of tumor necrosis factor-α in angiogenesis. *Am. J. Pathol.* 140:539–544.

Farber, E.M., Nall, M.L., Watson, W. (1974) Natural history of psoriasis in 61 twin pairs. *Arch. Dermatol.* 109:207–211.

Folkman, J. (19772a) Angiogenesis in psoriasis: Therapeutic implications. *J. Invest. Dermatol.* 59:40–48.

Folkman, J. (1972b) Anti-angiogenesis: a new concept for therapy of solid tumors. *Ann. Surg.* 175:409–416.

Folkman, J., Cotran, R.S. (1976) Relation of vascular proliferation to tumor growth. *Int. Rev. Exp. Pathol.* 16:207–248.

Folkman, J., Klagsbrun, M. (1987) Angiogenic factors. *Science* 235:442–447.

Ford, H., Hoffman, R.A., Wing, E.J., Magee, M., McIntyre, L., Simmons, R.L. (1987) Characterization of wound cytokines in the sponge matrix model. *Arch. Surg.* 124:1422–1428.

Frazier, W.A. (1987) Thrombospondin: a modular adhesive glycoprotein of platelets and nucleated cells. *J. Cell Biol.* 105:625–632.

Frazier, W.A. (1991) Thrombospondin. *Curr. Opinions in Cell Biol.* 3:792–799.

Gell, P.G.H. (1959) Cytological events in hypersensitivity reactions. In: *Cellular and Humoral Aspects of the Hypersensitivity States.* Lawrence (Ed) New York: Harper and Row, p. 43.

Gillespie, G.Y., Estes, J.E., Pledger, W.J. (1986) Macrophage derived growth factors for mesenchymal cells. *Lymphokines* 2:213–242.

Good, D.J., Polverini, P.J. Rastinejad, F., Le Beau, M.M., Lemons, R.S., Frazier, W.A., Bouck, N.P. (1990) A tumor suppressor-dependent inhibitor of angiogenesis is immunologically and functionally indistinguishable from a fragment of thrombospondin. *Proc. Natl. Acad. Sci. USA* 87:6624–6628.

Gordon, S. (1986) The biology of the macrophage. *J. Cell Sci. Suppl.* 4:267–286.

Gordon, S., Unkeless, J.C., Cohn, Z.A. (1974) Induction of macrophage plasminogen activator by endotoxin stimulation and phagocytosis. *J. Exp. Med.* 140:995–1010.

Graham, R.C., Shannon, S. (1972) Peroxidase arthritis. II. Lymphoid cell-endothelial interactions during developing immunologic inflammatory responses. *Am. J. Pathol.* 69:7–14.

Grotendorst, G., Grotendorst, C.A., Gilman, T. (1988) Production of growth factors (PDGF and TGF-β) at the site of tissue repair. *Prog. Clin. Biol. Res.* 266:131–145.

Harris, E.D. Jr. (1974) Recent insights into the pathogenesis of the proliferative lesions in rheumatoid arthritis. *Arthritis Rheum.* 19:48–72.

Hunt, T.K., Knighton, D.R., Thakral, K.K., Goodson, W.H., Andrews, W.S. (1984) Studies on inflammation and wound healing: angiogenesis and collagen synthesis stimulated *in vivo* by resident and activated wound Mϕ. *Surgery* 96:48–54.

Ingber, D.E., Folkman, J. (1989) Mechanochemical switching between growth and differentiation during fibroblast growth factor-stimulated angiogenesis *in vitro*. Role of the extracellular matrix. *J. Cell Biol.* 109:317–330.

Ingber, D.E. (1991) Extracellular matrix and cell shape: Potential control points for the inhibition of angiogenesis. *J. Cell Biochem.* 47:236–241.

Iruela-Arispe, M., Bornstein, P., Sage, H. (1991) Thrombospondin exerts an antiangiogenic effect on cord formation by endothelial cells *in vitro*. *Proc. Natl. Acad. Sci. USA* 88:-6026–5030.

Jaffe, E.A., Ruggiero, J.T., Falcone, D.J. (1985) Monocytes and Mϕ synthesize and secrete thrombospondin. *Blood* 65:79–84.

Jensen, J.A., Hunt, T.K., Scheuenstuhl, H., Banda, M. J. (1986) Effect of lactate, pyruvate, and pH on secretion of angiogenesis and mitogenesis factors by macrophages. *Lab. Invest.* 54:574–578.

Johnston, R.B., Jr. (1988) Monocytes and Mϕ. *New Eng. J. Med.* 318:747–752.

Kahaleh, M.B., DeLustro, F., Bock, W., LeRoy, E.C. (1986) Human monocyte modulation of endothelial cell and fibroblast growth: Possible mechanisms for fibrosis. *Clin. Immunol. Immunopathol.* 39:242–255.

Kaminski, M.J., Majewski, S., Jablonska, S., Pawinska, M. (1984) Lowered angiogenic capability of peripheral blood lymphocytes in progressive systemic sclerosis (scleroderma). *J. Invest. Dermatol.* 82:239–243.

Klagsbrun, M., D'Amore, P.A. (1991) Regulators of angiogenesis. *Ann.. Rev. Physiol.* 53:217–239.

Knighton, D.R., Hunt, T.K., Scheuenstuhl, N., Halliday, B.T., Werb, Z., Bunda, M.J. (1983) Oxygen tension regulates the expression of angiogenesis factor by macrophages. *Science* 221:1283–1285.

Koch, A.E., Polverini, P.J., Leibovich, S.J. (1986) Stimulation of neovascularization by human rheumatoid synovial tissue Mϕ. *Arth. Rheum.* 29:471–479.

Koch, A.E., Litvak, M.A., Burrows, J.C., Polverini, P.J. (1992) Decreased monocyte-mediated angiogenesis in scleroderma. *Clin. Immunol. Immunopath.* 64:153–160.

Koch, A.E., Polverini, P.J. Kunkel, S.L., Harlow, L.A., DiPietro, L.A., Elner, V.M., Elner, S.G., Strieter, R.M. (1992b) Interleukin-8 (IL-8) is a potent macrophage-derived mediator of angiogenesis that is blocked by IL-8 antibody and antisense oligonucleotides. *Science* 258:179–1801.

Koch, A.E., Halloran, M.M., Haskell, C.J., Shah, M.R., Polverini, P.J. (1995) Angiogenesis mediated by soluble forms of E-selectin and vascular cell adhesion molecule-1. *Nature* 376:517–519.

Lawler, J. (1986) The structural and functional properties of thrombospondin. *Blood* 67: 1191–1209.

Leibovich, S.J., Ross, R. (1975) The role of the macrophage in wound repair: a study with hydrocortisone and antimacrophage serum. *Am. J. Pathol.* 78:71–91.

Leibovich, S.J., Ross, R. (1976) A macrophage-dependent factor that stimulates the proliferation of fibroblasts *in vivo*. *Am. J. Pathol.* 84:501–514.

Leibovich, S.J., Polverini, P.J., Shepard, H.M., Wiseman, D.M., Shirely, V., Nuseir, N. (1987) Macrophage-induced angiogenesis is mediated by tumor necrosis factor-α. *Nature* 329: 430–432.

Leibovich, S.J., Wiseman, D.M. (1988) Mϕ, wound repair and angiogenesis. *Proc. Clin. Biol. Res.* 266:131–145.

Malhotra, R., Stenn, K.S., Fernandez, L.A., Braverman, I.M. (1989) Angiogenic properties of normal and psoriatic skin associated with epidermis, not dermis. *Lab. Invest.* 61: 162–165

Mantovani, A. (1994) Biology of disease. Tumor-associated Mϕ in neoplastic progression: A paradigm for the *in vivo* function of chemokines. *Lab. Invest.* 71:5–16.

Marczak, M., Majewski, S., Skopinska-Rozewska, E., Polakowski, I., Jablonska, S. (1986) Enhanced angiogenic capability of monocyte-enriched mononuclear cell suspensions from patients with systemic sclerosis. *J. Invest. Dermatol.* 86:355–358.

Maricq, H.R., Spencer-Green, G., LeRoy, E.C. (1976) Skin capillary abnormalities as indicators of organ involvement in scleroderma (systemic sclerosis), Raynaud's syndrome and dermatomyositis. *Am. J. Med.* 61:862–870.

Maricq, H.R. (1981) Widenfield capillary microscopy. Technique and rating scale for abnormalities seen in scleroderma and related disorders. *Arthritis Rheum.* 24:1159–1165.

Moore, J.W., Sholley, M.M. (1985) Comparison of the neovascular effects of stimulated Mϕ and neutrophils in autologous rabbit corneas. *Am. J. Pathol.* 120:87–98.

Mostafa, L.K., Jones, D.B., Wright, D.H. (1980a) Mechanism of induction of angiogenesis by human neoplastic lymphoid tissue: studies on the chorioallentoic membrane (CAM) of the chick embryo. *J. Pathol.* 132:197–205.

Mostafa, L.K., Jones, D.B., Wright, D.H. (1980b) Mechanism of induction of angiogenesis by human neoplastic lymphoid tissue: studies employing bovine aortic endothelial cells *in vitro*. *J. Pathol.* 132:207–216.

Nathan, C. (1987) Secretory products of Mϕ. *J. Clin. Invest.* 79:319–326.

Nickoloff, B.J., Riser, B.L., Mitra, R.S., Dixit, V.M., Varani, J. (1988) Inhibitory effect of gamma interferon on cultured keratinocytes thrombospondin production, distribution, and biological activity. *J. Invest. Dermatol.* 91:213–218.

Nickoloff, B.J. (1991) The cytokine network in psoriasis. *Arch. Dermatol.* 127:871–884.

Nickoloff, B.J., Mitra, R.S., Varani, J., Dixit, V.M., Polverini, P.J. (1994) Aberrant production of interleukin-8 and thrombospondin-1 by psoriatic keratinocytes mediates angiogenesis. *Am. J. Pathol.* 144:820–828.

Orchard, P.J., Smith, C.M., Woods, W.G., Day, D.L., Dehner, L.P. (1989) Treatment of hemangioendotheliomas with α-interferon. *The Lancet* 2:565–567.

O'Shea, K.S., Dixit, V.M. (1988) Unique distribution of the extracellular matrix component thrombospondin in the developing mouse embryo. *J. Cell. Biol.* 107:2737–2748.

Polverini, P.J., Cotran, R.S., Sholley, M.M. (1977a) Endothelial proliferation in the delayed hypersensitivity reaction: An autoradiographic study. *J. Immunol.* 118:529–532.

Polverini, P.J., Cotran, R.S., Gimbron, M.A., Jr., Unanue, E.R. (1977b). Activated Mϕ induce vascular proliferation. *Nature* 269:804–806.

Polverini, P.J., Leibovich, S.J. (1984) Induction of neovascularization *in vivo* and endothelial cell proliferation *in vitro* by tumor-associated Mϕ. *Lab. Invest.* 51:635–642.

Polverini, P.J., Leibovich, S.J. (1987) Effect of macrophage depletion on growth and neovascularization of hamster buccal pouch carcinomas. *J. Oral Pathol.* 16:436–441.

Polverini, P.J. (1989) Macrophage-Induced Angiogenesis: A Review. In: *Macrophage-Derived Regulatory Factors*, Sorg, C., (Ed.), Basel: S. Karger, pp. 54–73.

Polverini, P.J. (1995) The pathophysiology of angiogenesis. *Crit. Rev. Oral Biol. Med.* 6:230–247.

Polverini, P.J., DiPietro, L.A., Dixit, V.M., Hynes, R.O., Lawler, J. (1995) Thrombospondin 1 knockout mice show delayed organization and prolonged neovascularization of skin wounds. *FASEB J.* 9:272a.

Rappolee, D.A., Mark, D., Banda, M.J., Werb, Z. (1988) Wound Mφ express TGF-α and other growth factors *in vivo*: analysis by mRNA phenotyping. *Science* 241:708–712.

Rastinejad, F., Polverini, P.J., Bouck, N.P. (1989) Regulation of the activity of a new inhibitor of angiogenesis by a cancer suppressor gene. *Cell* 56:345–355.

Riches, D.W. (1988) The multiple roles of Mφ in wound healing. In: *The Molecular and Cellular Biology of Wound Repair*, Clark, R.A.F. and Henson, P.M. (eds.), New York: Plenum, pp. 4023–36

Ross, R., Odland, G.F. (1968) Human wound repair, II. Inflammatory cells, epithelial-mesenchymal interrelations and fibrogenesis. *J. Cell Biol.* 39:152–168.

Sage, H., Bornstein, P. (1982) Endothelial cells from umbilical vein and a hemangioendothelioma secrete basement membrane largely tot he exclusion of interstitial procollagens. *Arteriosclerosis* 2:27–36.

Sage, H., Bornstein, P. (1991) Extracellular proteins that modulate cell-matrix interactions. *J. Biol. Chem.* 266:14831–14834.

Sager, R. (1989) Tumor suppressor genes: the puzzle and the promise. *Science* 246:1406–1410.

Schrader, J.W. (1973) Mechanism of activation of the bone marrow-derived lymphocyte. III. A distinction between a macrophage-produced triggering signal and the amplifying effect on triggered B lymphocytes of all agenetic interactions. *J. Exp. Med.* 138:1466–1480.

Schreiber, A.B., Winkler, M.E., Derynck, R. (1986) Transforming growth factor-α: a more potent angiogenic mediator than epidermal growth factor. *Science* 232:1250–1253.

Sidkey, Y.A., Auerbach, R. (1975) Lymphocyte-induced angiogenesis (LIA): A quantitative and sensitive assay of the graft vs. host reaction. *J. Exp. Med.* 141:1084–1110.

Seibold, J.R., Scleroderma. In: *Textbook of Rheumatology*. Third ed. (Kelley, W.E., Harris, E.D., Ruddy, S., Sledge, C.B., eds.) Saunders, Philadelphia, PA 1985.

Silverstein, R.L., Nachman, R.L. (1987) Thrombosponin binds to monocytes and Mφ and mediates platelet-monocyte adhesion. *J. Clin. Invest.* 79:867–874.

Sorg, C., Lohmann-Matthes, M.L. (1989) Macrophages and accessory cells of the immune system. *Immunol. Today* 10:27–29.

Stein, M., Keshav, S. (1992) The versitility of Mφ. *Clin. Exp. Allergy* 22:18–27.

Stenzinger, W., Bruggen, J., Macher, E., Sorg, C. (1983) Tumor angiogenic activity (TAA) production *in vivo* and growth in the nude mouse by human malignant melanoma. *Eur. J. Cancer Clin. Oncol.* 19:649–656.

Strieter, R.M., Kunkel, S.L., Elner, V.M., Martonyi, C.L., Koch, A.E., Polverini, P.J., Elner, S.G. (1992) Interleukin-8: a corneal factor that induces neovascularization. *Am. J. Pathol.* 141:1279–1284.

Sunderkotter, C., Goebeler, M., Schultze-Osthoff, K., Bhardwaj, R., Sorg, C. (1991) Macrophage-derived angiogenesis factors. *Pharmac. Ther.* 51:195–216.

Sunderkotter, C., Steinbrink, K., Goebeler, M., Bhardwaj, R., Sorg, C. (1994) Macrophages and angiogenesis. *J. Leu. Biol.* 55:410–422.

Skekanecz, Z., Haines, G.K., Lin, T.R., Harlow, L.A., Goerdt, S., Rayan, G., Koch, A.E. (1994) Differential distribution of intercellular adhesion molecules (ICAM-1, ICAM-2, and ICAM-3) and the MS-1 antigen in normal and diseased human synovia. Their possible pathogenetic and clinical significance in rheumatoid arthritis. *Arthritis Rheum.* 37:221–231.

Thakral, K.K., Goodson, W., Hunt, T.K. (1979) Stimulation of wound blood vessel growth by wound Mφ. *J. Surg. Res.* 26:430–436.

Unanue, E.R. (1976) Secretory function of mononuclear phagocytes. *Am. J. Pathol.* 83:396–417.

Unanue, E.R., Kiely, J.-M., Calderon, J. (1976) The modulation of lymphocyte functions of molecules secreted by Mφ. II. conditions leading to increased secretion. *J. Exp. Med.* 144:155–166.

Van Furth, R. (1988) Phagocytic cells: development and distribution of mononuclear phagocytes in normal steady state and in inflammation. In: *Inflammation: Basic Principles and Clinical Correlates*. Gallin, J.I., Goldstein, I.M., Snyderman, R. (eds.) Raven Press, New York, pp. 281–295.

Vilette, D., Setiadi, H., Wautier, M.-P., Caen, J., Wautier, J.-L. (1990) Identification of an endothelial cell growth inhibitory activity produced by human monocytes. *Exp. Cell Res.* 188:219–225.

Weinberg, R. (1989) Oncogenes, antioncogenes, and the molecular basis of multistep carcinogenesis. *Cancer Res.* 49:3713–3721.

White, C.W., Wolf, S.J., Korones, D.N., Sondheimer, H.M., Tosi, M.F., Yu, A. (1991) Treatment of childhood angiomatous diseases with recombinant interferon-α. *J. Pediatr.* 118:59–66.

Wolfe, J.E., Jr., Hubler, W.R., Jr. (1976) Angiogenesis in psoriasis. In: *Psoriasis: Proceedings of the Second International Symposium.* Edited by Farber, E., Cox, A. New York: York Medical Books, pp. 40:375–377.

Wolfe, J.E. (1989) Angiogenesis in normal and psoriatic skin. *Lab. Invest.* 61:139–142.

Regulation of Angiogenesis
ed. by I.D. Goldberg & E.M. Rosen
© 1997 Birkhäuser Verlag Basel/Switzerland

Angiogenesis in human gliomas:
Prognostic and therapeutic implications

J. P. Johnson and J. N. Bruce

Department of Neurological Surgery, Neurological Institute of New York, College of Physicians and Surgeons of Columbia University, New York, New York 10032, USA

Introduction

Angiogenesis is fundamental to the growth of all solid tumors, and efforts at elucidating the critical elements of this process have yielded a wealth of information about the biology, clinical behavior and potential treatment of these neoplasms. This information is particularly relevant for brain tumors, which are among the most highly vascular of human neoplasms, and undoubtedly much of what is known about systemic cancers also applies to brain tumors. However, these factors must be evaluated differently with regard to tumors of the central nervous system and within the context of the unique character of cerebrovascular physiology and glioma biology. The nature and significance of glioma angiogenesis are yet to be fully explored, but already important insights into its clinical implications have been developed. Further clarification of the association between angiogenesis and clinical behavior holds exciting promise for novel anti-angiogenic strategies for glioma therapy.

Gliomas – overview

Neoplasms derived from glial precursor make up the bulk of primary brain tumors, accounting for approximately half of all such lesions (Walker et al., 1985). Although the term glioma is often taken to be synonymous with the astrocytoma, particularly its most malignant and common subtype, the glioblastoma, other tumors of neuroepithelial origin exist under this classification as well. These tumors share the critical characteristics of being intraparenchymal primary tumors of the brain with poorly defined margins, a tendency to recur locally after removal, and a resistance to virtually all standard adjuvant treatments. They are grouped by their presumed cell of origin into the astrocytoma, ependymoma, and oligodendroglioma.

Astrocytomas are presumed to arise from mature astrocytes and account for the vast majority of gliomas (Burger et al., 1985; Walker et al., 1985;

Adams and Victor, 1993; Kaye and Laws, 1995). They tend to occur in middle aged and elderly individuals, although all ages are affected. They may be seen throughout the neuraxis, with most adult tumors occurring in the cerebral hemispheres. Growth pattern are characterized by infiltrating margins that blend into the surrounding white matter, frustrating attempts at local therapy such as surgical resection or focused radiation.

A spectrum of malignancy exists within the category of astrocytomas. While several grading systems with slightly different criteria have been formulated to score this potential, generally gliomas can be classified into three somewhat distinct groups (low grade astrocytoma, anaplastic astrocytoma, and glioblastoma multiforme) (Daumas-Duport et al., 1988; Burger et al., 1991). However, these tumors exist along a continuous spectrum of malignancy, and attempts to classify them into groups are somewhat artificial. The low grade astrocytoma is characterized by hyperplasia of relatively normal appearing astrocytes. This relatively indolent tumor typically affects younger adults. Although median survival approaches 10 years, these tumors often progress to a higher grade with passing time. The anaplastic astrocytoma is a more aggressive lesion characterized by the appearance of more frequent mitoses and cellular atypia (Fulling and Garcia, 1985; Daumas-Duport et al., 1988). Median survival is approximately 2 years. The glioblastoma represents the most malignant, and unfortunately most common, of the astrocytomas, accounting for roughly 90 percent of all adult primary tumors of the cerebral hemisphere. Pathologically, these tumors show areas of necrosis and vascular proliferation in addition to marked anaplasia. Glioblastomas grow relentlessly and tend to recur locally despite any treatment modality employed (Burger et al., 1985; Russell and Rubinstein, 1989; Kaye and Laws, 1995). Although surgical resection and radiotherapy may prolong survival somewhat, the median survival after diagnosis is still only 9 months.

The pilocytic astrocytoma is another type of astrocyte-derived tumor which exhibits very different behavior to the more common fibrillary astrocytoma described above. These tumors are much more indolent, almost hamartomatous lesions that grow slowly and rarely progress to more aggressive histopathology (Russell and Rubinstein, 1989). In adults they often occur in the hypothalamus or in other eloquent areas around the third ventricle. Although the anatomical location may limit surgical resection, long term survival is the general rule. In children these tumors typically develop in the cerebellar hemisphere (Fulchiero et al., 1977; Winston et al., 1977). Complete resection is often possible and curative in these tumors.

The ependymoma and oligodendroglioma are relatively rare glial neoplasms derived from ependymal cells and oligodendrocytes, respectively. Although in their pure forms they are often less aggressive in appearance and behavior, they share the low grade astrocytoma's tendency towards infiltration, local recurrence and eventual progression to glioblastoma

(Mork et al., 1986; Burger et al., 1987; Ross and Rubinstein, 1989; Russell and Rubinstein, 1989; Burger et al., 1991; Kaye and Laws, 1995).

Tumor angiogenesis – general characteristics

It is a central tenet in tumor biology that neoplasms must incite angiogenesis in order to grow to clinical significance. Folkman proposed in 1971 that avascular collections of neoplastic cells grow to only 2–3 millimeters in diameter before reaching the limits of passive diffusion for nutrient delivery and waste removal (Folkman, 1971; Folkman and Hochberg, 1973; Folkman, 1975). Tumors may remain arrested at this size indefinitely until a change occurs which allows them to induce the development of new microvasculature. Following neovascularization, a tumor may grow exponentially and rapidly reach clinical significance. This growth pattern is the primary mechanism by which gliomas exert their deleterious effects, either by direct invasion or compression of local neural structures or cerebrospinal fluid pathways or by general increase in intracranial pressure in the rigid cranial compartment. Unlike gliomas, this rapid growth in extracranial tumors is critical with regard to the development of distant metastases, which is a process crucially linked to angiogenesis (Blood and Zetter, 1990; Liotta et al., 1991).

This concept has been refined and expanded along several fronts. It is apparent that the capacity for angiogenesis exists in all tissues and is tightly regulated. Tumor angiogenesis represents a loss of this control and a subversion of the process by the neoplastic cells (Blood and Zetter, 1990; Liotta et al., 1991). The sequence of events in tumor angiogenesis has been elucidated (Folkman, 1986; D'Amore and Thompson, 1987; Blood and Zetter, 1990; Folkman, 1990; Liotta et al., 1991), and it appears that the tumor cells themselves attract the development of new capillary networks from existing microvasculature. A proteolytic process is first initiated which disrupts the local host microvasculature. Endothelial cells are stimulated to migrate toward the tumor through these breaks and to proliferate. The development of a lumen and maturation to a capillary network follow. Apart from the endothelial cells themselves, circulating cells, notably macrophages and mast cells, also assist in this angiogenesis (D'Amore and Thompson, 1987).

The microvessels that result from the process of tumor angiogenesis resemble normal vessels in many respects but have several important differences (D'Amore and Thompson, 1987). Notably, pericytes are supporting cells which are closely associated with normal capillary beds but are absent from most tumor vessels. The basement membrane composition may also be altered in tumor capillaries. The endothelial cells themselves are also frequently larger and more irregular in tumor vessels (Blood and Zetter, 1990). These changes may account for the abnormal physiology of these capillary networks, especially the increased microvascular permeability, which leads to peritumoral fluid such as edema or ascites and probably contributes to metastasis as well (Jain, 1988).

While these concepts form a theoretical framework for the clinical importance of tumor angiogenesis, more recently the association between a tumor's propensity for angiogenesis and its clinical behavior has been established (Weidner, 1993). Demonstration of the correlation between angiogenesis and clinical behavior has been aided by the development of quantitative methods, including sensitive and accurate assays for the number of microvessels in a tumor. The number of vessels has been shown to be a strong predictor of the likelihood of metastasis in invasive ductal carcinoma of the breast (Weidner et al., 1991), prostate carcinoma (Wakui et al., 1992; Weidner, et al., 1993), and malignant melanoma (Srivastava et al., 1986; Srivastava et al., 1988). It is also one of the strongest predictors of long term survival in breast cancer (Bosari et al., 1992; Weidner et al., 1992). Microvessel assays are being incorporated into the routine evaluations of these cancers for clinical decision making because of their strong prognostic implications (Gasparini and Harris, 1995).

Glioma angiogenesis

Glial tumors are characteristically hypervascular as evidenced by their clinical behavior and their histologic appearance. Positron emission tomographic (PET) imaging of regional cerebral blood flow in individual with gliomas has demonstrated increased, although highly variable, perfusion of these tumors compared to surrounding brain (Ito et al., 1982). This is likely a reflection of their increased vascularity, although loss of cerebral autoregulatory control mechanisms may also contribute to this hyperperfusion. The progression of malignancy from low grade and anaplastic astrocytoma to glioblastoma is, by definition, associated with histological evidence of vascular proliferation (Daumas-Duport et al., 1988; Russell and Rubinstein, 1989; Burger et al., 1991). This neovascularization typically produces a striking pattern of capillary tufts resembling renal glomeruli which are found throughout the tumor (Burger et al., 1985; Russell and Rubinstein, 1989; Haddad, 1992b). The number of capillary channels is increased compared to normal brain (Brem, 1976; Russell and Rubinstein, 1989). Ultrastructurally, the basement membrane is often disrupted (Luse, 1960). The endothelial cells themselves are plump and occasionally hyperplastic and even redundant. Mitotic figures are sometimes seen within the endothelial cells themselves, and S-phase labeling with antibody to bromodeoxyuridine (BrdU) demonstrates a significantly higher rate of tumor endothelial cell growth than in normal brain vessels (Nagashima et al., 1985). In one study, the DNA labeling index of these tumor endothelial cells exceeded that of the glioblastoma cells themselves (Groothius et al., 1980).

Although most systemic tumor vessels lack normal stromal cells, immu-nocytochemical studies have suggested that glioma neovascularity contains pericytes and vascular smooth muscle cells (Wesserling et al., 1995). Haddad et al. (1992) have even suggested that smooth muscle cell prolife-ration accounts for a majority of the hypervascular changes associated with these glomeruloid structures. In some instances the degree of endothelial and perivascular anaplasia can be suggestive of sarcomatous degeneration; such tumors are designated gliosarcomas and carry a particularly grim prognosis (Burger et al., 1991; Haddad et al., 1992a). However, the func-tional significance of these cells is unknown.

Glioma vasculature is functionally abnormal as well and contributes heavily to the morbidity of these tumors. The blood-brain barrier is incom-petent, and microvascular permeability is increased. This leads to the development of peritumoral edema, which in the closed intracranial compartment leads ultimately to increased intracranial pressure (Adams and Victor, 1993). Furthermore, normal cerebral blood flow autoregulatory control is lost within glial tumors, producing local hyperemia which both contributes to fluid transudation and leads to peritumoral steal and brain ischemia (Kaye and Laws, 1995). These neovascular beds often display abnormal fragility and are prone to spontaneous hemorrhage, producing rapid clinical deterioration (Liwnicz et al., 1987).

The prognostic significance of neovascularity in gliomas is complex due to several unique aspects of their clinical behavior. While the large body of evidence linking angiogenesis and prognosis in non-central nervous system tumors is compelling, caution must be exercised in generalizing these results to glial neoplasms. In general, non-central ner-vous system tumors create clinical morbidity, and ultimately mortality, through the development of metastatic disease. Local cancer is of little significance in terms of survival in most cases of breast, prostate, or skin cancer. Angiogenesis is integral to the metastatic pathway, and the quan-titative link between the two has been established (Folkman, 1975; D'Amore and Thompson, 1987; Blood and Zetter, 1990; Liotta et al., 1991; Weidner et al., 1991; Horak et al., 1992; Wakui et al., 1992; Weidner et al., 1993).

In sharp contrast, glial tumors only rarely metastasize outside of the central nervous system (Liwnicz and Rubinstein, 1979; Cerame et al., 1985; Burger et al., 1991). Gliomas tend to spread via direct extension within the parenchyma of brain or spinal cord. Growth typically occurs along white matter tracts and respects natural planes such as pial or epen-dymal margins. Less commonly, tumors may spread throughout the neuraxis via cerebrospinal fluid dissemination. When removed, gliomas tend to recur locally at the margins of resection. Although they occasional-ly recur at multiple sites in the brain, suggestive of blood-borne metastasis within the CNS, glial tumors are nearly always fatal due to relentless, progressive local growth (Liwnicz and Rubinstein, 1979).

Therefore, the presumed causal link which exists for extracranial tumors between angiogenesis and prognosis via metastasis does not exist for gliomas. Other mechanisms by which angiogenesis may directly affect the clinical course of gliomas must be postulated. Foremost among these is the likelihood that aggressive local growth is facilitated by intense angiogenesis. However, an absolute link between angiogenesis and growth potential is somewhat unclear, as will be discussed below. Other mechanisms related to functionally abnormal vasculature have also been presented, including increased permeability leading to peritumoral edema, ischemic steal, and capillary fragility and hemorrhage. Each of these may also contribute to the link between angiogenesis and poorer prognosis. Finally, glial tumor cells which are capable of stimulating more abundant angiogenesis may also possess more aggressive growth characteristics. In this view, increasing angiogenesis would be only a marker for a higher degree of malignancy. However, these explanations are not supported by studies described below which suggest a specific connection between inhibition of angiogenesis and tumor regression.

The link between increasing degrees of angiogenesis and prognosis in gliomas is apparent in general, although several interesting exceptions exist. In fibrillary astrocytomas, angiogenesis is, by definition, associated with increasing tumor grade and poorer prognosis (Daumas-Duport et al., 1988). But because neovascularity is required for a diagnosis of glioblastoma, it is difficult to separate the exact contribution of angiogenesis to outcome. Brem et al. addressed this issue by scoring astrocytomas with a microscopic grading system which assessed vasoproliferation, endothelial hyperplasia, and endothelial cytology (Brem et al., 1972; Brem, 1976). They found that this score correlated with conventional tumor grade, and a higher score was associated with shorter survival. However, the strongest correlations were seen with endothelial hyperplasia and cytology, while capillary density alone did not correlate with length of survival. Thus, this study does not fully resolve whether angiogenesis is an independent predictor of prognosis in gliomas.

Within the subcategory of anaplastic astrocytoma, one study found that vascular endothelial proliferation was prognostically significant, while the degree of tumor cell anaplasia and necrosis were not (Fulling and Garcia, 1985). However, two larger studies, while also demonstrating shorter survival with increasing degrees of neovascularity, could not identify this as a prognostic factor independent of other variables, particularly increasing age (Burger et al., 1985; Cohadon et al., 1985). The presence of necrosis in glioblastomas is also as strong a prognostic factor as neovascularity (Nelson et al., 1983). It is unclear how necrosis, which suggests that tumor cells are outgrowing their capacity to induce angiogenesis, relates to the causal link of angiogenesis to outcome.

The strength of the link between neovascularity and outcome is further complicated by the behavior of glial neoplasms other than those of the fibrillary astrocytoma/glioblastoma lineage. The most puzzling of these are pilocytic astrocytomas. These relatively indolent tumors are frequently highly vascular, showing the same pattern of glomeruloid vascular-proliferation as glioblastomas. The degree of vasculogenesis in these tumors does not alter their prognosis (Fulchiero et al., 1977; Winston et al., 1977; Russell and Rubinstein, 1989). Vascular proliferation also failed to predict survival in oligodendrogliomas (Mork et al., 1986; Burger et al., 1987) and ependymomas (Ross, 1989). Clearly, the issue of the significance of angiogenesis in intrinsic glial neoplasms is complex and in need of further study.

Angiogenesis factors

A number of factors have been implicated in glioma angiogenesis. Each of these is in the family of polypeptide growth factors, is presumably produced by neoplastic cells and acts either upon endothelial cells themselves to cause migration or mitosis or upon intermediate cells such as macrophages to effect the release of other growth factors (Folkman and Klagsburn, 1987a,b).

Vascular endothelial growth factor (VEGF)

The importance of this glycoprotein, also known as vascular permeability factor (VPF), in glioma angiogenesis has become apparent through a large body of work in this area. This factor was initially found by Bruce et al. (1987) to be produced by glioma cells growing in culture and to be responsible for glioma-associated brain edema. The factor is now known to be produced by glioma cells both in culture and *in vivo* and is a potent endothelial cell mitogen. Levels of VEGF and its specific receptors Flt-1 and Flk-1, as demonstrated by immunohistochemistry, correlate closely with degree of angiogenesis in human gliomas (Plate et al., 1992; Plate et al., 1994). Production of VEGF has been shown to be inducible by hypoxia, and highest expression is seen associated with the boundaries of necrotic areas of these tumors (Shweiki et al., 1995). It may also mediate the actions of other angiogenesis factors, as it is upregulated by epidermal growth factor (EGF), platelet-derived growth factor (PDGF) (with which it shares significant homology), and basic fibroblast growth factor (bFGF) (Goldman et al., 1993; Tsai et al., 1995). Most convincingly, the administration of neutralizing antibody to VEGF produces a striking reduction in tumor vascularity and size in an animal glioma model (Kim et al., 1993). This factor is also unique in that it appears to mediate vasogenic brain edema, further supporting its role in targeting glioma vasculature (Bruce et al., 1987; Strugar et al., 1995).

Fibroblast growth factors (FGF)

Two types of this protein factor, which is also a potent endothelial cell mitogen, are relevant to gliomas. Acidic FGF has been shown by immunohistochemistry to be present in neoplastic but not normal astrocytes (Lieberman et al. 1987; Stefanik et al., 1991). Basic FGF is found in the matrix of glioma vessels but is not seen in normal brain vasculature. Moreover, the degree of staining for basic FGF has been shown to correlate with histological grade and tumor vascularity as assessed by angiography in human gliomas (Takahashi et al., 1992). Nguyen et al. found that urinary levels of basic FGF were generally elevated in patients with tumors of several types (Nguyen et al., 1994). Twenty percent of glioma patients demonstrated levels greater than the 90th percentile for healthy controls.

In an even more convincing demonstration of the potential importance of this factor, Li et al. (1994) measured basic FGF levels in the cerebrospinal fluid (CSF) of 26 children with brain tumors of various types. They found that the factor was detectable in 62% of the tumor patients but in none of the controls. The CSF of the tumor patients had endothelial mitogenicity which correlated with basic FGF levels and was eliminated by neutralizing antibody to basic FGF. Furthermore, microvascular counts from the surgical specimens from these patients correlated with CSF basic FGF levels, as did prognosis. It is interesting to note, however, that, although glioblastomas were of course underrepresented in this pediatric population, the highest basic FGF level, mitogenic activity, and neovascular grade were seen in a patient with a pilocytic astrocytoma. The potential utility of this type of assay as well as the role of basic FGF in glioma angiogenesis clearly warrants further study.

Epidermal growth factor (EGF)

Urinary levels of this factor were found to be elevated in patients with high grade gliomas as compared with other tumors, including low grade gliomas (Kanno et al., 1993). Furthermore, levels correlated with the volume of tumor and fell in response to treatment with either surgery or radiation. As EGF is also an endothelial cell mitogen capable of inducing angiogenesis *in vitro* (Folkman and Klagsburn, 1987a), a link may exist between this factor and glioma angiogenesis.

Transforming growth factors

The transforming growth factors (TGFs) are polypeptides which have the ability to alter the phenotype of several cell types to transformed cells (Folkman and Klagsburn, 1987a). They are strong endothelial cell mitogens, and stimulate angiogenesis *in vitro*. TGF-alpha has been demonstrated to be overexpressed in glioma cell lines and tumor specimens (Paulus et al., 1995; Yamada et al., 1995). This factor is purported to affect tumor cell-extracellular matrix interactions. In glioma cell lines it was found to facilitate tumor cell adhesion to the extracellular matrix and to inhibit vessel invasion (Paulus et al., 1995). The increased tumor cell adhesion might explain the unusual lack of metastatic potential despite intense angiogenesis in glial neoplasms.

Therapeutic strategies

Ever since the recognition of the importance of angiogenesis in tumor growth, a great deal of effort has been directed at the development of therapeutic strategies to block the development of tumor vascular proliferation (Folkman, 1971; Folkman, 1975; Langer and Murray, 1983; Blood and Zetter, 1990; Folkman, 1990; Sipos et al., 1994). In theory, if this response could be blocked, tumors would be unable to maintain a significant growth rate or to metastasize.

Anti-angiogenesis therapy for brain tumors offers several additional advantages, as well as significant limitations not found with extracranial neoplasms. One advantage is that, as mentioned previously, gliomas generally do not metastasize. A therapy targeted to local disease could theoretically be effective. Secondly, as an immunologically privileged site, these therapies might be expected to escape interference from the host immune system. Also, the adult brain contains no cells which are routinely dividing (except perhaps endothelial cells themselves) and does not, under routine circumstances, express normal angiogenesis (Russell and Rubinstein, 1989). Anti-angiogenesis therapies directed at highly vascular gliomas might therefore be expected to be less toxic to the surrounding brain. Finally, with sensitive modern imaging studies, tumor growth and response to therapy can be followed noninvasively.

In order to realize these potential benefits, however, drug delivery to the glioma must first be overcome. The blood-brain barrier limits effective delivery of most large molecules to these tumors via systemic routes (Kaye and Laws, 1995). Several alternative methods for direct delivery of compounds to the brain have been devised. Brem et al. have designed polymer wafers which are impregnated with the therapeutic agent (Langer and Murray, 1983; Jain, 1989; Brem et al., 1991; Tamargo and Brem, 1992; Brem et al., 1995). These may be implanted directly into the tumor; they are degraded over time and slowly release the drug, where it permeates the interstitium via passive diffusion. Another approach is high flow microinfusion, whereby a catheter implanted in the tumor delivers the compound at a constant rate of several microliters per minute (Bobo et al., 1994; Laske et al., 1994; Morrison et al., 1994; Lieberman et al., 1995). Large molecules may be delivered at a high and relatively uniform concentration to the entire tumor. Both of these delivery systems are currently being evaluated in clinical trials.

The following represents a list of agents with anti-angiogenic potential that have received attention for the treatment of gliomas.

Anti-VEGF

VEGF expression is elevated in glial tumors, and this factor appears to have a central role in glioma angiogenesis. As such, this may present a particularly inviting target for therapy. The systemic administration of anti-VEGF monoclonal antibody to nude mice with subcutaneous glial tumors dramatically suppressed both tumor growth and angiogenesis (Kim et al., 1993). Furthermore, the growth of intracerebral gliomas in nude mice could be prevented by retroviral infection with dominant-negative Flk-1/VEGF receptor (Millauer et al., 1994). These results elegantly reveal a specific role for VEGF in glioma growth which is probably mediated by angiogenesis and which may be inhibited selectively. The anti-VEGF antibody also represents a compound amenable to direct local delivery into gliomas, although this has not yet been examined.

Suramin

This polyanionic compound is a potent antiangiogenic and antitumoral compound. It appears to act by blocking the receptor binding of several growth factors, including basic FGF, PDGF, EGF, TGF, and VEGF (Stein, 1993). It has been shown to limit the growth of several glial tumor lines in cell culture, possibly by interrupting autocrine stimulation by these and other growth factors (Takano et al., 1994). In an *in vitro* angiogenesis assay, suramin also blocked neovascularization induced by glioma cells (Coomber, 1995). When administered to rats harboring intracerebral gliomas, suramin produced significant inhibition of growth and angiogenesis, although the survival benefit was eliminated by intratumoral hemorrhage seen in most cases (Takano et al., 1994).

Angiostatic steroids

Heparin-steroid conjugates have marked antiangiogenic effects, both in standard *in vitro* assays and against tumors growing *in situ* (Folkman et al., 1983; Crum et al., 1985; Harada et al., 1992; McNatt et al., 1992; Thorpe et al., 1993). Cortisone, given in conjunction with heparin, has been shown to cause tumor regression and prevent metastasis in neurofibrosarcoma-bearing nude mice (Folkman et al., 1983). The mechanism of this appears to involve inhibition of the proteolytic process which is necessary to the initiation of neovascularization (Folkman, 1987; Tamargo et al., 1990; Blei et al., 1993). Different steroids have very different effects with regard to angiogenesis, and glucocorticoid activity itself is not required for this potential. Several novel compounds have appeared with strong anti-angiogenic effects but lacking in the potential for glucocorticoid side effects (Harada et al., 1992; Norrby et al., 1993; Yamamoto et al., 1994).

Glucocorticoids are a mainstay of treatment for glioma-induced cerebral edema (Kaye and Laws, 1995). This implies an effect on tumor vasculature, although the mechanism of this well-known clinical effect remains poorly understood. Evidence exists, however, for a modulating effect of steroids on tumor-induced endothelial cell growth *in vitro* (Wolff et al., 1992). Wolff et al. (1993) found that dexamethasone given systematically to rats bearing intracerebral gliomas resulted in growth inhibition associated with decreased tumor microvascular density (1990). Tamargo et al. (1990) extended this finding to glial tumors in the rat flank model. Significant reduction in growth was observed when heparin and cortisone were incorporated into a degradable polymer wafer and implanted directly into these tumors.

AGM-1470 (TNP-470)

This agent is a synthetic analog of fumagillin, a fungal product with significant antiangiogenic properties (Ingber et al., 1990). It appears to have very specific effects on endothelial cell growth, producing cytostatic growth arrest in cell culture (Kusaka et al., 1994). This is associated with marked reduction in angiogenesis and tumor growth associated with various types of primary and metastatic tumors (Morita et al., 1994; Mori et al., 1995; O'Reilly et al., 1995; Tanaka et al., 1995a; Tanaka et al., 1995b; Tsujimoto et al., 1995; Yano et al., 1995; Yazaki et al., 1995). The compound also appears to have direct growth suppression effects on tumor cells themselves (Yanase et al., 1993), particularly glioma cells in culture (Takamiya et al., 1994), as well as nerve sheath tumor cells (Takamiya et al., 1993) and meningioma cells (Yazaki et al., 1995).

AGM-1470 has shown considerable promise against gliomas in animal models. Given systemically, it caused growth inhibition in flank or subrenally-implanted gliomas, which was associated with decreased angiogenesis and altered vessel morphology (Taki et al., 1994). Growth of intracranial glimas was inhibited and survival prolonged in one study of systemically-administered AGM-1470 (Takamiya et al., 1994), although another study demonstrated no benefit (Wilson and Penar, 1994). The decreased efficacy may be a reflection of difficulties with delivery of this large molecule to the intracerebral tumors. This may be overcome by some of the novel local delivery approaches that are being developed for gliomas. Additionally, it appears to be possible to augment the antitumoral effects of AGM-1470 by concomitantly increasing tumor oxygenation during treatment (Teicher et al., 1995).

Minocycline

Tetracycline antibiotics inhibit the degradation of basement membranes necessary in the initiation of angiogenesis, possibly via an effect on matrix metalloproteinases (Tamargo et al., 1991; Sipos et al., 1994), although other independent mechanisms may exist (Gilbertson Beadling et al., 1995). Although they do not affect the growth of tumor cells in culture, they do appear to inhibit angiogenesis *in vitro* assays (Tamargo et al., 1991). Minocycline has received attention as the most lipid soluble of these compounds. It has been shown to cause tumor regression and to prolong survival when injected directly into growing rat glial tumors in the brain but not when given systemically (Weingart et al., 1995).

Penicillamine

Penicillamine inhibits endothelial cell proliferation and angiogenesis *in vitro* (Brem et al., 1990). Although the mechanism is unclear, it may involve chelation of copper, a known modulator of angiogenesis. Gliomas have been demonstrated to be relatively rich in copper, with levels that correlated closely with tumor grade (Yoshida, 1993a). Systemically administered penicillamine appears to limit the growth of glial tumors in the rat flank, particularly in animals with dietary copper deficiency (Yoshida, 1993b; Yoshida et al., 1995). This is accompanied by diminished tumor vascularity as well. Penicillamine also inhibited tumor growth and angiogenesis in intracerebral carcinomas in the rabbit (Brem et al., 1990). This response was also augmented by systemic copper depletion.

Anti-FGF

Stan et al. (1995) demonstrated that the administration of polyclonal antibody to basic FGF by direct injection into intracerebral glial tumors in nude mice produced significant inhibition of tumor growth and angiogenesis. This direct injection is similar to microinfusion, and demonstrates the potential of this delivery system for macromolecular compounds.

Tumor Necrosis Factor (TNF)

Niida et al. (1995) have reported that in a co-culture system of endothelial and C6 glioma cells, the induction of capillary-like structures could be blocked by the administration of TNF-alpha. Further study will be necessary to elucidate the significance of this finding.

Conclusions

Angiogenesis is a critical process in tumor growth and an inviting target for therapeutic strategies. This is particularly true for gliomas, which are among the most highly vascular of neoplasms. Neovascularity is associated with worsening prognosis for the most common and devastating type of gliomas, the glioblastoma. Whether this is due to an effect of the tumor vasculature itself, as is probably the case for metastasis-prone systemic cancers, or simply a marker for increasing malignancy is not yet clear. Nevertheless, therapeutic strategies aimed at angiogenesis remain conceptually appealing for gliomas. Several agents have shown promise in preclinical testing. If these results are borne out in clinical trials, then anti-angiogenic treatments could form an important part of a multimodal approach to the treatment of these difficult tumors.

References

Adams, R. and Victor, M. (1993) *Principles of Neurology*. McGraw-Hill, New York.

Blei, F., Wilson, E.L., Mignatti, P. and Rifkin, D.B. (1993) Mechanism of action of angiostatic steroids: suppression of plasminogen activator activity via stimulation of plasminogen activator inhibitor synthesis. *J. Cell Phys.* 155 (3):568–578.

Blood, C. and Zetter, B. (1990) Tumor interactions with the vasculature: Angiogenesis and tumor metastasis. *Biochem. Biophys. Acta* 1032:89–118.

Bobo, R.H., Laske, D.W., Akbasak, A., Morrison P.F., Dedrick R.L. and Oldfield E.H. (1994) Convection-enhanced delivery of macromolecules in the brain. *PNAS (USA)* 91 (6): 2076–2080.

Bosari, S., Lee, A., DeLellis, R., Wiley, B., Heatley, G. and Silverman, M. (1992) Microvessel quantitation and prognosis in invasive breast carcinoma. *Hum. Path.* 23:755–761.

Brem, H., Mahaley, M. and Vick, N. (1991) Interstitial chemotherapy with drug polymer implants for the treatment of recurrent gliomas. *J. Neurosurg.* 74 (3):441–446.

Brem, H., Piantadosi, S., Burger, P.C., Walker, M., Selker, R., Vick, N.A., Black, K., Sisti, M., Brem, S. and Mohr, G. (1995) Placebo-controlled trial of safety and efficacy of intraoperative controlled delivery by biodegradable polymers of chemotherapy for recurrent gliomas. The Polymer-brain Tumor Treatment Group. *Lancet* 345 (8956):108–1012.

Brem, S. (1976) The role of vascular proliferation in the growth of brain tumors. *Clin. Neurosurg.* 23:440–453.

Brem, S., Cotran, R. and Folkman, J. (1972) Tumor angiogenesis: A quantitative method for histologic grading. *J. NCI* 48:347–356.

Brem, S., Zagzag, D., Tsanaclis, A., Gately, S., Elkouby, M. and Brien, S. (1990) Inhibition of angiogenesis and tumor growth in the brain: Suppression of endothelial cell turnover by penicillamine and the depletion of copper, a angiogenic cofactor. *Am. J. Path.* 137 (5): 1121–1142.

Bruce, J., Criscuolo, G., Merrill, M., Moquin, R., Blacklock, J. and Oldfield, E. (1987) Vascular permeability induced by protein product of malignant brain tumors: inhibition by dexamethasone. *J. Neurosurg.* 67:880–884.

Burger, P., Rawlings, C., Cox, E., McLendon, R., Schold, S. and Bullard, D. (1987) Clinicopathologic correlations in the oligodendroglioma. *Cancer* 59:1345–1352.

Burger, P., Scheithauer, B. and Vogel, F. (1991) *Surgical Pathology of the Nervous System and its Coverings*. New York, Churchill Livingstone.

Burger, P., Vogel, F., Green, S. and Strike, T. (1985) Glioblastoma multiforme and anaplastic astrocytoma; Pathologic criteria and prognostic implications. *Cancer* 56:1106–1111.

Cerame, M. Guthikonda, M. and Kohli, C. (1985) Extraneural metastases in gliosarcoma: A case report and review of the literature. *Neurosurg.* 17 (3):413–418.

Cohadon, F., Aouad, N., Rougier, A., Vital, C., Rivel, J. and Dartigues, J. (1985) Histologic and non-histologic factors correlated with survival time in supratentorial astrocytic tumors. *J. Neuro. Onc.* 3:105–111.

Coomber, B. (1995) Suramin inhibits C6 glioma-induced angiogenesis *in vitro. J. Cell Biochem.* 58 (2):199–207.

Crum, R., Szabo, S. and Folkman, J. (1985) A new class of steroids inhibits angiogenesis in the presence of heparin or a heparin fragment. *Science* 230:1375–1378.

D'Amore, P. and Thompson, R. (1987) Mechanisms of angiogenesis. *Ann. Rev. Phys.* 49: 453–464.

Daumas-Duport, C., Sheithauer, B., O'Fallon, J. and Kelly, P. (1988) Grading of Astrocytomas; a simple and reproducible method. *Cancer* 62:2152–2165.

Folkman, J. (1971) Tumor angiogenesis: Therapeutic implications. *NEJM* 285 (21):1182–1186.

Folkman, J. (1975) Tumor angiogenesis: A possible control point in tumor growth. *Ann. Internal Med.* 82 (1):96–100.

Folkman, J. (1986) How is blood vessel growth regulated in normal and neoplastic tissue? *Cancer Res.* 46:467–473.

Folkman, J. (1990) What is the evidence that tumors are angiogenesis dependent? *J. NCI* 82 (1):4–6.

Folkman, J. and Hochberg, M. (1973) Self regulation of growth in three dimensions. *J. Exp. Med.* 138:745–753.

Folkman, J. and Ingber, D. (1987) Angiostatic steroids: Method of discovery and mechanism of action. *Ann. Surg.* 206 (3):374–383.

Folkman, J. and Klagsbrun, M. (1987a) Angiogenic factors. *Science* 235:442–447.

Folkman, J. and Klagsbrun, M. (1987b) A family of angiogenic peptides. *Nature* 329: 671–672.

Folkman, J., Langer, R., Linhardt, C., Haudenschild, C. and Taylor, S. (1983) Angiogenesis inhibition and tumor regression caused by heparin or a heparin fragment in the presence of cortisone. *Science* 221:719–725.

Fulchiero, A., Winston, K., Leviton, A. and Gilles, F. (1977) Secular trends of cerebellar gliomas in children. *J. NCI* 58 (4):839–843.

Fulling, K. and Garcia, D. (1985) Anaplastic astrocytoma of the adult cerebrum; Prognostic value of histologic features. *Cancer* 55:928–931.

Gasparini, G. and Harris, A. (1995) Clinical importance of the determination of tumor angiogenesis in breast carcinoma: Much more than a new prognostic tool. *J. Clin. Onc.* 13:765–782.

Gilbertson-Beadling, S., Powers, E.A., Stamp-Cole, M., Scott, P.S., Wallace, T.L., Copeland, J., Petzold, G., Mitchell, M., Ledbetter, S. and Poorman, R. (1995) The tetracycline analogs minocycline and doxycycline inhibit angiogenesis *in vitro* by a non-metalloproteinase-dependent mechanism. *Cancer Chemo. Pharm.* 36 (5):418–424.

Goldman, C., Kim, J., Wong, W., King, V., Brock, T. and Gillespie, G. (1993) Epidermal growth factor stimulates vascular endothelial growth factor production by human malignant glioma cells: A model of glioblastoma multiforme pathophysiology. *Mol. Biol. Cell* 4:121–133.

Groothius, D., Fischer, D., Vicks, J. and Bigner, D. (1980) Experimental gliomas: an autoradiographic study of the endothelial component. *Neurology* 30:297.

Haddad, S., Moore, S., Schelper, R. and Goeken, J. (1992a) Vascular smooth muscle hyperplasia underlies the formation of glomeruloid vascular structures of glioblastoma multiforme. *J. Neuropath. Exp. Neurol.* 51 (5):488–492.

Haddad, S., Moore, S., Schelper, R. and Goeken, J. (1992b) Smooth muscle can comprise the sarcomatous component of gliosarcomas. *J. Neuropath. Exp. Neurol.* 51 (5):493–498.

Harada, I., Kikuchi, T., Shimonura, Y., Yamamoto, M., Ohno, H. and Sato, N. (1992) The mode of action of anti-angiogenic steroid and heparin. In: *Angiogenesis: Key principles – Science – Technology – Medicine.* EXS 61, Steiner, R., Weisz, P.B. and Langer, R. (eds). Basel: Birkhäuser Verlag 1992, 445–458.

Horak, E., Leek, R., Klenk, N., LeJeune, S., Smith, K., Stuart, N., Greenall, M., Stepniewska, K. and Harris, A. (1992) Angiogenesis, assessed by platelet/endothelial cell adhesion molecule antibodies, as indicator of node metastasis and survival in breast cancer. *Lancet* 340:1120–1124.

Ingber, D., Fujita, T., Kishimoto, S., Sudo, K., Kanamaru, T., Brem, H. and Folkman, J. (1990) Synthetic analogues of fumagillin that inhibit angiogenesis and suppress tumour growth. *Nature* 348:555–557.

Ito, M., Lammertsma, A. and Wise, R. (1982) Measurement of regional cerebral blood flow and oxygen utilisation in patients with cerebral tumors using O-15 and positron emission tomography; analytical techniques and preliminary results. *Neuroradiology* 23:63–74.

Jain, R. (1988) Determinants of tumor blood flow: A review. *Cancer Res.* 48:2641–2658.

Jain, R. (1989) Delivery of novel therapeutic agents in tumors: physiological barriers and strategies. *J. NCI* 81 (8):570–576.

Kanno, H., Chiba, Y., Kyuma, Y., Hayashi, A., Abe, H., Takada, H., Kim, I. and Yamamoto, I. (1993) Urinary epidermal growth factor in patients with gliomas: significance of the factor as a glial tumor marker. *J. Neurosurg.* 79 (3):408–413.

Kaye, A. and Laws, E., Ed. (1995) *Brain Tumors: An Encyclopedic Approach.* New York, Churchill Livingstone.

Kim, K., Li, B., Winer, J., Arrmanini, M., Gillett, N., Phillips, H. and Ferrara, N. (1993) Inhibition of vascular endothelial growth factor induced angiogenesis suppresses tumor growth *in vivo. Nature* 362:841–844.

Kusaka, M., Sudo, K., Matsutani, E., Kozai, Y., Marui, S., Fujita, T., Ingber, D. and Folkman, J. (1994) Cytostatic inhibition of endothelial cell growth by the angiogenesis inhibitor TNP-470 (AGM-1470). *Brit. J. Cancer* 69 (2):212–216.

Langer, R. and Murray, J. (1983) Angiogenesis inhibitors and their delivery systems. *Appl. Biochem. Biotech.* 8:9–24.

Laske, D.W., Ilercil, O., Akbasak, A., Youle, R.J. and Oldfield, E.H. (1994) Efficacy of direct intratumoral therapy with targeted protein toxins for solid human gliomas in nude mice. *J. Neurosurg.* 80 (3):520–526.

Li, V., Folkerth, R., Watanabe, H., Yu, C., Rupnick, M., Barnes, P., Scott, R., Black, P., Sallan, S. and Folkman, J. (1994) Microvessel count and cerebrospinal fluid basic fibroblast growth factor in children with brain tumors. *Lancet* 344:82–86.

Lieberman, D.M., Laske, D.W., Morrison, P.F., Bankiewicz, K.S. and Oldfield, E.H. (1995) Convection-enhanced distribution of large molecules in gray matter during interstitial drug infusion. *J. Neurosurg* 82 (6):1021–1029.

Lieberman, T., Friesel, R., Jaye, M., Lyall, R., Westermark, B., Drohan, W., Schmidt, A., Maciag, T. and Schlessinger, J. (1987) An angiogenic growth factor is expressed in human glioma cells. *EMBO* 6:1627–1632.

Liotta, L., Steeg, P. and Stetler-Stevenson, W. (1991) Cancer metastasis and angiogenesis: An imbalance of positive and negative regulation. *Cell* 64:327–336.

Liwnicz, B. and Rubinstein, L. (1979) The pathways of extraneural spread in metastasizing gliomas: A report of three cases and critical review of the literature. *Hum. Path.* 10 (4):453–467.

Liwnicz, B., Wu, S. and Tew, J. (1987) The relationship between the capillary structure and hemorrhage in gliomas. *J. Neurosurg.* 66:536–541.

Luse, S. (1960) Electron microscopic studies of brain tumors. *Neurology* 10 (10):881–905.

McNatt, L.G., Lane, D. and Clark, A.F. (1992) Angiostatic activity and metabolism of cortisol in the chorioallantoic membrane (CAM) of the chick embryo. *J. Steroid Biochem. Mol. Biol.* 42 (7):687–693.

Millauer, B., Shawver, L., Plate, K., Risau, W. and Ullrich, A. (1994) Glioblastoma growth inhibited *in vivo* by a dominant-negative Flk-1 mutant. *Nature* 367 (6463):576–579.

Mori, S., Ueda, T., Kuratsu, S., Hosono, N., Izawa, K. and Uchida, A. (1995) Suppression of pulmonary metastasis by angiogenesis inhibitor TNP-470 in murine osteosarcoma. *Int. J. Cancer* 61 (1):148–152.

Morita, T., Shinohara, N. and Tokue, A. (1994) Antitumour effect of a synthetic analogue of fumagillin on murine renal carcinoma. *Brit. J. Urol.* 74 (4):416–421.

Mork, S., Halvorsen, T., Lindegaard, K. and Eide, G. (1986) Oligodendroglioma: Histologic evaluation and prognosis. *J. Neuropath. Exp. Neurol.* 45 (1):65–78.

Morrison, P.F., Laske, D.W., Bobo, H., Oldfield, E.H. and Dedrick, R.L. (1994) High-flow microinfusion: tissue penetration and pharmacodynamics. *Am. J. Physiol.* 266:R292–305.

Nagashima, T., DeArmond, S., Murovic, S. and Hoshino, T. (1985) Immunocytochemical demonstration of S-phase cells by anti-bromodeoxyuridine monoclonal antibody in human brain tumor tissues. *Acta. Neuropath.* 67:155.

Nelson, J., Tsukada, Y., Schoenfeld, D., Fulling, K., Lamarche, J. and Peress, N. (1983) Necrosis as a prognostic criterion in malignant, supratentorial, astrocytic gliomas. *Cancer* 52: 550–554.

Nguyen, M., Watanabe, H., Budson, A.E., Richie, J.P., Hayes, D.F. and Folkman, J. (1994) Elevated levels of an angiogenic peptide, basic fibroblast growth factor, in the urine of patients with a wide spectrum of cancers [see comments]. *J. NCI* 86 (5):356–361.

Niida, H., Takeuchi, S., Tanaka, R. and Minakawa, T. (1995) Angiogenesis in microvascular endothelial cells induced by glioma cells and inhibited by tumor necrosis factor *in vitro*. *Neurol. Med. Chirurg.* 35 (4):209–214.

Norrby, K., Jakobsson, A. and Nilsson, C.L. (1993) Two potentially angiostatic factors, a steroid and L-azetidine-2-carboxylic acid, antagonize one another. *Int. J. Microcirc. Clin. Exp.* 13 (2):113–124.

O'Reilly, M.S., Brem, H. and Folkman, J. (1995) Treatment of murine hemangioendotheliomas with the angiogenesis inhibitor AGM-1470. *J. Ped. Surg* 30 (2):325–329; discussion 329–330.

Paulus, W., Baur, I., Huettner, C., Schmausser, B., Roggendorf, W., Schlingensiepen, K.H. and Brysch, W. (1995) Effects of transforming growth factor-beta 1 on collagen synthesis, integrin expression, adhesion and invasion of glioma cells. *J. Neuropath. Exp. Neurol.* 54 (2):236–244.

Plate, K., Breier, G., Weich, H., Mennel, H. and Risau, W. (1992) Vascular endothelial growth factor is a potent tumour angiogenesis factor in human gliomas *in vivo*. *Nature* 359:845–848.

Plate, K., Breier, G., Weich, H. and Risau, W. (1994) Vascular endothelial growth factor and glioma angiogenesis: Coordinate induction of VEGF receptors, distribution of VEGF protein and possible *in vivo* regulatory mechanisms. *Int. J. Cancer* 59 (4):520–529.

Ross, G. and Rubinstein, L. (1989) Lack of histopathological correlation of malignant ependymomas with postoperative survival. *J. Neurosurg.* 70:31–36.

Russell D. and Rubinstein, L. (1989) *Pathology of Tumors of the Nervous System*. Baltimore, Williams & Wilkins.

Shweiki, D., Neeman, M., Itin, A. and Keshet, E. (1995) Induction of vascular endothelial growth factor expression by hypoxia and by glucose deficiency in multicell spheroids: Implications for tumor angiogenesis. *PNAS (USA)* 92 (3):768–772.

Sipos, E.P., Tamargo, R.J., Weingart, J.D. and Brem, H. (1994) Inhibition of tumor angiogenesis. *Ann. NY. Acad. Sci.* 732:263–272.

Srivastava, A., Laidler, P., Hughes, L., Woodcock, J. and Shedden, E. (1986) Neovascularization in human cutaneous melanoma: A quantitative morphological and doppler ultrasound study. *Euro J. Cancer Clin. Onc.* 22 (10):1205–1209.

Srivastava, A., Laidler, P., Davies, R., Horgan, K. and Hughes, L. (1988) The prognostic significance of tumor vascularity in intermediate-thickness skin melanoma. *Am. J. Path.* 133 (2):419–423.

Stan, A., Nemati, N., Pietsch, T., Walter, G. and Dietz, H. (1995) *In vivo* inhibition of angiogenesis and growth of the human U-87 malignant glial tumor by treatment with an antibody against basic fibroblast factor. *J. Neurosurg.* 82 (6):1044–1052.

Stefanik, D., Rizkalla, L., Soi, A., Goldblatt, S. and Rizkalla, W. (1991) Acidic and basic fibroblast growth/factors are present in glioblastoma multiforme. *Cancer Res.* 51 (20): 5760–5765.

Stein, C. (1993) Suramin: A novel antineoplastic agent with multiple potential mechanisms of action. *Cancer Res.* 53 (10):2239–2248.

Strugar, J., Criscuolo, G., Rothbart, D. and Harrington, W. (1995) Vascular endothelial growth/permeability factor expression in human glioma specimens: Correlation with vasogenic brain edema and tumor-associated cysts. *J. Neurosurg.* 83 (4):682–689.

Takahashi, J., Fukumoto, M., Igarashi, K., Oda, Y., Kikuchi, H. and Hatanaka, M. (1992) Correlation of basic fibroblast growth factor expression levels with the degree of malignancy and vascularity in human gliomas. *J. Neurosurg.* 76 (5):792–798.

Takamiya, Y., Brem, H., Ojeifo, J., Mineta, T. and Martuza, R.L. (1994) AGM-1470 inhibits the growth of human glioblastoma cells *in vitro* and *in vivo*. *Neurosurg.* 34 (5):869–875; discussion 875.

Takamiya, Y., Friedlander, R.M., Brem, H., Malick, A. and Martuza, R.L. (1993) Inhibition of angiogenesis and growth of human nerve-sheath tumors by AGM-1370. *J. Neurosurg.* 78 (3):470–476.

Takano, S., Gately, S., Engelhard, H., Tsanaclis, A.M. and Brem, S. (1994) Suramin inhibits glioma cell proliferation in vitro and in the brain. *J. Neuro. Onc.* 21 (3): 189–201.

Taki, T., Ohnishi, T., Arita, N., Hiraga, S., Saitoh, Y., Izumoto, S., Mori, K. and Hayakawa, T. (1994) Anti-proliferative effects of TNP-470 on human malignant glioma *in vivo*: potent inhibition of tumor angiogenesis. *J. Neuro Onc.* 19 (3):251–258.

Tamargo, R. and Brem, H. (1992) Drug delivery to the central nervous system: a review. *Neurosurg. Quarterly* 2 (4):259–279.

Tamargo, R., Leong, K. and Brem, H. (1990) Growth inhibition of the 9L glioma using polymers to release heparin and cortisone acetate. *J. Neuro. Onc.* 9:131–138.

Tamargo, R.J., Bok, R.A. and Brem, H. (1991) Angiogenesis inhibition by minocycline. *Cancer Res.* 51 (2):672–675.

Tanaka, T., Konno, H., Matsuda, I., Nakamura, S. and Baba, S. (1995a) Prevention of hepatic metastasis of human colon cancer by angiogenesis inhibitor TNP-470. *Cancer Res.* 55 (4):836–839.

Tanaka, H., Taniguchi, H., Mugitani, T., Koishi, Y., Masuyama, M., Higashida, T., Koyama, H., Suganuma, Y., Miyata, K. and Takeuchi, K. (1995b) Intra-arterial administration of the angiogenesis inhibitor TNP-470 blocks liver metastasis in a rabbit model. *Brit. J. Cancer* 72 (3):650–653.

Teicher, B.A., Dupuis, N.P., Emi, Y., Ikebe, M., Kakeji, Y. and Menon, K. (1995) Increased efficacy of chemo- and radio-therapy by a hemoglobin solution in the 9L gliosarcoma. *In Vivo* 9 (1):11–18.

Teicher, B.A., Holden, S.A., Ara, G., Dupuis, N.P., Liu, F., Yuan, J., Ikebe, M. and Kakeji, Y. (1995) Influence of an anti-angiogenic treatment on 9L gliosarcoma: oxygenation and response to cytotoxic therapy. *Int. J. Cancer* 61 (5):732–737.

Thorpe, P.E., Derbyshire, E.J., Andrade, S.P., Press, N., Knowles, P.P., King, S., Watson, G.J., Yang, Y.C. and Rao-Bette, M. (1993) Heparin-steroid conjugates: new angiogenesis inhibitors with antitumor activity in mice. *Cancer Res.* 53 (13):3000–3007.

Tsai, J., Goldman, C. and Gillespie, G. (1995) Vascular endothelial growth factor in human glioma cell lines: Induced secretion by EGF, PDGF-BB, and bFGF. *J. Neurosurg.* 82 (5):864–873.

Tsujimoto, H., Hagiwara, A., Osaki, K., Ohyama, T., Sakakibara, T., Sakuyama, A., Ohgaki, M., Imanishi, T., Watanabe, N. and Yamazaki, J. (1995) Therapeutic effects of the angiogenesis inhibitor TNP-470 against carcinomatous peritonitis in mice. *Anti-Cancer Drugs* 6 (3):438–442.

Wakui, S., Furusato, M., Itoh, T., Sasaki, H., Akiyama, A., Kinoshita, I., Asano, K., Tokuda, T., Aizawa, S. and Ushigome, S. (1992) Tumor Angiogenesis in prostatic carcinoma with and without bone marrow metastasis: a morphometric study. *J. Path.* 168:257–262.

Walker, A., Robins, M. and Weinfeld, F. (1985) Epidemiology of brain tumors: the national survey of intracranial neoplasms. *Neurology* 35:219–226.

Weidner, N. (1993) Tumor angiogenesis: Review of current applications in tumor prognostication. *Seminars in Diagnostic Path.* 10 (4):302–213.

Weidner, N., Carroll, P., Flax, J., Blumenfeld, W. and Folkman, J. (1993) Tumor angiogenesis correlates with metastasis in invasive prostate carcinoma. *Am. J. Path.* 143 (2): 401–409.

Weidner, N., Folkman, J., Pozza, F., Bevilacqua, P., Allred, E., Moore, D., Meli, S. and Gasparini, G. (1992) Tumor angiogenesis: A new significant and independent prognostic indicator in early-stage breast carcinoma. *J. NCI* 84 (24):1875–1887.

Weidner, N., Semple, J., Welch, W. and Folkman, J. (1991) Tumor angiogenesis and metastasis – Correlation in invasive breast carcinoma. *NEJM* 324 (1):1–8.

Weingart, J., Sipos, E. and Brem, H. (1995) The role of monocycline in the treatment of intracranial 9L glioma. *J. Neurosurg.* 82 (4):635–640.

Wesserling, P., Schlingemann, R., Rietveld, F., Link, M., Burger, P. and Ruiter, D. (1995) Early and extensive contribution of pericytes/vascular smooth muscle cells to microvascular proliferation in glioblastoma multiforme: An immuno-light and immuno-electron microscopic study. *J. Neuropath. Exp. Neurol.* 54 (3):304–310.

Wilson, J.T. and Penar, P.L. (1994) The effect of AGM-1470 in an improved intracranial 9L gliosarcoma rat model. *Neurological Res.* 16 (2):121–124.

Winston, K., Gilles, F., Leviton, A. and Fulchiero, A. (1977) Cerebellar gliomas in children. *J. NCI* 58 (4):833–838.

Wolff, J., Guerin, C., Laterra, J., Bressler, J., Induri, R., Brem, H. and Goldstien, G. (1993) Dexamethasone reduces vascular density and plasminogen activator activity in 9L rat brain tumors. *Brain Res.* 604:79–85.

Wolff, J.E., Laterra, J. and Goldstein, G.W. (1992) Steroid inhibition of neural microvessel morphogenesis *in vitro*: receptor mediation and astroglial dependence. *J. Neurochem.* 58 (3):1023–1032.

Yamada, N., Kato, M., Yamashita, H., Nister, M., Miyazono, K., Heldin, C. and Funa, K. (1995) Enhanced expression of transforming factor-beta and its type-I and type-II receptors in human glioblastoma. *Int. J. Cancer* 62 (4):386–392.

Yamamoto, T., Terada, N., Nishizawa, Y. and Petrow, V. (1994) Angiostatic activities of medroxyprogesterone acetate and its analogues. *Int. J. Cancer* 56 (3):393–399.

Yanase, T., Tamura, M., Fujita, K., Kodama, S. and Tanaka, K. (1993) Inhibitory effect of angiogenesis inhibitor TNP-470 on tumor growth and metastasis of human cell lines *in vitro* and *in vivo. Cancer Res.* 53 (11):2566–2570.

Yano, T., Tanase, M., Watanabe, A., Sawada, H., Yamada, Y., Shino, Y., Nakano, H. and Ohnishi, T. (1995) Enhancement effect of an anti-angiogenic agent, TNP-470, on hyperthermia-induced growth suppression of human esophageal and gastric cancers transplantable to nude mice. *Anticancer Res.* 15 (4):1355–1358.

Yazaki, T., Takamiya, Y., Costello, P.C., Mineta, T., Menon, A.G., Rabkin, S.D. and Martuza, R.L. (1995) Inhibition of angiogenesis and growth of human non-malignant and malignant meningiomas by TNP-470. *J. Neuro. Onc.* 23 (1):23–29.

Yoshida, D., Ikeda, Y. and Nakazawa, S. (1993a) Quantitative analysis of copper, zinc, and copper/zinc ratio is selected human brain tumors. *J. Neuro. Onc.* 16 (2):109–115.

Yoshida, D., Ikeda, Y. and Nakazawa, S. (1993b) Suppression of 9L gliosarcoma growth by copper depletion with copper-deficient diet and D-penicillamine. *J. Neuro. Onc.* 17 (2): 91–97.

Yoshida, D., Ikeda, Y. and Nakazawa, S. (1995) Suppression of tumor growth in experimental 9L gliosarcoma model by copper depletion. *Neurol. Med. Chirurg.* 35 (3):133–135.

Regulation of Angiogenesis
ed. by I.D. Goldberg & E.M. Rosen
© 1997 Birkhäuser Verlag Basel/Switzerland

Regulation of angiogenesis in malignant gliomas

C. Guerin[1] and J. Laterra[2]

[1] *Department of Neurosurgery, National Naval Medical Center, Bethesda, Maryland
20889-5000, USA and*
[2] *Departments of Neurology, Oncology, and Neuroscience, The Johns Hopkins School of
Medicine and The Kennedy Krieger Institute, Baltimore, Maryland 21205, USA*

*Summary. Gliomas are highly resistant to conventional therapeutic measures, requiring the
development of novel treatments. Since gliomas are particularly vascular tumors, one approach
involves treatments directed at inhibiting angiogenic mechanisms. Although multiple factors
contribute to the ultimate vascularization of any tumor, some are especially relevant to gliomas. Early
experimental work directed at inhibiting angiogenic pathways has shown promise toward achieving
control of tumor growth. This article focuses on the evidence that angiogenesis and related vascular
cell responses play important roles in glioma biology, and reviews those biochemical pathways
known through experimentation to be involved in the vascular response to gliomas. Finally, con-
temporary vessel-targeted approaches that have been used to inhibit glioma growth are discussed.*

Introduction

Gliomas have long posed a formidable challenge to patients and physicians.
Despite advances in the treatment of other malignancies, glioma therapy
remains disappointing. Recent studies show that the overall incidence and
mortality due to brain tumors is increasing (Devesea et al., 1995). The vast
majority of these tumors are malignant, and most are of the astrocytic type.
Therefore, this review will focus on malignant astrocytoma.

The median survival for malignant astrocytoma patients remains less
than one year despite aggressive treatment with surgery, radiation and
chemotherapy (Black, 1991a, 1991b). One notable difference from other malignancies is that gliomas tend to recur locally, at the site of the original disease
(Hochberg and Pruitt, 1980). Death occurs from disease at this site, and metastases are distinctly unusual. Despite this primarily local growth pattern, invasiveness at the tumor border prevents surgical resection, as cells infiltrate into
normally functioning peritumoral brain that cannot be excised without risking
a substantial neurological deficit (Kelly et al., 1987). Likewise, preservation of
blood-brain barrier function in these areas of infiltrated brain likely inhibits
delivery of chemotherapeutic agents (Neuwelt et al., 1986). Clearly, new
approaches are needed in the treatment of these malignant tumors.

Vascular responses to malignant gliomas

Malignant astrocytomas are characterized by the heterogeneity of tumor cells
at multiple levels – e.g., appearance, antigenicity, genetic stability, chemo-

sensitivity, and radiosensitivity (McComb and Bigner, 1984; Shapiro 1986; Shapiro et al., 1991). Therefore, tumor cells are difficult to target directly with cytotoxic modalities, as some of them are frequently resistant to each therapeutic agent used. Another characteristic feature is the prominent vascular proliferation and blood-brain barrier dysfunction that accompanies malignant glioma growth. These vascular responses have long been recognized as a potential key to understanding glioma behavior (Burger, 1991; Scherrer, 1935). Like other neoplasms, gliomas are dependent on vascularization for growth (Brem et al., 1988; Folkman, 1990). Since the neovascular cells are not malignant themselves and are more homogeneous than the glioma cells, they may be more easily targeted therapeutically.

Astrocytoma vessels are not simply brain vessels within tumor tissue. Evidence suggests that microvessel endothelial cells, including those in brain, differentiate in tissue-specific patterns in response to their particular microenvironmental cues (Guerin et al., 1990; Stewart et al., 1987). The endothelial cells that comprise differentiated brain microvessels normally express a number of specializations that are collectively termed the blood-brain barrier (BBB) (Goldstein and Betz, 1986). These brain-specific endothelial properties consist of structural specializations, such as tight junctions, and unique biochemical specializations like transporters and enzymes. Brain microvessels are normally completely ensheathed by perivascular astrocytes that are strongly implicated in regulating the phenotype of brain microvessel endothelial cells (Laterra et al., 1991). Astrocytes *in vitro* can induce similar phenotypic changes in endothelial cells, as well as morphological changes consistent with a role in regulating the morphogenesis of microvascular tubules (DeBault and Cancilla, 1980; Rubin et al., 1991; Laterra et al., 1990; Tao-Cheng and Brightman, 1988).

As malignancy progresses, so do vascular changes such as endothelial proliferation, as well as the loss of interendothelial junctions, blood-brain barrier-specific endothelial transporters, and enzymes (Guerin et al., 1990; Harik and Roessmann, 1991; Long, 1970; Weller et al., 1977). With excessive vasoproliferation, glomeruloid structures consisting of hypertrophied endothelial cells appear (Burger, 1991). The enhanced permeability of these vessels accounts for the tumor-associated vasogenic edema that can cause substantial patient morbidity. The defective BBB expression that is most characteristic of endothelial cells within the more malignant gliomas appears to result from the replacement of environmental cues normally derived from the perivascular astroglial sheaths with those derived from abnormal perivascular tumor cells.

Gliomas are considered to be endothelial-rich tumors, and prognostic relevance can be assigned to the vascular changes themselves (Brem et al., 1972; Brem, 1976). Similar prognostic information may be found in other malignancies, with poor survival correlating with increasing vascular density (Weidner et al., 1991). Cerebrospinal fluid (CSF) from brain tumor patients is known to contain endothelial growth factors and can stimulate the proliferation of endothelial cells *in vitro* (Brem et al., 1983; and see below). Vessel dysfunction is directly responsible for some of the tumor-associated morbidity, through hemorrhage and the edema that results from increased permeability. In this sense, the routine treatment of tumor-associated brain edema with corticosteroids that increase endothelial cell expression of blood-brain barrier is a form of vessel-targeted therapy (Guerin et al., 1992a).

Regulation of angiogenesis in gliomas

With ever increasing information on the roles of specific growth factors, proteolytic systems, and cell-matrix interactions in the angiogenic process, it is important to specify which have been shown to be applicable to glioma vascular biology *in vivo*. Such information will increase the likelihood of successfully choosing a methodology or molecular target for anti-angiogenic glioma therapy. In addition, by targeting mechanisms that are unique to the glioma-vessel interaction, the risk of disrupting normal microvascular events within brain and other systemic organs can be minimized.

Glioma angiogenesis, like that occurring elsewhere, involves an intertwining of multiple factors. Glioma cells, endothelial cells, extracellular matrix, growth factors, growth factor receptors, proteases, protease inhibitors, and their complex interactions all contribute to the ultimate vascular response. We review these systems below, acknowledging that each acts in a dynamic environment affected by other such systems.

Vascular Endothelial Growth Factor/Vascular Permeability Factor (VEGF/VPF)

VEGF, also known as VPF, is a secreted, dimeric, heparin-binding glycoprotein. It is the only growth factor known to exhibit both endothelial mitogenic and vascular permeability-inducing activity (Ferrara et al., 1992 and Senger et al., 1993). It is expressed in a wide variety of tissues and is implicated in the enhanced angiogenesis occurring in embryonic tissue including brain, and in wound healing, inflammatory conditions, and neoplasia. Four isoforms result from alternative RNA splicing. The 121 and 165 amino acid forms are soluble, while the 189 and 206 amino acid forms tend to be cell-associated or bound to matrix. Expression of the two VEGF receptor tyrosine kinases *flt*-1 and KDR is essentially limited to endothelial cells, making it the only known endothelial-specific growth factor.

In vitro studies have shown that VEGF induces plasminogen activators (uPA and tPA) as well as plasminogen activator inhibitor-1 in endothelial cells (Pepper et al., 1991). RNA ratios suggest a balance favoring proteolysis, consistent with the known angiogenic activity of VEGF. Another study showed mild induction of interstitial collagenase without effects on type IV collagenase, stromelysin, or tissue inhibitor of metalloproteinase (Unemori et al., 1992).

VEGF expression is increased in a variety of vascular human brain tumors, including astrocytic gliomas, meningiomas, hemangioblastomas and metastases, but only rarely is VEGF elevated in less vascular tumors (Berkman et al., 1993). Brain tumor cyst fluid also contains VEGF (Berkman et al., 1993; Weindel et al., 1994). VEGF expression in human and experimental gliomas *in vivo* is not uniform, but is clustered in cells along capillaries and especially adjacent to the necrotic foci characteristic of the most malignant gliomas (Plate et al., 1992 b; Shweiki et al., 1992). In contrast, basic FGF expression is not increased in perinecrotic areas, although increased release of matrix-bound or intracellular FGF in these locations may occur. Such necrotic areas are likely to represent the most hypoxic regions of the tumor and the VEGF promoter is inducible by hypoxia. Indeed, VEGF expression is induced under hypoxic conditions *in vitro* in a variety of glial tumor lines, whereas the expression of basic and acidic fibroblast growth factors and platelet derived growth factor B chain RNAs are not similarly regulated (Plate et al., 1993).

Of the four isoforms, the soluble 165 amino acid form predominates in gliomas (Berkman et al., 1993). This isoform appears to diffuse readily to target endothelial cells. In support of this possibility, immunohistochemical studies show accumulation of the protein at microvessels, although the source of VEGF localizes to tumor cells by *in situ* hybridization (Plate et al., 1992 b and 1994). In a study comparing VEGF, bFGF, and TGF-α and -β, only the level of VEGF expression specifically correlated with vascular density of gliomas and meningiomas (Samoto et al., 1995). In contrast to its correlation with glioma angiogenesis, VEGF expression by glioma cells *in vivo* does not appear to be associated with expression of p53 or epidermal growth factor receptor, or with Ki67 labeling of glioma cells, a marker of glioma cell proliferation (Plate et al., 1994). Non VEGF-producing tumor cells, when transfected with a VEGF expression vector, showed no change in growth *in vitro*, but were found to grow faster in nude mice and the resulting solid tumors were more vascular (Zhang et al., 1995). Likewise, VEGF-transfected Hela cells are more tumorigenic, and tumor growth can be inhibited by antibodies to VEGF (Kondo et al., 1993). Thus, VEGF expression does not appear to act at the tumor cell level, but rather in a paracrine fashion to promote the vascular supply required for growth. However, another study demonstrated that although VEGF expression provides a growth advantage *in vivo*, this response may be determined by the multifactorial environmental context of individual tumor types (Ferrara et al., 1993).

VEGF receptors are upregulated in glioma vessels (Plate et al., 1992 b, 1993, and 1994). *In situ* hybridization reveals expression in tumor vessels but not in adjacent normal brain. As malignancy increases, both receptors are coexpressed; *flt*-1 initially, with the addition of KDR in the most malignant glioblastomas. VEGF expression also increases with tumor grade. Thus, gliomas may use VEGF to aid their growth by enhanced recruitment of vascular supply as malignancy progresses.

Although only one-tenth as potent as its mitogenic activity, VEGF's vascular permeability-inducing activity is still 1000 times that of histamine. Neutralizing anti-VEGF antibodies inhibit by up to 75–99% the ability of human tumor cyst fluids and conditioned medium obtained from glioma cell lines to permeabilize subcutaneous vessels *in vivo* (Berkman et al., 1993). In addition to the pathophysiologic role of vessel permeability in glioma edema formation, vascular hyperpermeability is also thought to be critical to the formation of tumor stroma, which is essential for tumor growth (Senger et al., 1993). While VEGF's permaeability-inducing activity is likely to be of physiologic importance for human tumors, this remains to be shown definitively in gliomas.

Platelet Derived Growth Factor (PDGF)

PDGF is a dimeric growth factor with three isoforms composed of disulfide-linked A and B chains (AA, BB, AB). PDGF is mitogenic for multiple cell types (Hedlin, 1992). The A chain is transcribed from the c-*sis* protooncogene. Receptor affinity differs among isoforms: PDGF-Rα binds all isoforms with high affinity, while PDGF-Rβ binds PDGF-BB with high affinity and PDGF-AB with lesser affinity.

Studies have shown expression of both PDGF chains and PDGF receptors in human astrocytomas. Whereas normal brain expressed predominantly A chain mRNA, tumor cells also prominently express B chain forms (Hermanson et al., 1988; Maxwell et al., 1990). Levels of both A and B chains are increased in the higher grade gliomas (Hermanson et al., 1992). In addition, the expression of both the α and β type receptors are increased in gliomas, by both gene amplification and other mechanisms (Fleming et al., 1992). In contrast to VEGF, immunohistochemistry has shown that the PDGF A chain and its α-receptor preferentially localize to tumor cells, supporting a role for PDGF A chains in the autocrine stimulation of tumor growth. Of more interest, B chains and β-receptor localize to endothelium, especially the hyperplastic capillaries of glioblastomas (Plate et al., 1992; Hermanson et al., 1992). Interestingly, PDGF-BB induces VEGF in a glioma cell line (Tsai et al., 1995). Thus, there may be a significant contribution to glioma neovascularization by the PDGF β-receptor system, with autocrine as well as paracrine stimulation by B chain isoforms that are produced by glioma cells.

Fibroblast Growth Factors (FGF)

Acidic and basic FGF (aFGF and bFGF) were the first angiogenic growth factors isolated. Now there are nine members of the FGF family and four receptors have been identified, each with multiple isoforms (Friesel and Maciag, 1995).

Malignant astrocytomas have elevated expression of both aFGF and bFGF (Maxwell et al., 1991; and see below). The CSF from patients with several types of brain tumors also contains bFGF (Li et al., 1994). Basic FGF levels correlate with glioma grade, with FGF immunopositivity in tumor cells, tumor vessels, and vessels in peritumoral brain (Brem et al., 1992; Zagzag et al., 1990). Since FGF is known to be stored in perivascular matrix, such immunopositive vessels may or may not express FGF receptors (Flaumenhaft and Rifkin, 1991). In studies of receptors, FGF-R2 is expressed in normal brain, while FGF-R1 has been found in glioblastoma tissue (Morrison et al., 1994a, 1994b). Glioblastomas can also express an alternatively spliced form of FGF-R1. Morrison and colleagues found FGF receptor expression to be limited to tumor cells, arguing for an autocrine tumor growth mechanism and against an angiogenic role. Nevertheless, another study of FGF-R1 did identify endothelial cell expression in gliomas and meningiomas, especially in the highly proliferative endothelia of glioblastomas (Ueba et al., 1994).

Scatter Factor/Hepatocyte Growth Factor (SF/HGF)

SF/HGF is a secreted, heterodimeric, heparin-binding member of the family of kringle-containing proteins (Rosen et al., 1993). Depending on its environmental context, SF/HGF functions as a mitogen, morphogen, and invasion/motility factor and is believed to mediate a number of developmental, regenerative, and neoplastic processes *in vivo*. It is also a potent angiogenic factor *in vivo*, induces endothelial cell production of proteases, and also modulates intercellular tight junctions of epithelial cells. The SF receptor is the c-*met* protooncogene.

The genes encoding human SF/HGF and its c-*met* receptor are on chromosome 7, which is very frequently multiploid in human gliomas, including the most aggressive glioblastoma multiforme. Most normal brain c-*met* appears to be expressed by phagocytic cells, but immunohistochemical studies stain neurons and endothelium (DiRenzo et al., 1993; Zarnegar and DeFrances, 1993). Glioma cells have been shown to co-express SF and c-*met* (Moriyama et al., 1995). Recent studies from our laboratory in collaboration with Eliot Rosen (Long Island Jewish Medical Center) show that human gliomas and glioma cell lines can produce active SF/HGF, which can enhance, by autocrine and paracrine mechanisms, the malignant behavior of glioma cells and endothelial cells *in vitro* (Rosen et al., 1996). We have found that the highest levels of SF/HGF are in the malignant gliomas. Furthermore, glioma cell-derived SF/HGF increases the proliferation of glioma cells and brain endothelial cells *in vitro*. SF-inducing factors are also present within human gliomas and are produced by glioma cell lines. Finally, experimental rat gliomas appear to become more aggressive *in vivo* after human SF gene transfer. The mechanism for this alteration is not yet known. SF has several properties that make it an attractive therapeutic target, but its role in glioma angiogenesis remains to be defined.

Plasminogen activators and inhibitors

The balance between proteases and their inhibitors plays a significant role in the regulation of angiogenesis (Liotta et al., 1991). Common to the angiogenic endothelial cell mitogens discussed so far is their ability to induce endothelial cells to produce proteolytic enzymes such as plasminogen activators and matrix metalloproteases (e.g. collagenases). An early phase of the angiogenic response involves perivascular proteolysis, to allow migration of endothelial cells and formation of neovasculature. One important system that influences the net perivascular proteolytic balance is the plasminogen activator system that is principally regulated by the serine proteases urokinase-type plasminogen activator (UPA) and tissue-type PA (TPA), and specific plasminogen activator inhibitors (PAIs) (Yamamoto et al., 1994a). The proteolytic activation of the proenzyme plasminogen generates plasmin, a broad spectrum serine protease that may also activate collagenases. Tissue plasminogen activator is thought to be principally involved in intravascular thrombolysis, while uPA is more important for tissue remodeling, including neovascularization and tumor cell invasion. The PA inhibitors include PAI-1, PAI-2, and protease nexin-1 (PN-1), also designated glia-derived nexin. PN-1 is produced by astroglial cells and normally localizes to blood-vessels in brain (Choi et al., 1990). *In vitro* studies suggest that the net balance between endothelial cell-derived uPA and astroglial (or glioma) cell-derived PA inhibitors functions to regulate morphogenic glial-endothelial interactions that lead to microvessel formation (Laterra et al., 1994) Urokinase receptors also play an important regulatory role by localizing enzyme to specific cell-surface microdomains involved in cell migration and invasion.

Both uPA and PAI-1 are known to increase as glioma grade increases (Hsu et al., 1995; Yamamoto et al., 1994a; Yamamoto et al., 1994c). In contrast, tPA levels are normal in lower grade tumors, but absent in glioblastomas (Bindal et al., 1994). Urokinase receptors are also present in increased amounts in higher grade gliomas, and are seen in tumor cells and endothelial cells, especially in areas of vascular proliferation and at the invasive tumor edge (Yamamoto et al., 1994b). PAI-1 is localized at vessels and necrotic areas, which suggests a role for microthrombosis in necrosis formation (Kono et al., 1994; Sawaya et al., 1995). PAI-2 in only rarely detected in gliomas (Landau et al., 1994).

Matrix Metalloproteinases (MMPs) and inhibitors

Another important protease system involves the MMPs, a large family of Zn^{2+} dependent proteases secreted by a variety of tumor cells and capable of degrading most components of the extracellular matrix (Woessner, 1991). They are secreted as inactive proenzymes requiring activation, and are further regulated by tissue inhibitors of metalloproteinase (TIMPs). Several reports of experimental manipulation of TIMP expression *in vivo* show significant effects on tumor growth and metastasis, with elevated TIMP expression being inhibitory and depressed levels stimulatory (Alvarez et al., 1990; DeClerk et al., 1992; Khoka et al., 1989).

Increased levels of the 92 kD type IV collagenase have been demonstrated in malignant gliomas (Rao et al., 1993). Elevated expression of type IV collagenases and TIMP-1 has been associated with malignancy in human gliomas, whereas interstitial collagenase, stromelysin, and TIMP-2 were not (Nakano et al., 1995).

Cathepsins

Cathepsins, especially the cysteine protease cathepsin B, are another class of proteolytic enzymes associated with malignancy in several organs. In gliomas, cathepsin B expression also increases with tumor grade. It localizes particularly to invading cells and endothelial proliferations (Rempe 1 et al., 1994; Mikkelsen et al., 1995; Sivaparvathi et al., 1995). No correlations between cathepsin B expression and glioma angiogenesis have been made. However, its broad substrate specificity, which includes extracellular matrix proteins, its ability to activate receptor-bound pro-urokinase, and its localization at sites of endothelial proliferation within human gliomas suggest that cathepsin B influences the glioma vascular response.

Extracellular Matrix (ECM)

ECM is also important to the control of angiogenesis (Ingber and Folkman, 1989). Several ECM components that could serve as therapeutic targets are specifically altered in gliomas. Integrins are transmembrane cell surface receptors that mediate cell-matrix interactions and cell-cell communication (Hynes, 1992). Integrin expression is altered in astrocytomas when compared to normal brain and other tumor histologies (Paulus et al., 1993). The proliferating vessels in glioblastoma also express novel integrins not seen in normal brain endothelium. Expression of α_v and β_3 integrin chains are increased in some gliomas, similar to their increased expression at proliferating vessels at sites of injury (Brooks et al., 1994).

The ECM glycoprotein tenascin is associated with tissue remodeling. The "glioma-mesenchymal extracellular matrix antigen" originally described as a potential marker of new vessels in gliomas has been identified as tenascin (McComb and Bigner, 1985). This antigen localizes predominantly around tumor vessels, and is only rarely found in normal brain. Its expression increases with glioma grade and is particularly increased in hyperplastic glioma capillaries (Zagzag et al., 1995). The expression of other perivascular ECM glycoproteins also differs in gliomas. Fibronectin expression decreases as tumor grade increases, and it tends to be seen in vessels near more differentiated astroglial cells that express glial fibrillary acidic protein (GFAP). The more malignant astrocytomas tend to have fibronectin-negative, tenascin-positive vessels that are surrounded by poorly differentiated GFAP-negative cells. The specificity of these changes is supported by studies showing that laminin is expressed in all vessels, normal and within tumors (Higuchi et al., 1993). Studies with specific glycoconjugates as probes confirm that glioblastoma vessels differ from normal vessels in lectin content (Debbage et al., 1988). These matrix glycoproteins are known to mediate many cell-matrix interactions involved in cell motility and tissue morphogenesis; however, the mechanistic relationships between their altered expression or tissue distribution and glioma vessel pathophysiology are not yet known.

Control of glioma growth by anti-angiogenic mechanisms

It is clear that the vasculature plays a prominent role in the pathobiology of the gliomas and is thus an attractive target for novel therapies. Moreover, the mechanistic similarities between angiogenesis and tumor cell invasion suggest that angiogenesis-directed treatments might simultaneously inhibit glioma cell invasion of distant sites within brain (Thorgeirsson et al., 1994). The problems of anti-glioma drug delivery across the blood-brain barrier may also be minimized, since the target endothelia can be directly exposed to agents administered by standard intravascular means (Guerin et al., 1992 b; Madrid et al., 1991). Although no antiangiogenic clinical trials have been undertaken for brain tumors, several lines of experimental evidence show promise. These approaches can be divided into those intended to target specific growth factors, or those that target proteases or extracellular matrix molecules.

Growth factor related approaches

Most work on gliomas has focused on bFGF and VEGF. Growth of a subcutaneous glioma in nude mice was inhibited by local injection of anti-bFGF antibody (Takahashi et al., 1991). Although vascularization was not assessed in this study, the antibody also inhibited *in vitro* growth, so some portion of its effect results from a direct action on tumor cells rather than vessels. In an intracranial glioma model, intratumoral treatment via an implanted device also inhibited growth (Stan et al., 1995) and treated tumors were significantly less vascular. In a third study, C6 gliomas were modestly inhibited by intravenous administration of anti-bFGF antibodies (Gross et al., 1992). The effective antibody was also capable of inhibiting angiogenesis in a kidney capsule model, whereas a less effective antibody was not. *In vitro* studies show that C6 glioma cells engineered to express anti-sense bFGF mRNA proliferate less rapidly than do control C6 cells (Redekop and Nans, 1995). Thus, bFGF appears to be an appropriate target in glioma treatment, although it remains unclear to what degree its *in vivo* effects are anti-angiogenic or tumoristatic.

Antibodies to VEGF were also found to inhibit glioblastoma growth in a nude mouse model after systemic, intraperitoneal treatment (Kim et al., 1993). In addition, this study showed that the treated tumors had a lower vascular density, and the anti-VEGF antibodies had little effect on glioma cell growth *in vitro*. Anti-VEGF antibodies also inhibited metastases as well as tumor growth in a colon cancer model (Warren et al., 1995). In contrast to receptor-negative quiescent endothelial cells, sprouting endothelial cells that participate in tumor-associated neovascularization express VEGF receptors. This is the basis for targeting VEGF-specific cell signaling pathways. C6 tumors in nude mice were treated with a retrovirus expressing a dominant negative mutant of *flk*-1, the murine homologue of the human KDR VEGF receptor (Millauer et al., 1994). Transduction with the dominant negative *flk*-1 leads to heterodimerization between mutant and endogenous receptor subunits and results in a signaling-defective receptor. C6 tumor growth *in vivo* was strongly inhibited by this approach, and histological analysis revealed noninvasive, largely necrotic nodules lacking capillaries. C6 cell growth *in vitro* was not affected after transduction by the same viral vector. These studies establish the applicability of VEGF-mediated therapies that apparently act via a purely anti-angiogenic mechanism.

Protease system/ECM related approaches

Angiogenic endothelial cell mitogens induce expression of proteases such as plasminogen activators and collagenases in endothelial cells. Early anti-angiogenic approaches were based on the inhibition of these proteolytic pathways, which remain promising targets for novel vessel-targeted treatments.

Brem and Folkman (1975) and Brem et al. (1980) initially inhibited the growth of brain tumors transplanted to the rabbit cornea by preventing vascularization with interstitial cartilage implants. Cartilage extracts were found to contain collagenase inhibitors responsible for this anti-angiogenic effect (Langer et al., 1976) and a specific cartilage-derived anti-angiogenic protein has since been characterized (Moses et al., 1990). The combination of heparin and cortisone was found to inhibit neovascularization strongly (Folkman et al., 1983) by a mechanism proposed to involve the dissolution of vascular basement membranes (Ingber et al., 1986). Although, initial systemic treatment with heparin/cortisone was ineffective against gliomas, subsequent approaches that involved local delivery of these agents did inhibit intracranial growth in a glioma model (Tamargo et al., 1990).

Glucocorticoid therapy alone is widely used in human gliomas for its ability to reduce tumor vessel permeability and tumor-associated interstitial edema. Standard therapy with glucocorticoids is not believed to affect glioma growth, although glucocorticoids are known to slow the growth of experimental gliomas. This tumor response is associated with concurrent decreases in tumor vascular density and tumor-associated plasminogen activator activity (Guerin et al., 1992 c; Wolff et al., 1993). *In vitro* studies suggest that inhibition of periendothelial plasminogen activation enhances astroglial cell-induced endothelial differentiation (Laterra et al., 1994). The induction of endothelial capillary-like structures by glioma cells *in vitro* is also inhibited by steroids, including glucocorticoids, probably via a receptor-mediated astroglioma cell response (Wolff et al., 1992).

Minocycline, a tetracycline antibiotic that also inhibits collagenases, can inhibit tumor-induced neovascularization *in vivo* (Tamargo et al., 1991). Interstitial treatment of intracranial gliomas with minocycline significantly prolongs animal survival (Weingart et al., 1995). When used in conjunction with the standard chemotherapeutic agent BCNU, animal survival was nearly double that of BCNU treatment alone. Tetracyclines are also selective inhibitors of endothelial cell proliferation, with pericytes, astrocytes, and glioma cells being relatively resistant (Guerin et al., 1992 a). This endothelial cell-specific anti-proliferative effect is non-cytotoxic, correlates with collagenase inhibition, and is independent of antibiotic action.

Copper can induce matrix glycoproteins that are associated with proliferating vessels. Systemic copper depletion with the chelating agent penicillamine did inhibit the growth of experimental gliomas concurrent with a decrease in endothelial BUdR labeling index. However, penicillamine therapy did not improve animal survival, probably due to an concomitant increase in treatment-induced brain edema (Brem et al., 1990). Copper depletion might alter endothelial proliferation via its effects on metal-binding angiogenic proteins such as SPARC (Lane et al., 1994).

An interesting recent approach used a monoclonal antibody to fibronectin, TV-1, in a non-glial *in vivo* model (Epstein et al., 1995). TV-1 only recognizes fibronectin isoforms associated with tumor vessels. In this study, interleukin-2 linked to TV-1 selectively increased tumor vessel permeability, which tripled the parenchymal uptake of a tumor-specific antibody. Of note, an anti-bFGF antibody linked to interleukin-2 was without effect. This approach could be modified to target the microvessels themselves directly.

Other approaches

Several other approaches that do not fit well into the categories above deserve mention. For AGM-1470, a synthetic antibiotic derivative, the mechanism of angiogenesis inhibition appears to be related to inhibition of cell cycle control pathways in endothelial cells (Abe et al., 1994; Antoine et al., 1994). It is a potent inhibitor of endothelial proliferation *in vitro* (Ingber et al., 1990). Studies *in vivo* show antitumor responses in extracranial glioma models, but conflicting results using intracranial models (Taki et al., 1994; Takamiya et al., 1994; Wilson and Penar, 1994).

Other interesting approaches involve novel uses of the tumor endothelium, but are not anti-angiogenic per se. Regression of established neuroblastomas was induced by treatment with an anti-endothelial immunotoxin that binds to a tumor-specific endothelial cell antigen (Burrows and Thorpe, 1993). Tumor regression appeared to result from vascular occlusion. A conventional anti-tumor cell immunotoxin had only minor effects. A study in which endothelial cells were injected intravenously into whole animals showed that they accumulate at angiogenic sites, raising the possibility of targeting a tumor with engineered endothelium (Ojeifo et al., 1995). Coimplantation of endothelial cells and glioma cells into rat brain showed that the implanted endothelial cells survive and proliferate within the growing experimental glioma (Lal et al., 1994). Using this approach, endothelial cells engineered to secrete interleukin-2 have been used to induce an anti-glioma response within rat brain (Nam et al., 1996).

Conclusions

Defining the mechanisms of glioma angiogenesis has led to several promising new approaches to treatment. With further elucidation of the relative importance of specific factors, our ability to manipulate the biological behavior of even the most malignant neoplasms should improve. Anti-angiogenic and/or endothelial cell-based therapies hold promise for providing novel tools against an entirely new set of tumor targets. Early evidence suggest that conventional chemotherapeutic approaches together with anti-angiogenic agents may act synergistically to produce potent therapeutic effects. Since conventional cytotoxic therapies alone have limited utility against malignant glioma, the development of new treatments such as those targeting the vasculature must continue.

Acknowledgments.
We thank Dr. Pamela Talalay for editorial assistance and Ms. Angela T. Williams for help in manuscript preparation. This work was funded in part by NIH grants NS32148 and NS33728.

References

Abe, J., Zhou, W., Takuwa, N., Taguchi, J., Kurokawa, K., Kumada, M. and Takuwa, Y. (1994) A fumagillin derivative angiogenesis inhibitor, AGM-1470, inhibits activation of cyclin-dependent kinases and phosphorylation of retinoblastoma gene product but not protein tyrosyl phosphorylation or protooncogene expression in vascular endothelial cells. *Cancer Res.* 54:3407–3412.

Alvarez, O.A., Carmichel, D.F. and DeClerk, Y.A. (1990) Inhibition of collagenolytic activity and metastasis of tumor cells by a recombinant human tissue inhibitor of metalloproteinases. *J. Natl. Cancer Inst.* 82:589–595.

Antoine, N., Greimers, R., DeRoanne, C., Kusaka, M., Heinen, E., Simar, L.J. and Castronovo, V. (1994) AGM-1470, a potent angiogenesis inhibitor, prevents the entry of normal but not transformed endothelial cells into the G1 phase of the cell cycle. *Cancer Res.* 54: 2073–2076.

Berkman, R.A., Merrill, M.J., Reinhold, W.C., Monacci, W.T., Saxena, A., Clark, W.C., Robertson, J.T., Ali, I.U. and Oldfield, E.H. (1993) Expression of the vascular permeability factor/vascular endothelial growth factor gene in central nervous system neoplasms. *J. Clin. Invest.* 91:153–159.

Bindal, A.K., Hammoud, M., Shi, W.M., Wu, S.Z., Sawaya, R. and Rao, J.S. (1994) Prognostic significance of proteolytic enzymes in human brain tumors. *J. Neuro-Oncol.* 22:101–110.

Black, P.M. (1991 a) Brain tumors. Part 1. *N. Engl. J. Med.* 324:1471–1476.

Black, P.M. (1991 b) Brain tumors. Part 2. *N. Engl. J. Med.* 324:1555–1564.

Brem, S. (1976) The role of vascular proliferation in the growth of brain tumors. *Clinical Neurosurg.* 23:440–453.

Brem, S., Cotran, R. and Folkman, J. (1972) Tumor angiogenesis: A quantitative method for histologic grading. *J. Natl. Cancer Inst.* 48:347–356.

Brem, H. and Folkman, J. (1975) Inhibition of tumor angiogenesis mediated by cartilage. *J. Exp. Med.* 141:427–439.

Brem, H., Thompson, D., Long, D.M. and Patz, A. (1980) Human brain tumors: Differences in ability to stimulate angiogenesis. *Surg. Forum* 31:471–473.

Brem, H., Patz, J. and Tapper D (1983) Detection of human central nervous system tumors: Use of migration-stimulating activity of the cerebrospinal fluid. *Surg. Forum* 34:532–534.

Brem, H., Tamargo, R.J., Guerin, C., Brem, S.S. and Brem, H. (1988) Brain tumor angiogenesis. *In*: Kornblith, P.L. and Walker, M.D. (eds.): *Adv. Neuro-Oncol.* Futura, Mount Kisco, New York, pp 89–102.

Brem, S.S., Zagzag, D., Tsanaclis, A.M.C., Gately, S., Elkouby, M.P. and Brien, S.E. (1990) Inhibition of angiogenesis and tumor growth in the brain – Suppression of endothelial cell turnover by penicillamine and the depletion of copper, an angiogenic cofactor. *Am. J. Pathol.* 137:1121–1142.

Brem, S.S., Tsanaclis, A.M.C., Gately, S., Gross, J.L. and Herblin, W.F. (1992) Immunolocalization of basic fibroblast growth factor to the microvasculature of human brain tumors. *Cancer* 70:2673–2680.

Brooks, P.C., Clark, R.A.F. and Cheresh, D.A. (1994) Requirement of vascular integrin $\alpha_v\beta_3$ for angiogenesis. *Science* 264:569–571.

Burger, P.C. (1991) *Surgical Pathology of the Nervous System and Its Coverings.* Third Edition. Churchill Livingstone, New York, pp 194–234.

Burrows, F.J. and Thorpe, P.E. (1993) Eradication of large solid tumors in mice with an immunotoxin directed against tumor vasculature. *Proc. Natl. Acad. Sci. USA* 90:8996–9000.

Choi, B.H., Suzuki, M., Kim, T., Wagner, S.L. and Cunningham, D.D. (1990) Protease nexin-1. Localization in the human brain suggests a protective role against extravasated serine proteases. *Am. J. Pathol.* 137:741–747.

DeBault, L.E. and Cancilla, P.A. (1980) γ-glutamyl transpeptidase in isolated brain endothelial cells: Induction by glial cells *in vitro*. *Science* 207:653–655.

Debbage, P.D., Gabius, H., Bise, K. and Marguth, F. (1988) Cellular glycoconjugates and their potential endogenous receptors in the cerebral microvasculature of man: A glycohistochemical study. *Euro. J. Cell Biol.* 46:425–434.

DeClerck, Y.A., Perez, N., Shimada, H., Boone, T.C., Langley, K.E. and Taylor, S.M. (1992) Inhibition of invasion and metastasis in cells transfected with an inhibitor of metalloproteinases. *Cancer Res.* 52:701–708.

Devesa, S.S., Blot, W.J., Stone, B.J., Miller, B.A., Tarone, R.E. and Fraumeni, J.F. Jr (1995) Recent cancer trends in the United States. *J. Natl. Cancer Inst.* 87:175–182.

DiRenzo, M.F., Bertolotto, A., Olivero, M., Putzolu, P., Crepaldi, T., Schiffer, D., Pagni, C.A. and Comoglio, P.M. (1993) Selective expression of the Met/HGF receptor in human central nervous system microglia. *Oncogene* 8:219–222.

Epstein, A.L., Khawli, L.A., Hornick, J.L. and Taylor, C.R. (1995) Identification of a monoclonal antibody, TV-1, directed against the basement membrane of tumor vessels, and its use to enhance the delivery of micromolecules to tumors after conjugation with interleukin-2. *Cancer Res.* 55:2673–2680.

Ferrara, N., Houck, K., Jakeman, L. and Leung, D.W. (1992) Molecular and biological properties of the vascular endothelial growth factor family of proteins. *Endocrine Reviews* 13:18–32.

Ferrara, N., Winer, J., Burton, T., Rowland, A., Siegel, M., Phillips, H.S., Terrell, T., Keller, G.A. and Levinson, A.D. (1993) Expression of vascular endothelial growth factor does not promote transformation but confers a growth advantage *in vivo* to Chinese hamster ovary cells. *J. Clin. Invest.* 91:160–170.

Flaumenhaft, R. and Rifkin, D.B. (1991) Extracellular matrix regulation of growth factor and protease activity. *Curr. Op. Cell. Biol.* 3:817–823.

Fleming, T.P., Saxena, A., Clark, W.C., Robertson, J.T., Oldfield, E.H., Aaronson, S.A. and Ali, I.U. (1992) Amplification and/or overexpression of platelet-derived growth factor receptors and epidermal growth factor receptor in human glial tumors. *Cancer Res.* 52:4550–4553.

Folkman, J. (1990) What is the evidence that tumors are angiogenesis dependent? *J. Natl. Cancer Inst.* 82:4–6.

Folkman, J., Langer, R., Linhardt, R., Haudenschild, C. and Taylor, S. (1983) Angiogenesis inhibition and tumor regression caused by heparin or a heparin fragment in the presence of cortisone. *Science* 221:719–725.

Friesel, R.E. and Maciag, T. (1995) Molecular mechanisms of angiogenesis: Fibroblast growth factor signal transduction. *FASEB J.* 9:919–925.

Goldstein, G.W. and Betz, A.L. (1986) The Blood Brain Barrier. *Sci. Am.* 254:74–83.

Gross, J.L., Herblin, W.F., Eidsvoog, K., Horlick, R. and Brem, S.S. (1992) Tumor growth regulation by modulation of basic fibroblast growth factor. *EXS* 61:421–427.

Guerin, C., Laterra, J., Hruban, R.H., Brem, H., Drewes, L.R. and Goldstein, G.W. (1990) The glucose transporter and blood brain barrier of human brain tumors. *Ann. Neurol.* 28:758–765.

Guerin, C., Laterra, J., Masnyk, T., Golub, L.M. and Brem, H. (1992a). Selective endothelial growth inhibition by tetracyclines that inhibit collagenase. *Biochem. Biophys. Res. Commun.* 188:740–745.

Guerin, C., Tamargo, R.J., Olivi, A. and Brem, H. (1992b) Brain tumor angiogenesis: Drug delivery and new inhibitors. *In*: Margoudakis, M.E. (ed.): *Angiogenesis in Health and Disease*. Plenum Press, New York, pp 265–274.

Guerin, C., Wolff, J.E., Laterra, J., Drewes, L.R., Brem, H. and Goldstein, G. (1992c) Vascular differentiation and glucose transporter expression in rat gliomas: Effects of steroids. *Ann. Neurol.* 31:481–487.

Harik, S.I. and Roessmann, U. (1991) The erythrocyte-type glucose transporter in blood vessels of primary and metastatic brain tumors. *Ann. Neurol.* 29:487–491.

Hedlin, C.H. (1992) Structural and functional studies on platelet-derived growth factor. *EMBO J.* 11:4251–4259.

Hermanson, M., Nister, M., Betsholtz, C., Heldin, C.H., Westermark, B. and Funa, K. (1988) Endothelial cell hyperplasia in human glioblastoma: Coexpression of mRNA for platelet-derived growth factor (PDGF) B chain and PDGF receptor suggests autocrine growth stimulation. *Proc. Natl. Acad. Sci. USA* 85:7748–7752.

Hermanson, M., Funa, K., Hartman, M., Claesson-Welsh, L., Heldin, C.H., Westermark, B. and Nister, M. (1992) Platelet-derived growth factor and its receptors in human tissue: Expression of messenger RNA and protein suggests the presence of autocrine and paracrine loops. *Cancer Res.* 52:3213–3219.

Higuchi, M., Ohnishi, T., Arita, N., Hirata, S. and Hayakawa, T. (1993) Expression of tenascin in human gliomas: Its relation to histological malignancy, tumor dedifferation and angiogenesis. *Acta Neuropathol. Berl.* 85:481–487.

Hochberg, F.H. and Pruitt, A. (1980) Assumptions in the radiotherapy of glioblastoma. *Neurology* 30:907–911.

Hsu, D.W., Efird, J.T. and Hedleywhyte, E.T. (1995) Prognostic role of urokinase-type plasminogen activator in human gliomas. *Am. J. Pathol.* 147:114–123.

Hynes, R.O. (1992) Integrins: Versatility, modulation, and signaling in cell adhesion. *Cell* 69:11–25.

Ingber, D., Madri, J.A. and Folkman, J. (1986) A possible mechanism for inhibition of angiogenesis by angiostatic steroids: Induction of capillary basement membrane dissolution. *Endocrinology* 119:1768–1775.

Ingber, D.E. and Folkman, J. (1989) How does extracellular matrix control capillary morphogenesis? *Cell* 58:803–805.

Ingber, D., Fujita, T., Kishimoto, S., Sudo, K., Kanamaru, T., Brem, H. and Folkman, J. (1990) Synthetic analogues of fumagillin that inhibit angiogenesis and suppress tumour growth. *Nature* 348:555–557.

Kelly, P.J., Daumas-Duport, C., Kispert, D.B., Kall, B.A., Scheithauer, B.W. and Illig, J.J. (1987) Imaging-based stereotaxic serial biopsies in untreated intracranial glial neoplasms. *J. Neurosurg.* 66:865–874.

Khokha, R., Waterhouse, P., Yagel, S., Lala, P.K., Overall, C.M., Norton, G. and Denhardt, D.T. (1989) Antisense RNA-induced reduction in murine TIMP levels confers oncogenicity on Swiss 3T3 cells. *Science* 243:947–950.

Kim, K.J., Li, B., Winer, J., Armanini, M., Gillett, N., Phillips, H.S. and Ferrara, N. (1993) Inhibition of vascular endothelial growth factor-induced angiogenesis suppresses tumour growth *in vivo*. *Nature* 362:841–844.

Kondo, S., Asano, M. and Suzuki, H. (1993) Significance of vascular endothelial growth factor/vascular permeability factor for solid tumor growth, and its inhibition by the antibody. *Biochem. Res. Commun.* 194:1234–1241.

Kono, S., Rao, J.S., Bruner, J.M. and Sawaya, R. (1994) Immunohistochemical localization of plasminogen activator inhibitor type 1 in human brain tumors. *J. Neuropath. Exp. Neurol.* 53:256–262.

Lal, B., Indurti, R.R., Couraud, P., Goldstein, G.W. and Laterra, J. (1994) Endothelial cell implantation and survival within experimental gliomas. *Proc. Natl. Acad. Sci. USA* 91:9695–9699.

Landau, B.J., Kwaan, H.C., Verrusio, E.N. and Brem, S.S. (1994) Elevated levels of urokinase-type plasminogen activator and plasminogen activator inhibitor type-1 in malignant human brain tumors. *Cancer Res.* 54:1105–1108.

Lane, T.F., Iruela-Arispe, M.L., Johnson, R.S. and Sage, E.H. (1994) SPARC is a source of copper-binding peptides that stimulate angiogenesis. *J. Cell Biol.* 125:929–943.

Langer, R., Brem, H., Falterman, K., Klein, M. and Folkman, J. (1976) Isolations of a cartilage factor that inhibits tumor neovascularization. *Science* 70:70–72.

Laterra, J., Guerin, C. and Goldstein, G.W. (1990) Astrocytes induce neutral microvascular endothelial cells to form capillary like structures *in vitro*. *J. Cell. Physiol.* 144:204–215.

Laterra, J., Stewart, P. and Goldstein, G. (1991) Development of the blood-brain barrier. *In*: Polin, R.A. and Fox, W.W. (eds.): *Neonatal and Fetal Medicine – Physiology and Pathophysiology*. W.B. Saunders, pp 1525–1531.

Laterra, J., Indurti, R.R. and Goldstein, G.W. (1994) Regulation of *in vitro* glia-induced microvessel morphogenesis by urokinase. *J. Cell. Physiol.* 158:317–324.

Li, V.W., Folkerth, R.D., Watanabe, H., Yu, C.N., Rupnick, M., Barnes, P., Scott, R.M., Black, P.M., Sallan, S.E. and Folkman, J. (1994) Microvessel count and cerebrospinal fluid basic fibroblast growth factor in children with brain tumours. *Lancet* 344 2915:82–86.

Liotta, L.A., Steeg, P.S. and Stetlerstevenson, W.G. (1991) Cancer metastasis and angiogenesis – An imbalance of positive and negative regulation. *Cell* 64:327–336.

Long, D.M. (1970) Capillary ultrastructure and the blood-brain barrier in human malignant
brain tumors. *J. Neurosurg.* 32:127–144.

Madrid, Y., Langer, L.F., Brem, H. and Langer, R. (1991) New directions in the delivery of drugs
and other substances to the CNS. *Adv. in Pharmacol.* 22:299–324.

Maxwell, M., Naber, S.P., Wolfe, H.J., Galanopoulos, T., Hedley-Whyte, E.T., Black, P.M. and
Antoniades, H.N. (1990) Coexpression of platelet-derived growth factor (PDGF) and PDGF-
receptor genes by primary human astrocytomas may contribute to their development and
maintenance. *J. Clin. Invest.* 86:131–140.

Maxwell, M., Naber, S.P., Wolfe, H.J., Hedleywhyte, E.T., Galanopoulos, T., Nevillegolden, J.
and Antoniades, H.N. (1991) Expression of angiogenic growth factor genes in primary
human astrocytomas may contribute to their growth and progression. *Cancer. Res.* 51:
1345–1351

McComb, R.D. and Bigner, D.D. (1984) The biology of malignant gliomas – A comprehensive
survey. *Clin. Neuropathol.* 3:93–106.

McComb, R.D. and Bigner, D.D. (1985) Immunolocalization of monoclonal antibody-defined
extracellular matrix antigens in human brain tumors. *J. Neuro-Oncol.* 3:181–186.

Mikkelsen, T., Yan, P.S., Ho, K.L., Sameni, M., Sloane, B.F. and Rosenblum, M.L. (1995)
Immunolocalization of cathepsin B in human glioma: Implications for tumor invasion and
angiogenesis. *J. Neurosurg.* 83:285–290.

Millauer, B., Shawver, L.K., Plate, K.H., Risau, W. and Ullrich, A. (1994) Glioblastoma growth
inhibited *in vivo* by a dominant negative Flk-1 in mutant. *Nature* 367:576–579.

Moriyama, T., Kataoka, H., Tsubouchi, H. and Koono, M. (1995) Concomitant expression of
hepatocyte growth factor (HGF), HGF activator and c-*met* genes in human glioma cells
in vitro. *FEBS Lett.* 372:78–82.

Morrison, R.S., Yamaguchi, F., Saye, H., Bruner, J.M., Yahanda, A.M., Donehower, L.A. and
Berger, M. (1994a) Basic fibroblast growth factor and fibroblast growth factor receptor I are
implicated in the growth of human astrocytomas. *J. Neuro-Oncol.* 18:207–216.

Morrison, R.S., Yamaguchi, F., Bruner, J.M., Tang, M., McKeehan, W. and Berger, M.S.
(1994b) Fibroblast growth factor receptor gene expression and immunoreactivity are elevated in human glioblastoma multiforme. *Cancer Res.* 54:2794–2799.

Moses, M.A., Sudhalter, J. and Langer R. (1990) Identification of an inhibitor of neovascularization from cartilage. *Science* 248:1408–1410.

Nakano, A., Tani, E., Miyazaki, K. (1995) Matrix metalloproteinases and tissue inhibitors of
metalloproteinases in human gliomas. *J. Neurosurg.* 83:298–307.

Nam, M., Johnston, P., Lal, B., Indurti, R., Wilson, M.A. and Laterra, J. (1996) Endothelial
cell-based antitumor gene delivery to the brain. *Brain Res.,* in press.

Neuwelt, E.A., Howieson, J., Frenkel, E.P., Specht, H.D., Weigel, R., Buchan, C.G. and Hill,
S.A. (1986) Therapeutic efficacy of multiagent chemotherapy with drug delivery enhancement by blood-brain barrier modification in glioblastoma. *Neurosurgery* 19:573–582.

Ojeifo, J.O., Forough, R., Paik, S., Maciag, T. and Zwiebel, J.A. (1995) Angiogenesis-directed
implantation of genetically modified endothelial cells in mice. *Cancer Res.* 55:2240–2244.

Paulus, W., Baur, I., Schuppan, D. and Roggendorf, W. (1993) Characterization of integrin
receptors in normal and neoplastic human brain. *Am. J. Pathol.* 143:154–163.

Pepper, M., Ferrara, N., Orci, L. and Montesano, R. (1991) Vascular endothelial growth factor
(VEGF) induces plasminogen activators and plasminogen activator inhibitor 1 in microvascular endothelial cells. *Biochem. Biophys. Res. Commun.* 181:902–906.

Plate, K.H., Breier, G., Farrel, C.L. and Risau, W. (1992a) Platelet-derived growth factor
receptor-β is induced during tumor development and upregulated during tumor progression
in endothelial cells in human gliomas. *Lab. Invest.* 67:529–534.

Plate, K.H., Breier, G., Weich, H.A. and Risau, W. (1992b) Vascular endothelial growth factor
is a potential tumour angiogenesis factor in human gliomas *in vivo*. *Nature* 359:845–848.

Plate, K.H., Breier, G., Millauer, B., Ullrich, A. and Risau, W. (1993) Up-regulation of vascular
endothelial growth factor and its cognate receptors in a rat glioma model of tumor
angiogenesis. *Cancer Res.* 53:5822–5827.

Plate, K.H., Breier, G., Weich, H.A., Mennel, H.D. and Risau, W. (1994) Vascular endothelial
growth factor and glioma angiogenesis: Coordinate induction of VEGF receptors, distribution
of VEGF protein and possible *in vivo* regulatory mechanisms. *Intl. J. Cancer* 59:520–529.

Rao, J.S., Steck, P.A., Mohanam, S., Stetler-Stevenson, W.G., Liotta, L.A. and Sawaya, R. (1993) Elevated levels of Mr 92,000 type IV collagenase in human brain tumors. *Cancer Res.* 53:2208–2211.

Redekop, G.J. and Naus, C.C.G. (1995) Transfection with bFGF sense and antisense cDNA resulting in modification of malignant glioma growth. *J. Neurosurg.* 82:83–90.

Rempel, S.A., Rosenblum, M.L., Mikkelsen, T., Yan, P.S., Ellis, K.D., Golembieski, W.A., Sameni, M., Rozhin, J., Ziegler, G. and Sloane, B.F. (1994) Cathepsin B expression and localization in glioma progression and invasion. *Cancer Res.* 54:6027–6031.

Rosen, E.M., Zitnik, R.J., Elias, J.A., Bhargava, M.M., Wines, J. and Goldberg, I.D. (1993) The interaction of HGF-SF with other cytokines in tumor invasion and angiogenesis. *EXS* 65:301–310.

Rosen, E.M., Laterra, J., Joseph, A., Jin, L., Way, D., Witte, M., Weinarnd, M. and Goldberg, I. (1996) Scatter factor expression and regulation in human glial tumors. *Int. J. Cancer* 67:1–8.

Rubin, L.L., Hall, D.E., Porter, S., Barbu, K., Cannon, C., Horner, H.C., Ianatpour, M., Liaw, C.W., Manning, K., Morales, J., Tanner, L.I., Tomaselli, K.J. and Bard, F. (1991) A cell culture model of the blood brain barrier. *J. Cell Biol.* 115:1725–1735.

Samoto, K., Ikezaki, K., Ono, M., Shono, T., Kohno, K., Kuwano, M. and Fukui, M. (1995) Expression of vascular endothelial growth factor and its possible relation with neovascularization in human brain tumors. *Cancer Res.* 55:1189–1193.

Sawaya, R., Yamamoto, M., Ramo, O.J., Shi, M.L., Rayford, A. and Rao, J.S. (1995) Plasminogen activator inhibitor-1 in brain tumors: Relation to malignancy and necrosis. *Neurosurgery* 36:375–380.

Scherrer, H.J. (1935) Gliomstudien. III. Angioplastische Gliome. *Virchows Arch.* 294:823–861.

Senger, D.R., VanDeWater, L., Brown, L.F., Nagy, J.A., Yeo, K.T., Yeo, T.K., Berse, B., Jackman, R.W., Dvorak, A.M. and Dvorak, H.F. (1993) Vascular permeability factor (VPF, VEGF) in tumor biology. *Cancer Met. Rev.* 12:303–324.

Shapiro, J.R. (1986) Biology of gliomas: Heterogeneity, oncogenes, growth factors. *Semin. Oncol.* 13:4–15.

Shapiro, J.R., Mehta, B.M., Ebrahim, S.A., Scheck, A.C. Moots, P.L. and Fiola, M.R. (1991) Tumor heterogeneity and intrinsically chemoresistant subpopulations in freshly resected human malignant gliomas. *Basic Life Sci.* 57:261–262.

Shweiki, D., Itin, A., Soffer, D. and Keshet, E. (1992) Vascular endothelial growth factor induced by hypoxia may mediate hypoxia-initiated angiogenesis. *Nature* 843–845.

Sivaparvathi, M., Sawaya, R., Wang, S.W., Rayford, A., Yamamoto, M., Liotta, L.A., Nicolson, G.L. and Rao, J.S. (1995) Overexpression and localization of cathepsin B during the progression of human gliomas. *Clin. Exp. Metastasis* 13:49–56.

Stan, A.C., Nemati, M.N., Pietsch, T., Walter, G.F. and Dietz, H. (1995) *In vivo* inhibition of angiogenesis and growth of the human U-87 malignant glial tumor by treatment with an antibody against basic fibroblast growth factor. *J. Neurosurg.* 82:1044–1052.

Stewart, P., Hayakawa, K., Farrell, C.L. and Del, M.R. (1987) Quantitative study of microvessel ultrastructure in human peritumoral brain tissue. Evidence for a blood brain barrier defect. *J. Neurosurg.* 67:697–705.

Takahashi, J.A., Fukumoto, M., Kozai, Y., Ito, N., Oda, Y., Kikuchi, H. and Hatanaka, M. (1991) Inhibition of cell growth and tumorigenesis of human glioblastoma cells by a neutralizing antibody against human basic fibroblast growth factor. *FEBS Lett.* 288:65–71.

Takamiya, Y., Brem, H., Ojeifo, J., Mineta, T. and Martuza, R.L. (1994) AGM-1470 inhibits the growth of human glioblastoma cells *in vitro* and *in vivo*. *Neurosurgery* 34:869–875.

Taki, T., Ohnishi, T., Arita, N., Hiraga, S., Saitoh, Y., Izumoto, S., Mori, K. and Hayakawa, T. (1994) Anti-proliferative effects of TNP-470 on human malignant glioma *in vivo*: Potent inhibition of tumor angiogenesis. *J. Neuro-Oncol.* 19:251–258.

Tamargo, R., Leong, K. and Brem, H. (1990) Growth Inhibition of the 9L glioma using polymers to release heparin and cortisone acetate. *J. Neuro-Oncol.* 9:131–138.

Tamargo, R., Bok, R.A. and Brem, H. (1991) Angiogenesis inhibition by minocycline. *Cancer Res.* 51:672–675.

Tao-Cheng, J. and Brightman, M. (1988) Development of membrane interactions between brain endothelial cells and astrocytes *in vitro*. *Int. J. Dev. Neurosci.* 6:25–37.

Thorgeirsson, U.P., Lindsay, C.K., Cottam, D.W. and Gomez, D.E. (1994) Tumor invasion, proteolysis and angiogenesis. *J. Neuro-Oncol.* 18:89–103.

Tsai, J.C., Goldman, C.K. and Gillespie, G.Y. (1995) Vascular endothelial growth factor in human glioma cell lines: Induced secretion by EGF, PDGF-BB and bFGF. *J. Neurosurg.* 82:864–873.

Ueba, T., Takahashi, J.A., Fukumoto, M., Ohta, M., Ito, N., Oda, Y., Kikuchi, H. and Hatanaka, M. (1994) Expression of fibroblast growth factor receptor-1 in human glioma and meningioma tissues. *Neurosurgery* 34:221–225.

Unemori, E.N., Ferrara, N., Bauer. E.A. and Amento, E.P. (1992) Vascular endothelial growth factor induces interstitial collagenase expression in human endothelial cells. *J. Cell. Physiol.* 153:557–562.

Warren, R.S., Yuan, H., Matli, M.R., Gillett, N.A. and Ferrara, N. (1995) Regulation by vascular endothelial growth factor of human colon cancer tumorigenesis in a mouse model of experimental liver metastastis. *J. Clin. Invest.* 95:1789–1797.

Weidner, N., Semple, J.P., Welch, W.R. and Folkman, J. (1991) Tumor angiogenesis and metastasis-correlation in invasive breast carcinoma. *N. Engl. J. Med.* 324:1–8.

Weindel, K., Moringlane, J.R., Marme, D. and Weich, H.A. (1994) Detection and quantification of vascular endothelial growth factor vascular permeability factor in brain tumor tissue and cyst fluid: The key to angiogenesis? *Neurosurgery* 35:439–448.

Weingart, J.D., Sipos, E.P. and Brem, H. (1995) The role of minocycline in the treatment of intracranial 9L glioma. *J. Neurosurg.* 82:635–640.

Weller, R.O., Foy, M. and Cox, S. (1977) The development and ultrastructure of the microvasculature in malignant gliomas. *Neuropathol. Appl. Neurobiol.* 3:307–322.

Wilson, J.T. and Penar, P.L. (1994) The effect of AGM-1470 in an improved intracranial 9L gliosarcoma rat model. *Neurol. Res.* 16:121–124.

Woessner, J.F. (1991) Matrix metalloproteinases and their inhibitors in connective tissue remodeling. *FASEB J.* 5:2145–2154.

Wolff, J.E.A., Laterra, J. and Goldstein, G. (1992) Steroid inhibition of neutral microvessel morphogenesis *in vitro*: Receptor mediation and astroglial dependence. *J. Neurochem.* 58:1023–1032.

Wolff, J.E.A., Guerin, C., Laterra, J., Bressler, J., Indurti, R.R., Brem, H. and Goldstein, G.W. (1993) Dexamethasone reduces vascular density and plasminogen activator activity in 9L rat brain tumors. *Brain Res.* 604:79–85.

Yamamoto, M., Sawaya, R., Mohanam, S., Rao, V.H., Bruner, J.M., Nicolson, G.L., Ohshima, K. and Rao, J.S. (1994a) Activities localization, and roles of serine proteases and their inhibitors in human brain tumor progression. *J. Neuro-Oncol.* 22:139–151.

Yamamoto, M., Sawaya, R., Mohanam, S., Rao, V.H., Bruner, J.M., Nicolson, G.L. and Rao, J.S. (1994b) Expression and localization of urokinase-type plasminogen activator receptor in human gliomas. *Cancer Res.* 54:5016–5020.

Yamamoto, M., Sawaya, R., Mohanam, S., Loskutoff, D.J., Bruner, J.M., Rao, V.H., Oka, K., Tomonaga, M., Nicolson, G.L. and Rao, J.S. (1994c) Expression and cellular localization of messenger RNA for plasminogen activator inhibitor type 1 in human astrocytomas *in vivo*. *Cancer Res.* 54:3329–3332.

Zagzag, D., Miller, D.C., Sato, Y., Rifkin, D.B. and Burstein, D.E. (1990) Immunohistochemical localization of basic fibroblast growth factor in astrocytomas. *Cancer Res.* 50:7393–7398.

Zagzag, D., Friedlander, D.R., Miller, D.C., Dosik, J., Cangiarella, J., Kostianovsky, M., Cohen, H., Grumet, M. and Greco, M.A. (1995) Tenascin expression in astrocytomas correlates with angiogenesis. *Cancer Res.* 55:907–914.

Zarneger, R. and DeFrances, M.C. (1993) Expression of HGF-SF in normal and malignant human tissues. *EXS* 65:181–199.

Zhang, H.T., Craft, P., Scott, P.A.E., Ziche, M., Weich, H.A., Harris, A.L. and Bicknell, R. (1995) Enhancement of tumor growth and vascular density by transfection of vascular endothelial cell growth factor into MCF-7 human breast carcinoma cells. *J. Natl. Cancer Inst.* 87:213–219.

Regulation of Angiogenesis
ed. by I.D. Goldberg & E.M. Rosen
© 1997 Birkhäuser Verlag Basel/Switzerland

Lymphangiogenesis: Mechanisms, significance and clinical implications

M.H. Witte, D.L. Way, C.L. Witte and M. Bernas

Department of Surgery, University of Arizona College of Medicine, Tucson, Arizona 85724, USA.

... it is clear that current descriptions of the reactions occurring in injured tissues should be expanded to include the lymphatics on the same footing as the blood vessels. The presence of the lymphatic proliferation may well be as important for successful repair as that of the blood vessels. B.D. Pullinger and Nobel laureate H.W. Florey, 1937.

Introduction

Great attention has been directed toward understanding angiogenesis over the past several decades since this phenomenon was reproduced *in vitro* in endothelial cell and mixed vascular tissue cultures (Folkman and Hauden-schild, 1980; Folkman, 1995). The focus, however, has been almost entirely on blood vessel growth, or what we have termed "*hem*angiogenesis" (M. Witte and Witte, 1987c). Yet a vast interstitial fluid circulation suffuses the tissues, bathes the parenchymal cells and interconnects with the blood vasculature via the lymphatic vasculature (Fig. 1). This arc of the circulatory system on the dark side of the blood-tissue interface and its growth ("*lymph*angiogenesis") has received scant attention even though lymphatic (re)generation is both vigorous and essential, and disorders of lymphatic dynamics are common, often disfiguring, and occasionally life- and limb-threatening.

In this chapter on lymphangiogenesis, we examine the lymphatic vasculature and in particular lymphangiogenesis in the larger context of the tissue fluid, matrix, immune and parenchymal cell populations and lymphoid aggregates (the extracellular milieu) in which they function. Initially, we review what is currently known about lymphangiogenesis *in vivo* largely from elegant descriptions, dating from the first decades of this century, of lymphatic regrowth after injury including inflammation and wounding, and also the longstanding embryologic controversy (unresolved to the present day) as to whether the lymphatic vasculature arises independently from tissue mesenchyme or "buds" centrifugally from a pre-existing central venous system. Thereafter, we present newer insights based on the first successful cell cultures of lymphatic endothelium (1984–1985) and the proliferation and migration of these cultured cells to form tubular structures ("lymphangiogenesis *in vitro*"). Despite the dearth of specific mechanistic information, we next examine physiologic and pathologic

Blood-Lymph Loop

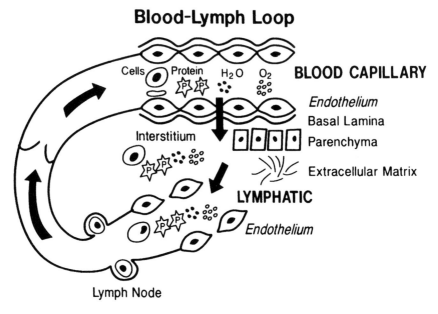

Figure 1. Blood-lymph circulatory loop. Within the bloodstream, liquid flows rapidly as a plasma suspension of erythrocytes; outside the bloodstream, it flows slowly as a tissue fluid-lymph suspension of immunocytes through lymphatics and lymph nodes. Small and large molecules including plasma protein (P), trafficking cells, particulates and respiratory gases cross the blood capillary endothelial barrier, percolate through the tissues, and also re-enter the lymph stream to complete the loop. (From M. Witte and Witte, 1987b).

processes in which lymphatic vessels arise, grow, and malfunction. These include a spectrum of lymphedema-angiodysplasia syndromes, some clearly genetic in origin and others strongly influenced by female sex hormones, which are characterized by limb or visceral swelling, scarring, immunodysregulation, nutritional disturbances, and deranged lymphangiogenesis including angiotumorigenesis. Finally, important areas of ignorance are highlighted, along with ways in which unanswered questions might be explored.

Structural and functional considerations

Discovered by Gasper Aselli (1627) of Padua virtually at the same time that William Harvey described the blood circulation, the lymphatic circulation has for a variety of reasons (inaccessibility, misconceptions, and confusion) remained relatively neglected until the present day. Even the most up-to-date vascular texts separate the "vasculature" (blood vessels) from the lymphatics, or focus narrowly on peripheral lymphedema and disregard the visceral lymphatic system and chyle. Other texts consider exclusively disorders of lymph nodes (primarily neoplasia), as though these nodes were held together by "strings".

Over the centuries since Aselli's discovery, knowledge about the lymphatic system has accumulated slowly, depending almost entirely on advances in imaging. Initial observations were made in cadavers and depended on the naked eye and intralymphatic injection of preservatives such as metallic mercury. Later, lymphatics were visualized by local or systemic infusion of colored dyes or fixed by instillation of latex compounds. Only in the past 40 years have patients regularly been studied, first using oil contrast agents for plain roentgenograms after direct lymphatic injection (conventional lymphography, reviewed in Kinmonth, 1972) and currently by interstitially instilled radioactive macromolecular tracers such as albumin, dextran, and sulfacolloid (isotope lymphography or lymphangioscintigraphy, reviewed in C. Witte and Witte, 1996). Accordingly, the normal and deranged anatomic variants of the lymphatic trunks, large collectors and regional lymph nodes have now been well delineated by these progressively less invasive techniques both in patients and experimental animals. Gradual refinements in microscopy using greater magnification and better resolution, including light and fluorescent, transmission and scanning microscopy, and the development of immunohistochemical staining, have also allowed greater topographic insight into the features distinctive of lymphatics as well as their similarities to blood vessels in health and disease.

Largely within the past hundred years, the physiology of the "absorbent vessels" has also been elucidated, primarily by lymphatic cannulation and sampling. Experimental conditions that accelerate, decrease, or block lymph flow or alter its composition have provided fundamental knowledge about diverse clinical disorders characterized by edema and effusion. The major breakthrough in understanding lymph formation and lymph flow was the demonstration by Ernest Starling (1896) that transcapillary movement of fluid and small solutes was governed by the balance of blood capillary and tissue hydrostatic and protein osmotic forces acting at the blood-tissue interface. Indeed, Starling's hydrodynamic law of transcapillary exchange and the recognition that edema (dropsy) represents an imbalance between lymph formation and absorption rested upon wide-ranging investigations documenting predictable alterations in central lymph flow under varied experimental conditions. Later lymphatic investigators (lymphologists) such as Cecil Drinker, F.C. Courtice, and H. Mayerson (Drinker and Yoffey, 1941; Courtice and Steinbeck, 1950; Mayerson et al., 1960), corroborated the unique role of the lymphatics in absorption of injected or leaked proteins and other macromolecules from the interstitium and clarified that the absorption of cholesterol and long-chain triglycerides from the digestive tract as chylomicra accounted for the distinctive milky appearance of mesenteric lacteal and thoracic duct lymph after a fatty meal. During the past 50 years, beginning with Yoffey (1929) and continuing with Gowans et al. (1961), the lymphocyte, previously thought an inert, mature cell, has come to be recognized as the crucial circulating player in the immune system. This continuous circulation and recirculation of lymphoid cell populations from blood to lymph and within the lymphoid organs (spleen, tonsils, thymus, Peyer patches, lymph nodes) and the bone marrow has been better defined, including the recognition of endothelial receptors, markers and binding integrins involved in trafficking of specific lymphoid cell populations. Furthermore, the intrinsic contractility of lymphatic vessels (Pippard and Roddie, 1987) as a critical pumping force popelling lymph centrally toward the great veins has also been established.

In 1987, we (M. Witte and Witte, 1987a) edited a monograph entitled "Are Lymphatics Different From Blood Vessels?" The consensus was that although differences between lymphatics and blood vessels exist, they also share many common features. Both vascular systems are lined by endothelium with the larger vessel wall supported by a smooth muscle framework, particularly around luminal valves (veins and lymphatics). Both have a vasa vasorum composed of the other vessel type. Lymph flowing in the lymphatic system is generally a clear fluid, platelet-free, containing very few red blood cells and hence much less coagulable than blood. Intrinsic and extrinsic clotting factors are present in lymph albeit in lower concentrations due to molecular sieving. Thus, lymph coagulates as a gel whereas blood clots as an organized thrombus. Both vasculatures synthesize and respond to a wide variety of vasoactive substances (Sinzinger et al. 1984a,b; 1986a,b; Sjöberg and Steen, 1991). In contrast to blood flow, which depends on the pumping action of the heart (*vis a tergo*) and skeletal muscle contraction (venous flow), lymph propulsion is generated by intrinsic contractility of the lymphatic truncal wall, which beats rhythmically and independently of respiration and cardiac activity, representing a phylogenetic vestige of amphibian lymph hearts. In tissue section, lymphatics closely resemble blood vessels but are generally more thin-walled, irregular, attenuated structures containing bloodless fluid. Immunohistochemical staining patterns can be indistinguishable and, despite expectations to the contrary, differences are generally quantitative rather than qualitative (e.g. Factor VIII-related antigen, von Willebrand factor (vWf; Nagle et al., 1985, 1987); even these distinctions seem to disappear in tissue culture as phenotypic differences are muted. Ultrastructurally, lymphatics exhibit discontinuous or absent basal lamina (blood vessels have a continuous or fenestrated basement membrane), complex overlapping intercellular junctional complexes, and specialized anchoring filaments which hold the vessel open as tissue pressure rises (Fig. 2) (Casley-Smith and Florey, 1961; Leak, 1970, 1972). Luminal and abluminal membrane net charge appears to be reversed in lymphatics compared to blood vessels. Permeability, surface charge distribution, vesicular-macromolecular movement, lipid absorption and transport, and vasoresponsiveness of the two vasculatures are distinct in some respects, vary from organ to organ and in different segments of the

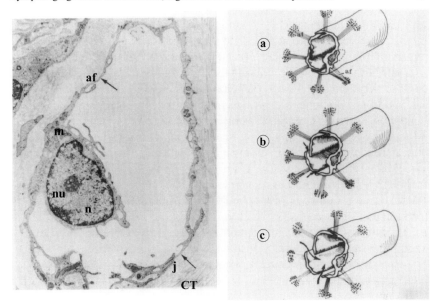

Figure 2. Distinctive ultrastructure of lymphatic capillary. Left: A survey electron micrograph demonstrating a lymphatic capillary in cross section. The close association of the adjoining connective tissue components (CT) with the lymphatic wall is maintained by numerous anchoring filaments (af) which join the abluminal surface to the adjacent interstitium as a meshwork. The endothelium is extremely attenuated at various points (arrows), and the nucleus (n) with its nucleolus (nu) protrudes into the lumen. Several intercellular junctions (j) are visible. Mitochondria (m) appear in the juxtanuclear region, as well as in the thin cytoplasmic rims. ×11,000. Right: Response of lymphatic capillaries to an increase in interstitial fluid volume. As the matrix expands, the tension on the anchoring filaments (af) rises and the lymphatic capillaries open widely to allow more rapid entry of liquid and solute. In contrast to the stretching of the lymph capillaries, a rise in tissue matrix pressure collapses the blood capillaries thereby restricting further plasma filtration. (From Leak, 1970 and 1972, respectively).

circulation, but also may overlap. Blood vascular endothelium, however, is normally a relatively leak-proof, non-thrombogenic surface exposed to high flow and high intraluminal pressure (in the arteries) and is to some extent shielded by the vessel wall from parenchymal-tissue matrix events. Lymphatics, in contrast, are a low flow, low pressure system in intimate contact or even in continuity with the extracellular matrix and ongoing chemical processes and mechanical forces in the interstitium and surrounding parenchymal cells (Ryan, 1989; Ryan and DeBerker, 1996).

The concept of the blood-lymph circulatory loop as a functional unit (Fig. 1) with clear division of labor as the unifying primary circulation of the body is that of extracellular liquid which moves rapidly as plasma as a suspension of red blood cells in the bloodstream, and more sluggishly as lymph-tissue fluid as a suspension of immunocytes in the interstitial space outside the blood vascular compartment (M. Witte and Witte, 1987c). Liquid, macromolecules, and migrating cells pass through the blood capillary endothelial interface, enter the tissues and are gradually absorbed into the lymphatic system, percolate through lymph nodes, and eventually return to the blood circulation via connections with the central veins (i.e. primarily at the right and left thoracic duct-subclavian vein junction but also at lymph nodal sites). When this finely tuned and delicately balanced blood-lymph loop breaks down, the subtle but critical role of the lymphatic system is uncovered with characteristic brawny swelling, fibrosis, immunodysregulation, and angiogenesis-angiotumorigenesis involving both lymphatics and blood vessels (e.g. Stewart-Treves syndrome, Kaposi sarcoma) as the outcome (vide infra Fig. 20) (M. Witte et al. 1990a). Although pathologic processes similar to those affecting blood vessels also occur within lymphatics (i.e. thrombosis, inflammation, dissemination of malignant cells, wall hypertrophy and sclerosis), the slow flow of the tissue fluid-lymph stream renders lymphatic disorders more chronic disabling conditions associated with intense fibrosis and overgrowth rather than the often calamitous ischemic or congestive events associated with sudden interruption of blood flow.

Lymphangiogenesis and hemangiogenesis *in vivo*

Whereas the recent explosion of interest in angiogenesis has focused almost entirely on the regulation of blood vessel growth and what turns this process on and off, it is becoming clear that similar processes and probably similar underlying physiochemical signals and mediators must be involved in normal and pathologic lymphangiogenesis.

In 1896, Thoma postulated a set of principles governing the growth of blood vessels, which includes what we now consider both non-sprouting (elongation and widening) as well as sprouting (budding) angiogenesis (Sholley et al., 1984). He proposed that increased luminal size of a vessel or its wall surface area depends on the rate of blood flow and then stabilizes unless flow increases or decreases. The thickness of the vessel wall depends upon its tension, which also depends upon the diameter of the lumen and blood pressure, with increased capillary pressure leading to the formation of new capillaries (changes in microvascular density). Though generally consistent with current knowledge, the mechanism by which these biophysical parameters are translated into biochemical events remains unclear. Thoma's principles can be reframed in the modern language of vascular remodeling (whether blood vessels or lymphatics) namely endothelial cell proliferation, migration, rearrangement, or programmed cell death (apoptosis) along with the production, degradation, and realignment of extracellular matrix. Indeed, conditions which greatly increase lymph production and flow rate, such as portal hypertension from hepatic cirrhosis, and systemic venous hypertension in congestive heart failure, strikingly dilate the hepatosplanchnic lymphatic network, heighten intraluminal pressure, increase the thickness of the wall (sclerotic plaques may appear), and stimulate formation of new lymphatic pathways (reviewed in M. Witte et al., 1969; C. Witte et al., 1980 a, b; C. Witte et al., 1984).

Phylogenetically, body fluid turnover or circulation can be traced back to single cell organisms, becoming more developed in the coeloms of coelenterates and forming specialized lymph hearts (*cor lymphaticum*) which beat independently of the "blood heart" in amphibia or reptiles (Kampmeier, 1969). When these hearts are slowed or paralyzed, frogs quickly drown in their rapidly forming lymph. As species evolved away from an aquatic environment, lymph nodes first developed in birds, posing a potential site of restriction to the free flow of lymph. Eventually both features are combined in humans with well-defined lymphatic segments termed "lymphangions" capable of rhythmic intrinsic contractions interrupted by innumerable nodules with highly immunoreactive cells (lymph glands).

Embryologically, controversy has persisted since early in this century about the origin of lymphatics (reviewed in Yoffey and Courtice, 1970; van der Putte, 1975). According to Sabin (1902), lymphatics and veins arise from a common primordium in the venous system. After contrast injections into pig skin, she observed gradually extending lymphatic plexuses starting first at the base of the neck in direct communication with the venous system. She concluded that lymphatics derive from central veins and their growth progresses centrifugally by sprouting toward the periphery. On the other hand, Kampmeier (1969) and others (Huntington, 1914), after study of serial tissue sections and from phylogenetic considerations, proposed the centripetal theory: that the lymphatic system arises independently from tissue mesenchyme and only later joins with the venous system. According to this latter view, fetal cystic hygroma and similar lymphangiomatous malformations reflect failure of the jugular lymph sac to establish venous connection. On the other hand, prelymphatic tissue pathways and perivascular sinuses (e.g. the spaces of Disse in the liver and Virchow-Robin in the brain, and the canal of Schlemm in the eye) are properly considered part of the lymphatic drainage apparatus and under pathologic conditions can even become endothelial-lined.

Lymphatic regeneration is a rapid and vigorous phenomenon (reviewed in Yoffey and Courtice, 1970; Casley-Smith, 1983). After experimental circumferential transection (skeletonization) of hindlimb structures (except the femoral artery, vein and bone), new lymphatics cross the incision site as early as the fourth day postoperatively, and by the eighth day lymphatic reconnection is anatomically (by distribution of India ink) and physiologically (transient edema disappears) restored (Reichert, 1926). More recently, Bellmann and Odén (1958–1959; Odén, 1960) (Fig. 3, right panel) meticulously documented by thoratrast microlymphangiography the time course and extensiveness of newly formed lymphatic vessels in circumferential wounds in the rabbit ear, including lymphatic bridging through the newly formed scar. As the lymph vessels increased in caliber,

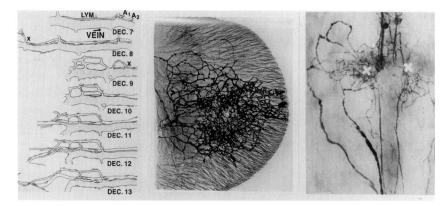

Figure 3. Classical observations on lymphangiogenesis *in vivo*. Left: Series of camera-lucida oil-immersion records, showing growth of an individual lymphatic capillary in the rabbit ear, (LYM). Corresponding parts of a lymphatic and vein have been placed below one another in drawings of Dec (December) 7, 8, and 9 to 13. The drawings of Dec 9 to 13 have been moved to the right so that X represents the same spot on Dec 8 and 9. During growth of the lymphatic there has been conspicuous retraction and disappearance of branches and loops of the blood vessel. Retraction of a short lymphatic tip is seen between Dec 11 and 13. Lymphatic endothelial nuclei are stippled. A1 and A2 are two daughter nuclei after a mitotic division. ×177 (Magnification of original drawing, ×700). Middle: Twenty-one days after making a turpentine abscess, which perforated a mouse ear, a dense new network of lymphatic capillaries surrounds the hole. ×8. Right: *In vivo* micro-lymphangiogram, 24 days after a short transverse incision in the skin of a rabbit ear. The extent of the incision is depicted by the crosses. The distal part of the ear is uppermost. Several arcading vessels around the incision and numerous fine connections through the scar are seen. (From Clark and Clark, 1932; Pullinger and Florey, 1937; and Bellman and Odén, 1958–1959, respectively).

intraluminal valves and sinuous dilatations appeared. Subsequent studies (Magari and Asano, 1978; Satomua et al., 1978; Leak, 1985; Magari, 1987), including induced peritonitis and after autotransplantation of the small intestine, documented restoration of distinctive ultrastructural features in the newly regenerated lymphatic vessels, including the characteristic overlapping junctional complexes and Weibel-Palade bodies (storage depots for vWf).

In now classical studies, Clark and Clark (1932) demonstrated the extension of lymphatic capillaries by outgrowth from pre-existing lymph vessels in rabbit ear transparent chambers (Fig. 3, left panel). The living mammalian lymphatic capillaries were seen as clear vessels with walls of delicate endothelium. There was no tendency toward anastomosis with blood capillaries. Abundant new lymphatic capillaries in the area under observation appeared somewhat later (approximately 30 days) than blood capillaries (5–10 days). Pullinger and Florey (1937) (Fig. 3, middle panel), after examining the proliferation of lymphatics in inflammation and abcesses in the living mouse ear, emphasized that unless opaque or colored materials were used to identify lymphatics, they might otherwise be overlooked. These workers also noted that, despite the similar appearance of each vascular endothelium, lymphatics consistently connected to lymphatics, veins to veins, and arteries to arteries without intermingling.

In adult mammals, the endometrium, ovary, uterus, and breast cyclically exhibit dynamic periodic growth and regression regulated by sex hormones and accompanied by dramatic changes in hemangiogenesis and lymphangiogenesis (Yoffey and Courtice, 1970; Otsuki et al., 1986; Reynolds et al., 1992). The female breast at puberty in particular displays a great upsurge in discrete lymphatics and labyrinths (analogous to prelymphatic pathways) as the mamma hypertrophies; conversely, with aging and menopause, these breast lymphatics involute. Whether estradiol and other sex steroids directly stimulate the proliferation of blood vessel and lymphatic endothelial cells by modulating their responses to growth factors such as basic FGF, IGF-I, and EGF (Reynolds et al., 1992) or to vasoactive substances such as endothelin and prostacyclin (Berge et al., 1990) remains, however, unclear.

Two recent studies provide the first direct quantitative and qualitative information on the proliferation rate of lymphatic endothelium *in vivo*. Like the more than a trillion blood vascular endothelial cells which normally lie dormant in the adult human body (less than 1% incorporate tritiated thymidine, and cell life spans several years) (Jaffe, 1987; Denekamp, 1993), lymphatic endothelial turnover is minuscule. In the rat cornea, basal levels of 0.6% were found in lymphatic endothelium with similar values in adjacent blood capillaries and venules (Junghans and Collin, 1989). Within 36 hours of cautery of the cornea, 6.8% of lymphatic endothelial cells incorporated tritiated thymidine compared to 13.2–13.6% of blood vascular endothelial cells, with rapid diminution thereafter. During studies of the developmental stages of the sheep, Trevella and Dugan (1990) observed greatly increased radiolabeling with tritiated thymidine in the endothelial lining of fetal lymphatic compared to neonatal vessels, which in turn showed greater labeling than adult lymphatics. Mesenteric lymphatic endothelial cells also showed higher proliferative rates than peripheral lymphatic endothelium.

Lymphatic endothelium and lymphangiogenesis *in vitro*

Far from an inert "cellophane wrapper", endothelium is now recognized as the mentor of the blood circulation, a vast endocrine organ and metabolic factory that maintains an intact nonthrombogenic vascular surface, modulates its own structure and function and that of various other cell types including neutrophils, lymphocytes, monocytes, macrophages, pericytes, mast cells, and eosinophils, while masterminding cell trafficking into, through, and out of tissues (Fig. 4). Endothelium produces and responds to a wide variety of autocrine, paracrine, and endocrine signals, thereby mediating inflammation, infection, wound repair, immune reactions,

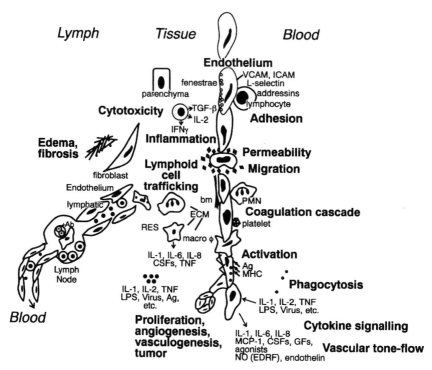

Figure 4. Schematic illustration depicting the postulated role of endothelial processes in microcirculatory events bearing on angiogenesis in the blood-lymph loop. These include macromolecular and liquid permeability, vasoreponsiveness, leukocyte adhesion and transmigration, coagulation cascading, particulate phagocytosis, antigen presentation and cytokine activation, lymphoid cell trafficking, and proliferative events leading to new vessel or tumor growth. Although scarcely studied, processes corresponding to those implicated at the blood vascular endothelial surface probably also occur at the lymphatic endothelial interface. The relative anatomic and dynamic relationships between blood and lymph vascular endothelium, parenchymal and extravascular connective tissues and transmigrating leukocytes are shown. Ag = antigen, bm = basement membrane, CSFs = colony-stimulating factors, ECM = extracellular matrix, EDRF = endothelium derived relaxing factor, GFs = growth factors, ICAM = intercellular adhesion molecule, IFN = interferon, IL = interleukin, LPS = lipopolysaccharide, macro φ = macrophage, MCP-1 = monocyte chemoattractant protein 1, MHC = major histocompatibility complex, NO(EDRF) = nitric oxide (endothelium derived relaxing factor), PMN = polymorphonuclear leukocyte, RES = reticuloendothelial system, TGF = transforming growth factor, TNF = tumor necrosis factor, VCAM = vascular cell adhesion molecule, ● = exogenous particulates, ■ = macromolecules, drops represent fluid (plasma, interstitial or lymph), (From Borgs et al., 1995).

mechanical stress, coagulopathy, oncogenesis, and tumor dissemination. The specific role of soluble chemicals and solid-state regulators (e.g. growth factors, cytokines, proteases/inhibitor balance, procoagulants, anticoagulants, matrix components), mechanical forces (shear stress-tension and "grip and stick"), and genetic and environmental factors that transform the usually quiescent endothelial cell into a proliferating and organizing tissue of vascular structures surrounded by extracellular matrix is nonetheless poorly understood (Carey, 1991; D'Amore and Braunhut, 1987; Gibbons and Dzau, 1990; Baldwin, 1982; Montesano et al., 1983; Tsuboi et al., 1990; Ziche et al., 1992; Ryan and Barnhill, 1983; Ryan, 1990; Friedlander et al., 1995). Furthermore, the once simple barrier function concept of pores governing microvascular permeability and exchange has now evolved into a complex ultrastructural domain of multiple cell-cell junctional complexes (gap junctions, tight junctions, and adhering junctions) with transcellular vesicular shuttling, intricately regulated by a molecular cascade of adhesive (tethering) and contractile elements closely linked to vasomotion, proteolytic balance and tissue matrix interactions. Endothelium is phenotypically diverse and heterogeneous (Turner et al., 1987; Zetter, 1987) varying in morphology and markers by organ, species, vessel size (large caliber vs. capillary), and location (arterial, venous, capillary, and lymphatic). Indeed, microvascular endothelial cultures, for example, from omentum and skin, are usually indeterminate mixtures derived from both lymphatic and blood microvessels abundant in such tissues. *In vitro*, endothelial phenotypic expression is influenced by diverse chemical and mechanical mediators. But exactly how signals are transduced into alterations in endothelial cell morphology, antigenic markers, metabolic products, permeability, migration, and cell cycle activity remains elusive. Nonetheless, under a wide variety of conditions, endothelium retains its chameleon pluripotentiality, perhaps traceable to its angioblastic origin from which the reticuloendothelial system, circulating immune cells, smooth muscle, and even fibroblasts also arise in the primitive blood island primordia. Even adult "differentiated" endothelium may retain this stem cell potential (Hay, 1984). Transdifferentiation may occur in response to altered culture conditions (e.g. basement membrane components, cytokines, growth factors, hypoxia, cyclic AMP), and endothelial cells may convert to a regenerating mode, develop procoagulant activity, and express cellular attachment sites while losing distinctive endothelial markers (Bissell et al., 1982; Lipton et al., 1991).

In 1984, pure cultures of lymphatic endothelium were first isolated from bovine mesenteric lymphatic duct by Johnston and Walker (1984) (Fig. 5) and by our group from a patient with a massive cervicomediastinal lymphangioma (Bowman et al., 1984) (Fig. 6). In 1985, Gnepp and Chandler (Fig. 7) isolated lymphatic endothelium from canine and human thoracic duct, and subsequently others have cultured and studied bovine, ovine, rat, and ferret lymphatic collector endothelium through multiple passages (Leak, 1991; Leak and Jones, 1993; Weber et al., 1994; Yong and Jones, 1991; Borron and Hay, 1994). Routine isolation procedures and culture methods as for blood macro- and microvessel endothelium were applied. Lymphatic endothelial cells have been grown in monolayer and on microcarrier beads using fetal bovine serum with and without heparin and endothelial cell growth supplement. In each instance, lymphatic endothelium grows in sheets with a cobblestone morphology, shows typical endo-

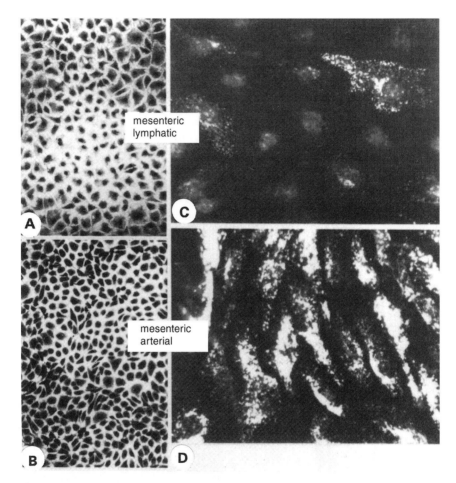

Figure 5. Lymphatic endothelium *in vitro*. Appearance of bovine lymphatic endothelial (A) compared with bovine mesenteric arterial endothelial (B) in tissue culture and after staining for vWf (C, D). Endothelial cells (A, B) were stained with Hemacolour (Harleco). A, Passage 9; (×141). B, Passage 4; (×141). Assay for vWf shows greater intensity in positive staining in arterial (D) compared with lymphatic endothelium (C). (From Johnston and Walker, 1984).

thelial structural features, and is generally positive for vWf. Anti-thrombin 3 and MHC1 are also expressed on the cell surface. Characteristic lymphatic-like overlapping cell junctions have been noted in normal and lymphangioma-derived endothelium, and tight junctions are rare. Fibronectin, F-actin, *Ulex europaeus* ligand, and Weibel-Palade bodies have been documented. Chemical enzyme markers such as 5′ nucleotidase and adenylate and guanylate cyclase enzymes are lost in tissue culture and cannot be used to distinguish endothelial cells of lymphatic from blood vessel origin (Weber et al., 1994). Matrix metalloproteinases (Meade-Tollin et al., 1995), uPA, tPA and a variety of cytokines (Witte, personal observations and Pepper

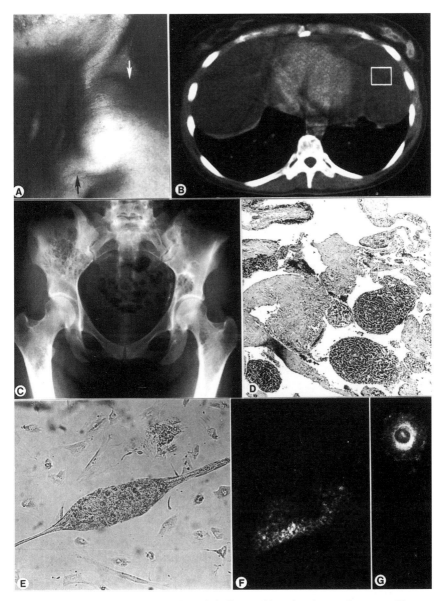

Figure 6. Dysplastic human lymphatic endothelium and lymphangiogenesis *in vitro*. Twenty-nine-year-old woman with massive cervico-mediastinal-intrathoracic cystic hygroma (Gorham disease) (A, B). Lytic soap bubble lesions of pelvis consistent with bony lymphangiomatosis (C). Histopathology of resected neck and chest mass shows complex endothelium-lined cysts with intervening lymphoid aggregates (D). *In vitro* culture of the resected mass sustained long-term a mixture of cell types including fibroblasts and endothelial cells, the latter occasionally forming tubular structures (lymphangiogenesis) (E) strongly positive for Factor VIII related antigen (von Willebrand factor) (vWf) (F) as well as polygonal cells brightly decorated with *Ulex europaeus* lectin (G). (From Bowman et al., 1984).

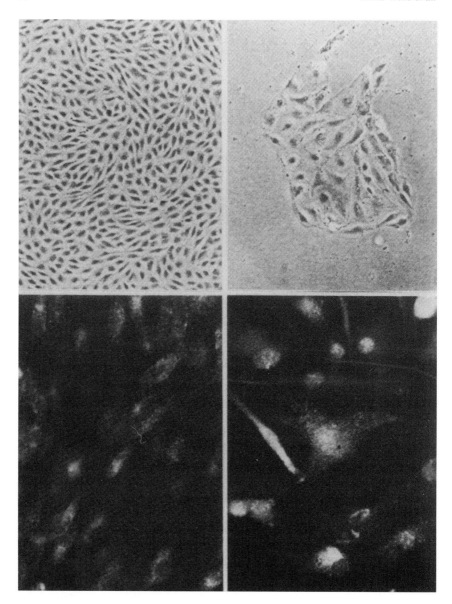

Figure 7. Lymphatic endothelium *in vitro*. Upper left: *In vitro* culture of canine thoracic duct demonstrating a sheet of uniform contact inhibited, non-overlapping endothelial cells with typical cobblestone morphology. (×131). Lower left: Indirect immunofluorescence of canine thoracic duct endothelial cells demonstrating positivity for vWf (granular cytoplasmic staining) (×189). Upper right: Cultured human thoracic duct demonstrating a nest of typical endothelial cells, Day 9. (×129). Lower right: Indirect immunofluorescence of cultured human duct endothelial cells demonstrating granular cytoplasmic staining and perinuclear fluorescence. (×189). (From Gnepp and Chandler, 1985).

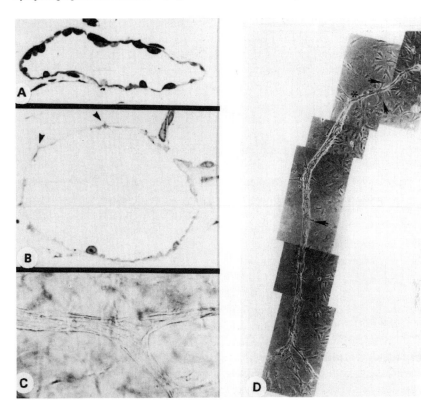

Figure 8. Lymphangiogenesis *in situ* and *in vitro*. (A) Hematic-like vascular channel lined by plump endothelial cells with hump-shaped, cross-sectional profiles in a 9-day-old plasma clot culture of a rat thoracic duct. Toludine blue stain. (×1000). (B) Newly formed, branching vascular channel in a 12-day-old living plasma clot culture of rat thoracic duct. (×333). (C) Light micrograph of a lymphatic-like channel in a 25-day-old plasma clot culture of a rat thoracic duct. The highly attenuated endothelium is anchored to the surrounding fibrin by abluminal cytoplasmic filaments (arrows). Toludine blue stain. (×928). (D) One to 3 days after treatment with collagen type I, adjacent cells (arrows) continue to migrate and adhere to tubular structures with an increase in their dimension and length. Adhesion of cells (asterisk) at various points along the length of the tubular structures is accompanied by branching and the subsequent formation of elaborate capillary-like networks in the culture dishes. This lymphatic capillary tube that formed *in vitro* was derived from bovine lymphatic endothelial cells (×155). (A, B and C taken from Nicosia, 1987; D taken from Leak and Jones, 1994).

et al., 1994) are produced in varying amounts, and the lymphatic cells also actively metabolize acetylated low density lipoproteins (Borron and Hay, 1994). Proliferation rate depends on varying growth conditions in primary and passage level and has been reported to be as high as 36 – 48 hour doubling time (Yong and Jones, 1991). Cell density is not clearly influenced by basement membrane components (collagen, laminin, and fibronectin) (Djonedi and Brodt, 1991; Leak and Jones, 1994). Growth factor responsiveness (Pepper et al., 1994) and production are under study by several groups but definitive reports are not yet available. Migratory behavior, phagocytic activity and transendothelial permeability have been examined under varying conditions although suitable comparisons with control blood vessel endothelium are not interpretable because of the latter's heterogeneity. Tumor cells (Bevacqua et al., 1990) and lymphocytes (Borron and Hay, 1994; Stazzone et al., 1995) readily bind to lymphatic endothelium as measured by adhesion assays. Junctional complex proteins characteristic of lymphatics *in vivo* such as desmoplakin (Schmelz et al., 1995) have recently been identified *in vitro* and may relate to the anchoring filamentous structure, a feature determined in early development.

The phenomenon of "lymphangiogenesis *in vitro*" was first demonstrated in 1984 (Bowman et al.) in lymphatic endothelial cultures from a cervicomediastinal lymphangioma (Fig. 6) where tubule formation and associated cyst-like structures spontaneously formed. This phenomenon has been repeatedly observed by us *in vitro* in other human lymphangioma cell culture isolates (later Figs. 14 and 15). Nicosia (1987) (Fig. 8 A–C) used rat thoracic duct collagen-embedded explants to isolate both lymphatic (intimal extension) and more hematic (adventitial origin) endothelial cells. Lymphatic sprouting was detected in the collagen plasma clot assay. Leak and Jones (1994) demonstrated lymphangiogenesis *in vitro* in normal lymphatic cultures in a delayed post-confluent state (weeks) or earlier after collagen I was added to the culture medium (Fig. 8 D). On the other hand, fibronectin or complete basement membrane replacement as matrigel failed to stimulate lymphatic (in contrast to blood vessel) tube formation *in vitro*. Extensive and spontaneous lymphangiogenesis *in vitro* has also consistently occurred in lymphatic endothelial cultures early on (after days) in a tumorigenic lymphatic cell line HTDEC-1 derived from human thoracic duct (later Fig. 19) (Way et al., 1996).

Disorders of lymphangiogenesis

Whereas interruption of arterial or venous blood flow produces sudden, dramatic, often life-threatening disorders, interference with lymph flow typically results in more insidious, indolent, but nonetheless inexorable, sequelae, manifesting as chronic disabling, occasionally life-threatening, conditions (C. Witte and Witte, 1987). Whether congenital or acquired, involving the limbs or the viscera, a constellation of swelling, scarring, immune dysregulation, malnutrition, and disordered lymphangiogenesis (and hemangiogenesis) emerges, not uncommonly becoming a central feature (M. Witte et al., 1990a). These illnesses (Fig. 9) range from strangulating cystic hygroma, vascular birthmark (Klippel Trenaunay-Weber) syndrome, Kaposi sarcoma, and Stewart-Treves (lymph)angiosarcoma to grotesque elephantine limb overgrowth.

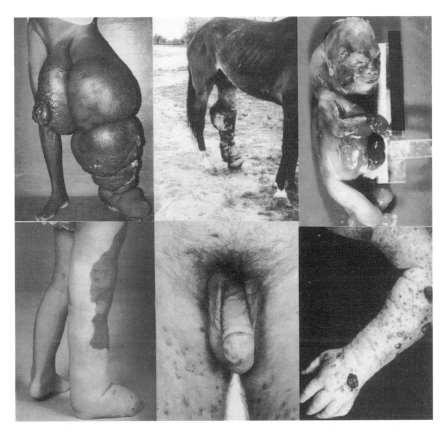

Figure 9. Vivid examples of various genetic, congenital, and acquired lymphologic disorders associated with disturbed (defective, dysplastic, exuberant, and neoplastic) lymphangiogenesis. Upper: Lymphatic filariasis (left), post-traumatic equine lymphedema (middle), fetal Turner (XO) syndrome (ovarian dysgenesis) with strangulating cervical cystic hygroma and hydrops (right). Bottom: Klippel-Trenaunay syndrome (left), acquired immunodeficiency syndrome-associated Kaposi sarcoma (middle), and Stewart-Treves lymphangiosarcoma from post-mastectomy lymphedema. (From M. Witte et al., 1990a except for the last from Stewart and Treves, 1948).

Lymphedema (low output failure of the lymph circulation)

The prototype disorder for lymph stasis is chronic peripheral lymphedema, a potent stimulator of lymphangiogenesis (reviewed in Földi, 1983; Olszewski, 1991; C. Witte et al., 1984; C. Witte and Witte, 1995). Typically, a lengthy latent period transpires before evolution of unremitting edema, sometimes interspersed with intermittent episodes of inflammation or infection which finally culminate in brawny induration and diffuse tissue fibrosis. For many years, the link between lymphatic insufficiency and the external manifestations was obscure because of the variable swelling and often considerable delay between lymphatic ablation and symptoms, the frequent superimposition of cellulitis, and seemingly limitless capacity for

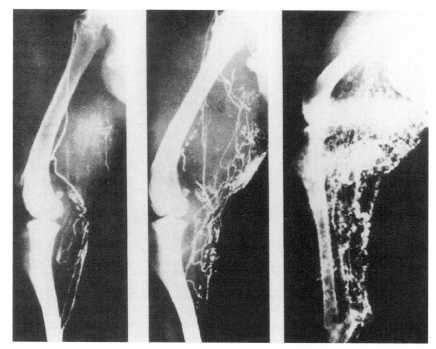

Figure 10. Neolymphangiogenesis following experimental lymphatic obstruction. Sequential conventional lymphograms of dog hind limb 9 (left), 14 (middle), and 48 (right) months after operative manipulation to reproduce peripheral lymphedema. During the first year, there is gradual dilation and collateralization of lymphatic channels with dermal backflow despite lack of clinical edema. By 4 years, however, edema is well-established, and lymphography discloses extensive macrolymphatic proliferation and dysplasia. (From Olszewski, 1973).

lymphatic regeneration. It is now clear, based on pioneering long-term experimental studies by Danese (Danese et al., 1968), Olszewski (1973), and Clodius (Clodius and Altofer, 1977) and their colleagues, that obstruction to lymphatic drainage alone is sufficient to cause unremitting lymph stasis (Fig. 10). Not only is there tissue protein "build-up" but also insufficient removal of newly sequestered plasma protein by tissue macrophages and macrophage proteases. Once lymph flow is impaired, truncal contractions first quicken, then over time the intraluminal valves gradually give way, the mechanical advantage of lymphangion contractility weakens (fails), the lymph fluid column becomes more continuous, and refractory lymphedema supervenes. Hydrostatic pressure in the draining tissue watersheds and lymphatics increases as intrinsic truncal contraction fails to expel lymph completely. In contrast to the normal situation, the fluid column in the lymphatics is now fully "primed", and skeletal muscle contraction or forced external compression becomes an effective mechanism to propel lymph cephalad. The same sequence of events probably also takes place in the abdominal viscera and underlies intestinal lymphangiectasia and chylous reflux syndromes.

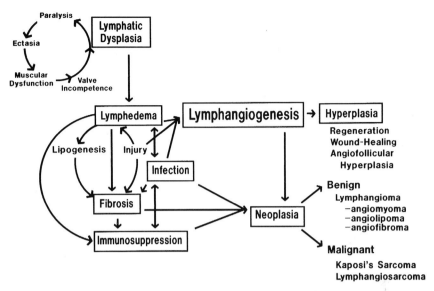

Figure 11. Hypothesized links between the pathophysiologic events and sequelae of lymph stasis and disturbed lymphangiogenesis, hyperplasia, dysplasia, and neoplasia. During a latent period (no overt edema), lymphatic truncal dysplasia is accompanied by lymphatic ectasia, muscular dysfunction, lymphangioparalysis, and intraluminal valve incompetence culminating in persistent protein-rich (more than 1.5 g/dl) lymphedema. Superimposed opportunistic infection, immunosuppression, and injury with repeated lymphangitis promote intractable lymphedema, intense (lymph)angiogenesis, lipid deposition and fibrosis, and on rare occasions malignant vascular neoplasia (note controversy as to whether Kaposi sarcoma is a true neoplasm or cancer). This sequence of events may evolve as a local phenomenon as in peripheral lymphedema ("local AIDS") or as a multitude of isolated or generalized visceral disorders characterized by scar formation and angiohyperplasia and neoplasia including AIDS-related syndromes. (From M. Witte and Witte, 1986).

Similarly, in hindlimb lymph stasis and lymphedema from experimental filarial infection, a condition that closely mimics the human counterpart, nests of wriggling live adult worms residing in peripheral lymphatics stimulate endothelial overgrowth, massive lymphatic dilatation ("lakes"), striking and tortuous lymphatic neovascularization, and racemose collateralization (Fig. 12) (Case et al., 1991, 1992a; Crandall et al., 1987). Later in the course of infection, adult worms die, usually after a year or more, and inflammatory immunoreactive obliteration of the lymphatics is superimposed.

Vigorous lymphangiogenesis and lymphatic dysplasia characteristic of chronic experimental lymph stasis and lymphedema from surgically-induced lymphatic obstruction or filarial infection is accompanied by a concomitant hemangiogenesis as evidenced by an increase in blood vessel cross-sectional area on tissue section and hypervascularity on arteriography (Casley-Smith and Casley-Smith, 1986; Daroczy, 1996; Snowden and Hammerberg, 1986). Clinically, hemangiogenesis in the lymphedematous limb is typically displayed in the dermal hyperemia of the swollen

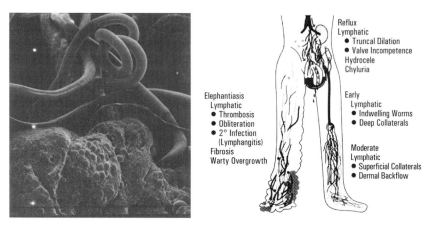

Figure 12. Lymphatic growth abnormalities in filariasis. Left: Scanning electron micrograph of emerging adult filarial nematode (*Brugia malayi*) from a markedly dilated lymphatic cistern with hyperplastic heaped-up endothelium in an experimentally infected ferret. Right: Proposed scheme of lymphatic growth abnormalities in human filariasis as depicted by lymphangioscintigraphy. Early in the disease, impedance to lymph truncal transport promotes collateralization within the deep lymphatics. With further progression, collaterals form prominently in the superficial cutaneous lymphatics, and dermal diffusion (backflow) becomes more apparent. At this stage, regional nodes and central lymphatics are usually still visualized. With repeated infections and recurrent lymphadenitis, lymphatic insufficiency becomes more diffuse (elephantiasis) although regional nodes are occasionally still seen. With more central lymphatic blockade (e.g. retroperitoneum), lymph may flow retrograde (e.g. as chyluria or hydrocele) even without peripheral lymphedema. The process outlined in the legs may occur similarly in the upper extremities. Compare with Normal scheme in Figure 16. (Right from M. Witte et al., 1993).

limb. Accordingly, in both primary and secondary lymph stasis, the proliferative blood vascular response and heightened surface area for microvascular filtration of liquid and macromolecules into the interstitium further burdens the already limited lymphatic transport capacity. Indiscriminate vascular hyperplasia may take on bizarre angiomatoid distortions and may even progress to highly aggressive malignant vascular tumors (i.e. the Stewart-Treves syndrome) in longstanding lymphedema, whether congenital or acquired, in the limbs, irradiated breast, or at other sites. This striking overgrowth phenomenon is not, however, limited to endothelial and vascular proliferation and oncogenesis. The tissue remodeling also involves the laying down of dense fibrous tissue (fibroblasts) and fat accumulation (adipocytes) (Ryan, 1996), the sheer bulk of which further distorts tissue biomechanics and interferes with lymph and tissue fluid circulation.

Thus, in the setting of intractable protein-rich lymphedema (in contrast to longstanding low protein edema of renal or heart failure), dermal life processes are gradually extinguished or so distorted as to be unrecognizable. Yet the subfascial (muscle) compartment with its tight fascial binding is characteristically spared. The following sequence of events probably takes place: anatomical or functional lymphatic obstruction leads to a heightened intralymphatic transmural pressure, which is transmitted to the endothelial lining cells as mechanical stress. This shear/strain effect then signals the lymphatic endothelial cells to release vasoactive angiogenic chemical mediators similar to those produced by blood vascular endothelium. These, in turn, initiate mitogenesis and cell migration in lymphatics and perhaps indiscriminately also in nearby blood vessels (lymphangiogenesis and hemangiogenesis). The subsequent response may further enhance lymph formation despite a reduced lymphatic transport capacity. Continued accumulation of protein-rich edema fluid and trapped immune cells in the interstitium progressively disrupts attachments and adhesions of a variety of cell types, promoting lymphatic remodeling and collateral formation. The angiogenic process involves release of growth factors not only from endothelial and other neighboring cell types but also from ground substance stores. Exacerbated by poor lymphatic drainage and extracellular matrix alterations, an array of different cell types is stimulated to proliferate in the poorly structured milieu, culminating in generalized dysplastic soft tissue overgrowth of the limb or body part and occasionally in highly aggressive angiosarcomas (Fig. 11).

Tumor angiogenesis and lymphangiogenesis

Lymphatic (lymphogenous) spread of cancer cells by direct invasion or tumor embolization is a crucial prognostic indicator of tumor aggressiveness, forms the basis for histologic staging of many solid organ cancers, and underlies the rationale for operative and radiation treatment directed at the draining regional lymph nodes (Willis, 1952). Yet, despite an upsurge of interest in tumor angiogenesis involving blood vessels (Gullino, 1978; reviewed in Folkman, 1995) and the use of antiangiogenic agents to control tumor growth, the lymphatic vessels have generally been disregarded. Indeed, the very existence of lymphatics in tumors has been called into question by several workers (Zeidman et al., 1955; Jain, 1989; Folkman, 1995).

Despite the strong experimental evidence that tumor angiogenesis drives tumor growth (Folkman, 1995), clinical evidence is based largely on cross-sectional counts of microvessels in solid organ cancers (Weidner, 1991) assuming that these represent or correlate with nutritive blood vessels. Under most conditions, macrocirculatory blood flow measurements provide a reliable index of nutritive flow and correlate with organ hyperfunction (e.g. in the spleen and endocrine glands); conversely, reduction in blood flow ("ischemic therapy") is a reasonable approach to controlling hyperfunction (C. Witte et al., 1980a). On the other hand, tumor blood flow has not been clearly correlated with either microvascular cross-sectional index or tumor aggressiveness. Tumor blood vessels branch abnormally, are hyperpermeable, and even sinusoidal in structure. It is accordingly unclear how a microvascular index can be developed that would distinguish arterial, venous, blood capillary, and lymphatic structures so that the designation of "nutritional blood flow" is justified. The formation of granulation tissue, wound healing, and Kaposi sarcoma is relevant here. Can static microvessel cross-section in these processes (lesions) be legitimately correlated with nutritive blood flow?

Recent conclusions about the absence of lymphatic vessels in tumors may in part relate to the difficulty of demonstrating them by injection techniques and also by inference from the elevated interstitial fluid pressure reported in some tumors (Jain, 1989). Yet lymphatics have been described in a variety of tumors, as first reviewed by Reichert in 1926, and many tumors do not appear overtly or microscopically edematous. Although intratumor lymphatics have been described in 10% of breast cancers, it is unclear whether these are residual lymphatic channels that have been compressed, or obliterated, or newly formed. Initial lymphatics are particularly well-adapted by ultrastructure to permit or promote the entry and exit of cells, such as tumor cells, by virtue of their lack of basement membrane, overlapping intercellular junctions, and anchoring retracting filaments which open rather than collapse (as do blood vessels) in response to elevated tissue pressure (Fig. 2). Whether hyperpermeable tumor microvasculature (presumed to be hematic in origin) and leaky sinusoids, from action by VEGF (vascular endothelial growth factor) (Dvorak et al., 1992), come to resemble lymphatic vasculature has not been considered, nor has the possibility that bloody lymphatic-venous communications may arise in tumors, rendering specific differentiation of lymphatics from blood vessels difficult if not impossible.

Therefore, the question of the existence of lymphatics in tumors and their origin *de novo* should be considered as still open along with the "nutritive equivalent" of a static microvascular cross-sectional index. As recently shown in Kaposi sarcoma tissue sections (Dictor et al., 1991), tumor microvessels need to be better defined and distinguished ultrastructurally and immunohistochemically in a variety of benign and malignant tumors. Comparative responses of lymphatic and blood vessel endothelial cells to putative angiogenic growth factors need to be documented, particularly those implicated in tumor angiogenesis. Further, the pathophysiology of edema and particularly serous effusions in patients with cancer should be examined to see whether the lymphatic network is overloaded ("high output failure of the lymph circulation") due to blood capillary hyperpermeability and enhanced surface area, or else obstructed from tumor infiltration or tumor emboli ("low output failure of the lymph circulation"), or a combination of both.

Lymphangiomas and mixed angiotumors

Whereas intense interest has focused on "tumor angiogenesis", scant attention has been paid to "angiotumorigenesis" (M. Witte and Witte, 1986), the growth and development of benign and malignant tumors of lymphatic and blood vessel origin. Even their classification is a matter of controversy.

Lymphangiomas, like their blood vessel counterparts (hemangiomas), are common disfiguring tumors of childhood and may also arise and enlarge rapidly in adulthood (Mullikan and Young, 1988; Papendieck, 1992; Leu and Lie, 1995). Some angioma syndromes follow Mendelian inheritance (usually autosomal dominant) whereas others are dysmorphogenic conditions that may represent somatic rather than germ-line mutations (Table 1). *In vivo* behavior and aggressiveness varies, and at times a fine line exists between a ductal malformation (e.g. cystic hygroma or intestinal lymphangiectasia) and true neoplasia ("aggressive" lymphangioma) as these tumors insinuate into adjacent tissues and encroach on vital organs (*viz.* Fig. 6) (Goetsch, 1938; Bowman et al., 1984). Some (lymph) angiomas may be multifocal or multicentric, and questions have been raised as to whether the process is "metastatic" (so-called "benign metastasizing lymphangioma") (Aristizabal et al., 1977) and "vanishing bone disease" (Gorham and Stout, 1955) (Fig. 6). Do cells from the primary tumor actually invade lymphatics and blood vessels and travel to distant sites or do they arise from a common proliferative-neoplastic stimulus? Occasionally, malignant transformation takes place (e.g. Maffucci syndrome). Sex hormones, such as exogenous or endogenous estrogen (oral contraceptives and puberty) appear to stimulate lymphangioma and other angiotumor growth. Lymphangiomas, moreover,

Table 1. Chromosomal aneuploidy and hereditary-dysmorphogenic disorders associated with lymphatic growth abnormalities

Chromosomal Aneuploidy		Hereditary-Dysmorphogenic Disorders	
Syndrome	Lymphatic-Vascular Growth Abnormalities	Syndrome	Lymphatic-Vascular Growth Abnormalities
Turner (XO)	CH(WN), LE, ILC, CA	Lymphedema I: Nonne-Milroy[*]	LE, ILC
Trisomy 18	CH(WN), LE	Lymphedema II: Meige	LE
Trisomy 13	CH, LE	Distichiasis-Lymphedema[*]	LE
Trisomy 21	CH, LE	Lymphedema-Hypo-parathyroidism	LE
Duplication 11b	CH, LE	Noonan[*]	LE, ILC, CA
Triploidy	CH, LE	Neurofibromatosis[*] (Type I)	LE
Klinefelter (XXY)	CH, LE	Bannayan Overgrowth[*]	LA
Trisomy 22	CH(WN)	Proteus	LA, HA
13q-	CH	Maffucci	LA, HA, M
11q-	CH	Klippel-Trenaunay-Weber	LE, LA, HA, AVM

CH = cystic hygroma; WN = webbed neck; LE = lymphedema; ILC = intestinal lymphangiectasia; CA = cardiac abnormalities; LA = lymphangioma; HA = hemangioma; M = malignancies; AVM = arteriovenous malformations; [*] = autosomal dominant trait.

are often mixed with hemangiomatous components (e.g. Klippel-Trenaunay and Maffucci syndrome) suggesting that both vasculatures are stimulated (or inhibited) together, or alternatively, that stem cells differentiate or fail to differentiate indiscriminately.

The *in vivo* behavior of these vascular tumors, characterized by formation of tubes, cysts, and other dysplastic structures, can be simulated *in vitro* now that long-term propagation of endothelial cultures from common benign lymphangiomas and rarer vascular tumors is possible (Bowman et al., 1984; Way et al., 1987, 1994a). Whereas control endothelium from vein, artery, lymphatic, and microvasculature *in vitro* exhibits typical contact-inhibited homogeneous cobblestone morphology, dysplastic and neoplastic endothelium characteristically exhibits a more heterogeneous cell population of multiple and at times focal phenotypes occasionally with

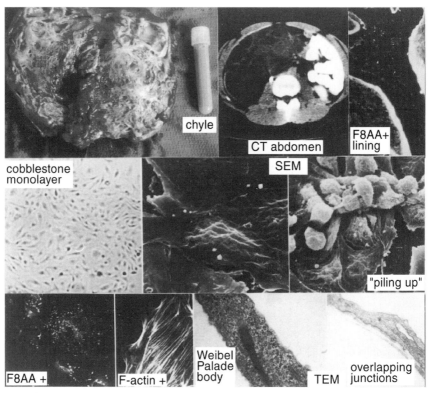

Figure 13. Lymphatic endothelial cell line (CH3) derived from a recurrent retroperitoneal lymphangioma in a 23-year-old woman (raising questions about the link between lymphatic malformations and neoplasia). Operative specimen with chylous contents, radiolucent mass on abdominal computer tomography (CT), and cyst lining immunofluorescence for endothelial marker vWf (F8AA). CH3 displays cobblestone morphology and on scanning electron microscopy is composed of polygonal cells, in some areas piling up, that are positive for vWf and F-actin and on transmission electron microscopy (TEM) contain Weibel-Palade bodies and exhibit typical lymphatic overlapping cell junctions (From Way et al., 1987).

areas of "piling up" (i.e. lack of contact-inhibition), probably denoting a transformed phenotype (Fig. 13). The percentage of the cell population exhibiting vWf positivity is variable. Lymphangioma and cystic hygroma-derived cells also form endothelial lined cyst-like spaces surrounded by dense, thickened, almost tubular aggregates of bipolar elongated, spindle-like cells (Figs. 6, 14 and 15) similar to AIDS-KS cell isolates (*vide infra*). Distinctive branched structures formed by lymphangioma cells, spontaneous "lymphangiogenesis *in vitro*", is not exhibited by normal endothelium under similar growth conditions but does appear in protracted confluence states (Leak and Jones, 1994).

These findings signify that relatively pure endothelial populations can be isolated from vascular tumors and propagated for long periods *in vitro* while retaining the morphologic characteristics of lymphatic or blood vascular endothelium, albeit with greater heterogeneity and tendency to exhibit spontaneous angiogenesis-angiodysmorphogenesis *in vitro*. These *in vitro* models should not only help to delineate endothelial cell structure and function and to define the biologic behavior of transdifferentiated or transformed endothelium, but also provide test systems for modulating these dysplastic and neoplastic processes.

Kaposi sarcoma (KS) – the lymphatic connection

More than a century after Kaposi's original description, the cell of origin and the nature of KS (multicentric vascular hyperplasia vs. neoplasm) remain mired in controversy. An AIDS-defining illness, KS is present in 20% of HIV-infected individuals and is found at death in more than 50%. The "KS cell" is considered to arise from lymphatic (most often mentioned) (Beckstead et al., 1985) or blood vascular endothelial variants. The KS lesion displays intense endothelial cell proliferation, disorganized angiogenesis with migratory spindle cells, particulate phagocytosis, interstitial sclerosis, and tumor formation in rare cases progressing to an angiosarcoma-like stage resembling the Stewart-Treves syndrome associated with longstanding lymphedema.

Figure 14. Spontaneous lymphangiogenesis *in vitro* in cell culture derived from a lymphangioma excised behind the knee in a young child. Note loose clusters of lymphatic endothelial cells sprouting into branches (left) which become more prominent (middle) and eventually evolve into a sheet-like aggregate with intense lymphatic-like sprouting branches (right). (From Witte and Witte, 1987b).

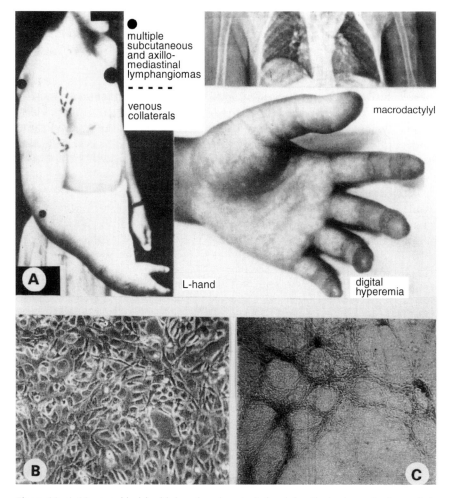

Figure 15. A 10-year-old girl with lymphangiomatosis involving the back, mediastinum, left axilla, retroperitoneum and cervical paraspinal space associated with muscular-bony arm overgrowth, macrodactyly, digital hyperemia and venous ectasia (A). Lymphatic endothelial cells isolated from a resected portion of the lymphangioma display typical cobblestone morphology (B), and spontaneous lymphangiogenesis (tube/vessel formation) *in vitro* (C). (A taken from Witte and Witte, 1987b).

The close linkage of KS to the lymphatic system and to lymphedema (extremity, facial, torso, and pulmonary) has long been recognized in classical (Mediterranean) and endemic (African) and more recently in epidemic AIDS-KS (Fig. 16) (Dorfman, 1988; M. Witte et al., 1989). AIDS-KS patients commonly die with massive pulmonary edema, serous effusions, and anasarca. We and others (M. Witte et al., 1990b, 1990c; Dictor and Andersson, 1988; Dictor et al., 1995) have documented venous and lymphatic-lesional communications, lymphatic dilatation and

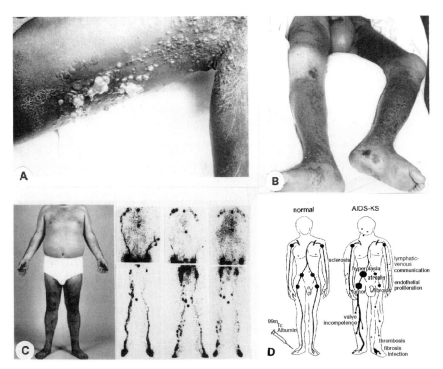

Figure 16. (A) Lymphangiomatous form of African (endemic) Kaposi sarcoma on upper leg. (B) 35-year-old homosexual man with AIDS-associated Kaposi sarcoma and massive whole body lymphedema. (C) 43-year-old homosexual man with moderately severe lower extremity lymphedema from AIDS-associated Kaposi sarcoma. Lymphangioscintigrams of the legs (lower panels) show prompt lymphatic transport of radioactive tracer (99 mTc-albumin) to large regional nodes (clinically palpable) with several focal accumulations of radioactive intensity in the calf and mid-thigh corresponding to the distribution of cutaneous Kaposi lesions seen in the patient photograph. The arms (upper panels) show a similar but less dramatic pattern. Midline rounded densities are markers on sternal notch and xiphoid (upper panels) and pubis and knees (lower panels). (D) Schematic diagram of normal lymphatic anatomy (left) and abnormal patterns of lymphatic growth visible on isotope lymphangioscintigraphy (LAS) in AIDS-KS (right) operating singly or in combination in different patients or different limbs of the same patient, namely: opening of lymphatic-venous communications with bidirectional filling of lesions, lymphatic truncal sclerosis, proximal nodal obstruction from KS replacement or nodal hyperplasia, and distal lymphatic obliteration from exuberant endothelial proliferation, intraluminal thrombosis, and tissue fibrosis with or without local infection. (A, courtesy of R. Dorfman; C taken from M. Witte et al., 1990c; D taken from M. Witte et al., 1990b).

sclerosis, leaky blood and lymph vessels, major basement membrane profile resembling lymphatics (including absence of heparan proteoglycan sulfate), and lymphatic truncal obliteration occasionally in conjunction with an externally striking lymphangitic-lymphangiomatous appearance (Fig. 16). Various lymphadenopathies are also associated with KS, including secondary lymphoid malignancies.

The coalescence of disordered lymphangiogenesis, tumor angiogenesis, and angiotumorigenesis probably lies at the heart of KS (*vide infra*

Fig. 21). By ultrastructural analysis, Dictor et al. (1991) have followed the evolution of KS lesions from radial lymphatic-venous communications. The process is characterized by endothelial cell migration from preexisting vessels (lymphatics, veins, and blood capillaries), intense proliferation, dysmorphogenesis (resembling angiomas), sclerosis and ultimately, in late stages, oncogenesis. As the surrounding matrix undergoes cycles of dissolution and deposition, growth factors and other biologic signals are likely released further modulating endothelial proliferation and disrupting matrix structure. These events involve not only small lesional vessels but also larger collecting lymphatics and veins. Of interest in this regard is a 15-year-old patient seen in 1968 and reported in 1973 (Elvin-Lewis et al.), who developed progressive whole-body lymphedema associated with cachexia, severe immunodeficiency and systemic chlamydial infection, and ultimately died with disseminated Kaposi sarcoma. He was subsequently shown two decades later to have HlV-1 infection by Western blot immunoanalysis and antigen capture assay of preserved serum and tissues (Garry et al., 1988). Originally the skin and internal tumors could not be distinguished from lymphangiosarcoma arising in the setting of unremitting lymphedema (Stewart-Treves syndrome), which may bear a striking histologic resemblance to advanced KS. At autopsy, he displayed the entire spectrum of KS lesions ranging from benign lymphangiectasia, nodules of lymphangioma, and frank angiosarcoma, that is, widespread KS in all stages of evolution and closely linked to the lymphatic system. On the other hand, KS may remain quiescent for many years as a solitary minute lesion, or even as multicentric or bulky lesions, only, like certain benign hemangiomas and lymphangiomas (Fig. 22), to undergo sudden and explosive growth or spontaneously regress, or involute after immunorestoration in renal transplant patients.

We and others have successfully isolated AIDS-KS cultured cell lines from skin, gingiva, and palatine KS lesions using standard culture media, either without supplemental growth factor or supplemented only with endothelial cell growth factor (Fig. 17) (reviewed in Way et al., 1993; Nicosia and Pietra, 1994). Neither HTLV-II-conditioned media nor matrigel is necessary. These spindle-stellate cells fail to generate tumor nodules or transient mouse-derived angiogenic lesions in nude mice, are morphologically predominantly endothelial in appearance, exhibit positivity for endothelial markers at sustained confluence (but variably before confluence), display prominent extracellular vWf deposition in tissue culture, survive for long periods (more than 5 years) in culture without apparent senescence, lack specific surface antigens for hybridoma monoclonal antibody development (Bernas et al., 1993), secrete constitutive 72D matrix metalloproteinases (Meade-Tollin et al., 1995), and fail to demonstrate spontaneous angiogenesis *in vitro*. On the other hand, our AIDS-KS cells inconsistently secrete high concentrations of proteases (tPA and uPA) and inconsistently produce elevated cytokine IL-6 and scatter factor

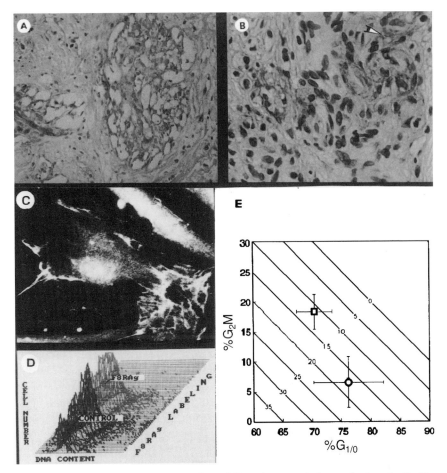

Figure 17. Endothelial-vascular features of AIDS-Kaposi sarcoma lesions and cultured cells. (A, B) Immunohistochemical staining of paraffin sections of original tissue from AIDS-Kaposi sarcoma skin biopsies. (A) Glomeruloid structure with variably stained endothelial lining cells to von Willebrand factor (vWf). (B) Characteristic diffuse extracellular deposition of vWf (arrow). (C) Fluorescein-stained micrograph of cultured Kaposi sarcoma cells also displaying intense vWf deposition in extracellular matrix. (D) Flow cytometric histogram corresponding to vWf (F8RAg) marker decoration shown in C. Peak left front is control propidium iodide and secondary antibody with fluorescein isothiocynate (FITC) staining of cells but without primary antibody. Peak right rear is vWf stained cell population showing 88% positivity above control. (E) DNA cell cycle analysis of quiescent cultured KS (square) and non-KS cells (circle). Diagonal lines represent S phase percentages. Data points represent mean ±SE. (From Way et al., 1993).

(E. Rosen, personal communication) in the supernatant. No viruses (human papillomavirus, cytomegalovirus, human immunodeficiency virus, human herpes virus 8) have been detected as particles by ultrastructural analysis or as genetic sequences by polymerase chain reaction (Way et al., 1993; Dictor et al., 1996). Cell-cell contact and communications appear deranged, with disorganization of junctional complexes (loss of desmoplakin, plakoglobin) (Schmelz et al., 1995). Cultured KS cells also demonstrate cell cycle abnormalities of elevated G_2M phase at contact-induced quiescence in conjunction with increased proliferating cell nuclear antigen (PCNA) in the G_2 phase, reflecting an increased proliferative potential (Way et al., 1993). They also display peculiar signal transduction events in early proliferation related to an abnormal intracellular Ca^{++} pool (Martinez-Zaguilan et al., 1996). Thus, KS cells may be capable of outcompeting normal cells for cell cycle entry and tissue remodeling during a post-injury response or other analogous perturbations. By responding more rapidly to growth-promoting and migratory stimuli and proceeding through mitosis to the final stage of division while inhibiting the exit from quiescence and entrance of normal cells into G_1, cells of KS endothelial lineage (transdifferentiated, if not yet transformed) would gradually replace normal vascular endothelial cells.

X chromosome, sex hormones, and lymphangiodysplasias

Whereas blood vascular disorders do not show a predilection for females, a variety of lymphatic vascular disorders show such a sexual dimorphism (Table 1) (Greenlee et al., 1993). In XO Turner syndrome (Fig. 9) or ovarian dysgenesis (with the female phenotype), a common chromosomal aneuploid state (approx. 1.5% of conceptuses, 99% of which are lethal) (Jacobs, 1975), the lymphatic ductal system fails to develop properly resulting in a high incidence of strangulating fetal cystic hygroma early in pregnancy, cardiac atresia, and hydrops, and post-natally in webbed neck (pterygium colli-involuted cystic hygroma) and hypoplastic peripheral lymphedema (Chervenak et al., 1983). Meige's syndrome of congenital lymphedema is associated with ovarian dysfunction and amenorrhea. Moreover, the preponderance of female patients with lymphedema precox suggests that the onset of puberty unmasks a defective lymphatic drainage system which has probably operated marginally since birth.

Cyclic surges in female sex hormone levels relating to puberty, the menstrual cycle, and pregnancy, and exogenous administration of estrogen replacement and oral contraceptives, not only aggravate the severity of lymphedema but also promote lymphatic and blood vascular proliferation and neoplasia (phlebothrombosis, ruptured arterial aneurysms, vascular malformations, lymphangioma-hemangiomas) as well as non-endothelial tumors and life-threatening coagulopathies. Another syndrome exclusively afflicting fertile women is pulmonary lymphangioleiomyomatosis, which produces striking pulmonary lymphangiogenesis and lymphangiectasia with perilymphatic and perivenous pericyte proliferation (Graham et al., 1984; Clemm et al., 1987; Taylor et al., 1990). Endothelial estrogen receptors have been implicated in promoting the prominent pericyte proliferation around lymphatic and venous vessels typical of this disorder. Accordingly, ovarian ablation and the anti-estrogen hormone tamoxifen have been used clinically as treatment, albeit with limited success. Lung transplantation (3% of the total number performed) is the last resort to avert death from progressive pulmonary fibrosis and pulmonary hypertension.

The pathophysiologic link between female sex hormones (particularly estrogen), vascular overgrowth, and lymph stasis has not been studied but may relate either to a heightened microvascular permeability-surface area product or to increased growth factor release-responsiveness, stimulating lymphatic and blood vessel endothelial cell cycle kinetics. These effects would tip the balance in favor of lymph formation over lymph absorption, thereby promoting progressive lymphedema while at the same time stimulating lymphangiogenesis. There may also be alterations in tissue matrix associated with induced changes in endothelial phenotype, particularly cell shape (Hendrix et al., 1994). On the other hand, a sufficient complement of female genetic material or conversely the XY genotype and adequate female hormone production appears necessary for proper development of the lymphatic vascular system, as inadequacies lead to hypoplastic lymphatic growth, fetal hydrops, and even death.

Genotypic influences and "lymphangiogenesis gene(s)"

Growing evidence suggests that lymphangiogenesis is at least in part genetically controlled (Table 1) (Greenlee et al., 1993). The inactivated second X chromosome or else the Y chromosome is clearly necessary to prevent XO Turner syndrome. It has been hypothesized that a gene or genes expressed from non-activated portions of the inactive X chromosome or from the Y chromosome are involved in the development of lymphatics and that a deficiency of the product of this gene(s) is responsible for the Turner syndrome phenotype. Interestingly, the peripheral lymphedema and the webbed neck, typically noted at birth in XO ovarian dysgenesis, tend to improve with time (in contrast to other lymphedemas, which either stabilize or worsen) suggesting that defective lymphatic growth *in utero* may be influenced by unbalanced vascular growth inhibiting factors of maternal-placental origin. Other chromosomal aneuploidy disorders, including the common trisomy 21 (Down syndrome), also occasionally present with strangulating fetal cystic hygroma, lymphedema, and intestinal lymphangiectasia in addition to cardiac anomalies.

A variety of hereditary lymphedema-angiodysplasia syndromes (generally with autosomal dominant inheritance) also provide clues to the potential sites of "lymphangiogenesis genes". Most notable is Milroy disease (Fig. 18), a familial disorder characterized by progressive lymphedema from birth or early childhood and associated with lymphatic hypoplasia and occasional intestinal lymphangiectasia.

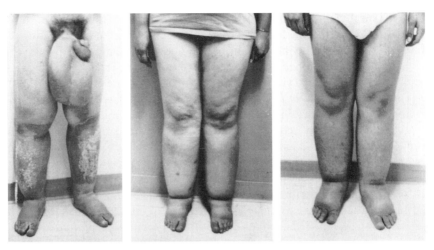

Figure 18. Hereditary lymphedema (Milroy disease) affecting both legs since early childhood in brothers (right and left) and sister (center). Father, but not mother, had lymphedema and one of the two daughters of the sister also has bilateral lymphedema (other family members unaffected). Lymphangioscintigram confirmed deficient lower limb lymphatic collectors (lymphatic hypoplasia). (From Witte and Witte, 1991).

Alterations in endothelial genotypic and phenotypic expression should be sought in family cohorts by both the classical (bottom up) approach, carefully studying clinical pedigrees for phenotypic variations including simultaneous occurrence of other rare inherited conditions in a search for genes involved in angiodysplasia, and the reverse genetics approach (top down-positional cloning with translocation patients) using fluorescent *in situ* hybridization (FISH) searches of the human genome. Should specific genes be identified and the protein abnormalities associated with them delineated, specific mechanisms may be uncovered which could be used as targets to modulate lymphangiogenesis, either to increase it where defective or suppress it where excessive.

In vivo and *in vitro* models of lymphangiogenesis

Unraveling the pathomechanisms underlying angiogenic and angiotumorigenic disorders, particularly those involving the lymphatic system, depends upon developing experimental counterparts and devising or adapting methods to evaluate and quantitate endothelial proliferation and vascular morphogenesis and oncogenesis *in vivo*. Also critical is the development of long-term cultures and cocultures of lymphatic endothelium from normal and diseased microvessels and large lymphatic collectors. If lymphatic and blood vascular elements in microvascular cultures cannot be distinguished or if macrovascular cultures also contain such mixtures (Fig. 8), this issue needs to be addressed.

In vivo lymphangiogenesis

Several experimental setups either newly developed or worthy of being revisited with modern techniques, hold promise in the elucidation of the mechanisms of lymphangiogenesis and its disorders. The cellular and molecular events during embryonic development and remodeling of lymphatics during adult life after wounding and regeneration, obstruction, inflammation, and oncogenesis should be the target of such future studies.

Several animal models are available to produce lymph stasis, which is associated with exuberant (compensatory) or defective (causative) lymph-angiogenesis: 1) Acute lymphatic collector (thoracic duct) obstruction in rats can be maintained with the animal under anesthesia for up to 6 hours without serious complication. Alternatively, the incision can be closed and the rat restudied. 2) Chronic lymphedema can be produced by segmental excision of deep inguinal lymphatics of the rat groin (Kanter et al., 1990) The hindlimb superficial lymphatics are interrupted by circumferential excision of skin and subcutaneous tissue around the upper thigh, and the skin edges sewn so as to establish a permanent narrow integumental gap. Postoperative irradiation to the groin is added. 3) Chronic lymph stasis (filariasis) in the ferret (*Mustela putorius furo*) is induced by subcutaneous inguinal inoculation of *Brugia malayi* third-stage larvae in one hindlimb while the contralateral limb acts as a control (Crandall et al., 1987; M. Witte et al., 1988). Lymphatic dysfunction, disturbed lymphangioscinti-graphic patterns and impaired interstitial lymphatic albumin absorption kinetics can be documented within 3–4 weeks. The earliest clinically detectable abnormality in filariasis is hindlimb lymphatic dilatation and collateralization related to nests of live adult worms thrashing inside the peripheral lymphatics (Fig. 12, left). These lymphangiogenic events can not only be seen invasively (Case et al., 1992a) and imaged non-invasively (M. Witte et al., 1988) but can also be recreated *in vitro*. This early lymphatic abnormality has been used to screen for productive (microfilaremic) in-fection (M. Witte et al., 1993) and also exploited to follow the efficacy of therapeutic intervention in patients in areas such as northeastern Brazil where infection is endemic (Amaral et al., 1994). 4) Hydrops fetalis in the ovine fetus (Andres and Brace, 1990) by ligation and excision *in utero* of the fetal left thoracic, cervical and brachycephalic lymph ducts during the last trimester of pregnancy in ewes simulates fetal hydrops and cystic hygroma of Turner syndrome. 5) Congenital hereditary lymphedema occurs spora-dically in dogs and several other species. The condition, thought to be carried as an autosomal dominant, is characterized by edema of varying severity, usually limited to the hindlimbs but occasionally more generalized (Patterson et al., 1967).

Most models of *in vivo* angiotumorigenesis have been directed at KS look-alike lesions. KS lesions and cultured cells (with one possible exception of a pleural effusion-derived highly malignant tumor cell line

KSY1, reported by Lunardi-Islander et al., 1995) fail to propagate as tumors although transient mouse-derived hyperpermeable angiogenic lesions resembling granulation tissue have been described by some workers (Salahuddin et al., 1988) though not others (M. Witte et al., 1994a) in nude mice. Viral infection or transfection of murine endothelial cells, on the other hand, can induce angiomas and sarcomas in mice. As the cell of origin arises from an undetermined heterogeneous mixture of lymphatic and blood microvessels, these growths [(polyomavirus middle T oncogene-induced multicentric hemangiomas (Bautch et al., 1987) and middle T SV40 antigen-transformed vascular sarcomas (O'Connell and Rudmann, 1993)] can be considered mixed angiotumors rather than exclusively hemangiotumors. Interestingly, increased proteolytic activity appears to be pivotal for the aberrant morphogenetic behavior of endothelial cells expressing the middle T oncogene (Montesano et al., 1990). Two avian models are also valuable in angiotumorigenesis research, namely the chicken comb and wattle (which represent congenital angiomas) and avian leukosis viral infection. The latter virus induces rapidly growing angiomas and angiosarcomas associated with immunodeficiency and death (Järplid, 1961; Dictor, 1988). This retrovirus inserts next to the gene that regulates fibroblast growth factor, a potent stimulator of endothelial proliferation and angiogenesis.

In vitro lymphangiogenesis

Based on our earlier success in establishing RSE-1, a tumorigenic cell line derived from rat hepatic sinusoidal endothelium (Way et al., 1994b;

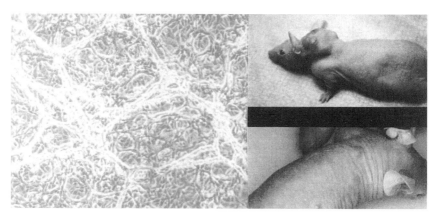

Figure 19. Tumorigenic human lymphatic endothelial cell line HTDEC-1. Left: Confluent cultured lymphatic endothelial cells derived from a human thoracic duct displaying cobblestone morphology and spontaneous lymphangiogenesis *in vitro* without added growth factors or matrix components. Right: Lymphatic vascular tumors arising in nude mice after intradermal or subcutaneous inoculation with these cells. Note Kaposi sarcoma-like nodules (bottom), which revert to original lymphatic endothelial phenotype when excised and returned to culture.

Borgs et al., 1994), an immortal non-transfected human thoracic duct endothelial cell line (HTDEC-1) has been isolated and propagated (Way et al., 1996) (Fig. 19). HTDEC-1 proliferates rapidly in culture without added growth factor supplements, displays spontaneous lymphangio-genesis *in vitro*, exhibits extended lymphatic-like intercellular junctional complexes which stain positively for a junctional protein, desmoplakin, (Schmelz et al., 1994) characteristic of lymphatics, and binds lymphocytes in an adhesion assay (Stazzone et al., 1995). Strikingly, subcutaneously or intradermally inoculated HTDEC-1 cells uniformly produce rapidly growing transplantable non-metastatic endothelial-derived tumors in nude (nu/nu) mice similar to the RSE-1 derived tumors. These tumors typically display an endothelial phenotype on return to culture. Using normal lymphatic endothelium derived from lymphatic collectors as a control and human lymphangioma-derived cells for comparison, HTDEC-1 is poten-tially an important resource for investigating mechanisms of lymphangio-genesis (lymphatic dysplasia and lymphangiotumorigenesis) and for modulating lymphatic growth *in vivo* and *in vitro*. Several assay systems currently available to quantify hemangiogenesis are more or less suited to such studies of lymphangiogenesis: disk assay (Nelson et al., 1993), chick embryo chorioallantoic membrane, vessel segment organ culture in collagen gels (Nicosia, 1987), *in vitro* cell monoculture post-confluence assays (Leak and Jones, 1993), *in vivo* microsphere tumor angiogenesis assay (Fotsis et al., 1994), and transdermal transparent window assay (Yuan et al., 1994). Of these, the vessel segment organ culture and post-confluence assays are directly applicable to lymphangiogenesis. The post-fixation transdermal window assay, however, could be modified for microinjection of vital dye and subsequent evaluation of lymphangiogenesis *in vivo* and *in vitro*.

Evaluation and treatment of clinical problems of lymphangiogenesis

Evaluation

As described earlier, some approaches and techniques for evaluating both normal and pathologic lymphangiogenesis have been available for a long time. These methods need to be revisited and refined using state-of-the-art techniques to pinpoint and dissect underlying pathomechanisms in lymphatic disorders in patients in order to improve understanding and develop rational treatment.

The proliferative potential of lymphatic growth can be defined and followed serially by further extension of DNA incorporation studies with Brd/U (Dolbeare et al., 1985; Borgs et al., 1992) and similar markers either *in vivo* or from excised lymphatics or lymphatic tumors *in vitro*. Profiling the autocrine and paracrine growth factor environment within the lymphatic endothelium, surrounding extracellular matrix, and non-endothelial cells, as has been done in aortic atherosclerosis (Ross, 1993), would further delineate the sequence of abnormal lymphangiogenic events and the cell types involved. Using serial non-invasive or invasive imaging with high resolution dynamic lymphangioscintigraphy, fluorescent microlymphangiography (Bollinger et al., 1981), and magnetic resonance imaging (Case et al., 1992 c), the number, size, and pattern of lymphatic growth can be delineated and followed (Figs. 12 and 16) (C. Witte and Witte, 1996). These dynamic structural changes need to be correlated clinically with physiologic parameters of lymphatic function revealed by these same techniques, including lymphatic absorption and macromolecular transport kinetics, intralymphatic pressure measurements, and intrinsic lymphatic contractility. Magnetic resonance imaging and nuclear magnetic resonance spectroscopy can also be used to define and quantitate associated structural and biochemical matrix events to pinpoint the alterations that accompany and aggravate abnormal lymphatic growth and lymph stasis. With greater magnification, *in vivo* microscopy including videomicroscopy can provide other structural and functional details. From tissue sections, quantitative cross-sectional indices with immunohistochemical differentiation is possible, and ultrastructural definition of lymphatics using transmission and scanning-tunnel electron microscopy can further define the structural-molecular basis of lymphatic vessel abnormalities and associated matrix disturbances. The focus needs to be on quantitating where possible the steps in angiogenesis beginning with proliferation rate before and after angiogenic stimulation or during pathologic processes, describing the evolution of lymphatic growth and pattern changes under different conditions, and then finally evaluating the response to treatment.

Treatment

Because of the close association of angiogenic disturbances with swelling, scarring, malnutrition, and immunodysregulation in disorders of the lymphatic arc of the blood-lymph loop, therapy should target these clinical manifestations, i.e. attempt to alleviate or remove the conditions associated with and possibly provoking or perpetuating abnormal lymphatic and soft tissue growth (Fig. 20) (M. Witte et al., 1990a). These approaches (not all feasible or demonstrably effective at this time) include mobilization of edema fluid by enhancing lymph drainage through compressive physical therapy (Földi, 1983), using collagenolytic or other anti-fibrotic agents to reduce scar tissue, immunorestoration, operative relief of underlying lymphatic obstruction (lymphatic-venous shunts), benzopyrones to facilitate macrophage protease digestion of high protein edema fluid (an important link in lymphatic and general tissue overgrowth: Casley-Smith and Casley-

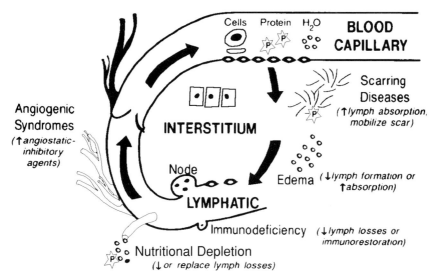

Figure 20. Clinical disorders of the blood-lymph circulatory loop (see Fig. 1) manifesting as swelling, scarring, immunodeficiency, nutritional depletion, and uncontrolled lymphangiogenesis and angiotumorigenesis (see also Fig. 9). The phrases in italics designate current and potential future therapeutic approaches. (From M. Witte et al., 1990a).

Smith, 1986; Casley-Smith et al., 1993), and gene replacement or hormone manipulation to moderate abnormal vessel growth.

Furthermore, until specific growth factor modulation of lymphatic compared with blood vascular endothelium is better defined experimentally and *in vitro*, approaches similar to those outlined for angiogenic or angiodefective disorders of blood vessels (i.e. hemangiogenesis) should be pursued in disorders of lymphangiogenesis. An array of angioinhibitory agents has been identified for this purpose, and these are being tested for their ability to restrain tumor angiogenesis associated with solid organ cancers such as breast, prostate, lung, and colon, hemangiomas, and also AIDS-KS (Burrows and Thorpe, 1994; Folkman, 1995). The putative angioinhibitor repertoire includes agents which are thought to inhibit endothelial cell proliferation and/or migration; block or manipulate enzymatic steps in basement membrane-extracellular matrix biosynthesis and assembly; block key growth factors; modulate the inflammatory cytokine balance; eradicate angiogenic microorganisms; promote protease inhibition and a procoagulant microenvironment; and/or restore the immune system. Alternatively, where angiogenesis is defective (Milroy disease), (lymph)angiogenic stimulation would seem appropriate, including the potential to deliver gene therapy *in utero* to prevent fetal wastage from life-threatening congenital lymphatic malformations.

Targeting the lymphatic system for selective or local delivery of lymphangiomodulatory agents, analogous to non-invasive imaging by intradermal administration of macromolecular tracers or invasive imaging by direct lymphatic cannulation of oily contrast agents, is desirable. Drug formulations in macromolecular complexes such as liposomes or in aerosomes (Unger et al., 1992), deposited interstitially or directly infused into lymphatic collectors, should be explored. In this way, angioinhibitory or angiogenic agents, including future gene therapy, could conceivably be delivered selectively to prevent or correct the underlying lymphatic growth disturbance.

Medical ignorance, questions, and future challenges

Many vexing yet intriguing questions (those known and unknown) regarding lymphangiogenesis and its disorders can be framed in the context of the four components of the lymphatic system (lymphatic channels, lymph, lymphoid aggregates, and lymphocytes) and their interrelationships to the blood vasculature through the blood-lymph loop (Figs. 1, 20, and 21) (reviewed in M. Witte and Witte, 1996). Such is the extent of medical ignorance (reviewed in M. Witte et al., 1990a) – the things we know we don't know, don't know we don't know, and think we know but don't!

How different are lymphatics from blood macro- and microvessels? Which differences are truly distinctive and immutable? For example, which junctional proteins or adhesion molecules are maintained through multiple

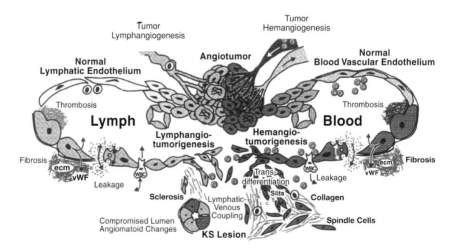

Figure 21. Pathologic processes common to and linking blood vessels with lymphatics. Normally, these two vasculatures remain separate and connect directly only at a few strategic sites. In a variety of disorders, however, which are characterized by sprouting and non-sprouting lymph-hemangiogenesis and angiotumorigenesis, the two vasculatures come to resemble one another, interdigitate, and even merge indistinguishably. ecm = extracellular matrix; KS = Kaposi sarcoma; vWf = von Willebrand factor; WBC = white blood cell.

passages in tissue culture, and on what basis are they altered in pathologic angiogenic processes *in vivo*? To what extent is endothelium a pluripotential stem cell, and into what cell types other than endothelium can it transdifferentiate in embryonic and even in adult life? Which signals and mediators act on both lymphatic and blood vessel endothelium, and in what ways do the responses differ? Are there specific monoclonal antibodies to different types of endothelium? If cultured endothelial cells are exposed to altered environmental conditions, can their distinctive markers be modulated and will they transdifferentiate indistinguishably? Do lymphatic and blood vessel endothelial cells differ in their growth factor requirements, responses, and production *in vivo* and *in vitro*?

How does the lymphatic system maintain its separateness from the blood vessels *in vivo*? What happens and what are the consequences when the distinctions become merged in dysangiogenesis and angiotumorigenesis as in Kaposi sarcoma? How important are altered junctional complexes or pericyte loss and gain in these processes? Are they cause or effect? Does heightened macromolecular permeability, for instance, of tumor blood vessels or a sinusoidal circulation such as is found in the liver represent a merging toward the lymphatic capillary?

How is the obstructive or obliterative process in the lymphatic (or blood) vasculature translated, and from which physicochemical signals, into a proliferative stimulus generating collateral formation and dissipation of hydrostatic pressure? How is this phenomenon linked to vascular and matrix remodeling as both a "recapitulation" of embryonic development and a response to wounding or blockage in post-natal life? What signals perpetuate these events, and how do sex hormones or the female sex exaggergate or otherwise modulate these processes to the benefit and/or harm of the host? What is missing when estrogens are insufficient rather than excessive? What genetic, dysmorphogenic, and environmental influences lead to maldevelopment, malformation, or malfunction of the intestinal lacteals and large collecting trunks including massive lymphangiectasia, reversal of lymph flow, impaired lymphangion contraction, and intraluminal valvular incompetence?

Which growth factors are present in lymph and do they differ in type or amounts from those in blood? Are they altered by lymphatic obstruction? What determines non-sprouting vs. sprouting angiogenesis? What is the genetic control of lymphangiogenesis – are there lymphangiogenesis genes and how do they relate to the Turner XO syndrome and other hereditary non-sex-linked lymphedema-angiodysplasia syndromes? Do these genes modulate known (or yet to be discovered) growth factors and how? What is the influence of viral infection and somatic non-germ line mutations in producing lymphangiodysplastic syndromes at times associated with hemangiodysplasias? How do these conditions relate to lymphatic endothelial tube, cyst, and cavernous formation *in vitro*? Can these angiogenic, angiodysplastic, and angiotumorigenic processes be controlled by anti-angiogenic drugs, gene manipulation, physical therapy, and operations? What mechanisms are involved in spontaneous angioinvolution of massive grotesque lymphatic growth disturbances (Fig. 22)?

In lymphedema or other lymphatic growth syndromes, what factors limit the capacity of the lymphatic network, including pre-lymphatic tissue channels, to grow and handle an increased load of microvascular filtrate (i.e. high output failure of lymph flow) or even the normal load of tissue fluid (low output failure of lymph flow)? What determines the distribution of lymph flow and growth between contracting deep lymphatic collectors and valveless superficial lymphatic channels? Are the limits set genetically or perhaps in the local environment or even directly by an altered endothelial phenotype? What approaches – drugs, physical and other measures – are available or possible to enhance lymphatic growth if needed or to stimulate lymphatic function and promote beneficial tissue modeling?

As for tumor lymphangiogenesis, does it exist and under what conditions? Does it occur only in certain tumors? Can intratumor microvessels be distinguished as lymphatic vs. "hematic", and which are nutritive? What is the connection between vascular hyperpermeability and hem- and lymphangiogenesis? Can newer imaging techniques such as NMR spectroscopy be used to assess nutritive blood flow and depict lymphatic pathways? How do lymphatic endothelial cells respond to tumor angiogenic and tumor antiangiogenic agents (transforming growth factor-β, taxol, and protease inhibitors) in terms of proliferation, migration, and angiogenesis? Can hemangiogenesis be stimulated selectively, that is, without simultaneously stimulating lymphatic growth? Is the tumor lymphatic system obstructed or functionally overloaded or both? How does the tumor interstitial fluid drain into or connect with the peritumor host lymphatics, and what if any is the connection to tumor immunity and tumor spread?

What mechanisms underlie the soft tissue overgrowth phenomena including striking scarring (fibrogenesis) and fat deposition (lipogenesis) as lymph stasis and lymphedema progress? Do alterations in tissue tension, cell adhesion, and proteolytic balance play a crucial role in development of the elephantine limb? Can this skin and subcutaneous overgrowth be prevented, modulated, or reversed? What is the link between high-protein lymphedema fluid (and secreted cytokines or attracted cell types) and the subsequent accumulation of adhesive multimers (e.g. fibronectin and vWf), lymph coagulation, lymphatic obliteration, and progressive rigidity of tissue scaffolding? Are matrix-bound growth factors released into the lymphedematous interstitium to act indiscriminately on heterogeneous cell populations? How can these non-endothelial growth processes be assessed, prevented, and reversed? And which drugs, physical methods, and operations might slow the process?

In addition to the increased incidence of opportunistic bacterial infections in the lymphedematous limb or body part, what accounts for the heightened occurrence of benign and rare malignant vascular and other

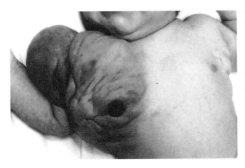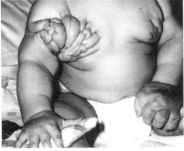

Figure 22. Left: Massive and deforming lymphangioma of the right chest wall, shoulder gird-le and upper arm (lymphangiogenesis and lymphangiotumorigenesis *in vivo*) in a 6-month-old girl. Right: Spontaneous regression 4 months later of the shoulder-chest wall lesion (lymphangioinvolution *in vivo*). Hemangiomas on left hand persist unchanged.

opportunistic neoplasms (e.g. Stewart-Treves angiosarcoma and squamous and basal cell carcinomas) in this pathologic terrain and does it relate to immunodysregulation? Is the defect primarily in trafficking of immune cells between blood and lymph streams or is it related to impaired clearance of particulates, microorganisms and carcinogens that penetrate the skin barrier with or without defective lymph node function – or all of them operating together? When there is rapid growth of lymphatics and blood vessels from any cause, what is the oncogenic switch (e.g. mechanical forces, loss of adhesion, or matrix alterations) that transforms these proliferating endothelial cells into a vascular malignancy?

Epilogue

The derivation of the Latin word *lymphaticus*, first introduced by Thomas Bartholin in 1653 (Yoffey and Courtice, 1970), signifies "distracted and confused". This designation seems an appropriate starting point to venture into the *terra incognita* of the lymphatic system and lymphangiogenesis in health and disease. What little is known about lymphangiogenesis and the plethora of information about hemangiogenesis *in vivo* and *in vitro* should set us on a clear yet necessarily meandering path of investigation. In this journey, a special reverence for the lymphatic vasculature, indeed lympho-mania, is appropriate following so many years of neglect. After all, in the poetry of Homer and in Greek mythology (Kampmeier, 1969), whereas blood fills the veins of mere mortals, the ichor or clear lymph – gathered by nymphs (the goddesses of moisture) – flows through the vessels of the gods.

Acknowledgments
We are indebted to Peter Borgs, Ph.D. and Grace Wagner for their help in preparing this chapter. This research was supported in part by the Arizona Disease Control Research Commission Contracts 8277-000000-1-1-AT-6625, -ZB-7492, and I-103 and NIH R01 HL48493.

References

Amaral, F., Dreyer, G., Figueredo-Silva, J., Noroes, J., CaValcanti, A., Samico, S.C., Santos, A. and Continho, A. (1994) Live adult worms detected by ultrasonography in human Bancroftian filariasis. *Am. J. Trop. Med. Hyg.* 50 (6):753–757.

Andres, R.L. and Brace, R.A. (1990) The development of hydrops fetalis in the ovine fetus after lymphatic ligation or lymphatic excision. *Am. J. Obstet. Gynecol.* 162: 1331–1334.

Aristizabal, S.A., Galindo, J.H., Davis, J.R. and Boone, M.L. (1977) Lymphangiomas involving the ovary. *Lymphology* 10:219–223.

Aselli, G. (1627) *De Lactibus sive Lacteis Venis, Quarto Vasorum Mesarai corum Genere novo invento.* J.B. Biddelium, Mediolani, Milano.

Baldwin, W.M. III (1982) The symbiosis of immunocompetent and endothelial cells. *Immunol. Today* 3:267.

Bautch, V.L., Toda, S., Hassell, J.A. and Hanahan, D. (1987) Endothelial cell tumors develop in transgenic mice carrying polyoma virus middle T oncogene. *Cell* 51:529–537.

Beckstead, J.H., Wood, G.S. and Fletcher, V. (1985) Evidence for origin of Kaposi's sarcoma from lymphatic endothelium. *Am. J. Pathol.* 119:294–300.

Bellman, S. and Odén, B. (1958–1959) Regeneration of surgically divided lymph vessels. An experimental study on the rabbit's ear. *Acta Chir. Scand.* 116:99–117.

Berge, L.N., Hansen, J.B., Svensson, B., Lyngmo, V. and Nordoy, A. (1990) Female sex hormones and platelet/endothelial cell interactions. *Haemostasis* 20:313–320.

Bernas, M., Enriques, J., Way, D., Witte, M., Ragland, A., Bradley-Dunlop, D. and Fiala, M. (1993) Absence of monoclonal antibody detectable Kaposi sarcoma-specific antigens on lesion-derived cultured cells. *Life Sciences* 52:663–668.

Bevilacqua, S., Welch, D.R., Diez de Pinos, S.M., Shapiro, S.A., Johnston, M.G., Witte, M.H., Leong, S.P.L., Dorrance, T.L., Leibovitz, A. and Hendrix, M.J.C. (1990) Quantitation of human melanoma, carcinoma and sarcoma tumor cell adhesion to lymphatic endothelium. *Lymphology* 23:4–14.

Bissell, M.J., Hall, H.G. and Parry, G. (1982) How does the extracellular matrix direct gene expression? *J. Theoret. Biol.* 99:31–68.

Bollinger, A., Jäger, K., Sgier, F. and Seglias, J. (1981) Fluorescence microlymphography. *Circulation* 64:1195–1200.

Borgs, P., Way, D.L., Witte, M.H., Case, T.C., Ramirez, G. and Witte, C.L. (1992) Evaluating *In Vitro* Toxicity to Mammalian Endothelial Cells. *In*: R. Watson (ed.): *In Vitro Methods in Toxicology and Pharmacology.* CRC Press, Boca Raton, FL, pp 270–284.

Borgs, P., Ramirez Jr, G., Way, D.L., Bernas, M.J., Witte, M.H. and Witte, C.L. (1994) Models of Kaposi sarcoma and endothelial cell angiotumorigenicity in rodents. *Lymphology* 27 (Suppl): 749–752.

Borgs, P., Witte, M.H. and Witte, C.L. (1995) Alcohol, Endothelial Cells and Immunity. In: R.R. Watson (ed.); Alcohol, Drugs of Abuse, and Immune Function. CRC Press, Boca Raton, FL, pp 121–130.

Borron, P. and Hay, J.B. (1994) Characterization of ovine lymphatic endothelial cells and their interactions with lymphocytes. *Lymphology* 27:6–13.

Bowman, C., Witte, M.H., Witte, C.L., Way, D., Nagle, R., Copeland, J. and Daschbach, C. (1984) Cystic hygroma reconsidered: Hamartoma or neoplasm? Primary culture of an endothelial cell line from a massive cervicomediastinal cystic hygroma with bony lymphangiomatosis. *Lymphology* 17:15–22.

Burrows, F.J. and Thorpa, P.E. (1994) Review: Vascular targeting -a new approach to the therapy of solid tumors. *Pharmacol. Ther.* 64:155–174.

Carey, D.J. (1991) Control of growth and differentiation of vascular cells by extracellular matrix proteins [review]. *Ann. Rev. Physiol.* 53:161–177.

Case, T., Leis, B., Witte, M., Way, D., Bernas, M., Borgs, P., Crandall, C., Crandall, R., Nagle, R., Jamal, S., Nayar, J. and Witte, C. (1991) Vascular abnormalities in experimental and human lymphatic filariasis. *Lymphology* 24:174–183.

Case, T.C., Witte, M.H., Way, D.L. and Witte, C.L. (1992a) Videomicroscopy of intralymphatic-dwelling *Brugia malayi*. *Ann. Trop. Med.* 84:435–438.

Case, T.C., Unger, E., Bernas, M.J., Witte, M.H., Witte, C.L., McNeill, G., Crandall, C. and Crandall, R. (1992b) Lymphatic imaging in experimental filariasis using magnetic resonance. *Invest. Radiol.* 27:293–297.

Case, T.C., Witte, C.L., Witte, M.H., Unger, E.C. and Williams, W.H. (1992c) Magnetic resonance imaging in human lymphedema: Comparison with lymphangioscintigraphy. *Magnetic Resonance Imaging* 10:549–558.

Casley-Smith, J.R. and Florey, H.W. (1961) The structure of normal small lymphatics. *Quart J. Exptl. Physiol.* 46:1010–106.

Casley-Smith, J.R. (1983) The Regenerative Capacity of the Lymphovascular System. *In*: M. Földi and J.R. Casley-Smith (eds): *Lymphangiology*, Schattauer, Stuttgart, pp 276–278.

Casley-Smith, J.R. and Casley-Smith, Judith R. (1986) *High-Protein Oedemas and the Benzo-Pyrones*. J.B. Lippincott Company, Sydney.

Casley-Smith, J.R., Morgan, R. and Piller, N. (1993) Treatment of lymphedema of the arms and legs with 5,6 benzo α pyrone. *N. Engl. J. Med.* 329:1158–1163.

Chervenak, F.A., Isaacson, G., Blakemore, K.J., Breg, W.R. Hobbins, J.C., Berkowitz, R.L., Tortora, M., Mayden, K. and Mahoney, M.J. (1983) Fetal cystic hygroma: cause and natural history, *N. Engl. J. Med.* 309:822–825.

Clark, E.R. and Clark, E.L. (1932) Observations on the new growth of lymphatic vessels as seen in transparent chambers introduced into the rabbit's ear. *Am. J. Anat.* 51:49–87.

Clemm, C., Jehn, U., Wolf-Hornung, B., Siemon, G. and Walter, G. (1987) Lymphangiomyomatosis: A report of three cases treated with tamoxifen. *Klin. Wochenschr.* 65:391–393.

Clodius, L. and Altorfer, J. (1977) Experimental chronic lymphostasis of extremities. *Folia Angiolog.* 25:137.

Courtice, F.C. and Steinbeck, A.W. (1950) The lymphatic drainage of plasma from the peritoneal cavity of the cat. *Aust. J. exp. Biol. med. Sci.* 28:161–169.

Crandall, R.B., Crandall, C.A., Hines, S.A. and Nayar, N.K. (1987) Periperal lymphedema in ferrets infected with *Brugia malayi*. *Am. J. Trop. Med. Hyg.* 37:138–142.

D'Amore, P.A. and Braunhut, S.J. (1987) Stimulatory and Inhibitory Factors in Vascular Growth Control. *In*: U.S. Ryan (ed.) *Endothelial Cells, Vol II*. CRC Press, Boca Raton, pp 13–36.

Danese, C.A., Georgalas-Bertakis, M. and Morales, L.E. (1968) A model of chronic postsurgical lymphedema in dogs' limb. *Surgery* 64:814–820.

Daroczy, J. (1995) Pathology of Lymphedema. *In*: T. Ryan and P.S. Mortimer (ed): *Clinical Management of Lymphatic Disease. Cutaneous Lymphatic System. Clinics in Dermatology.* 13:433–444.

Denekamp, J. (1993) Angiogenesis, neovascular proliferation and vascular pathophysiology as targets for cancer therapy. *Br. J. Radiol.* 66:181–196.

Dictor, M. (1988) Vascular remodeling in Kaposi's sarcoma and avian hemangiogenesis: Relation to the vertebrate lymphatic system. *Lymphology* 21:53–60.

Dictor, M. and Andersson, C. (1988) Lymphaticovenous differentiation in Kaposi's sarcoma. *Am. J. Pathol.* 130:411–417.

Dictor, M., Carlén, B., Bendsöe, N. and Flamholc, N. (1991) Ultrastructural development of Kaposi's sarcoma in relation to the dermal microvasculature. *Virchows Archiv. A* 419:35–43.

Dictor, M., Bendsöe, N., Runke, S. and Witte, M. (1995) Major basement membrane components in Kaposi's sarcoma, angiosarcoma and benign vascular neogenesis. *J. Cutan. Pathol.* 22:423–441.

Dictor, M., Rambech, E., Way, D., Witte, M. and Bendsoe, N. (1996) Human herpesvirus 8 (KSHV) DNA in Kaposi sarcoma, cell lines and immunosuppressed patients: A molecular epidemiologic study. *Am. J. Pathol.* 148:2009–2016.

Djonedi, M. and Brodt, P. (1991) Isolation and characterization of rat lymphatic endothelial cells. *Microcirc. Endo. Lymph.* 7:161–182.

Dolbeare, F., Beisker, W., Pallavicini, M.G., Vanderlaan, M. and Gray, J.W. (1985) Cytochemistry for bromodeoxyuridine/DNA analysis: Stoichiometry and sensitivity. *Cytometry* 6:521–530.

Dorfman, R.F. (1988) Kaposi's sarcoma: Evidence supporting its origin from the lymphatic system. *Lymphology* 21:45–52.

Drinker, C.K. and Yoffey, J.M. (1941) *Lymphatics, Lymph and Lymphoid Tissue.* Harvard U.P., Cambridge, Mass.

Dvorak, H.F., Nagy, J.A., Berse, B., Brown, L.F., Yeo, K.T., Yeo, T.K., Dvorak, A.M., van de Water, L., Sioussat, T.M. and Senger, D.R. (1992) Vascular permeability factor, fibrin, and the pathogenesis of tumor stroma formation [review]. *Annals New York Acad. Sci.* 667:101–111.

Elvin-Lewis, M., Witte, M., Witte, C., Cole, W. and Davis, J. (1973) Systemic chlamydial infection associated with generalized lymphedema and lymphangiosarcoma. *Lymphology* 6:113–121.

Földi, M. (1983) Insufficiency of Lymph Flow. *In*: M. Földi and J.R. Casley-Smith (eds): *Lymphangiology*, Shattauer-Verlag, Stuttgart and New York: 195. (extensive review with references)

Folkman, J. and Haudenschild, C. (1980) Angiogenesis *in vitro. Nature* 288:551–556.

Folkman, J. (1995) Clinical applications of research on angiogenesis. *N. Engl. J. Med.* 26:1757–1763.

Fotsis, T., Zhang, Y., Pepper, M.S., Adlercreutz, Montesano, R., Nawroth, P.O. and Schweigerer, L. (1994) The endogeneous oestrogen metabolite 2-methoxyoestradiol inhibits angiogenesis and suppresses tumour growth. *Nature* 368:237–239.

Friedlander, M., Brooks, P.C., Shaffer, R.W., Kincaid, C.M., Varner, J.A. and Cheresh, D.A. (1995) Definition of two angiogenetic pathways by distinct a_v integrins. *Science* 270:1500–1502.

Garry, R.F., Witte, M.H., Gottlieb, A.A., Elvin-Lewis, M., Gottlieb, M.S., Witte, C.L., Cole, W.R. and Drake, W.L. (1988) Documentation of an AIDS virus infection in the United States in 1968. *JAMA* 260:2085–2087. Letter to Editor Reply (1989) *JAMA* 261:2198–2199.

Gibbons, G.H. and Dzau, V.J. (1990) Endothelial Function in Vascular Remodeling. *In*: J.B. Warren (ed) *The Endothelium: An Introduction to Current Research*, Wiley-Liss New York, pp 81–93.

Gnepp, D.R. and Chandler, W. (1985) Tissue culture of human and canine thoracic duct endothelium. *In Vitro* 21:200–206.

Goetsch, E. (1938) Hygroma colli cysticum and hygroma axillare. *Arch. Surg.* 36:394.

Gorham, L.W. and Stout, A.P. (1955) Massive osteolysis (acute spontaneous absorption of bone, phantom bone, disappearing bone). *J. Bone Jt. Surg.* 37:985.

Gowans, J.L., Gesner, B.M. and McGregor, D.D. (1961) The Immunological Activity of Lymphocytes. *In: Biological Activity of the Leucocyte.* Ciba Foundation Study Group No. 10, Churchill, London, pp 32–44.

Graham, M.L. II, Spelsberg, T.C., Dines, D.E., Payne, W.S., Bjornsson, J. and Lie, J.T. (1984) Pulmonary lymphangiomyomatosis: with particular reference to steroid-receptor assay studies and pathologic correlation. *Mayo Clin. Proc.* 59:3–11.

Greenlee, R., Hoyme, H., Witte, M., Crowe, P. and Witte, C. (1993) Developmental disorders of the lymphatic system. *Lymphology* 26:156–168.

Gullino, P.M. (1978) Angiogenesis and oncogenesis. *J. Natl. Cancer Inst.* 61:639–643.

Hay, E.D. (1984) Cell-Matrix Interaction in the Embryo: Cell Shape, Cell Surface, Cell Skeletons, and Their Role in Differentiation. *In*: R.L. Trelstad (ed.): *The Role of Extracellular Matrix in Development*, Alan R. Liss, New York, pp 1–31.

Hendrix, M.J.C., Seftor, E.A., Seftor, R.E.B.,Way, D.L., Bernas, M., Weinand, M., Witte, C.L. and Witte, M.H. (1994) Sex steroid modulation of endothelial cell interactions: Implications for angiodysplasia syndromes. *Lymphology* 27 (Suppl): 138–141.

Huntington, G.S. (1914) The development of the mammalian jugular lymph sac, of the tributary primitive ulnar lymphatic, and of the thoracic duct from the view point of recent investigations of vertebrate lymphatic oncogeny, together with a consideration of the genetic relations of lymphatic and haemal vascular channels in the embryos of amniotes. *Am. J. Anat.* 16:259–316.

Jacobs, P.A. (1975) The Load Due to Chromosome Abnormalities in Man. *In*: F.M. Salzano (ed.): *The Role of Natural Selection in Human Evolution*. North-Holland, Amsterdam, pp 337–352.

Jaffe, E.A. (1987) Cell biology of endothelial cells. *Hum. Pathol.* 18:234–239.

Jain, R.K. (1989) Delivery of novel therapeutic agents in tumors: Physiological barriers and strategies [review]. *J. Natl. Cancer Inst.* 81:570–576.

Järplid, B. (1961) Haemangioendotheliomas in poultry. *J. Comp. Path. Ther.* 71:370–376.

Johnston, M. and Walker, M. (1984) Lymphatic endothelial and smooth muscle cells in tissue culture. *InVitro* 20:566–572.

Junghans, B.M. and Collin, H.B. (1989) Limbal lymphangiogenesis after corneal injury: an autoradiographic study. *Current Eye Research* 8:91–100.

Kampmeier, O.F. (ed.) (1969) *Evolution and Comparative Morphology of the Lymphatic System.* Charles C. Thomas, Springfield, Illinois, pp 620.

Kanter, M.A., Slavin, S.A. and Kaplin, W. (1990) An experimental model for chronic lymphedema. *Plast. Reconst. Surg.* 85:573–580.

Kinmonth, J.B. (ed.) (1972) *The Lymphatics: Diseases, Lymphography and Surgery*. Edward Arnold, London.

Leak, L.V. (1970) Electron microscopic observations on lymphatic capillaries and the structural components of the connective tissue-lymph interface. *Microvasc. Res.* 2:361–391.

Leak, L.V. (1972) The Fine Structure and Function of the Lymphatic Vascular System. *In*: H. Meesen (ed.) *Handbook der Allgemeinen Pathologie.* Springer-Verlag, Berlin and New York, pp 149–196.

Leak, L. (1985) Interaction of the peritoneal cavity to intraperitoneal stimulation: a peritoneal nodal system to monitor cellular and extracellular events in the formation of granulation tissue. *Am. J. Anat.* 173:171–183.

Leak, L.V. (1991) Lymphatic endothelial cells *in vitro*: isolation and characterization. *J. Cell Biol.* 115:365A.

Leak, L.V. and Jones, M. (1993) Lymphatic endothelium isolation characterization and long-term culture. *Anat. Rec.* 236:641–652.

Leak, L.V. and Jones, M. (1994) Lymphangiogenesis *in vitro*: Formation of lymphatic capillary-like channels from confluent monolayers of lymphatic endothelial cells. *In Vitro Cell. Dev. Bio.* 30A:512–518.

Leu, H.J. and Lie, J.T. (1995) Diseases of the Veins and Lymphatic Vessels Including Angiodysplasias. *In*: W.E. Stehbens, W.E. and J.T. Lie (eds): *Vascular Pathology*, Chapman & Hall Medical, London, pp 489–516.

Lipton, B.H., Bensch, K.G. and Karasek, M.A. (1991) Microvessel endothelial cell transdifferentiation: Phenotype characterization. *Differentiation* 46:117–133.

Lunardi-Islander, Y., Bryant, J.L., Zeman, R.A., Lam, V.H., Samaniego, F., Besnier, J.M., Hermans, P., Thierry, A.R., Gill, P. and Gallo, R.C. (1995) Tumorigensis and metastasis of neoplastic Kaposi's sarcoma cell line in immunodeficient mice blocked by a human pregnancy hormone. *Nature* 375:64–68.

Magari, S. and Asano, S. (1978) Regeneration of the deep cervical lymphatics-light and electron microscopic observations. *Lymphology* 11:57–61.

Magari, S. (1987) Comparison of fine structure of lymphatics and blood vessels in normal conditions and during embryonic development and regeneration. *Lymphology* 20:189–195.

Martinez-Zaguilan, R.M., Chinnock, B.F., Wald-Hopkins, S., Bernas, M., Way, D., Weinand, M., Witte, M.H. and Gillies, R.J. (1996) Calcium and pH homeostasis in Kaposi sarcoma cells. *Cell Physiol. and Biochem., in press.*

Zaguilan, R.M., Chinnock, B.F., Hopkins, S.W., Bernas, M., Way, D.,Weinand, M., Witte, M.H. and Gillies, R.J. (1996) Calcium and pH homeostasis in Kaposi sarcoma cells. *Cell. Physiol. & Biochem.; in press.*

Mayerson, H.S., Wolfram, C.G., Shirley, H.H. and Wasserman, K. (1960) Regional differences in capillary permeability. *Am. J. Physiol.* 198:155–160.

Meade-Tollin, L., Way, D., Witte, M., Bernas, M., Fiala, M., Mallery, S. and Witte, C. (1995) Constitutive metalloproteinase secretion patterns in normal, Kaposi sarcoma, and angiotumor endothelial cell populations. *J. Invest. Med.* 43:311A.

Montesano, R., Orci, L. and Vassalli, P. (1983) *In vitro* rapid organization of endothelial cells into capillary-like networks is promoted by collagen matrices. *J. Cell Biol.* 97:1648–1652.

Montesano, R., Pepper, M.S., Möhle-Steinlein, U., Risau, W., Wagner, E.F. and Orci, L. (1990) Increased proteolytic activity is responsible for the aberrant morphogenetic behavior of endothelial cells expressing the middle T oncogene. *Cell* 62:435–445.

Mulliken, J.B. and Young, A.E. (1988) *Vascular Birthmarks: Hemangiomas and Malformations.* W.B. Saunders, Philadelphia.

Nagle, R., Witte, M.H., Witte, C.L. and Way, D. (1985) Factor VIII-associated antigen in canine lymphatic endothelium. *Lymphology* 18:84–85.

Nagle, R.B., Witte, M.H., Martinez, A.P.,Witte, C.L., Hendrix, M.J., Reed, K. and Way, D. (1987) Factor VIII-associated antigen in human lymphatic endothelium. *Lymphology* 20:20–24.

Nelson, M.J., Conley, F.K. and Fajardo, L.F. (1993) Application of the disc angiogenesis system to tumor-induced neovascularization. *Expl. Molec. Pathol.* 58:105–113.

Nicosia, R.F. (1987) Angiogenesis and the formation of lymphaticlike channels in cultures of thoracic duct. *In Vitro Cell. Devel. Biol.* 23:167–174.

Nicosia, R.F. and Pietra, G.G. (1994) Neoplasia. *In*: N.A. Mortillaro and A.E. Taylor (eds), *The Pathophysiology of the Microcirculation.* CRC Press, Boca Raton: Ann Arbor. London: Tokyo, pp 35–60.

Odén, B. (1960) A micro-lymphangiographic study of experimental wounds healing by second intention. *Acta Chir. Scandinav.* 120:100–114.

O'Connell, K.A. and Rudmann, A.A. (1993) Cloned spindle and epithelioid cells from murine Kaposi's sarcoma-like tumors are of endothelial origin. *J. Invest. Dermatol.* 100: 742–745.

Olszewski, W. (1973) On the pathomechanism of development of postsurgical lymphedema. *Lymphology* 6:35–52.

Olszewski, W.L. (ed.) (1991) *Lymph Stasis-Pathophysiology, Diagnosis and Treatment.* CRC Press, Boca Raton.

Otsuki, Y., Magari, S. and Sugimoto, O. (1986) Lymphatic capillaries in rabbit ovaries during ovulation: an ultrastructural study. *Lymphology* 19:55–64.

Papendieck, C.M. (1992) *Atlas Color Temas de Angiologia Pediatrica.* Editorial Medica Panamericana, Buenos Aires, pp 225.

Patterson, D.F., Medway, W., Luginbühl, H. and Chacko, S. (1967) Congenital hereditary lymphedema in the dog. Part I. Clinical and gentic studies. *J. Med. Genet.* 4:145–165.

Pepper, M.S., Wasi, S., Ferrara, N., Orci, L. and Montesano, R. (1994) *In vitro* angiogenic and proteolytic properties of bovine lymphatic endothelial cells. *Exptl. Cell Research* 210: 298–305.

Pippard, C. and Roddie, I.C. (1987) Comparison of fluid transport systems in lymphatic and veins. *Lymphology* 20:224–229.

Pullinger, D.B. and Florey, H.W. (1937) Proliferation of lymphatics in inflammation. *J. Pathol. Bacteriol.* 45:157–170.

Reichert, F.L. (1926) The regeneration of the lymphatics. *Arch. Surg.* 13:871–881.

Reynolds, L.P., Killilea, S.D. and Redmer, D.A. (1992) Angiogenesis in the female reproductive system. *FASEB J.* 6:886–892.

Ross, R. (1993) The pathogenesis of atherosclerosis: A perspective for the 1990s. *Nature* 362:801–809.

Ryan, T.J. and Barnhill, R.L. (1983) Physical Factors and Angiogenesis. *In*: J. Nugen, M. O'Connor (eds): *Development of the Vascular System.* Ciba Foundation Symposium 100, Pitman, London, UK, pp 80–94.

Ryan, T.J. (1989) Biochemical consequences of mechanical forces generated by distention and distortion. *J. Am. Acad. Dermatol.* 21:115–130.

Ryan, T.J. (1990) Grip and stick and the lymphatics. *Lymphology* 23:81–84.

Ryan, T.J. (1995) Lymphatics and Adipose Tissue. *In*: T. Ryan and P.S. Mortimer (eds.): *Cutaneous Lymphatic System. Clinics in Dermatology* 13:493–498.

Ryan, T.J. and DeBerker, D. (1995) The Interstitium and the Connective Tissue Environment of the Lymphatic in Human Skin. *In*: T. Ryan and P.S. Mortimer (eds.): *Cutaneous Lymphatic System. Clinics in Dermatology* 13:451–458.

Sabin, F.R. (1902) On the origin and development of the lymphatic system from the veins and the development of the lymph hearts and the thoracic duct in the pig. *Am. J. Anat.* 1:367–389.

Salahuddin, S.Z., Nakamura, S., Biberfeld, P., Kaplan, M.H., Markham, P.D., Larsson, L. and Gallo, R.C. (1988) Angiogenic properties of Kaposi's sarcoma-derived cells after long-term culture *in vitro. Science* 242:430–433.

Satomua, K., Tanigawa, N. and Magari, S. (1978) Microlymphangiographic study of lymphatic regeneration following intestinal anastomosis. *Surgery, Genec. & Obstet.* 146:415–418.

Schmelz, S., Moll, R., Kuhn, C. and Franke, W.W. (1994) *Complexus adhaerentes,* a new group of desmoplakin-containing junctions in endothelial cells: II. Different types of lymphatic vessels. *Differentiation* 57:97–117.

Schmelz, M., Way, D., Bernas, M., Witte, M., Witte, C., Moll, R. and Franke, W.W (1995) *Complexus adhaerentes* and Kaposi sarcoma. Abstract booklet, 15th Intl. Congress of Lymphology, São Paulo, Brazil, pp 50.

Sholley, M.M., Ferguson, G.P., Seibel, H.R., Montour, J.L. and Wilson, J.D. (1984) Mechanisms of neovascularization: vascular sprouting can occur without proliferation of endothelial cells. *Lab. Invest.* 51:624–634.

Sinzinger, H., Kaliman, J. and Mannheimer, E. (1984a) Regulation of human lymph contractility by prostaglandins and thromboxane. *Lymphology* 17:43–45.

Sinzinger, H., Kaliman, J. and Mannheimer, E. (1984b) Arachidonic acid metabolites of human lymphatics. *Lymphology* 17:39–42.

Sinzinger, H., Kaliman, J. and Mannheimer, E. (1986a) Effect of leukotrienes C_4 and D_4 on prostaglandin I_2-liberation from human lymphatics. *Lymphology* 19:79–81.

Sinzinger, H., Kaliman, J. and Mannheimer, E. (1986b) Enhanced prostaglandin I_2-formation of human lymphatics during pulsatile perfusion. *Lymphology* 19:153–156.

Sjöberg, T. and Steen, S. (1991) *In vitro* effects of a thromoboxane A_2-analogue U-46619 and noradrenaline on contractions of the human thoracic duct. *Lymphology* 24:113–115.

Snowden, K. and Hammerberg, B. (1986) Vascular patterns in the filaria-infected canine limb. *Lymphology* 19:77–78.

Starling, E.H. (1896) Physiologic factors involved in the causation of dropsy. *Lancet* 1:1267–1270.

Stazzone, A.M., Borgs, P., Way, D. and Witte, M. (1995) Flow cytometric assay of *in vitro* lymphocyte-endothelial cell binding. *J. Invest. Med.* 43:185A.

Stewart, F.W. and Treves, N. (1948) Lymphangiosarcoma in post-mastectomy lymphedema. *Cancer* 1:64–81.

Taylor, J.R., Ryu, J., Colby, T.V. and Raffin, T.A. (1990) Lymphangioleiomyomatosis; Clinical course in 32 patients. *N. Engl. J. Med.* 323(18):1254–1260.

Thoma, R. (1896) *Textbook of General Pathology.* (transl. by A. Bruce), Vol. 1, Adam and Charles Black, London.

Trevella, W. and Dugan, M.C. (1990) Endothelial Interactions. *In*: M. Nishi, S. Uchino and S. Yabuki (eds): *Progress in Lymphology-XII*, Elsevier Science Publisher B.V., p 269, Amsterdam, New York, Oxford.

Tsuboi, R., Sato, Y. and Rifkin, D.B. (1990) Correlation of cell migration, cell, invasion, receptor number, proteinase production, and basic fibroblast growth factor levels in endothelial cells. *J. Cell Biol.* 110:511–517.

Turner, R.R., Beckstead, J.H., Warnke, R.A. and Wood, G.S. (1987) Endothelial cell diversity: *in situ* demonstration of immunologic and enzymatic heterogeneity that correlates with specific morphologic subtypes. *Am. J. Clin. Pathol.* 87:569–575.

Unger, E.C., Lund, P.J., Shen, D.K., Fritz, T.A., Yellowhair, D. and New, T.E. (1992) Nitrogen-filled liposomes as a vascular US contrast agent: Preliminary evaluation. *Radiology* 185:453–456.

van der Putte, S.C.J. (1975) *The Development of the Lymphatic System in Man.* Springer-Verlag. Berlin, Heidelberg, New York, p 7.

Way, D., Hendrix, M., Witte, M., Witte, C., Nagle, R. and Davis, J. (1987) Lymphatic endothelial cell line (CH3) from a recurrent retroperitoneal lymphangioma. *In Vitro* 23:647–652.

Way, D.L., Witte, M.H., Fiala, M., Ramirez G., Nagle, R.B., Bernas, M.J., Dictor, M., Borgs, P. and Witte, C.L. (1993) Endothelial transdifferentiated phenotype and cell-cycle kinetics of AIDS-associated Kaposi sarcoma cells. *Lymphology* 26:79–89.

Way, D.,Witte, M., Bernas, M. Weinland, M., Scott, L., Shao, X., Witte, C. and Fiala, M. (1994a) *In vitro* models of angiotumorigenesis. *Lymphology* 27 (Suppl): 136–137.

Way, D., Borgs, P., Bernas, M., Ramirez, G., Shao, X., Witte, M., Weinand, M., Riedel, U., Johnson, V. and Witte, C. (1994b) Characterization of an established immortal endothelial cell line (RSE-1); Comparison to AIDS-Kaposi sarcoma cell cultures. *Lymphology* 27 (Suppl): 761–762.

Way, D., Borgs, P., Witte, M., Schmelz, M., Sobonya, R., Shao, X. and Witte, C. (1996) Isolation and characterization of a new established human lymphatic endothelial cell line (HTDEC-1). *Lymphology* 29 (Suppl); *in press*.

Weber, E., Lorenzoni, P., Lozzi, G. and Sacchi, G. (1994) Differentiation between blood and lymphatic endothelium: bovine blood and lymphatic large vessels and endothelial cells in culture. *J. Histochem. Cytochem.* 42:1109–1115.

Weidner, N., Semple, J.P., Welch, W.R. and Folkman, J. (1991) Tumor angiogenesis and metastasis – correlation in invasive breast carcinoma. *N. Engl. J. Med.* 324:1–8.

Willis, R.A. (1952) Metastasis via Lymphatics and the Cancerous Thoracic Duct. *In*: R.A. Ellis (ed.) *The Spread of Tumours in the Human Body.* C.V. Mosby Co., St. Louis, pp 18–35.

Witte, C.L., Corrigan, J.J., Witte, M.H., Van Wyck, D.B., O'Mara, R.E. and Woolfenden, J.M. (1980a) Circulatory control of splenic hyperfunction in children with peripheral blood dyscrasia. *Surgery, Gynec. & Obstet.* 150:75–80.

Witte, C.L., Witte, M.H. and Dumont, A.E. (1980b) Lymph imbalance in the genesis and perpetuation of ascites syndrome in hepatic cirrhosis. *Gastroenterology* 78:1059–1068.

Witte, C.L., Witte, M.H. and Dumont, A.E. (1984) Pathophysiology of Chronic Edema, Lymphedema, and Fibrosis. *In*: N.C. Staub and A.E. Taylor (eds): *Edema: Basis Science and*

Clinical Manifestations: A Comprehensive Treatise. Raven Press, New York, NY, pp 521–542.

Witte, C.L. and Witte, M.H. (1987) Contrasting patterns of lymphatic and blood circulatory disorders. *Lymphology* 20:257–266.

Witte, C.L. and Witte, M.H. (1991) Pathophysiology of Lymphatic Insufficiency and Principles of Treatment. *In:* W.L. Olszewski (ed.): *Lymph Stasis-Pathophysiology, Diagnosis and Treatment*, CRC Press, pp 327–344.

Witte, C.L. and Witte, M.H. (1995) Lymph Dynamics and Pathophysiology of Lymphedema. *In:* R.B. Rutherford (ed.) *Vascular Surgery*. Fourth Edition. W.B. Saunders Company, Philadelphia, PA, pp 1889–1899.

Witte, C.L. and Witte, M.H. (1996) Physiology and Imaging of the Periperal Lymphatic System. *In:* W.S. Moore (ed.): *Vascular Surgery; A Comprehensive Review,* 5th Edition, W.B. Saunders Company, Philadelphia, Pennsylvania; *in press*.

Witte, M.H., Dumont, A.E., Clauss, R.H., Rader, B., Levine N. and Breed, E.S. (1969) Lymph circulation in congestive heart failure: Effect of external thoracic duct drainage. *Circulation* 39:723–733.

Witte, M.H. and Witte, C.L. (1986) Lymphangiogenesis and lymphologic syndromes. *Lymphology* 19:21–28.

Witte, M.H. and Witte, C.L. (eds) (1987a) Are Lymphatics Different From Blood Vessels? *Lymphology* 20:167–276.

Witte, M.H. and Witte, C.L. (1987b) Lymphatics and blood vessels, lymphangiogenesis and hemangiogenesis: From cell biology to clinical medicine. *Lymphology* 20:171–178.

Witte, M., McNeill, G., Crandall, C., Case, T., Witte, C., Crandall, Hall, J. and Williams, W. (1988) Whole body lymphangioscintigraphy in ferrets chronically infected with *Brugia malayi. Lymphology* 21:251–257.

Witte, M.H., Stuntz, M. and Witte, C.L. (1989) Kaposi's sarcoma: A lymphologic perspective. *Int. J. Dermatol.* 28:561–570.

Witte, M.H., Witte, C.L. and Way, D.L. (1990a) Medical ignorance, AIDS-Kaposi sarcoma complex, and the lymphatic system. *Western J. Med.* 153:17–23.

Witte, M.H., Witte, C.L., Way, D.L. and Fiala, M. (1990b) AIDS, Kaposi sarcoma, and the lymphatic system: update and reflections. *Lymphology* 23:73–80.

Witte, M.H., Fiala, M., McNeill, G., Witte, C.L., Williams, W. and Szabo, J. (1990c) Lymphangioscintigraphy in AIDS-associated Kaposi sarcoma. *Am. J. Roent.* 155:311–315.

Witte, M.H., Jamal, S., Williams, W., Witte, C.L., Kumaraswami, V., McNeill, G., Case, T. and Panicker, T.M.R. (1993) Lymphatic abnormalities in human filariasis as depicted by lymphangioscintigraphy. *Arch. Intern. Med.* 153:737–744.

Witte, M.H., Borgs, P., Way, D.L., Bernas, M., Ramirez, G. Jr, and Witte, C.L. (1994a) Kaposi sarcoma, vascular permeability, and scientific integrity. *JAMA* 271:1769–1771. Letters to the Editor. *JAMA* 272:916–924.

Witte, M.H. and Witte, C.L. (1995) Epilogue: Beyond the sphere of knowledge in lymphology. In: T. Ryan and P.S. Mortimer (eds): *Cutaneous Lymphatic System. Clinics in Dermatology* 13:511–514.

Yoffey, Y.M. (1929) A contribution to the study of the comparative histology and physiology of the spleen, with reference chiefly to its cellular constitutents. I: In fishes. *J. Anat.* 63: 314–344.

Yoffey, J.M. and Courtice, F.C. (eds) (1970) *Lymphatics, Lymph and Lymphomyeloid Complex.* Academic Press, London, pp 942.

Yong, L.C. and Jones, B. (1991) A comparison study of cultured vascular and lymphatic endothelium. *Exp. Pathol.* 42:11–25.

Yuan, F., Salchi, H.A. Boucher, Y., Vasthare, U.S., Tuma, R.F. and Jain, R. (1994) Vascular permeability and microcirculation of glioma and mammary carcinomas transplanted in rat and mouse cranial windows. *Cancer Res.* 54:4564–4568.

Zeidman, I., Copeland, B.E. and Warren, S. (1955) Experimental studies on the spread and cancer in the lymphatic system. II. Absence of a lymphatic supply in carcinoma. *Cancer* 8:123–127.

Zetter, B.R. (1987) Endothelial Heterogeneity: Influence of Vessel Size, Organ Localization, and Species Specificity on the Properties of Cultured Endothelial Cells. *In:* U.S. Ryan (ed.): *Endothelial Cells, Vol II.* CRC Press, Boca Raton, pp 63–80.

Ziche, M., Morbidelli, L., Alessandri, G. and Guillino, P.M. (1992) Angiogenesis can be stimulated or repressed *in vivo* by a change in GM3:GD3 ganglioside ratio. *Lab. Invest.* 67:711–715.

Regulation of Angiogenesis
ed. by I.D. Goldberg & E.M. Rosen
© 1997 Birkhäuser Verlag Basel/Switzerland

Angiogenesis as a biologic and prognostic indicator in human breast carcinoma

K. Engels[1], S. B. Fox[1] and A. L. Harris[2]

[1] *Department of Cellular Science, University of Oxford, John Radcliffe Hospital, Oxford OX3 9DU, UK*
[2] *ICRF Molecular Oncology Laboratory, Institute of Molecular Medicine, John Radcliffe Hospital, Oxford OX3 9DU, UK*

Summary. In this review we describe angiogenesis pathways involved in the development of breast carcinoma. Different assessment techniques for angiogenesis and their optimisation are discussed. Angiogenesis is an important factor for prognosis and will be increasingly important in therapeutic decisions.

Introduction

Angiogenesis, the formation of new blood vessels from the existing vascular bed, is tightly controlled by a complex network of angiogenesis-promoting and -inhibiting factors and is observed only transiently during particular processes such as reproduction, development and wound healing. It is a complex multistep process involving extracellular matrix remodelling, endothelial cell (EC) migration and proliferation, capillary differentiation and anastomosis (Paweletz and Knierim, 1989; Blood and Zetter, 1990). Although it has been recognised for many centuries that neoplastic tissue is more vascular than its normal counterpart, it is only since Folkman's hypothesis on anti-angiogenesis that interest in angiogenesis has increased and its importance in many physiological and pathological conditions has been described. The number of known angio- and anti-angiogenic factors is steadily increasing and their correlation with pathological and clinical data in carcinomas has been studied. Until recently only a research subject, angiogenesis is now finding its way into the first clinical trials.

Nevertheless, there are still more questions than answers. The exact processes regulating angiogenesis, the genetic changes involved, host-tumor relations modulating the angiogenic response, and many other points are still not fully understood. In this review we will describe the process of angiogenesis using breast carcinoma as a model, demonstrate its value for prognostic evaluation, and discuss the currently available techniques for assessing tumor angiogenesis.

Angiogenesis in breast carcinoma

Evolution

Initially it was thought that carcinomas are caused simply by lack of growth control, but the interactions between tumor and host, the intra- and peri-tumoral environment, are much more complicated and a cell population has to acquire many new characteristics to develop into a malignant tumor with metastatic manifestations. Loss of the homeostasis between cell growth and death (apoptosis) increases the number of cells. The development of an invasive phenotype resulting from a combination of different tumor cell properties is the key step in tumor development, defining the malignant disease. Metastatic growth, usually terminating the patient's life, is the end of a long evolution. The multistepwise nature of the process called tumorigenesis or carcinogenesis has been assumed for a long time (Foulds, 1958). Certain changes in genes controlling cell functions were thought to cause this process, and were described by the concept of tumor suppressor genes and oncogenes introduced in 1914 and translated and published in 1929 (Boveri, 1929). But only since the work of Fearon and Vogelstein (1990) on genetic changes during the stepwise progression of colorectal carcinomas this model has been accepted. Malfunction of DNA repair and control mechanisms resulting in mutational activation of oncogenes coupled with the mutational deactivation of tumor suppressor genes leads to changes in cell morphology and physiology. Benign tumorigenesis was characterized by a small number of changes and malignant tumorigenesis accordingly by more changes. The genetic alterations often occurred in a preferred sequence, but the total accumulation of changes, rather than their order with respect to one another, was responsible for determining the tumor's biological properties (Fearon and Vogelstein, 1990). It has been assumed that changes in genes coding for DNA control and repair mechanisms are among the first alterations of the genome resulting in a genetic instability. These changes affect the control of a constant cell number via the control of cell cycle (e.g. cyclin D, p53, Rb, c-*myc*) and apoptosis (e.g. bcl-2, p53). Through further genetic changes transformed cells gain more abnormal properties before and after they switch to the malignant phenotype and the cells with the greatest capacity for growth, invasion, and survival will undergo a positive selection (Sato et al., 1991; Moffett et al., 1992) finally resulting in widespread tumor growth, metastasis, and therapy resistance.

Genes and angiogenesis in breast carcinoma

Genetic changes in many tumors, including breast carcinoma, have been described and in the human breast the progression from normal duct epithelium to hyperplasia without and with atypia, *in situ* and finally invasive carcinoma, describes the morphological aspects of a stepwise development. However, the exact genetic changes in each step have still to be characterised. A strict hierarchy of genetic changes, where the switch to an angiogenic phenotype has its exact and constant place, is very unlikely.

In breast carcinomas *p53* is one of the most widely examined tumor suppressor genes; mutations can be found in approximately 30–50% of all sporadic breast carcinomas, usually occurring early in malignant progression. It is located on 17p13 and if mutated in several sensitive regions the protein conformation changes and stability increases, so that finally it functions as an oncogene product. The reduced expression of thrombospondin-1 *(TSP-1)* after loss of the wild-type *p53* allele in fibroblasts of Li-Fraumeni patients coincided with the onset of angiogenesis. Transfection with wild-type *p53* could stimulate the endogenous *TSP-1* gene and positively regulated the *TSP-1* promoter sequences reversing the switch (Smith et al., 1993). A link between *p53* mutation and protein kinase C (PKC) function promoting the expression of vascular endothelial growth factor (VEGF) has been described in tumor cell lines (Kieser et al., 1994). Control of angiogenesis in glial cells is regulated by *p53* via a yet unknown angiogenesis inhibitor (Van et al., 1994).

Loss of heterozygosity (LOH) of various tumor suppressor genes, which can appear early or late during progression, has been described in a varying proportion of breast carcinomas. Even though a very small number of patients was studied, LOH on 16q (D16S413) and 17p (D17S796) was found in approximately 50% of cases with atypical ductal hyperplasia (ADH) (Lakhani et al., 1995) and ductal carcinoma *in situ* (DCIS) of low and high nuclear grade (Stratton et al., 1995). In lobular carcinoma *in situ* (CLIS) LOH on 16q was present in a comparable percentage, whereas the incidence of LOH on 17p was reduced to 8%. These findings underline the stepwise progression concept and the appearance of genetic alterations before the switch to an invasive phenotype (see also e.g. Smith et al., 1993).

Although in experiments using human and animal tissue some genetic changes associated with the switch to the angiogenic phenotype were described, it is not known which changes are responsible for the activation of angiogenesis in breast carcinomas. Some further correlations between genetic alterations and angiogenesis are described below.

In mouse prostate reconstitutions, transfection with the *myc* oncogene induced hyperplasia; transfection with *ras* induced dysplasia with angiogenesis; and combined transfection with both oncogenes resulted in the development of carcinomas (Thompson et al., 1989). The transformation of baby hamster kidney cells (BHK) by polyomavirus results in tumor formation and reduced expression of the angiogenic inhibitor gp140 (Bouck, 1990).

Transfection of the human epithelial mammary cell line MCF-10A with the *int-2* oncogene (FGF-3) demonstrated induced angiogenic activity in the chick chorioallentoic membrane (CAM) assay and on rat mesenterium (Costa et al., 1994).

Growth and death

The homeostasis of cell growth and death (apoptosis) in normal tissue is under a balanced control of a number of factors signalling growth induction and inhibition, survival and death. In tumors this balance is shifted towards proliferation not simply by activating stimulatory factors; growth inhibition has to be weakened as well.

Early genetic aberrations in cell cycle control might result in proliferation dependent on systemic or local growth factors. The transformed cells demonstrate a pathologically increased reaction to hormones regulating growth in normal breast tissue, possibly by upregulation of growth factor receptors or alteration of intracellular signalling mechanisms. Such hormones are endocrine steroids, peptides, and other molecules produced by secretory cells of ovaries, pituitary, endocrine pancreas, thyroid, and adrenal cortex. Estrogen and progesterone regulate genes for growth factor production and secretion and other cell-cycle associated genes (Clarke and Sutherland, 1990). But tumor cells not only show an abnormal reaction to a normal environment, they also create an abnormal environment by releasing growth factors primarily directed towards stromal cells, with consequent proliferation of stromal fibroblasts, histiocytes, and vascular endothelial cells, which again can all release growth factors acting in an autocrine, paracrine or endocrine fashion. The development of the tumor stroma is critical for continued growth and progression of the neoplasm (Ethier, 1995).

Some growth factors secreted by breast carcinoma cells may auto-stimulate growth of the cells secreting the growth factor (autocrine pathway) or stimulate other breast carcinoma cells which are not secreting this factor (paracrine pathway). Cells might even be stimulated via an intracrine pathway, where growth factors act in the cells without being secreted. The expression of the appropriate growth factor receptors and the dependence of the transformed cells on this specific factor are both essential.

Further genetic alterations finally serve to separate cells from the requirement for growth stimuli (Ethier, 1995) resulting in a situation where the cell cycle of the malignant cells is continuously switched to proliferation. During further progression, although independent of growth factors, tumor growth can be promoted by their presence via their affect on stromal components (Ethier, 1995). In this process endothelial and inflammatory cells are important regulatory entities.

Apoptosis represents the negative regulator in balancing cell number in normal tissues. Carcinogenesis may involve several steps at which apoptotic pathways are initially activated by controlling genes; if cell death is activated to the same degree as proliferation, tumor growth may remain stationary. In growing carcinomas these mechanisms are bypassed. Increased expression of the oncogene *bcl-2* can inhibit apoptosis (for reviews see Williams, 1991; Nunez and Clarke, 1994; Hockenbery, 1995) and its co-expression with c-*myc* could accelerate tumorigenesis in tumor cell lines and transgenic models (Strasser et al., 1990). The correct function of the tumor suppressor gene *p53* is correlated with apoptosis induction possibly via downregulation of *bcl-2* (Haldar et al., 1994) and its loss or mutation can cause a block in the apoptotic pathway (for review see Elledge and Lee, 1995).

The loss of the apoptotic ability of a tumor cell is a strong advantage in selection, since competing cells can still undergo apoptosis and hence be outgrown by the new cell clone. The exact apoptotic pathways and the means of their activation in breast carcinomas are still unknown.

In breast carcinogenesis these processes are represented by certain morphological stages. Dysregulation of growth control results in hyperplasia, accompanying morphological changes cause hyperplasia with cellular atypia, DCIS, and CLIS. The latter are defined as lesions with cells showing definitive malignant cytology but lacking invasive tumor growth.

All the above described changes can coincide with the onset of angiogenesis.

Invasion

The switch to an invasive phenotype is the keystone in carcinogenesis defining a malignant disorder. It is a complex process involving the secretion of proteolytic enzymes and the gain of migratory abilities by tumor cells. Proteolytic enzymes (proteases) allow tumor cells to dissolve components of extracellular stroma and basement membranes. Three different classes of proteases are known to be correlated with the malignant phenotype:

1. **Urokinase plasminogen activator** (uPA) activates the broad specificity serine protease plasmin which is central to invasion and metastasis since it is able to degrade most matrix components either directly or by activation of other latent enzyme systems. UPA, which is usually absent from resting endothelium (Larsson et al., 1984; Kristensen et al., 1994), is upregulated by many angiogenic factors including VEGF and basic fibroblast growth factor (bFGF) (Pepper et al., 1990; Pepper et al., 1991) which together also demonstrate potent synergism (Pepper et al., 1992). Several studies have shown that elevated levels of uPA and PAI-1 (plasminogen activator

inhibitor) are associated with a poor prognosis in several tumor types including breast (Duffy, 1990; Schmitt, 1990; Sumiyoshi, 1991; Foekens et al., 1992; Spyratos et al., 1992; Grondahl et al., 1993; Jänicke et al., 1993; Ganesh et al., 1994; Duggan et al., 1995), and colon (Ganesh et al., 1994). The natural inhibitor of uPA, plasminogen activator inhibitor 2 (PAI-2), is associated with a good prognosis (Foekens et al., 1995) whereas PAI-1 correlates with a poor prognosis in breast carcinoma (Foekens et al., 1994). We have demonstrated in a series of 103 breast carcinoma patients a significant correlation between microvessel density and both uPA and PAI-1 (Fox et al., 1993 b). Furthermore, expression of the components of the urokinase system was concentrated at the endothelial-tumor interface where angiogenic remodelling is most prominent. This orchestrated interplay between endothelial and tumor cells suggests that the reduction in survival in patients with high levels of uPA and PAI-1 is partly due to the angiogenic activity of these tumors. Although the significant correlation between high levels of uPA and poor prognosis might have been anticipated, a similar association with elevated PAI-1 appears paradoxical. Nevertheless, the immunohistochemical pattern of expression suggests that PAI-1 protects both endothelium and neoplastic tumor elements from active uPA and that tumors low in PAI-1 might be more susceptible to autolysis (Grondahl et al., 1993).

2. The imbalance between **matrix metalloproteinase** (MMP-1/2/3/9) and **tissue inhibitors of metalloproteinase** (TIMP-1/2/3) represents another factor in this process (e.g. Iwata et al., 1995). This group, including collagenases, gelatinases and stromelysins, will be discussed later.

3. **Cathepsin D** is a widely expressed lysosomal aspartyl protease which plays an important role in tumor invasion and metastasis (Rochefort, 1990). It is initially synthesised as an inactive precursor (pro-cathepsin D), before being converted by proteolytic processing via an intermediate to its mature forms (Erikson, 1989). Evidence that cathepsin D may play a role in tumor progression has come from *in vitro*, *in vivo* and clinical studies. *In vitro* cathepsin D is mitogenic for breast carcinoma cell lines (Vignon et al., 1986) and degrades basement membranes (Briozzo et al., 1988) following an endocytotic or phagocytotic process (Moutcourrier et al., 1990). It also facilitates the release of the potent angiogenic factor bFGF (Briozzo, 1988) and thus could potentially stimulate neovascularization. In breast carcinomas high cathepsin D expression is a predictor of a shorter recurrence-free survival (Briozzo et al., 1988; Tandon et al., 1990), metastasis-free survival (Spyratos et al., 1989), and overall survival (Tandon et al., 1990; Winstanley et al., 1993).

Migration

The gain of migratory abilities has been shown to be associated with changes in adhesion molecule expression, which also plays an important role in angiogenesis and will be discussed later. The loss of E-cadherin resulted in a more motile, fibroblastic morphology in breast carcinoma cells with an increased motility (Liotta et al., 1991). Transfection of carcinoma cell lines with E-cadherin cDNA could block the development of an invasive phenotype, and treatment of these cells with anti-E-cadherin antibodies reversed this effect (Frixen et al., 1991). During the invasive process tumor cells might be kept together by varying levels of E-cadherin and/or integrin expression (Glukhova et al., 1995).

Further motility-promoting factors are members of the fibroblast growth factor (FGF) family and hepatocyte growth factor (HGF, scatter factor), which acts through a tyrosine kinase-encoding product of the c-*met* oncogene.

Metastases

At this point the neoplastic cells have gained the ability to proliferate without control and leave their designated space, migrating through given structures. This involves access to the circulatory system of the host resulting in a potentially life-threatening tumor spread. Breast carcinomas and many other tumor types penetrate into lymphatics and develop metastases in local lymph nodes. After entering blood vessels, carcinoma cells have to survive a rough journey through the circulation. Again, mutations protect against the activation of the apoptotic pathway by the host immune system and poor environmental conditions. Abnormal adhesion molecule expression can result in formation of tumor cell aggregates and emboli which are entrapped in the capillary network of potential target organs, and also allows the attachment of tumor cells to vascular endothelial cells. Again the release of proteolytic enzymes and migratory abilities are crucial for penetration of the vessel wall and invasion of the target organ.

Having reached the target organ and successfully formed a tumor cell deposit the above described processes continue; the genetic instability of the malignant cells allows a further evolution resulting in further tumor spread, increasing aggressiveness and therapy resistence.

Role of angiogenesis

Angiogenesis in model systems
During the above described changes the tumor cell population faces competition with other, normal cells for oxygen, nutrition and growth factors. Transport of such substances into prevascular tumors happens by diffusion only. Without an adequate vascular supply tumors can only form a conglomerate of $1-2$ mm diameter, containing approximately 10^6 cells

(Sutherland et al., 1971; Folkman and Hochberg, 1973; Sutherland, 1988). The switch to the angiogenic phenotype (the induction of angiogenesis) solves this problem. In model systems this switch resulted in unlimited growth (Knighton et al., 1977). Tumor cells gain direct access to blood vessels and can enter the circulation, resulting in the formation of distant metastasis which is very unlikely in the prevascular state (Liotta et al., 1974). Since tumor cells, which have acquired all of the above described properties except angiogenesis, can profit from their angiogenic counterparts and enter the circulation with them, the distant metastasis might face the same starting situation as the primary tumor and enter a dormant state where a potential high proliferation rate is balanced by a high apoptotic rate, until again the change to an angiogenic phenotype is performed (O'Reilly et al., 1994; Holmgren et al., 1995).

Results from experimental and clinical studies suggest that this onset of angiogenesis can happen any time during carcinogenesis and is not always correlated with malignancy. A series of *in vitro* and *in vivo* experiments demonstrated the switch to the angiogenic phenotype during the progression from normal cells via hyperplasia to malignant cells. Using a transgenic mouse model, dermal fibrosarcomas developed from mild fibromatosis, in which basic fibroblast growth factor (bFGF/FGF2) was only found intracellularly, via aggressive fibromatosis (advanced preneoplastic form) to fibrosarcomas. The latter two secreted bFGF and both lesion types were highly angiogenic. Since the bFGF gene could be found in nonsecreting lesions and 80–85% of the bFGF synthesised by fibrosarcoma cells was excreted, it is likely that a transcriptional activation of this gene causes the switch to the angiogenic phenotype (Kandel et al., 1991). In transgenic mice carrying the SV40 large T oncogene the development of highly vascularized tumors of pancreatic b-cells arising from multifocal hyperplastic islets was studied. Angiogenic activity appeared prior to the formation of solid tumors, first in a subset of hyperplastic islets. The frequency correlated closely with the subsequent tumor incidence (Folkman et al., 1989).

Using the rabbit cornea assay the angiogenic potential of mouse and human mammary lesions has been examined. With increasing frequency normal, hyperplastic, and neoplastic murine breast tissue induced angiogenesis. The hyperplastic lesions with the highest angiogenic activity had the highest rate of neoplastic transformation (Gimbrone and Gullino, 1976a and b). Fragments of human benign hyperplasias of the breast induced angiogenesis in 30% in the same assay whereas *in situ* carcinomas and invasive carcinomas induced angiogenesis in 66% and 65%, respectively (Brem et al., 1978).

Implanting rat mammary tissue onto the rabbit iris gave similar results. Glands of virgin, pregnant and lactating mice gave an angiogenic response in approximately 5%, carcinogen-treated glands in 5%, hyperplastic glands were angiogenic in 13–25%, and carcinomas in 75–100% (Maiorana and

Gullino, 1978). In various multiply passaged cell lines the switch to the angiogenic phenotype could be observed before the switch to the neoplastic phenotype (Ziche and Gullino, 1982).

Angiogenesis in human tumors

Clinical data support these results. In cervix biopsies a high vascularity and a close rim of microvessels around the lesions was found with increasing frequency in benign, low-grade squamous intraepithelial, high-grade squamous intraepithelial, and invasive lesions. Corresponding results have been found for the mRNA expression of VEGF (Guidi et al., 1995).

Differences in angiogenic factor expression between superficial and invasive bladder carcinoma could be found. VEGF expression was four times higher in superficial than in invasive carcinoma and ten times higher than in normal bladder epithelium. The expression of platelet derived endothelial cell growth factor (Pd-ECGF) was 33 times higher in invasive than in superficial and 260 times higher than in normal tissue. These results indicate two different angiogenic pathways which are possibly associated with different genetic alterations. Indeed, mutations on chromosome 9 can be found in 70% of superficial bladder carcinomas, but are rare in invasive tumors; the opposite is the case for mutations on 17p. This indicates two different sequences of genetic alterations causing the above lesions (O'Brien et al., 1995). This correlation with angiogenic factor expression remains to be proven.

Increased mean vessel size in hyperplastic lesions and DCIS compared to normal breast tissue has been seen using an image analysis system for assessment. There was no difference in vascular density between these groups (Ottinetti and Sapino, 1988). A study including invasive carcinomas gave similar results: increased mean vessel size in neoplastic lesions compared to normal tissue, no difference in vascular density between these groups. The investigators described marked variations in vessel density even within the same diagnostic group despite using an image analysis system. The tumors with a high mean vascular size tended to be from younger patients, being smaller and more likely to be ER-negative (Porter et al., 1993). Using hematoxylin and eosin-stained tissue sections and sections stained immunohistochemically against laminin, a significant increase in stromal vascular density in DCIS and invasive carcinoma compared to benign and normal breast tissue was found. Interestingly vascular density was significantly lower in invasive carcinoma than in DCIS (Samejima and Yamazaki, 1988).

We found an increase in vascularity in 57% of 75 DCIS cases using antibodies against von Willebrand factor (FVIIIrAg). High grade and poorly differentiated lesions more advanced on the scale of carcinogenesis demonstrated the angiogenic phenotype significantly more often than low grade and well differentiated lesions (p = 0.005 and p = 0.003, respectively). a close rim of microvessels around the basement membrane of the involved ducts was present in 63 % of all cases and significantly correlated with increased vascularity (p = 0.0001) (Engels et al., 1996). Invasive breast carcinomas with a DCIS component demonstrated this rim pattern in 23 % of 22 patients (Weidner et al., 1991). Guidi et al. (1994) found a rim pattern in 35 % of 55 cases with pure DCIS, but no significant correlation to histologic features, *HER2/neu* oncogene expression, or Ki-S1 proliferation index was found. High vessel density (semiquantitative assessment) was significantly associated with comedo type (p = 0.04), high nuclear grade (p = 0.05), marked stromal dysplasia (p = 0.05), and *HER2/neu* overexpression (p = 0.03). However, in our series 29 % of lesions demonstrating no rim pattern showed an increase of vascularity in the stroma (Engels et al., 1996). This condition, which has been found in 6 out of 14 (43 %) by Guidi et al. (1994), suggests that angiogenesis is not induced only by the secretion of one single angiogenic factor by neoplastic cells. The different vascular patterns found in DCIS are shown in Figure 1.

The results draw a more complicated picture of angiogenesis than initially expected. The angiogenic environment is formed by a complex balance of angiogenesis-promoting and -inhibiting factors released by tumor cells, extracellular matrix (Vlodavsky et al., 1990; Houck et al., 1992), and inflammatory cells like macrophages (Hunt et al., 1984; Polverini and Leibovich, 1984; Koch et al., 1992; Sheid, 1992; Leek et al., 1994; Sunderkotter et al., 1994). Several angiogenic factors and inhibitors with different half lifes could be released simultaneously by tumor cells and meeting different inhibitors in the surrounding stroma causing different patterns of vascularity. Extratumoral elements modulate this situation by releasing and binding factors. An increased vascular density in the stroma without showing the rim pattern could be explained by the release of angiogenic factors by extratumoral elements that do not induce vascular proliferation close to the tumor itself. The presence of ducts

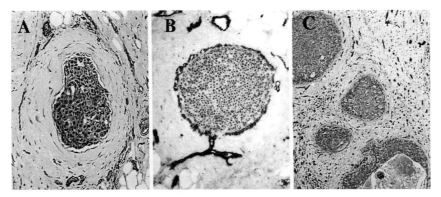

Figure 1. Different vessel patterns in DCIS. A) Distant rim and B) dense rim of microvessels around a DCIS duct close to the basement membrane. C) High vascular density between DCIS lesions. All vessels are highlighted using antibodies against FVIIIrAg (DAKO, UK).

showing the rim pattern, next to others not showing it in the same lesion, indicates variations in the angiogenic environment. The expression of angiogenic factors and their receptors can be regulated by changes in environmental factors such as oxygen (Shweiki et al., 1992; Stein et al., 1995) and glucose levels (Stein et al., 1995).

Even though no significant correlation between the different vessel patterns and relapse-free survival or overall survival could be found, vessel density was significantly higher in DCIS subtypes known to be associated with a higher relapse rate and a higher risk in developing an invasive tumor (comedo, $p = 0.005$; high nuclear grade, $p = 0.005$; poor differentiation, $p = 0.003$). A similar situation could be found comparing the presence of the dense vascular rim with survival where the rim pattern appeared slightly more often in the high risk groups, but no significant correlation to survival could be found.

Angiogenic factors

A large number of angiogenic factors has now been identified. Many angiogenic factors were initially identified by a variety of *in vivo* systems, including the chorioallantoic membrane assay (Auerbach et al., 1975), rabbit corneal implant assay (Gimbrone et al., 1974), hamster cheek pouch assay (Warren and Shubik, 1966) and rodent subcutaneous sponge model (Andrade et al., 1987). Their effect on the component steps of angiogenesis (i.e. proliferation, migration, tube formation) has been dissected using a variety of *in vitro* assay systems. Endothelial cell proliferation is studied using thymidine uptake or cell counts, migration in Boyden chambers (Postlethwaite et al., 1976) or wound healing models (Fan et al., 1992), and tube formation in three dimensional substrates like collagen or matrigel (Montesano, 1992). However, few studies have used each assay and each endothelial cell type (Kuzu et al., 1992), leaving a confused picture of the precise role of the different factors in tumor angiogenesis. Furthermore, many factors have pleiotropic effects and their contribution to tumor angiogenesis *in vivo* is unknown.

Angiogenic factors can act directly by stimulating endothelial cells and/or indirectly via accessory cells like macrophages and/or stromal cells. Thus candidate angiogenic factors have often been found to have opposite actions *in vivo* to that expected from *in vitro* experiments. Examples include TGF-β (Frater-Schroder et al., 1987) and TNF-α (Schweigerer et al., 1987), both of which inhibit endothelial cell growth *in vitro* but stimulate *in vivo*. The *in vivo* picture is further clouded by the release of many angiogenic factors from the extracellular matrix (Vlodavsky et al., 1990; Houck et al., 1992) which also contains soluble growth factor receptors (Hanneken et al., 1995) and might therefore give a further level of control.

In the following only factors which have been studied in breast carcinoma will be highlighted.

Vascular endothelial growth factor (VEGF), a member of the platelet derived growth factor (PDGF) family, is a 34–42 kD dimeric multi-functional glycoprotein which is angiogenic in many *in vitro* and *in vivo* model systems. Four isoforms, $VEGF_{121}$, $VEGF_{165}$, $VEGF_{189}$, and $VEGF_{206}$, generated by alternative splicing, which have different affinities for heparin but are equally active, have been identified. VEGF is a specific mitogen for endothelium, increases vascular permeability (its alternative name is vascular permeability factor, VPF), and induces proteolytic enzymes necessary for angiogenesis (Ferrara et al., 1992; Senger et al., 1993; Ferrara, 1995). Little is known about its regulation, but in a similar manner to erythropoetin it is upregulated by hypoxia via c-*src* activation (Shweiki et al., 1992; Goldberg and Schneider, 1994; Minchenko et al., 1994; Mukhopadhyay et al., 1995). Two tyrosine kinases, flt-1 (de Vries et al., 1992) and KDR (Terman et al., 1992) have been identified as high affinity receptors for VEGF. Flt-4 was also reported to be a VEGF receptor (Galland et al., 1992; Pajusola et al., 1992; Galland et al., 1993), although later studies demonstrated that flt-4 did not bind to VEGF (its ligand is currently unknown) (Pajusola et al., 1994). Its almost exclusive expression on lymphatics suggests that flt-4 might be a specific lymphatic marker and might play a role in tumor metastasis. *In situ* hybridization studies have demonstrated distinct expression profiles of VEGF and its receptors flt-1, flk-1 (mouse homologue for KDR) and flt-4. They showed specific spatio-temporal associations with blood islands and endothelial cells at all stages of development. Furthermore, although their pattern of expression was similar, the differences observed suggested that each has a specific biological function (Kaipainen et al., 1993; Millauer et al., 1993; Yamaguchi et al., 1993). This was recently confirmed by targeted mutations of flt-1 and flk-1 (KDR) in murine vascular development. A *flt-1* gene knockout caused poor capillary assembly, whilst a *flk-1* (KDR) knockout resulted in failure of blood island formation; both were incompatible with life, and death occurred *in utero* before 8 days *post coitum* (Fong et al., 1995; Shalaby et al., 1995).

Few studies have attempted to address which biological functions the different VEGF receptors might perform. The effects of VEGF appear to be mostly mediated by KDR. VEGF binding to transfected KDR but not flt-1 in porcine aortic endothelial cells caused cytoskeletal changes, chemotaxis and mitosis (Waltenberger et al., 1994). However, two isoforms of placenta growth factor ($PlGF_{131}$ and $PlGF_{152}$), a VEGF-related glycoprotein, are also ligands for flt-1 but not for KDR (Park et al., 1994) and although PlGF has little mitogenic or permeability activity for endothelium alone (K_d of 200 pM), it enhances the effects of low concentrations of VEGF. Thus, PlGF might be part of a redundancy mechanism by which angiogenesis can proceed with limited VEGF (Park et al., 1994) and angiogenic growth factors with different specificities for different receptors may act in concert with matrix elements such as heparan (Tessler et al., 1994) to modulate angiogenesis.

Expression of VEGF has been studied in many human carcinomas, including brain (Alvarez et al., 1992; Plate et al., 1992; Berkman et al., 1993; Weindel et al., 1994; Samoto et al., 1995), cervix (Guidi et al., 1995), ovary (Olson et al., 1994; Boocock et al., 1995), bladder and kidney (Brown et al., 1993a, Takahashi et al., 1994), gastrointestinal tract (Brown et al., 1993b), lung (Mattern et al., 1995), and lymph node (Dvorak et al., 1991), but very few studies describe the relevance of VEGF in breast carcinoma. Transplanting MCF-7 cells transfected with $VEGF_{121}$ in nude mice resulted in significantly faster growing tumors with a higher vascular density compared to wild-type cells (Zhang et al., 1995). *In situ* hybridization demonstrated a high level of VEGF mRNA in invasive ductal and *in situ* ductal (comedo-type) breast carcinoma and a low level in normal breast epithelium and invasive lobular carcinoma (Brown et al., 1992). In the same study *flt-1* and *KDR* mRNA were found in endothelium of small vessels adjacent to malignant cells in invasive ductal and DCIS, not in invasive lobular carcinoma and benign tissue. Using an immunohistochemical approach high VEGF expression was significantly correlated with a high vascular density and a reduced relapse-free survival in a multivariate analysis (Toi et al., 1994). Recently high expression of KDR, a VEGF receptor, has also been correlated with high vessel counts and advanced stage colon carcinomas (Takahashi et al., 1995).

Platelet derived endothelial cell growth factor (Pd-ECGF) has been shown to be thymidine phosphorylase (TP) (Barton et al., 1992; Furukawa et al., 1992; Moghaddam and Bicknell, 1992; Usuki et al., Finnis et al., 1993; Sumizawa et al., 1993) which catalyses the reversible phosphorylation of thymidine to deoxyribose-1-phosphate and thymine. It is reported as both chemotactic and mitogenic for endothelial cells and angiogenic in several model systems (Ishikawa et al., 1989; Haraguchi et al., 1994; Moghaddam et al., 1995). TP is not a classic growth factor and the mechanism by which TP promotes angiogenesis is unknown. Some evidence suggests that metabolites of TP might be responsible for its angiogenic activity (Morris et al., 1989; Moghaddam and Bicknell, 1992; Haraguchi et al., 1994). Using an immunohistochemical approach in 240 primary invasive breast carcinomas (PGF44c, mouse monoclonal) we found TP-expression in neoplastic cells often heterogenous and upregulated in both DCIS (Fig. 2A and B) and invasive (47% of all cases) elements. Tumor TP expression was usually both nuclear and cytoplasmic but occassionally only one of these was observed. Endothelial cell TP expression, although focal, was observed in 61% and, although positivity was also present within the tumor body, was most prominent at the tumor periphery (Fig. 2C) and associated with inflammatory cells, an area where tumor angiogenesis is most active (Fox et al., 1993 a), suggesting that TP also has an important role in endothelial cell metabolism. However, we observed no significant correlation between tumor TP expression and tumor vascularity. Although this might suggest that TP is not a significant angiogenic factor in breast carcinoma, TP could be important early in tumor angiogenesis through re-modelling of the existing vasculature. After this initial step other angiogenic factors might then assume more significance. The significant correlation of a high TP expression in a series of 75 DCIS cases with the presence of dense vascular rim around these lesions (Engels et al., unpublished data) and high TP expression in small tumors of low grade supports this sequence (Fox et al., 1995 c). Furthermore, this is in accordance with the increase in tumor size but not microvessel density in mouse xenografts of MCF-7 breast carcinoma cells transfected with TP over controls (Moghaddam et al., 1995). In both instances TP appears to alter the rate of vascularisation, consistent with TP being chemotactic but non-mitogenic for endothelium (Ishikawa et al., 1989; Miyazono et al., 1991; Haraguchi et al., 1994). In necrotic tumor regions where tumor cell immunoreactivity was absent, upregulation of TP in nonneoplastic elements was observed. Thus, although presently not demonstrated, TP, like vascular endothelial growth factor, might also be modulated by hypoxia (Shweiki et al., 1992). Nevertheless, in tumor cell lines it has been shown that tumor necrosis factor-α, interleukin-1 and interferon-γ upregulate TP (Ho et al., 1990; Eda et al., 1993) and therefore *in vivo* through autocrine and paracrine pathways, tumors might directly regulate their TP expression. Furthermore, cytokines might also recruit macrophages (O'Sullivan et al., 1993) rich in TP which may themselves also augment tumor cell TP through paracrine loops. Dedifferentiated tumors which show loss of cognate receptors (e.g. interleukin-4) would be unable to use these networks, which would also account for the low levels of TP observed in high grade tumors.

Although we observed no significant reduction in relapse-free survival (RFS) or overall survival (OS) in TP positive patients, in node positive patients RFS was significantly higher and OS showed a borderline significance with TP positive patients. This may be due to TP modulating the sensitivity of tumor cells to drugs widely used in adjuvant therapy in this patient group. It might both metabolise 5-fluorouracil to its active form and also, by degrading thymidine, enhance the sensitivity of the tumor cells to methotrexate. The former has been demonstrated *in vitro* (Patterson et al., 1994).

Thrombospondin-1 (TSP-1) is a 170 kD matrix glycoprotein which is secreted by several cell types including endothelium. It modulates platelet aggregation, wound healing, cell adherence and inhibits angiogenesis *in vivo* and *in vitro* (Good et al., 1990; Weinstat and Steeg, 1994). It is one of five members of a gene family which, by alternative splicing, post trans-

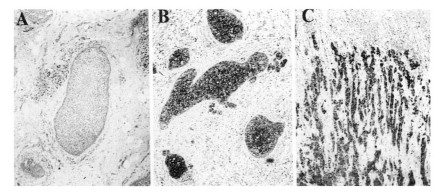

Figure 2. TP expression in breast carcinomas. DCIS lesions negative (A) and positive (B) for TP. (C) Invasive ductal breast carcinoma with higher TP expression at the invading border. Staining in all with monoclonal antibody PG44 c.

lational modification, and multimodular structure, has a diversity of effects, including on angiogenesis (Weinstat and Steeg, 1994). It is secreted by tumor cells and inflammatory cells like macrophages (DiPietro and Polverini, 1993).

The loss of an unknown oncogene in BHK hamster kidney cells is followed by a downregulation of the *TSP-1* gene and the secretion of TSP-1 below levels sufficient to inhibit angiogenesis (Rastinejad et al., 1989; Good et al., 1990). Furthermore, in fibroblasts from Li-Fraumeni patients (these individuals have an increased incidence of carcinomas due to the inheritance of a mutant p53 allele), the loss of the remaining p53 coincided with a reduced expression of TSP-1 and an angiogenic phenotype, re-introduction of wild-type p53 increased TSP-1 and produced an anti-angiogenic phenotype (Dameron et al., 1994 a, b). In some tumor cell lines malignant progression and angiogenesis are associated with low TSP-1 expression (Weinstat and Steeg, 1994; Zabrenetzky et al., 1994). In the human breast carcinoma cell line BT549 the reintroduction of p53 increases TSP-1 secretion and the cells lost their angiogenic phenotype. When TSP-1 was neutralised, angiogenic activity and bFGF-induced migration, previously blocked by TSP-1, was detectable (Volpert et al., 1995). The formation of hybrids from MCF-7 cells with nontumorigenic immortal mammary cells which expressed TSP-1 resulted in nontumorigenic cells lacking TSP-1 expression (Zajchowski et al., 1990). The forced expression of TSP-1 decreased *in vivo* tumor growth in human breast carcinoma cell lines. In breast tissue TSP-1 mRNA could be found in high concentration in normal ductal cells (Clezardin et al., 1993), whereas it was rarely detected in tumor cells from invasive ductal carcinomas, although TSP-1 protein could be found in neighbouring stromal cells (Wong et al., 1992; Clezardin et al., 1993; Tuszynski and Nicosia, 1994).

Adhesion molecules

Role of adhesion molecules
Intercellular relations are mediated by endothelial cell adhesion molecules
(CAMs) including selectins, integrins and immunoglobulin superfamilies,
which have physiological roles in immune cell trafficking and localisation
of the inflammatory response and also play a major role in angiogenesis
(Springer, 1990; Hogg and Landis, 1993). They maintain the integrity of
the normal breast gland. Changes in their expression are necessary for
invasion and angiogenesis. For the angiogenic process three types of
cellular interactions can be differentiated: first, plasma membranes of
adjacent endothelial cells interact during the extension of capillary blood
vessels; second, since the circumference of a capillary blood vessel consists of a single endothelial cell, intracellular adhesion between plasma
membrane regions of a single cell occurs during the formation of a vessel
lumen; third, the interaction between proliferating and migrating endothelial cells and extracellular matrix provide a scaffold for the growing
vessel.

The regulation of CAMs on normal endothelium has been extensively
studied *in vitro* and it has become apparent that several properties of
CAMs may affect their function. Some are constitutively expressed
(ICAM-2 and CD31) (de Fougerolles et al., 1991) whilst others are induced by cytokines (e.g. ICAM-1, VCAM-1 and E-selectin) (Pober, 1988).
The pattern of CAM expression also varies within different vascular beds
(Kuzu et al., 1992), reflecting the functional heterogeneity of endothelium.
Several of the major CAMs and their ligands are additionally subject to
post-translational modification such as glycosylation (Zetter, 1993). This
may further modulate their function *in vivo*, although there is yet little
evidence for site specific variation in glycosylation patterns. CAM
expression *in vivo* has been examined in several non-neoplastic diseases
(Steinhoff et al., 1991; van der Wal et al., 1992; Volpes et al., 1992;
Roy et al., 1993) with only data in a limited number of tumors (Denton
et al., 1992; Zocchi et al., 1993; Nelson et al., 1994).

Adhesion molecules in human tumors
Using an immunohistochemical approach, we examined the frequency and
pattern of expression of PECAM (CD31), ICAM-1 (CD54), ICAM-2,
(CD102) ICAM-3 (CD50), VCAM-1 (CD106), E-(CD62E, ELAM-1)
and P-selectins (CD62P, PADGEM, GMP-140) in a series of primary
invasive breast carcinomas (n = 64) and compared these with samples of
normal breast tissue (n = 14), with the aims of defining the phenotype of
tumor endothelium and assessing the induction of these molecules in the
various cellular elements within the tumor. Endothelium in normal breast
tissue showed widely distributed constitutive expression of the lineage-
specific markers PECAM, ICAM-2 and P-selectin, but more focal and

weak expression of ICAM-1 and E-selectin; no ICAM-3 or VCAM-1 expression was observed. ICAM-1 was only expressed by myoepithelial cells and no other cells in the breast demonstrated immunoreactivity for other markers.

In the breast tumors we observed a significant upregulation of ICAM-3 and E-selectin in breast tumor endothelium which was not restricted to any particular vessel type. Furthermore, in contrast to the pan-endothelial staining of PECAM and ICAM-2, we observed both upregulated and preferential endothelial cell expression of the selectins at the periphery of the tumor in 65% of cases, a pattern similar to previous reports of selectin expression in lung and renal cell carcinomas (Zocchi et al., 1993). This upregulation of selectins at the tumor periphery may have several roles. It has been suggested that the selectins are essential for capillary morphogenesis (Nguyen et al., 1993a; Kaplanski et al., 1994), and at the invading tumor margin neovascularization and endothelial cell division are most active (Fox et al., 1995b). Antibodies against bovine E-selectin but not P-selectin could inhibit tube formation in a two-dimensional assay (Bischoff, 1995). Furthermore, P-selectin might also be important in tumor progression by promoting fibrin deposition via interaction with leukocytes, thereby enhancing tumor stroma formation (Dvorak, 1987; Palabrica et al., 1992).

As an early event in most solid tumors before in-growth of the neovasculature, fibrinogen extravasates through leaky microvasculature under the influence of VEGF (Senger et al., 1993). Through the procoagulant properties of both the tumor cells and host tissue, fibrinogen is converted to fibrin which helps generate the tumor matrix. Fibrin itself is angiogenic, and together with other angiogenic factors released by the tumor results in an influx of endothelium and accessory cells leading to tumor angiogenesis.

Thus CAMs, like many others, probably perform several functions depending on their immediate environment. Tumor cell endothelial adhesion and transmigration, in analogy with lymphocyte-endothelial interactions, could be enabled and would be advantageous to the tumor by favouring metastasis (Springer, 1990). There is evidence that cytokines and/or growth factors originating from neoplastic and inflammatory cells (Ruco et al., 1990; Rubbert et al., 1991) modulate the expression of selectins and their ligands in different experimental systems (Aruffo et al., 1992; Roberts and Bevilacqua, 1994). This concept was supported by our observation of peripheral accentuation of both selectins in the presence and absence of inflammatory cell infiltrates. Furthermore, up-regulation of these CAMs by tumor autocrine or paracrine loops may confer a growth advantage and contribute to changes to an invasive phenotype. This might be augmented by recruited inflammatory cells, (O'Sullivan et al., 1993), platelets or components of the coagulation system.

Although increased expression of ICAM-2 has been described in endothelial cells in lymphomas (Renkonen et al., 1992), immunoreactivity in the endothelial cells of our normal breast specimens was too intense to detect any upregulation. Furthermore, we observed only focal and weak immunoreactivity of VCAM-1 in 10% of breast tumor endothelium, a significant difference to that reported for lung (7/8) and renal cell carcinomas (6/8) using the same antibody (Zocchi et al., 1993). This would argue against ICAM-2 or VCAM-1 playing a major role in breast carcinoma metastasis, as it may do in other tumor types, and supports the concept that differential expression of CAMs in tumors and the vascular beds of various tissues might account for site specific metastasis (Zetter, 1993). Differences between the major CAMs involved in development and metastasis in different tumor types may have important implications for designing potential treatment strategies.

On tumor cells we confirmed previous reports of expression of ICAM-1 (Denton et al., 1992), but we also identified ICAM-3 not only in lymphocytes (Fawcett et al., 1992) but also in endothelial cells and tumor cells. The induction of ICAM-3 on endothelium has been observed in lymphomas (Doussis et al., 1993) but has not previously been recorded in carcinomas. In one case tumor cell PECAM expression was seen; like ICAM-3, PECAM is not an antigen previously observed on tumor cells of epithelial origin *in vivo*. PECAM can act as a homotypic and heterotypic adhesion molecule in inflammation (Vaporiciyan et al., 1993) and may mediate tumor-endothelial cell adhesion in human neoplastic cell lines (Tang et al., 1993). However, the frequency of PECAM expression in our series of human tumors would suggest this adhesion molecule is important in only a small subset of breast tumors. The expression of endothelial cell- or haemopoietic-associated markers on tumor cells might be due either to a general or specific dysregulation of surface molecule expression. In the latter case, it might be one strategy that tumors use both to evade host surveillance and to regulate their adhesive properties. Furthermore, the increase in levels of soluble CAMs reported in many tumors (Banks et al., 1993) suggests their cellular origins are not solely endothelial.

Metalloproteinases and inhibitors
Metalloproteinases (MMP) and their inhibitors (TIMPs) are believed to play an independent role in controlling extracellular matrix remodelling (Alexander and Werb, 1991; Matrisian, 1992; Birkedal et al., 1993). TIMP-1 mRNA was detected in cells of breast (Polette et al., 1993b) and head and neck (Polette et al., 1993a) carcinomas. TIMP-2 transcripts were localised in stromal cells of breast (Poulsom et al., 1993), colorectal (Poulsom et al., 1992), and head and neck (Polette et al., 1993a) carcinomas. TIMP-3 was initially described in breast carcinoma cells (Uria et al., 1994), but it was recently shown to be predominantly expressed by fibroblasts within the bed of these tumors and was more intense close to carcinoma cells, which were found to express the TIMP-3 gene infrequently (Byrne et al., 1995). These expression patterns indicate an

area of intense proteolysis close to carcinoma cells which is regulated by the simultaneous release of proteolytic enzymes and their inhibitors.

Integrins
Integrins seem to play a role in endothelial cell migration and tube formation. The integrin heterodimers $\alpha_2\beta_1$, a receptor for collagens and laminin, and $\alpha_5\beta_1$, a receptor for fibronectin, have been found to regulate the integrity and permeability of a HUVECS monolayer (Lampugnani et al., 1991). Ca^{++}-independent migration of HUVECS on collagen was mediated by $\alpha_2\beta_1$ (Leavesley et al., 1993) and blocking its function with specific antibodies enhanced tube formation on collagen by blocking binding of $\alpha_2\beta_1$ to collagen (Gamble et al., 1993). The integrin $\alpha_v\beta_3$ was involved in Ca^{++}-dependent signalling and endothelial cell migration on vitronectin (Leavesley et al., 1993). Specific antibodies against $\alpha_v\beta_3$ significantly inhibited migration of HUVECS on vitronectin, but had no effect on collagen-dependent migration (Leavesley et al., 1993). Using the CAM assay antibodies against $\alpha_v\beta_3$ induced apoptosis in proliferative vascular endothelial cells (Brooks et al., 1994b) and thereby inhibited normal vessel growth and bFGF-stimulated or tumor-induced angiogenesis while leaving pre-existing vessels unaffected (Brooks et al., 1994a). Using a mouse model, angiogenic activity and growth of human breast carcinoma cells could be inhibited via systemic application of antibodies against $\alpha_v\beta_3$ (Brooks et al., 1995). These initially contradictory results might be explained by a different interaction of $\alpha_v\beta_3$ with collagen or vitronectin gels and the chorioallentoic membrane. $\alpha_v\beta_3$ is expressed on growing vessels and necessary for EC migration during angiogenesis; it is absent on normal mature vessels (Brooks et al., 1994a).

Finale

Tumor development is a neverending story. The continuously changing environmental conditions within and around the primary tumor and its secondaries will cause continuous reactions such as release of angiogenic and other growth factors, resulting in a further selection of more resistant and more aggressive carcinoma cells until the defense mechanisms of the host and the possibilities of the therapist are outrun.

Angiogenesis as a prognostic marker in breast carcinoma

Assessment of angiogenesis

Introducing angiogenesis into routine medicine requires reliable assessment techniques. Reproducible and affordable diagnostic procedures have to be established to produce prognostic information and to control the effects of anti-angiogenic drugs; some are currently in clinical trials (Scott and Harris, 1994). In the future angiogenic activity might provide a screening tool for some malignant diseases (e. g. ovarian carcinoma), give additional criteria for therapeutic decisions (e. g. anti-angiogenic neoadjuvant, adjuvant, and antimetastatic therapy), and be used for follow-up control of carcinoma patients.

Assessment of angiogenesis can be performed in tissue after biopsy and *in vivo*. In tumor biopsies vascular parameters like vascular density are widely used. Since this criteria gives only a single frame in the dynamic multistep process of angiogenesis the assessment of biochemical and physiological criteria might give other useful information. The expression of angiogenic and other growth factors and adhesion molecules, and the ratio of newly formed vessels to old vessels are currently being studied. The detection of (anti-)angiogenic factors in blood, urine, liquor, etc. might give information about the systemic effect of tumors. Tumor blood flow as a result of vascularity, altered vascular architecture, intratumoral tissue pressure, and host immune mechanisms can be assessed using imaging techniques.

All the above are used in studies, but optimisation is still necessary and no agreement has been reached on which technique is most useful in clinical practise or correlates with therapeutic response or outcome.

Methods for assessing tumor angiogenesis in tumor sections

Most studies have employed a method based on the pioneering work of Weidner et al. (1991). Tinctorial stains or an immunohistochemical approach have been used to highlight tumor blood vessels and microvessel density is then assessed in the most vascular areas of the tumor. These studies have shown that increased microvessel density as a measure of angiogenesis is a powerful prognostic tool in many human tumor types (Srivastava et al., 1988; Weidner et al., 1991; Barnhill et al., 1992b; Bosari et al., 1992; Horak et al., 1992; Wakui et al., 1992; Bigler et al., 1993; Toi et al., 1993; Weidner et al., 1993; Albo et al., 1994; Brawer et al., 1994; Bundred et al., 1994; Dickinson et al., 1994; Jaeger et al., 1994; Li et al., 1994; Macchiarini et al., 1994; Ogawa et al., 1994; Olivarez et al., 1994; Porschen et al. 1994; Simpson et al., 1994; Smith-McCune and Weidner, 1994; Vesalainen et al., 1994; Wiggins et al., 1995). Nevertheless, probably due to differences and problems in methodologies, not all investigators have been able to confirm that highly vascular tumors are associated with a reduction in survival (Hall et al., 1992; Van Hoef et al., 1993; Barnhill et al., 1994; Sightler et al., 1994; Vesalainen et al., 1994; Axelsson et al., 1995; Ohsawa et al., 1995; Rutgers et al., 1995). These differences highlight the problems originally recognised by Brem et al. (1972) with the MAGS score, who demonstrated that the quantitation of tumor angiogenesis is still limited by the methods used for capillary identification and quantitation.

The major considerations in quantifying tumor angiogenesis in tissue sections are 1) the choice of endothelial antibody, 2) the selection of tumor areas for assessment, 3) the vascular parameter measured, 4) methods for assessment, and 5) the cut-off used in correlation analysis with other clinicopathological variables and survival.

Selection of endothelial antibody
Studies vary in the method used to highlight tumor capillaries. Some have used tinctorial stains (Protopapa et al., 1993) whilst other have employed immunohistochemistry. It is well recognised that tinctorial stains only pick up a small proportion of the total tumor vasculature but it is less appreciated that endothelium is highly heterogenous (McCarthy et al., 1991; Kuzu et al., 1992). Thus the choice of antibody will influence the number of microvessels available for assessment. Initially, when selecting for the most reliable and sensitive endothelial antibody, we observed a significant variation in the number of highlighted vessels in hot spots (Horak et al., 1992). We have found that JC70, an anti-CD31 antibody, recognises the most capillaries and have obtained reliable microvessel immunostaining in routinely handled formalin-fixed paraffin-embedded tissues (Parums et al., 1990; Horak et al., 1992; Dickinson et al., 1994; Fox et al., 1994).

We have also assessed Factor VIII related antigen (FVIIIrAg) which has been used in many studies (Macchiarini et al., 1992; Ottinetti and Sapino, 1988; Bigler et al., 1993; Bosari et al., 1992; Bundred et al., 1994; Hall et al., 1992; Li et al., 1994; Ogawa et al., 1994; Sahin et al., 1992; Van Hoef et al., 1993; Weidner et al., 1991, 1992). This identifies only a proportion of vascular endothelium and therefore underestimates the true extent of the microvasculature. Moreover, antibody to FVIIIrAg, unlike JC70, also recognises lymphatics (McCarthy et al., 1991; Kuzu et al., 1992). Nevertheless, when FVIIIrAg and CD31 staining have been compared, both have provided prognostic information (Horak et al., 1992; Toi et al., 1993). Furthermore, there is a significant correlation between the mean and median vascular counts for both antigens (Horak et al., 1992; Weidner et al., 1992; Toi et al., 1993; Fox et al., 1994; Weidner and Gasparini, 1994) which suggests that differences in published data are mainly due to the different methods of quantifying tumor vascularity. A good alternative endothelial marker is anti-CD34 (Li et al., 1994).

Other sub-optimal markers which have also been employed are antibodies against vimentin (Wakui et al., 1992), lectin (Svrivastava et al., 1988; Carnochan et al., 1991; Barnhill et al., 1992), alkaline phosphatase (Mlynek et al., 1985), and type IV collagen (Visscher et al., 1993; Vesalainen et al., 1994; Visscher et al., 1994), which all suffer from lower specificity and sensitivity.

An alternative approach would be not to highlight all endothelium but to identify selectively only the vasculature induced by the tumor. This might be valuable not only in more accurately quantifying tumor angiogenesis but might also have important implications for anti-vascular targeting (Burrows and Thorpe, 1994). Several antibodies have been identified which recognise antigens that have been reported to be upregulated in tumor-associated endothelium compared to normal tissues, and include EN7/44, endoglin, endosialin and E-9. However, to date no studies assessing these antibodies have been performed.

Tumor area to assess
In selecting the area of tumor for quantification some studies have examined random fields (Protopapa et al., 1993), others have measured average vascularity (Svrivastava et al., 1988; Carnochan et al., 1991), but most have assessed three angiogenic "hot spots" (Weidner et al., 1991; Weidner et al., 1992). We use the hot spot technique since we believe it is these angiogenic areas that are likely to be biologically important.

However, the number of hot spots counted varies from 1 to 5 (Weidner et al., 1991; Barnhill et al., 1992 a; Hall et al., 1992; Sahin et al., 1992; Weidner et al., 1992; Van Hoef et al., 1993; Sightler et al., 1994) and the tumor field area which is currently assessed ranges from $0.12-0.74$ mm^2. The tumor field area measured will significantly effect the vascular index since too small an area will always give a high vascular index and too large an area will dilute out the "hot spot". Similarly, tumors have a limited number of hot spots, which would be diluted by counting a large number of tumor fields. Therefore, practical considerations aside, we examine three hot spots.

Vascular parameter to measure
The majority of studies quantifying tumor angiogenesis have measured microvessel density (Svrivastava et al., 1988; Carnochan et al., 1991; Weidner et al., 1991; Barnhill et al., 1992 a; Bosari et al., 1992; Horak et al., 1992; Macchiarini et al., 1992; Weidner et al., 1992; Bigler et al., 1993; Toi et al., 1993; Weidner et al., 1993; Bundred et al., 1994; Jaeger et al., 1994; Li et al., 1994; Macchiarini et al., 1994; Ogawa et al., 1994; Olivarez et al., 1994; Porschen et al., 1994; Sightler et al., 1994; Visscher et al., 1994; Weidner and Gasparini, 1994; Weidner et al., 1994). However, there are several problems associated with this technique making it unsuitable for a diagnostic pathology service. The counting itself is time consuming, particularly when the tumor is of high vascularity and there is significant

inter- and intra-observer variation. The latter has recently been illustrated in the study of Axelsson et al. (1995) who, after an initial training period with Weidner, did not observe any correlation between microvessel density and clinicopathological variables including survival. Although the experience of the observer has been shown to alter estimates of microvessel density significantly (Barbareschi et al., 1995), in our laboratory highly trained observers sometimes disagreed on counts even with strict adherence to guidelines. Furthermore, microvessel density might not be the most important vascular parameter; a large vascular perimeter or area might be a better measure of angiogenesis since it may reflect the endothelial surface and volume of blood available for interaction with the tumor.

We have examined different vascular parameters such as microvessel density, luminal area and perimeter (Fox et al., 1995a). We observed significant correlations between microvessel number, luminal perimeter and area (Simpson et al., 1994), suggesting that all these indices are satisfactory for quantifying tumor vascularity.

Methods for assessment

Due to the laborious nature of counting microvessel density and problems with its reproducibility, together with the conceptual difficulty of adjacent microvessels with significantly different vascular parameters assuming equal importance with the set criteria (Weidner et al., 1991; Weidner et al., 1992) we have tried to develop more rapid and objective methods to quantify tumor angiogenesis.

We and others have attempted to automate the procedure by using computer image analysis systems (Svrivastava et al., 1988; Wakui et al., 1992; Bigler et al., 1993; Visscher et al., 1993; Wesseling et al., 1993; Brawer et al., 1994; Furusato et al., 1994; Simpson et al., 1994; Williams et al., 1994; Barbareschi et al., 1994; Charpin et al., 1995; Fox et al., 1995a). This should improve reproducibility and overcome the observer variation associated with "manual" counting, and several of the above studies have used different endothelial markers and vascular parameters to confirm that quantitative angiogenesis gives independent prognostic information.

However, these systems also have several drawbacks, not least capital and running costs. As with manual assessment, technical and computer software considerations are paramount. An endothelial marker which will give sensitive and specific capillary staining should be used to reduce background signal. Those directed against basement membrane epitopes, lectins or intermediate filaments are sub-optimal since these will stain other tumor elements. The softwear employed to analyse the staining is another important factor. Most systems are not automated and require a high degree of operator interaction and thus, like manual counting, suffer from observer bias. No software is available for identifying hot spots, but when developed it will necessitate a motorized stage, adding to cost. Partially automated systems with area and shape filters using defined color tolerances are available, which require reduced human control. We have tried several methods to reduce operator control including a "pixel analysis" of the stained area. This involves selection by the computer of regions of the raw acquired image within set color tolerances to that of the set staining color. Unfortunately, due to the low tolerance threshold required the computer could not accurately quantitate the tumor vasculature. In our experience a computer image analysis system is costly, time consuming, unsuitable to routine diagnostic practice and no more accurate than a trained observer.

We have also assessed a system based on a microscope eyepiece graticule to overcome the time and observer variation problems associated with counting microvessels. After scanning the stained tumor section at low power ($\times 10 - 100$) to identify the three most vascular fields we have used a 25 point Chalkley eyepiece graticule (Chalkley, 1943) at high power ($\times 250$) over each hot spot. The eyepiece graticule is oriented so the maximum number of points at $\times 250$ (0.155 mm^2) are on or within areas of highlighted vessels. Thus interpretation difficulties of individual counts are bypassed and no assumptions are made as to a particular size or shape of vessel; this method is rapid ($2-3$ minutes) and reproducible. Chalkley counts give independent prognostic information in breast carcinoma patients (including the node negative subgroup) (Fox et al., 1994; Fox et al., 1995a) and in invasive transitional carcinomas of the bladder (Dickinson et al., 1994).

To simplify the process further, we are also assessing a vascular grading system based on the subjective appraisal by trained observers over a conference microscope. Weidner et al. (1991, 1992) used a similar system but stratified tumors into four groups. We were unable to reproduce this work but could define three groups of low, medium and high vascular categories. We have found excellent correlations of vascular grades with both microvessel density ($p = 0.002$) and Chalkley count ($p = 0.0001$), suggesting that this method can be used to quantify tumor angiogenesis (Fox et al., 1995a). We are currently validating this method to quantify tumor angiogenesis in a large series of randomised patients to determine its prognostic utility.

Criteria for assessment

A further confounding source of variation in published studies is the value used for stratification into different prognostic groups. This alone would result in different conclusions being drawn from the same data. Studies have used the highest and the mean count in node-negative patients with recurrence (Bosari et al., 1992), or variable cut-offs given as a function of tumor area (Weidner et al., 1992; Van Hoef et al., 1993) or microscope magnification (Sahin et al., 1992; Protopapa et al., 1993; Toi et al., 1993; Bundred et al., 1994; Guinebretiere et al., 1994; Olivarez et al., 1994). We have used the median (Horak et al., 1992; Dickinson et al., 1994;

Fox et al., 1994) or tertile groups (Fox et al., 1995a) since such categorisation avoids strong assumptions about the relationship between tumor vascularity and other variables including survival.

The above difficulties demonstrate the urgent need for a randomized prospective study using agreed methodological standards with strict guidelines. There are still reports published using a variety of sub-optimal endothelial markers, selection procedures and counting methods making comparisons between studies difficult. Perhaps it is a testimony to the prognostic power of this technique that even using a variety of techniques, studies still demonstrate that tumors with high vascular indices have a significantly poorer prognosis.

Detection of angiogenic factors and receptors

There are numerous reports documenting upregulation of several angiogenic factors and their receptors at the mRNA and protein level using a variety of techniques in a range of histological types (Zagzag et al., 1988; Gomm et al., 1991; Alvarez et al., 1992; Wong et al., 1992; Daa et al., 1993; Schultz-Hector and Haghayegh, 1993; Zarnegar and DeFrances, 1993; Anandappa et al., 1994; Garver et al., 1994; Reynolds et al., 1994a and b; Brown et al., 1995; Guidi et al., 1995; Janot et al., 1995; Moghaddam et al., 1995; Takahashi et al., 1995; Visscher et al., 1995). However, only a few have correlated these data to clinicopathological parameters or survival.

We have also examined the angiogenic factor thymidine phosphorylase by immunohistochemistry in 328 breast carcinomas. Although a significant correlation between tumor vascularity and thymidine phosphorylase expression in 100 breast carcinomas has been reported (Toi et al., 1995) we could not demonstrate a similar a relationship in 185 (Fox et al., 1995c).

The angiogenic factors bFGF and VEGF have been measured in patient serum (Fujimoto et al., 1991; Kondo et al., 1994), urine (Nguyen et al., 1993 and 1994; O'Brien et al., 1995), and cerebrospinal fluid (Li et al., 1994) in various tumor types. These have shown that bFGF is higher in tumors than controls and that increasing bFGF is associated with advanced tumor grade and stage. Similarly, soluble tyrosine kinase receptors such as flt-1 (Kendall and Thomas, 1993) or factor VIII related antigen (Blann and McCollum, 1994) might give an indication of the angiogenic activity of a tumor.

Although many of the above studies have shown significant relationships between angiogenic factor expression as a measure of tumor angiogenesis and patient survival, none of the current techniques are sensitive or specific enough to use for quantifying tumor angiogenesis. This is because angiogenesis is the result of a balance of positive and negative regulators and it is likely that different tumor types use several different angiogenic factors at different times during their development. Therefore even measuring several angiogenic stimulators or inhibitors might still not give an accurate assessment of the angiogenic activity of individual tumors. Nevertheless, in parallel with our growing understanding of angiogenic factors and the identification of specific profiles for individual tumor types, their measurement might play an increasing role in quantitative tumor angiogenesis.

Proteolytic enzymes

Measurement of proteases and their inhibitors, particularly components of the urokinase system, might give some indication of the angiogenic and invasive activity of a tumor and provide independent prognostic information. Several studies have shown that elevated levels of uPA and PAI-1 are associated with a poor prognosis in several tumor types including breast (Duffy et al., 1990; Schmitt et al., 1990; Sumiyoshi et al., 1991; Foekens et al., 1992; Spyratos et al., 1992; Grondahl et al., 1993; Jänicke et al., 1993; Ganesh et al., 1994; Duggan et al., 1995) and colon (Ganesh et al., 1994). Current assays using homogenates of primary tumors have shown potential, but it might be possible to measure PAI-1, uPA and its receptor in serum (Nils Brunner, personal communication).

Cell adhesion molecules

Tumor stroma including microvasculature is not just an inert support for the neoplastic element but, via unique structural features, it is a major regulator of many neoplastic processes including angiogenesis (Iozzo, 1995; Lochter and Bissell, 1995). The tumor co-ordinates the necessary cell-cell interactions for its development using a variety of molecules including several families of cell adhesion molecules (CAMs). Clinical studies showing that melanoma patients with upregulated selectins on endothelium have a significantly worse prognosis validate the interest in CAMs and their cognate ligands in tumor angiogenesis (Schadendorf et al., 1995) and soluble CAMs are readily identified in sera of carcinoma-bearing patients (Banks et al., 1993; Kageshita et al., 1993). However, the use of CAMs is not limited to its potential as a prognostic marker but might also have other important therapeutic applications (Brooks et al., 1994, 1995).

Tumor vascular architecture

The vascular morphology of tumors is different within tumors of similar and different histological types (Warren, 1979). Vascular architecture also varies depending on the anatomical site and host (Lauk et al., 1989; Paku and Lapis, 1993). However, in all these patterns the newly formed vessels are abnormal in several respects. The tumor vasculature can be dilated, saccular, sinusoidal and tortuous and it may contain multiple bifurcations, loops, and blind-ending sprouts (Warren, 1979). It has been suggested that particular vascular patterns might help distinguish benign from malignant lesions (Smolle et al., 1989; Cockerell et al., 1994) and also be a prognostic marker; in ocular melanomas, a closed back to back loop vascular pattern was associated with death from metastasis (Folberg et al., 1993).

Imaging

It is desirable to be able to perform repeated observations on human tumors non-invasively *in vivo* to assess novel anti-angiogenic agents, to correlate these measurements with pathology and to design appropriate treatment before surgery. The presently available *in vivo* techniques are mostly based on changes in tumor blood flow. Like the vascular architecture of tumors, tumor blood flow also shows significant variation (Vaupel et al., 1989). The abnormal microvessel morphology causes arteriovenous shunting of blood, thus bypassing the usual capillary nutrient exchange (Vaupel et al., 1989). This has been estimated to account for at least 30% of tumor blood flow in animal models, with similar shunts reported for some human tumors. In addition, blood flow is intermittent, with periods of reverse flow or stasis (Vaupel et al., 1989; Jain, 1994). Furthermore, a combination of high tumoral interstitial pressure, altered vessel rheology (which enhances erythrocyte sludging and platelet aggregation resulting in increased blood viscosity) also compromises blood flow (Vaupel et al., 1989; Jain, 1990). These flow defects are not uniform but occur in different tumor regions at different times. Moreover, since the new tumor microvessels are not innervated (Mitchell et al., 1994) and demonstrate abnormal responses to vasoactive stimuli they are unable to respond to the demands of the growing tumor (the average tumor perfusion rate also decreases as tumors enlarge). All these defects in perfusion explain why the average perfusion rate for some tumors is sometimes less than that observed in its normal tissue counterpart (Nystrom et al., 1969) and account for the common occurrence of necrosis in many tumors. Nevertheless, the perfusion rate of poorly differentiated tumors is often higher than of well differentiated tumors (Vaupel et al., 1989). Thermography, which was used in the past as a screening technique to detect hyperaemic regions within the breast, was possibly the first angiogenesis test.

It might be possible in the future to assess metabolic effects and functions of cellular enzymes using positron emission tomography (PET).

Color doppler ultrasound

The first studies using color doppler ultrasound to examine breast lesions gave inconsistent results and excluded routine application (Burns et al., 1982; Bamber et al., 1988; Jackson, 1988; Jellins, 1988; Srivastava et al., 1988; Britton and Couldon, 1990; Cosgrove et al., 1990; Dixon et al., 1992; Cosgrove et al., 1993; Dock, 1993). Recent studies could demonstrate significant differences in blood flow (Peters-Engl et al., 1995) and vascular parameters between benign and malignant lesions, and showed that vascular parameters differed significantly between high and low blood flow lesions (Grischke et al., 1994). But comparison between studies is still difficult due to the use of different ultrasound units and scanning techniques.

Despite the current drewbacks color doppler is very promising, especially in carcinomas which are difficult to reach with other screening techniques such as ovarian carcinomas (Wu et al., 1994; Antonic and Rakar, 1995). Furthermore, in ovarian tumors, blood flow demonstrated a significant relationship to tumor levels of the angiogenic factor TP (Reynolds et al., 1994a and b). Accuracy will be further improved by high resolution scanners. It is easy to learn and use and relatively cheap. Further advantages are the lack of radiation or need for contrast medium with the risk of allergic reaction. Its application can be seen in breast carcinoma screening where mammography gives doubtful results, in preoperative diagnosis, and in intra- and post-operative control.

Magnetic resonance imaging (MRI)

The high sensitivity of MRI in combination with the intravenous injection of Gadolinium as a contrast medium in recognising breast carcinomas has been demonstrated (Heywang et al., 1993; Kaiser, 1993; Kaiser et al., 1993; Heywang, 1994; Heywang et al., 1994). Although specific contrast enhancement patterns have been shown to be correlated with invasive carcinomas, the vascular changes causing these have still to be determined. It is unlikely that vascular density alone causes the typical contrast enhancement patterns, vascular shunts being responsible for 30% of tumor perfusion (Vaupel et al., 1989) and irregular and immature vascular architecture could be more important. Frouge et al. (1994) showed a correlation between MRI results and angiogenesis represented by vessel counts in tissue sections in a small number of patients. Technical difficulties might be responsible for the small number of studies comparing angiogenic activity with MRI results. Particularly troublesome is the direct comparison between the radiological planes and the histological sections.

The drawbacks of this technique include high costs, limiting its application to a few centres and patients where alternative techniques do not give satisfactory results. Furthermore, the intravenous use of contrast medium might be difficult in some cases.

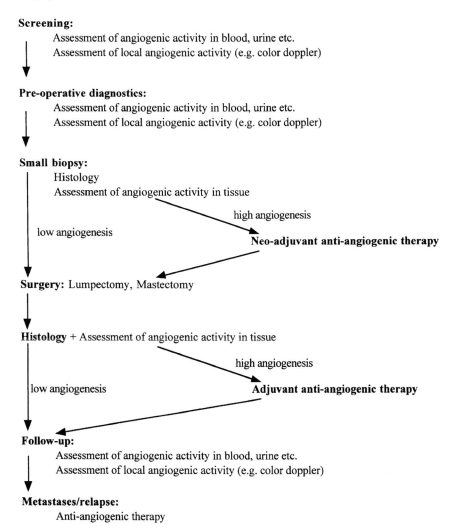

Screening:
  Assessment of angiogenic activity in blood, urine etc.
  Assessment of local angiogenic activity (e.g. color doppler)

Pre-operative diagnostics:
  Assessment of angiogenic activity in blood, urine etc.
  Assessment of local angiogenic activity (e.g. color doppler)

Small biopsy:
  Histology
  Assessment of angiogenic activity in tissue

                                          high angiogenesis

  low angiogenesis
                              Neo-adjuvant anti-angiogenic therapy

Surgery: Lumpectomy, Mastectomy

Histology + Assessment of angiogenic activity in tissue

                                          high angiogenesis

  low angiogenesis   **Adjuvant anti-angiogenic therapy**

Follow-up:
  Assessment of angiogenic activity in blood, urine etc.
  Assessment of local angiogenic activity (e.g. color doppler)

Metastases/relapse:
  Anti-angiogenic therapy

Figure 3. Role of angiogenesis in diagnosis and therapy

Others
Other techniques for assessing tumor angiogenesis which have not been fully evaluated include positron emission tomography (PET) and magnetic resonance spectroscopy (MRS). Scanning of patients using tumor endothelium-associated antigens could give *in vivo* information about growth factor receptors and possibly help to differentiate between normal and tumor-associated vessels. Development of appropriate markers and improvement of scanners are crucial for routine clinical application.

Currently at an experimental stage, techniques such as PET are very expensive and available only in a few centres.

Conclusion

The development of angiogenesis is a key step during carcinogenesis. The optimised assessment of angiogenesis provides information for prognosis and therapy decisions and will be important in future therapy plans, which include anti-angiogenic therapy options (Fig. 3).

Acknowledgements
The work described in this review has been funded by the Imperial Cancer Research Fund (ICRF) and the Deutsche Forschungsgemeinschaft (DFG; En 308/1-1).

References

Albo, D., Granick, M., Jhala, N., Atkinson, B. and Solomon, M. (1994) The relationship of angiogenesis to biological activity in human squamous cell carcinomas of the head and neck. *Ann. Plast. Surg.* 32:558–594.

Alexander, C.M. and Werb, Z. (1991) Extracellular matrix degradation. In: E.D. Hay, (ed.) *Cell biology of extracellular matrix.* Plenum Press, New York, 255–302.

Alvarez, J.A., Baird, A., Tatum, A., Daucher, J., Chorsky, R., Gonzalez, A.M. and Stopa, E.G. (1992) Localization of basic fibroblast growth factor in human glial neoplasms. *Mod. Pathol.* 5(3):303–307.

Anandappa, S.Y., Winstanley, J.H., Leister, S., Green, B., Rudland, P.S. and Barraclough, R. (1994) Comparative expression of fibroblast growth factor mRNAs in benign and malignant breast disease. *Br. J. Cancer* 69(4):772–776.

Andrade, S.P., Fan, T.P. and Lewis, G.P. (1987) Quantitative *in vivo* studies on angiogenesis in a rat sponge model. *Br. J. Exp. Pathol.* 68(6):755–766.

Antonic, J. and Rakar, S. (1995) Colour and pulsed doppler US and tumour marker CA125 in differentiation between benign and malignant ovarian masses. *Anticancer Res.* 15:1527–1532.

Aruffo, A., Dietsch, M.T., Wan, H., Hellstrom, K.E. and Hellstrom, I. (1992) Granule membrane protein 140 (GMP140) binds to carcinomas and carcinoma-derived cell lines. *Proc. Natl. Acad. Sci. USA* 89(6):2292–2296.

Auerbach, R., Arensman, R., Kubai, L. and Folkman, J. (1975) Tumor induced angiogenesis: lack of inhibition by irradiation. *Int. J. Cancer* 15(2):241–245.

Axelsson, K., Ljung, B.E., Moore, D.H. 2nd., Thor, A.D., Chew, K.L., Edgerton, S.M., Smith, H.S. and Mayall, B.H. (1995) Tumor angiogenesis as a prognostic assay for invasive ductal breast carcinoma. *J. Natl. Cancer Inst.* 87(13):997–1008.

Bamber, J.C., De, G.L., Cosgrove, D.O., Simmons, P., Davey, J. and McKinna, J.A. (1988) Quantitative evaluation of real-time ultrasound features of the breast. *Ultrasound Med. Biol.* 1(81):81–87.

Banks, R.E., Gearing, A.J., Hemingway, I.K., Norfolk, D.R., Perren, T.J. and Selby, P.J. (1993) Circulating intercellular adhesion molecule-1 (ICAM-1), E-selectin and vascular cell adhesion molecule-1 (VCAM-1) in human malignancies. *Br. J. Cancer* 68(1):122–124.

Barbareschi, M., Weidner, N., Gasparini, G., Morelli, L., Forti, S., Eccher, C., Fina, P., Caffo, O., Leonardi, E., Mauri, F., Bevilacqua, P. and Dalla Palma, P. (1995) Microvessel quantitation in breast carcinomas. *Appl. Immunochem.* 3(2):75–84.

Barnhill, R.L., Fandrey, K., Levy, M.A., Mihm, M.J. and Hyman, B. (1992a) Angiogenesis and tumor progression of melanoma. Quantification of vascularity in melanocytic nevi and cutaneous malignant melanoma. *Lab. Invest.* 67(3):331–337.

Barnhill, R.L., Mihm, M.J. and Ceballos, P.I. (1992b) Angiogenesis and regressing cutaneous malignant melanoma [letter]. *Lancet* 339(8799):991–992.

Barnhill, R., Busam, K., Berwick, M., Blessing, K., Cochran, A., Elder, D., Fandrey, K., Karaoli, T. and White, W. (1994) Tumour vascularity is not a prognostic factor for cutaneous melanoma. *Lancet* 344:1237–1238.

Barton, G.J., Ponting, C.P., Spraggon, G., Finnis, C. and Sleep, D. (1992) Human platelet-derived endothelial cell growth factor is homologous to *Escherichia coli* thymidine phosphorylase. *Protein Science* 1:688–690.

Berkman, R.A., Merrill, M.J., Reinhold, W.C., Monacci, W.T., Saxena, A., Clark, W.C., Robertson, J.T., Ali, I.U. and Oldfield, E.H. (1993) Expression of the vascular permeability factor/vascular endothelial growth factor gene in central nervous system neoplasms. *J. Clin. Invest.* 91(1):153–159.

Bigler, S., Deering, R. and Brawer, M. (1993) Comparisons of microscopic vascularity in benign and malignant prostate tissue. *Hum. Pathol.* 24:220–226.

Birkedal, H.H. Moore, W.G., Bodden, M.K., Windsor, L.J., Birkedal, H.B., De, C.A. and Engler, J.A. (1993) Matrix metalloproteinases: a review. *Crit. Rev. Oral Biol. Med.* 4(2): 197–250.

Bischoff, J. (1995) Approaches to studying cell adhesion molecules in angiogenesis. *Trends in Cell Biology* 5:69–74.

Blann, A.D. and McCollum, C.N. (1994) Circulating endothelial-cell leukocyte adhesion molecules in atherosclerosis. *Thromb. Haemost.* 72(1):151–154.

Blood, C.H. and Zetter, B.R. (1990) Tumor interactions with the vasculature: angiogenesis and tumor metastasis. *Biochim. Biophys. Acta.* 1032(1):89–118.

Boocock, C.A., Charnock, J.D., Sharkey, A.M., McLaren, J., Barker, P.J., Wright, K.A., Twentyman, P.R. and Smith, S.K. (1995) Expression of vascular endothelial growth factor and its receptors flt and KDR in ovarian carcinoma. *J. Natl. Cancer Inst.* 87(7): 506–516.

Bosari, S., Lee, A.K., DeLellis, R.A., Wiley, B.D., Heatley, G.J. and Silverman, M.L. (1992) Microvessel quantitation and prognosis in invasive breast carcinoma. *Hum. Pathol.* 23(7):755–761.

Bouck, N. (1990) Tumor angiogenesis: the role of oncogenes and tumor suppressor genes. *Cancer Cells* 2(6):179–185.

Boveri, T. (1929) *The origin of malignant tumors*. Williams and Wilkins, Baltimore

Brawer, M.K., Deering, R.E., Brown, M., Preston, S.D. and Bigler, S.A. (1994) Predictors of pathologic stage in prostatic carcinoma. The role of neovascularization. *Cancer* 73(3):678–687.

Brem, S., Cotran, R. and Folkman, J. (1972) Tumor angiogenesis: a quantitative method for histological grading. *J. Natl. Cancer Inst.* 48:347–356.

Brem, S.S., M., J.H. and Gullino, P.M. (1978) Angiogenesis as a marker of pre-neoplastic lesions of the human breast. *Cancer* 41:239–248.

Briozzo, P., Morisset, M., Capony, F., Rougeot, C. and Rochefort, H. (1988) *In vitro* degradation of extracellular matrix with Mr 52,000 cathepsin D secreted by breast cancer cells. *Cancer Res.* 48:3688–3692.

Britton, P.D. and Coulden, R.A. (1990) The use of duplex Doppler ultrasound in the diagnosis of breast cancer. *Clin. Radiol.* 42(6):399–401.

Brooks, P.C., Clark, R.A. and Cheresh, D.A. (1994a) Requirement of vascular integrin $\alpha v \beta 3$ for angiogenesis. *Science* 264(5158):569–571.

Brooks, P.C., Montgomery, A.M., Rosenfeld, M., Reisfeld, R.A., Hu, T., Klier, G. and Cheresh, D.A. (1994b) Integrin $\alpha v \beta 3$ antagonists promote tumor regression by inducing apoptosis of angiogenic blood vessels. *Cell* 79(7):1157–1164.

Brooks, P.C., Strömblad, S., Klemke, R., Visscher, D., Sarkar, F.H. and Cheresh, D.A. (1995) Antiintegrin $\alpha v \beta 3$ blocks human breast cancer growth and angiogenesis in human skin. *J. Clin. Invest.* 96:1815–1822.

Brown, L.F., Yeo, K.T., Berse, B., Yeo, T.K., Senger, D.R., Dvorak, H.F. and Van, d. W.L. (19992) Expression of vascular permeability factor (vascular endothelial growth factor) by epidermal keratinocytes during wound healing. *J. Exp. Med.* 176(5):1375–1379.

Brown, L.F., Berse, B., Jackman, R.W., Tognazzi, K., Manseau, E.J., Dvorak, H.F. and Senger, D.R. (1993a) Increased expression of vascular permeability factor (vascular endothelial growth factor) and its receptors in kidney and bladder carcinomas. *Am. J. Pathol.* 143(5):1255–1262.

Brown, L.F., Berse, B., Jackman, R.W., Tognazzi, K., Manseau, E.J., Senger, D.R. and Dvorak, H.F. (1993b) Expression of vascular permeability factor (vascular endothelial growth factor) and its receptors in adenocarcinomas of the gastrointestinal tract. *Cancer Res.* 53(19):4727–4735.

Brown, L.F., Berse, B., Jackman, R.W., Tognazzi, K., Guidi, A.J., Dvorak, H.F., Senger, D.R., Connolly, J.L. and Schnitt, S.J. (1995) Expression of vascular permeability factor (vascular endothelial growth factor) and its receptors in breast cancer. *Hum. Pathol.* 26(1): 86–91.

Bundred, N., Bowcott, M., Walls, J., Faragher, E. and Knox, F. (1994) Angiogenesis in breast cancer predicts node metastasis and survival. *Br. J. Surgery* 81:768 (Abstract)

Burns, P.N., Halliwell, M., Wells, P.N. and Webb, A.J. (1982) Ultrasonic Doppler studies of the breast. *Ultrasound Med. Biol.* 8(2):127–143.

Burrows, F.J. and Thorpe, P.E. (1994) Vascular targeting – a new approach to the therapy of solid tumors. *Pharmacol. Ther.* 64(1):155–74.

Byrne, J.A., Tomasetto, C, Rouyer, N., Bellocq, J.-P., Rio, M.-C. and Basset, P. (1995) The tissue inhibitor of metalloproteinases-3 gene in breast carcinoma: identification of multiple polyadenylation sites and a stromal pattern of expression. *Molecular Medicine* 1(4):418–427.

Carnochan, P., Briggs, J.C., Westbury, G. and Davies, A.J. (1991) The vascularity of cutaneous melanoma: a quantitative histological study of lesions 0.85–1.25 mm in thickness. *Br. J. Cancer* 64(1):102–107.

Chalkley, H. (1943) Method for the quantitative morphological analysis of tissues. *J. Nat. Cancer Inst.* 4:47–53.

Charpin, C., Devictor, B., Bergeret, D., Andrac, L., Boulat, J., Horschowski, N., Lavaut, M.N. and Piana, L. (1995) CD31 Quantitative immunocytochemical assays in breast carcinomas. *Am. J. Clin. Pathol.* 103:443–448.

Clarke, C.L. and Sutherland, R.L. (1990) Progestin regulation of cellular proliferation. *Endocr. Rev.* 11(2):266–301.

Clezardin, P., Frappart, L., Clerget, M., Pechoux, C. and Delmas, P.D. (1993) Expression of thrombospondin (TSP1) and its receptors (CD36 and CD51) in normal, hyperplastic, and neoplastic human breast. *Cancer Res.* 53(6):1421–1430.

Cockerell, C.J., Sonnier, G., Kelly, L. and Patel, S. (1994) Comparative analysis of neo-vascularisation in primary cutaneous melanoma and Spitz nevus. *Am. J. Dermatopathol.* 16:9–13.

Cosgrove, D.O., Bamber, J.C., Davey, J.B., McKinna, J.A. and Sinnett, H.D. (1990) Color Doppler signals from breast tumors. Work in progress. *Radiology* 176(1):175–180.

Cosgrove, D.O., Kedar, R.P., Bamber, J.C., Al, M.B., Davey, J.B., Fisher, C., McKinna, J.A., Svensson, W.E., Tohno, E., Vagios, E. et al. (1993) Breast diseases: color Doppler US in differential diagnosis [see comments]. *Radiology* 189(1):99–104.

Costa, M., Danesi, R., Agen, C., Di, P.A., Basolo, F., Del, B.S. and Del, T.M. (1994) MCF-10A cells infected with the int-2 oncogene induce angiogenesis in the chick chorioallantoic membrane and in the rat mesentery. *Cancer Res.* 54(1):9–11.

Daa, T., Kodama, M., Kashima, K., Yokoyama, S., Nakayama, I. and Noguchi, S. (1993) Identification of basic fibroblast growth factor in papillary carcinoma of the thyroid. *Acta Pathol. Jpn.* 43(10):582–589.

Dameron, K.M., Volpert, O.V. and Bouck, N. (1994a) The p53 tumor suppressor gene inhibits angiogenesis by stimulating the production of thrombospondin. *Cold Spring Harbor Symp. Quant. Biol.* 54:1–7.

Dameron, K.M., Volpert, O.V., Tainsky, M.A. and Bouck, N. (1994b) Control of angiogenesis in fibroblasts by p53 regulation of thrombospondin-1. *Science* 265(5178):1582–1584.

de Fougerolles, A., Stacker, S.A., Schwarting, R. and Springer, T.A. (1991) Characterization of ICAM-2 and evidence for a third counter-receptor for LFA-1. *J. Exp. Med.* 174(1): 253–267.

Denton, K.J., Stretch, J.R., Gatter, K.C. and Harris, A.L. (1992) A study of adhesion molecules as markers of progression in malignant melanoma. *J. Pathol.* 167(2):187–191.

deVries, C., Escobedo, J.A., Ueno, H., Houck, K., Ferrara, N. and Williams, L.T. (1992) The fms-like tyrosine kinase, a receptor for vascular endothelial growth factor. *Science* 255(5047):989–991.

Dickinson, A.J., Fox, S.B., Persad, R.A., Hollyer, J., Sibley, G.N. and Harris, A.L. (1994) Quantification of angiogenesis as an independent predictor of prognosis in invasive bladder carcinomas. *Br. J. Urol.* 76(6):762–766.

DiPietro, L.A. and Polverini, P.J. (1993) Angiogenic macrophages produce the angiogenic inhibitor thrombospondin 1. *Am. J. Pathol.* 143(3):678–684.

Dixon, J.M., Walsh, J., Paterson, D. and Chetty, U. (1992) Colour Doppler ultrasonography studies of benign and malignant breast lesions. *Br. J. Surgery* 79(3):259–260.

Dock, W. (1993) Duplex sonography of mammary tumors: a prospective study of 75 patients. *J. Ultrasound Med.* 12(2):79–82.

Doussis, A.I., Kaklamanis, L., Cordell, J., Jones, M., Turley, H., Pulford, K., Simmons, D., Mason, D. and Gatter, K. (1993) ICAM-3 expression on endothelium in lymphoid malignancy. *Am. J. Pathol.* 143(4):1040–1043.

Duffy, M., Reilly, D., O'Sullivan, C., O'Higgins, N., Fennelly, J.J. and Andreasen, P. (1990) Urokinase, plasminogen activator, a new and independent prognostic marker in breast cancer. *Cancer Res.* 50:6827–6829.

Duggan, C., Maguire, T., McDermott, E., O'Higgins, N., Fennelly, J.J. and Duffy, M.J. (1995) Urokinase plasminogen activator and urokinase plasminogen activator receptor in breast cancer. *Int. J. Cancer* 61(5):597–600.

Dvorak, H.F. (1987) Thrombosis and cancer. *Hum. Pathol.* 18(3):275–284.

Dvorak, H.F., Sioussat, T.M., Brown, L.F., Berse, B., Nagy, J.A., Sotrel, A., Manseau, E.J., Van, d. W.L. and Senger, D.R. (1991) Distribution of vascular permeability factor (vascular endothelial growth factor) in tumors: concentration in tumor blood vessels. *J. Exp. Med.* 174(5):1275–1278.

Eda, H., Fujimoto, K., Watanabe, S., Ura, M., Hino, A, Tanaka, Y, Wada, K and Ishitsuka, H. (1993) Cytokines induce thymidine phosphorylase expression in tumor cells and make them more susceptible to 5'-deoxy-5-fluorouridine. *Cancer Chemother. Pharmacol.* 32:333–338.

Elledge, R.M. and Lee, W.-H. (1995) Life and death by 53. *Bioessays* 17(11):923–930.

Engels, K., Fox, S.B., Gatter, K.C. and Harris, A.L. (1996) Distinct angiogenic patterns are associated with high grade in-situ ductal carcinoma of the breast. *J. Pathol.,* in press.

Erikson, A. (1989) Biosynthesis of lysosomal endopeptidases. *J. Cell Biochem.* 40:31–41.

Ethier, S.P. (1995) Growth factor synthesis and human breast cancer progression. *J. Natl. Cancer Inst.* 87(13):964–973.

Fan, T.-P.D. Frost, E.E. and Wren, A.D. (19992) A multichannel wounding device for the study of vascular repair *in vitro. In:* M.E. Maragoudakis, P. Gullino, and P.T. Lelkes, (eds): *Angiogenesis in health and disease.* Plenum Press, New York, pp 315–320.

Fawcett, J., Holness, C.L., Needham, L.A., Turley, H., Gatter, K.C., Mason, D.Y. and Simmons, D.L. (1992) Molecular cloning of ICAM-3, a third ligand for LFA-1, constitutively expressed on resting leukocytes. *Nature* 360(6403):481–484.

Fearon, E.R. and Vogelstein, B. (1990) A genetic model for colorectal tumorigenesis. *Cell* 61(5):759–767.

Ferrara, N. (1995) Leukocyte adhesion. Missing link in angiogenesis [news; comment]. *Nature* 376(6540).

Ferrara, N., Houck, K., Jakeman, L. and Leung, D.W. (1992) Molecular and biological properties of the vascular endothelial growth factor family of proteins. *Endocr. Rev.* 13(1):18–32.

Finnis, C., Dodsworth, N., Pollitt, C.E., Carr, G. and Sleep, D. (1993) Thymidine phosphorylase activity of platelet-derived endothelial cell growth factor is responsible for endothelial cell mitogenicity. *Eur. J. Biochem.* 212(1):201–210.

Foekens, J.A., Schmitt, M., van, P.W., Peters, H.A., Bontenbal, M., Jänicke, F. and Klijn, J.G. (1992) Prognostic value of urokinase-type plasminogen activator in 671 primary breast cancer patients. *Cancer Res.* 52(21):6101–6105.

Foekens, J.A., Schmitt, M., van, P.W., Peters, H.A., Kramer, M.D., Jänicke, F. and Klijn, J.G. (1994) Plasminogen activator inhibitor-1 and prognosis in primary breast cancer. *J. Clin. Oncol.* 12(8):1648–1658.

Foekens, J.A., Buessecker, F., Peters, H.A., Krainick, U., van, P.W., Look, M.P., Klijn, J.G. and Kramer, M.D. (1995) Plasminogen activator inhibitor-2: prognostic relevance in 1012 patients with primary breast cancer. *Cancer Res.* 55(7):1423–1427.

Folberg, R., Rummelt, V., Ginderdeuren, R.-V., Hwang, T., Woolsen, R., Pe'er, J. and Gruman, L. (1993) The prognostic value of tumor blood vessel morphology in primary uveal melanoma. *Ophthalmology* 100:1389–1398.

Folkman, J. and Hochberg, M. (1973) Self-regulation of growth in three dimensions. *J. Exp. Med.* 138(4):745–753.

Folkman, J., Weisz, P.B., Joullie, M.M., Li, W.W. and Ewing, W.R. 1989) Control of angiogenesis with synthetic heparin substitutes. *Science* 243(4897): 1490–1493.

Fong, G.-H., Rossant, J., Gertsenstein, M. and Breitman, M.L. (1995) Role of the Flt-1 receptor tyrosine kinase in regulating the assembly of vascular endothelium. *Nature* 376:66–70.

Foulds, L. (1958) The natural history of cancer. *J. Chronic Dis.* 8:2–37.

Fox, S.B., Gatter, K.C., Bicknell, R., Going, J.J., Stanton, P., Cooke, T.G. and Harris, A.L. (1993a) Relationship of endothelial cell proliferation to tumor vascularity in human breast cancer. *Cancer Res.* 53(18):4161–4163.

Fox, S.B., Stuart, N., Smith, K., Brunner, N. and Harris, A.L. (1993b) High levels of uPA and PAI-1 are associated with highly angiogenic breast carcinomas. *J. Pathol.* 170 (suppl.):388A.

Fox, S.B., Leek, R., Smith, K., Hollyer, J., Greenall, M. and Harris, A. (1994) Tumor angiogenesis in node negative breast carcinomas-relationship to epidermal growth factor receptor and survival. *Breast Cancer Res. Treat.* 29:109–116.

Fox, S.B., Leek, R.D., Weekes, M.P., Whitehouse, R.M., Gatter, K.C. and Harris, A.L. (1995a) Quantitation and prognostic value of breast cancer angiogenesis: comparison of microvessel density, Chalkley count and computer image analysis. *J. Pathol,. in press.*

Fox, S.B., Turner, G., Gatter, K. and Harris, A. (1995b) The increased expression of adhesion molecules ICAM-3, E and P selectin on breast cancer endothelium. *J. Pathol., in press*

Fox, S.B., Westwood, M., Moghaddam, A., Comley, M., Turley, H., Whitehouse, R.M., Bicknell, R., Gatter, K.C. and Harris, A.L. (1995c) The angiogenic factor platelet derived endothelial cell growth factor/thymidine phosphorylase is upregulated in breast cancer epithelium and endothelium. *Br. J. Cancer, in press.*

Frater-Schroder, M., Risau, W., Hallmann, R., Gautschi, P. and Bohlen, P. (1987) Tumor necrosis factor type alpha, a potent inhibitor of endothelial cell growth *in vitro*, is angiogenic *in vivo. Proc. Natl. Acad. Sci. USA* 84(15):5277–5281.

Frixen, U.H., Behrens, J., Sachs, M., Eberle, G., Voss, B., Warda, A., Lochner, D. and Birchmeier, W. (1991) E-cadherin-mediated cell-cell adhesion prevents invasiveness of human carcinoma cells. *J. Cell Biol.* 113(1):173–185.

Frouge, C., Guinebretiere, J.M., Contesso, G., Di, P.R. and Blery, M. (1994) Correlation between contrast enhancement in dynamic magnetic resonance imaging of the breast and tumor angiogenesis. *Invest. Radiol.* 29(12):1043–1049.

Fujimoto, K., Ichimori, Y., Kakizoe, T., Okajima, E., Sakamoto, H., Sugimura, T. and Terada, M. (1991) Increased serum levels of basic fibroblast growth factor in patients with renal cell carcinoma. *Biochem. Biophys. Res. Comm.* 180:386–392.

Furukawa, T., Yoshimura, A., Sumizawa, T., Haraguchi, M. and Akiyama, S.I. (1992) Angiogenic factor. *Nature* 356:668.

Furusato, M., Wakui, S., Sasaki, H., Ito, K. and Ushigome, S. (1994) Tumour angiogenesis in latent prostatic carcinoma. *Br. J. Cancer* 70(6):1244–1246.

Galland, F., Karamysheva, A., Mattei, M.G., Rosnet, O., Marchetto, S. and Birnbaum, D. (1992) Chromosomal localization of FLT4, a novel receptor-type tyrosine kinase gene. *Genomics* 13(2):475–478.

Galland, F., Karamysheva, A., Pebusque, M.J., Borg, J.P., Rottapel, R., Dubreuil, P., Rosnet, O. and Birnbaum, D. (1993) The FLT4 gene encodes a transmembrane tyrosine kinase related to the vascular endothelial growth factor receptor. *Oncogene* 8(5):1233–1240.

Gamble, J.R., Matthias, L.J., Meyer, G., Kaur, P., Russ, G., Faull, R., Berndt, M.C. and Vadas, M.A. (1993) Regulation of in vitro capillary tube formation by anti-integrin antibodies. *J. Cell Biol.* 121(4):931–943.

Ganesh, S., Sier, C.F.M., Heerding, M.M., Griffioen, G., Lamers, C.B.H.W. and Verspaget, H.W. (1994) Urokinase receptor and colorectal cancer survival. *Lancet* 344:401–402.

Garver, R.J., Radford, D.M., Donis, K.H., Wick, M.R. and Milner, P.G. (1994) Midkine and pleiotrophin expression in normal and malignant breast tissue. *Cancer* 74(5): 1584–1590.

Gimbrone, M.A.J., Cotran, R., Leapman, S. and Folkman, J. (1974) Tumour growth neovascularization: An experimental model using rabbit cornea. *J. Natl. Cancer Inst.* 52: 413–427.

Gimbrone, M.J. and Gullino, P.M. (1976a) Angiogenic capacity of preneoplastic lesions of the murine mammary gland as a marker of neoplastic transformation. *Cancer Res.* 36: 2611–2620.

Gimbrone, M.J. and Gullino, P.M. (1976b) Neovascularization induced by intraocular xeno-grafts of normal, preneoplastic, and neoplastic mouse mammary tissues. *J. Natl. Cancer Inst.* 56(2):305–318.

Glukhova, M., Koteliansky, V., Sastre, X. and Thiery, J.P. (1995) Adhesion systems in normal breast and in invasive breast carcinoma. *Am. J. Pathol.* 146(3):706–716.

Goldberg, M.A. and Schneider, T.J. (1994) Similarities between the oxygen-sensing mecha-nisms regulating the expression of vascular endothelial growth factor and erythropoietin. *J. Biol. Chem.* 269(6):4355–4359.

Gomm, J.J., Smith, J., Ryall, G.K., Ballic, R., Turnbull, L. and Coombes, R.C. (1991) Locali-sation of basic fibroblast growth factor and transforming growth factor b1 in the human mammary gland. *Cancer Res.* 51:4685–4692.

Good, D.J., Polverini, P.J.., Rastinejad, F., Le, B.M., Lemons, R.S., Frazier, W.A. and Bouck, N.P. (1990) A tumor suppressor-dependent inhibitor of angiogenesis is immunologically and functionally indistinguishable from a fragment of thrombospondin. *Proc. Natl. Acad. Sci. USA* 87(17):6624–6628.

Grischke, E.M., Kaufmann, M., Eberlein-Gonska, M., Mattfeld, T., Sohn, C. and Bastert, G. (1994) Angiogenesis as a diagnostic factor in primary breast cancer: microvessel quantita-tion by stereological methods and correlation with color doppler sonography. *Onkologie* 17:34–42.

Grondahl, H.J., Christensen, I.J., Rosenquist, C., Brunner, N., Mouridsen, H.T., Dano, K. and Blichert, T.M. (1993) High levels of urokinase-type plasminogen activator and its inhibitor PAI-1 in cytosolic extracts of breast carcinomas are associated with poor prognosis. *Cancer Res.* 53(11):2513–2521.

Guidi, A., Fischer, L., Harris, J. and Schnitt, S. (1994) Microvessel density and distribution in ductal carcinoma in situ of the breast. *J. Natl. Cancer Inst.* 86:614–619.

Guidi, A.J., Abu-Jawdeh, G., Berse, B., Jackman, R.W., Tognazzi, K., Dvorak, H.F. and Brown, F.F. (1995) Vascular permeability factor (vascular endothelial growth factor) expression and angiogenesis in cervical neoplasia. *J. Natl. Cancer Inst.* 87(16):1237–1245.

Guinebretiere, J.M., Le, M.G., Gavoille, A., Bahi, J. and Contesso, G. (1994) Angiogenesis and risk of breast cancer in women with fibrocystic disease [letter; comment]. *J. Natl. Cancer Inst.* 86(8):635–636.

Haldar, S., Negrini, M., Monne, M., Sabbioni, S. and Croce, C.M. (1994) Down-regulation of bcl-2 by p53 in breast cancer cells. *Cancer Res.* 53(8):2095–2097.

Hall, N.R., Fish, D.E., Hunt, N., Goldin, R.D., Guillou, P.J. and Monson, J.R. (1992) Is the relationship between angiogenesis and metastasis in breast cancer real? *Surg. Oncol.* 1(3):223–229.

Hanneken, A., Maher, P. and Baird, A. (1995) High affinity immunoreactive FGF receptors in the extracellular matrix of vascular endothelial cells-implications for the modulation of FGF-2. *J. Cell Biol.* 128:1221–1228.

Haraguchi, M., Kazutaka, M., Uemura, K., Sumizawa, T., Furukawa, T., Yamada, K. and Akiyama, S.-I. (1994) Angiogenic activity of enzymes. *Nature* 368:198.

Heywang, K.S. (1994) Contrast-enhanced magnetic resonance imaging of the breast. *Invest. Radiol.* 29(1):94–104.

Heywang, K.S., Schlegel, A., Beck, R., Wendt, T., Kellner, W., Lommatzsch, B., Untch, M. and Nathrath, W.B. (1993) Contrast-enhanced MRI of the breast after limited surgery and radiation therapy. *J. Comput. Assist. Tomogr.* 17(6):891–900.

Heywang, K.S., Haustein, J., Pohl, C., Beck, R., Lommatzsch, B., Untch, M. and Nathrath, W.B (1994) Contrast-enhanced MR imaging of the breast: comparison of two different doses of gadopentetate dimeglumine. *Radiology* 191(3):639–646.

Ho, C.K., Ou, B.R., Hsu, M.L., Su, S.N., Yung, C.H. and Wang, S.Y. (1990) Induction of thymidine kinase activity and clonal growth of certain leukemic cell lines by granulocyte-derived factor. *Blood* 75(12):2438–2444.

Hockenbery, D.M. (1995) bcl-2, a novel regulator of cell death. *Bioessays* 17 (7):631–638.

Hogg, N. and Landis, R.C. (1993) Adhesion molecules in cell interactions. *Curr. Opin. Immunol.* 5(3):383–390.

Holmgren, L., O'Reilly, M.S. and Folkman, J. (1995) Dormancy of micrometastasis: Balanced proliferation and apoptosis in the presence of angiogenesis suppression. *Nature Medicine* 1(2):149–153.

Horak, E.R., Leek, R., Klenk, N., LeJeune, S., Smith, K., Stuart, N., Greenall, M., Stepniewska, K. and Harris, A.L. (1992) Angiogenesis, assessed by platelet/endothelial cell adhesion molecule antibodies, as indicator of node metastases and survival in breast cancer. *Lancet* 240(8828):1120–1124.

Houck, K.A., Leung, D.W., Rowland, A.M., Winer, J. and Ferrara, N. (1992) Dual regulation of vascular endothelial growth factor bioavailability by genetic and proteolytic mechanisms. *J. Biol. Chem.* 267(36):26031–26037.

Hunt, T.K., Knighton, D.R., Thakral, K.K., Goodson, W.3. and Andrews, W.S. (1984) Studies on inflammation and wound healing angiogenesis and collagen synthesis stimulated in vivo by resident and activated wound macrophages. *Surgery* 96(1):48–54.

Iozzo, R.V. (1995) Tumor stroma as a regulator of neoplastic behaviour. *Lab. Invest.* 73:157–160.

Ishikawa, F., Miyazono, K., Hellman, U., Drexler, H., Wernstedt, C., Hagiwara, K., Usuki, K., Takaku, F., Risau, W. and Heldin, C.H. (1989) Identification of angiogenic activity and the cloning and expression of platelet-derived endothelial cell growth factor. *Nature* 33(6216):557–562.

Iwata, H., Kobayashi, S., Iwase, H. and Okada, Y. (1995) [The expression of MMPs and TIMPs in human breast cancer tissues and importance of their balance in cancer invasion and metastasis]. *Nippon Rinsho* 53(7):1805–1810.

Jackson, V.P. (1988) Duplex sonography of the breast. *Ultrasound Med. Biol.* 1(131):131–137.

Jaeger, T., Weidner, N., Chew, K., Moore, D., Kerschman, R., Carrol, P. and Waldman, F. (1994) Tumor angiogenesis predicts lymph node metastasis in invasive bladder cancer. *Proc. Am. Assoc. Cancer Res.* 35:66 (abstract 394).

Jain, R.K. (1990) Vascular and interstitial barriers to delivery of therapeutic agents in tumors. *Cancer Metastasis Rev.* 9(3):253–266.

Jain, R.K. (1994) Barriers to drug delivery in solid tumors. *Sci. Am.* 271(1):58–65.

Jänicke, F., Schmitt, M., Pache, L., Ulm, K., Harbeck, N., Hofler, H. and Graeff, H. (1993) Urokinase (uPA) and its inhibitor PAI-1 are strong and independent prognostic factors in node-negative breast cancer. *Breast Cancer Res. Treat.* 24(3):195–208.

Janot, F., el-Naggar, A.K., Morrison, R.S., Liu, T.J., Taylor, D.L. and Clayman, G.L. (1995) Expression of basic fibroblast growth factor in squamous cell carcinoma of the head and neck is associated with degree of histologic differentiation. *Int. J. Cancer* 64(2):117–123.

Jellins, J. (1988) Combining imaging and vascularity assessment of breast lesions. *Ultrasound Med. Biol.* 1(121):121–130.

Kageshita, T., Yoshii, A., Kimura, T., Kuriya, N., Ono, T., Tsujisaki, M. Imai, K. and Ferrone, S. (1993) Clinical relevance of ICAM-1 expression in primary lesions and serum of patients with malignant melanoma. *Cancer Res.* 53:4927–4932.

Kaipainen, A., Korhonen, J., Pajusola, K., Aprelikova, O., Persico, M.G., Terman, B.I. and Alitalo, K. (1993) The related FLT4, FLT1, and KDR receptor tyrosine kinases show distinct expression patterns in human fetal endothelial cells. *J. Exp. Med.* 178(6):2077–2088.

Kaiser, W.A. (1993) [MR mammography]. *Radiologie* 33(5):292–299.

Kaiser, W.A., Diedrich, K., Reiser, M. and Krebs, D. (1993) [Modern diagnostics of the breast]. *Geburtshilfe Frauenheilkd.* 53(1):1–14.

Kandel, J., Bossy, W.E., Radvanyi, F., Klagsbrun, M., Folkman, J. and Hanahan, D. (1991) Neovascularization is associated with a switch to the export of bFGF in the multistep development of fibrosarcoma. *Cell* 66(6):1095–104.

Kaplanski, G., Farnarier, C., Benoliel, A., Foa, C., Kaplanski, S. and Bongrand, P. (1994) A novel role for E- and P-selectins: shape of endothelial cell monolayers. *J. Cell Sci.* 107:2449–2457.

Kendall, R.L. and Thomas, K.A. (1993) Inhibition of vascular endothelial cell growth factor activity by an endogenously encoded soluble receptor. *Proc. Natl. Acad. Sci. USA* 90(22):10705–10709.

Kieser, A., Weich, H.A., Brandner, G., Marme, D. and Kolch, W. (1994) Mutant p53 potentiates protein kinase C induction of vascular endothelial growth factor expression. *Oncogene* 9(3):963–969.

Knighton, D., Ausprunk, D., Tapper, D. and Folkman, J. (177) Avascular and vascular phases of tumour growth in the chick embryo. *Br. J. Cancer* 35(3):347–356.

Koch, A.E., Cho, M., Burrows, J.C., Polverini, P.J. and Leibovich, S.J. (1992) Inhibition of production of monocyte/macrophage-derived angiogenic activity by oxygen free-radical. scavengers. *Cell Biol. Int. Rep.* 16(5):415–425.

Kondo, S., Asano, M., Matsuo, K, Ohmori, I. and Suzuki, H. (1994) Vascular endothelial growth factor/(vascular permeability factor is detectable in the sera of tumor-bearing mice and cancer patients. *Biochim. Biophys. Acta* 1221(2):211–214.

Kristensen, P., Larsson, L., Nielsen, L., Grondahl-Hansen, J., Andeasen, P. and Dano, K. (1994) Human endothelial cells contain one type of plasminogen activator. *FEBS Lett.* 168:33–37.

Kuzu, I., Bicknell, R., Harris, A.L., Jones, M., Gatter, K.C. and Mason, D.Y. (1992) Heterogeneity of vascular endothelial cells with relevance to diagnosis of vascular tumours. *J. Clin. Pathol.* 45(2):143–148.

Lakhani, S.R., Collins, N., Stratton, M.R. and Sloane, J.P. (1995) Atypical ductal hyperplasia of the breast: clonal proliferation with loss of heterozygosity on chromosomes 16q and 17p. *J. Clin. Pathol.* 48:611–615.

Lampugnani, M.G., Resnati, M., Dejana, E. and Marchisio, P.C. (1991) The role of integrins in the maintenance of endothelial monolayer integrity. *J. Cell Biol.* 112(3): 479–490.

Larsson, K., Skriver, L., Nielsen, J., Grondahl-Hansen, J., Kristensen, P. and Dano, K. (1984) Distribution of urokinase-type activator immunoreactivity in the mouse. *J. Cell Biol.* 98:894–903.

Lauk, S., Zietman, A., Skates, S., Fabian, R. and Suit, H.D. (1989) Comparative morphometric study of tumor vasculature in human squamous cell carcinomas and their xenotransplants in athymic nude mice. *Cancer Res.* 49(16):4557–4561.

Leavesley, D.I., Schwartz, M.A., Rosenfeld, M. and Cheresh, D.A. (1993) Integrin beta 1- and beta 3-mediated endothelial cell migration is triggered through distinct signaling mechanisms. *J. Cell Biol.* 121(1):163–170.

Leek, R.D., Harris, A.L. and Lewis, C.E. (1994) Cytokine networks in solid human tumors: regulation of angiogenesis. *J. Leukoc. Biol.* 56(4):423–435.

Li, V., Folkerth, R., Watanabe, H., Yu, C., Rupnick, M., Barnes, P., Scott, R., Black, P., Sallan, S. and Folkman, J. (1994) Microvessel count and cerebrospinal fluid basic fibroblast growth factor in children with brain tumours. *Lancet* 344:82–86.

Liotta, L.A., Kleinerman, J. and Saidel, G.M. (1974) Quantitative relationships of intravascular tumor cells, tumor vessels, and pulmonary metastasis following tumor implantation. *Cancer Res.* 34:997–1004.

Liotta, L.A., Steeg, P.S. and Stetler, S.W. (1991) Cancer metastasis and angiogenesis: an imbalance of positive and negative regulation. *Cell* 64(2):327–336.

Lochter, A. and Bissell, M.J. (1995) Involvement of extracellular matrix constituents in breast cancer. *Cancer Biol.* 6:165–1773.

Macchiarini, P., Fontanini, G., Hardin, M.J., Squartini, F. and Angeletti, C.A. (1992) Relation of neovascularization to metastasis of non-small-cell lung cancer. *Lancet* 240(8812):145–146.

Macchiarini, P., Fontanini, G.., Dulmet, E., de Montepreville, V., Chapelier, A., Cerrina, J., Ladurie, F.-R. and Darteville, P. (1994) Angiogenesis: an indicator of metastasis in non-small cell lung cancer invading the thoracic inlet. *Ann. Thor. Surg.* 57:1534–1539.

Maiorana, A. and Gullino, P.M. (1978) Acquisition of angiogenic capacity and neoplastic transformation in the rat mammary gland. *Cancer Res.* 38(12):4409–4414.

Matrisian, L.M. (1992) The matrix-degrading metalloproteinases. *Bioessays* 14(7):455–463.

Mattern, J., Koomägi, R. and Volm, M. (1995) Vascular endothelial growth factor expression and angiogenesis in non-small cell lung carcinomas. *Int. J. Oncology* 6:1059–1062.

McCarthy, S.A., Kuzu, I., Gatter, K.C. and Bicknell, R. (1991) Heterogeneity of the endothelial cell and its role in organ preference of tumour metastasis. *Trends Pharmacol. Sci.* 12(12):462–467.

Millauer, B., Wizigmann, V.S., Schnurch, H., Martinez, R., Moller, N.P., Risau, W. and Ullrich, A. (1993) High affinity VEGF binding and developmental expression suggest Flk-1 as a major regulator of vasculogenesis and angiogenesis. *Cell* 72(6):835–846.

Minchenko, A., Bauer, T., Salceda, S. and Caro, J. (19994) Hypoxic stimulation of vascular endothelial growth factor expression *in vitro* and *in vivo*. *Lab. Invest.* 71:374–379.

Mitchell, B.S. Schumacher, U. and Kaiserling, E. (1994) Are tumours innervated? Immuno-histological investigations using antibodies against the neuronal marker protein gene product 9.5 (PGP 9.5) in benign, malignant and experimental tumours. *Tumour Biol.* 15(5):269–274.

Miyazono, K., Usuki, K. and Heldin, C.H. (1991) Platelet-derived endothelial cell growth factor. *Prog. Growth Factor Res.* 3(3):207–217.

Mlynek, M., van Beunigen, D., Leder, L.-D. and Streffer, D. (1985) Measurement of the grade of vascularisation in histological tumour tissue sections. *Br. J. Cancer* 52:945–948.

Moffett, B.F., Baban, D., Bao, L. and Tarin, D. (1992) Fate of clonal lineages during neoplasia and metastasis studied with an incorporated genetic marker. *Cancer Res.* 52(7):1737–1743.

Moghaddam, A. and Bicknell, R. (1992) Expression of platelet-derived endothelial cell growth factor in *Escherichia coli* and confirmation of its thymidine phosphorylase activity. *Biochemistry* 31(48):12141–12146.

Moghaddam, A., Zhang, H.T., Fan, T.P., Hu, D.E., Lees, V.C., Turley, H., Fox, S.B., Gatter, K.C., Harris, A.L. and Bicknell, R. (1995) Thymidine phosphorylase is angiogenic and promotes tumor growth. *Proc. Natl. Acad. Sci. USA* 92(4):998–1002.

Montesano, R. (1992) 1992 Mack Forster Award Lecture. Review. Regulation of angiogenesis *in vitro. Eur. J. Clin. Invest.* 22(8):504–515.

Morris, P.B., Ellis, M.N. and Swain, J.L. (1989) Angiogenic potency of nucleotide meta-bolites: potential role in ischemia-induced vascular growth. *J. Mol. Cell. Cardiol.* 21(4): 351–358.

Moutcourrier, P., Manreat, P.H., Salazar, G., Morisset, M., Sahuquet, A. and Rochefort, H. (1990) Cathepsin D in breast cancer cells can digest extracellular matrix in large acidic vesicles. *Cancer Res.* 50:6045–6054.

Mukhopadhyay, D., Tsiokas, L., Zhou, X.M., Foster, D., Brugge, J.S. and Sukhatme, V.P. (1995) Hypoxic induction of human vascular endothelial growth factor expression through c-Src activation. *Nature* 375(6532):577–581.

Nelson, H., Ramsey, P., Donohue, J. and Wold, L. (1994) Cell adhesion molecule expression within the microvasculature of human colorectal malignancies. *Clin. Immunol. Immun-pathol.* 72:129–136.

Nguyen, M., Strubel, N.A. and Bischoff, J. (1993a) A role for sialyl Lewis-X/A glycoconjuga-tes in capillary morphogenesis. *Nature* 365(6443):267–269.

Nguyen, M., Watanabe, H., Budson, A.E., Richie, J.P. and Folkman, J. (1993b) Elevated levels of the angiogenic peptide basic fibroblast growth factor in urine of bladder cancer patients. *J. Natl. Cancer Inst.* 85(3):241–242.

Nguyen, M., Watanabe, H., Budson, A.E., Richie, J.P., Hayes, D.F. and Folkman, J. (1994) Elevated levels of an angiogenic peptide, basic fibroblast growth factor, in the urine of patients with a wide spectrum of cancers [see comments]. *J. Natl. Cancer Inst.* 86(5):356–361.

Nunez, G. and Clarke, M.F. (1994) The bcl-2 family of proteins: regulators of cell death and survival. *Trends in Cell Biology* 4:399–403.

Nystrom, C., Forssman, L. and Roos, B. (1969) Myometrial blood flow studies in carcinoma of the corpus uteri. *Acta Radiol. Ther.* 8:193–198.

O'Brien, T., Cranston, D., Fuggle, S., Bicknell, R. and Harris, A.L. (1995) Different angiogenic pathways characterize superficial and invasive bladder cancer. *Cancer Res.* 55(3):510–513.

O'Reilly, M.S., Holmgren, L., Shing, Y., Chen, C., Rosenthal, R.A., Moses, M., Lane, W.S., Cao, Y., Sage, E.H. and Folkman, J. (1994) Angiostatin: a novel angiogenesis inhibitor that mediates the suppression of metastases by a Lewis lung carcinoma [see comments]. *Cell* 79(2):315–328.

O'Sullivan, C., Lewis, C.E., Harris, A.L. and McGee, J.O. (1993) Secretion of epidermal growth factor by macrophages associated with breast carcinoma. *Lancet* 342(8864): 148–149.

Ogawa, Y., Chung, Y., Nakata, B., Takatsuka, S., Yamashita, Y., Maeda, K., Sawada, T., Kato, Y., Fujimoto, Y., Yoshikawa, K. and Sowa, M. (1994) An evaluation of angiogenesis in breast cancer with factor VIII related antigen staining. *Proc. Am. Assoc. Cancer Res.* 35:186 (abstr. 1113).

Ohsawa, M., Tomita, Y., Kuratsu, S., Kanno, H. and Aozasa, K. (1995) Angiogenesis in malignant fibrous histiocytoma. *Oncology* 52:51–54.

Olivarez, D., Ulbright, T., DeRiese, W., Foster, R., Reister, T., Einhorn, L. and Sledge, G. (1994) Neovascularization in clinical stage A testicular germ cell tumor: prediction of metastatic disease. *Cancer Res.* 54:2800–2802.

Olson, T.A., Mohanraj, D., Carson, L.F. and Ramakrishnan, S. (1994) Vascular permeability factor gene expression in normal and neoplastic human ovaries. *Cancer Res.* 54(1):276–280.

Ottinetti, A. and Sapino, A. (1988) Morphometric evaluation of microvessels surrounding hyperplastic and neoplastic mammary lesions. *Breast Cancer Res. Treat.* 11:241–248.

Pajusola, K., Aprelikova, O., Korhonen, J., Kaipainen, A., Pertovaara, L., Alitalo, R. and Alitalo, K. (1992) FLT4 receptor tyrosine kinase contains seven immunoglobulin-like loops and is expressed in multiple human tissues and cell lines [published erratum appears in *Cancer Res.* 1993 Aug 15, 53 (16):3845]. *Cancer Res.* 52(20):5738–5743.

Pajusola, K., Aprelikova, O., Pelicci, G., Weich, H., Claesson, W.L. and Alitalo, K. (1994) Signalling properties of FLT4, a proteolytically processed receptor tyrosine kinase related to two VEGF receptors. *Oncogene* 9(12):3545–3555.

Paku, S. and Lapis, K. (1993) Morphological aspects of angiogenesis in experimental liver metastases. *Am. J. Pathol.* 143(3):926–936.

Palabrica, T., Lobb, R., Furie, B.C., Aronovitz, M., Benjamin, C., Hsu, Y.M., Sajer, S.A. and Furie, B. (1992) Leukocyte accumulation promoting fibrin deposition is mediated in vivo by P-selectin on adherent platelets. *Nature* 359(6398):848–851.

Park, J.E., Chen, H.H., Winer, J., Houck, K.A. and Ferrara, N. (1994) Placenta growth factor. Potentiation of vascular endothelial growth factor bioactivity, *in vitro* and *in vivo*, and high affinity binding to Flt-1 but not to Flk-1/KDR. *J. Biol. Chem.* 269(41):25646–25654.

Parums, D., Cordell, J., Micklem, K., Heryet, A., Gatter, K. and Mason, D. (1990) JC70: a new monoclonal antibody that detects vascular endothelium associated antigen on routinely processed tissue sections. *J. Clin. Pathol.* 43:752–757.

Patterson, A., Zhang, H., Moghaddam, A., Bicknell, R., Talbot, D., Stratford, I. and Harris, A. (1994) Increased sensitivity to the pro-drug 5′-deoxy-5-fluorouridine and modulation of 5-fluoro-2′-deoxyuridine sensitivity in MCF-7 cell transfected with thymidine phosphorylase. *Cancer Res.* submitted.

Paweletz, N. and Knierim, M. (1989) Tumor-related angiogenesis. *Crit. Rev. Oncol. Hematol.* 9(3):197–242.

Pepper, M.S., Belin, D., Montesano, R., Orci, L. and Vassalli, J.D. (1990) Transforming growth factor-beta 1 modulates basic fibroblast growth factor-induced proteolytic and angiogenic properties of endothelial cells in vitro. *J. Cell Biol.* 111(2):743–755.

Pepper, M.S., Ferrara, N., Orci, L. and Montesano, R. (1991) Vascular endothelial growth factor (VEGF) induces plasminogen activators and plasminogen activator inhibitor-1 in microvascular endothelial cells. *Biochem. Biophys. Res. Commun.* 181(2):902–906.

Pepper, M.S., Ferrara, N., Orci, L. and Montesano, R. (1992) Potent synergism between vascular endothelial growth factor and basic fibroblast growth factor in the induction of angiogenesis in vitro. *Biochem. Biophys. Res. Commun.* 189(2):824–831.

Peters-Engl, C., Medl, M. and Leodolter, S. (1995) The use of colour-coded and spectral Doppler ultrasound in the differentiation of benign and malignant breast lesions. *Br. J. Cancer* 71(1):137–139.

Plate, K.H., Breier, G., Weich, H.A. and Risau, W. (1992) Vascular endothelial growth factor is a potential tumour angiogenesis factor in human gliomas *in vivo*. *Nature* 359(6398):845–848.

Pober, J.S. (1988) Warner-Lambert/Parke-Davis award lecture. Cytokine-mediated activation of vascular endothelium. Physiology and pathology. *Am. J. Pathol.* 133(3):426–433.

Polette, M., Clavel, C., Birembaut, P. and De, C.Y. (1993a) Localization by in situ hybridization of mRNAs encoding stromelysin 3 and tissue inhibitors of metallo-proteinases TIMP-1 and TIMP-2 in human head and neck carcinomas. *Pathol. Res. Pract.* 189(9):1052–1057.

Polette, M., Clavel, C., Cockett, M., Girod, d. B.S., Murphy, G. and Birembaut, P. (1993b) Detection and localization of mRNAs encoding matrix metalloproteinases and their tissue inhibitor in human breast pathology. *Invasion Metastasis* 13(1):31–37.

Polverini, P.J. and Leibovich, S.J. (1984) Induction of neovascularization *in vivo* and endothelial proliferation *in vitro* by tumor-associated macrophages. *Lab Invest.* 51(6):635–642.

Porschen, R., Classen, S., Piontek, M. and Borchard, F. (1994) Vascularization of carcinomas of the esophagus and its correlation with tumor proliferation. *Cancer Res.* 54(2):587–591.

Porter, P.L., Patton, K.F., Self, S.A., Gown, A.M. and Schmidt, R.A. (1993) A quantitative study of blood vessel size and density in normal and neoplastic breast tissue. *Mod. Pathol.* 6:18; USCAP Abstract 87.

Postlethwaite, A.E., Snyderman, R. and Ang, A.H. (1976) The chemotactic attraction of human fibroblasts to a lymphocyte-derived factor. *J. Exp. Med.* 144:1188–1203.

Poulsom, R., Pignatelli, M., Stetler, S.W., Liotta, L.A., Wright, P.A., Jeffery, R.E., Longcroft, J.M., Rogers, L. and Stamp, G.W. (1992) Stromal expression of 72 kda type IV collagenase (MMP-2) and TIMP-2 mRNAs in colorectal neoplasia. *Am. J. Pathol.* 141(2): 389–396.

Poulsom, R., Hanby, A.M., Pignatelli, M., Jeffery, R.E., Longcroft, J.M., Rogers, L. and Stamp, G.W. (1993) Expression of gelatinase A and TIMP-2 mRNAs in desmoplastic fibroblasts in both mammary carcinomas and basal cell carcinomas of the skin. *J. Clin. Pathol.* 46(5):429–436.

Protopapa, E., Delides, G.S. and Revesz, L. (1993) Vascular density and the response of breast carcinomas to mastectomy and adjuvant chemotherapy. *Eur. J. Cancer.*

Rastinejad, F., Polverini, P.J. and Bouck, N.P. (1989) Regulation of the activity of a new inhibitor of angiogenesis by a cancer suppressor gene. *Cell* 56(3):345–355.

Renkonen, R., Paavonen, T., Nortamo, P. and Gahmberg, C.G. (1992) Expression of endothelial adhesion molecules in vivo. Increased endothelial ICAM-2 expression in lymphoid malignancies. *Am. J. Pathol.* 140(4):763–767.

Reynolds, K., Farzaneh, F., Collins, W.P., Campbell, S., Bourne, T.H., Lawton, F., Moghaddam, A., Harris, A.L. and Bicknell, R. (1994a) Association of ovarian malignancy with expression of platelet-derived endothelial cell growth factor. *J. Natl. Cancer Inst.* 86(16): 1234–1238.

Reynolds, K., Farzaneh, F., Collins, W.P., Campbell, S., Bourne, T.H., Lawton, F., Moghaddam, A., Harris, A.L. and Bicknell, R (1994b) Correlation of ovarian malignancy with expression of platelet-derived endothelial cell growth factor. *J. Natl. Cancer Inst., in press.*

Roberts, W. and Bevilacqua, M. (1994) Vascular endothelial growth factor inhibits E-selectin and VCAM-1 expression in vitro and in vivo. *J. Cell. Biochem. suppl.* 18A:329 (abstract).

Rochefort, H. (1990) Biological and clinical significance of cathepsin D in breast cancer. *Semin. Cancer Biol.* 1:153–160.

Roy, J., Platt, J.L. and Weisdorf, D.J. (1993) The immunopathology of upper gastrointestinal acute graft-versus-host disease. Lymphoid cells and endothelial adhesion molecules. *Transplantation* 55(3):572–578.

Rubbert, A., Manger, B., Lang, N., Kalden, J.R. and Platzer, F. (1991) Functional characterization of tumor-infiltrating lymphocytes, lymph-node lymphocyts and peripheral-blood lymphocytes from patients with breast cancer. *Int. J. Cancer* 49(1):25–31.

Ruco, L.P., Pomponi, D., Pigott, R., Stoppacciaro, A., Monardo, F., Uccini, S., Boraschi, D., Tagliabue, A., Santoni, A., Dejana, E., Mantovani, A. and Baroni, C.D. (1990) Cytokine production (IL-1 α, IL-1 β, and TNF α) end endothelial cell activation (ELAM-1 and HLA-DR) in reactive lymphadenitis, Hodgkin's disease, and in non-Hodgkin's lymphomas. An immunocytochemical study. *Am. J. Pathol.* 137(5):1163–1171.

Rutgers, J.L., Mattox, T.F. and Vargas, M.P. (1995) Angiogenesis in uterine cervical squamous cell carcinoma. *Int. J. Gynecol. Pathol.* 14:114–118.

Sahin, A., Sneige, N., Singletary, E. and Ayala, A. (1992) Tumor angiogenesis detected by Factor-VIII immunostaining in node-negative breast carcinoma (NNBC): a possible predictor of distant metastasis. *Mod. Pathol.* 5:17A (abstract).

Samejima, N. and Yamazaki, K. (1988) A study on the vascular proliferation in tissues around the tumor in breast cancer. *Jpn. J. Surg.* 18:235–242.

Samoto, K., Ikezaki, K., Ono, M., Shono, T., Kohno, K., Kuwano, M. and Fukui, M. (1995) Expression of vascular endothelial growth factor and its possible relation with neovascularization in human brain tumors. *Cancer Res.* 55(5):1189–1193.

Sato, T., Akiyama, F., Sakamoto, G., Kasumi, F. and Nakamura, Y. (1991) Accumulation of genetic alterations and progression of primary breast cancer. *Cancer Res.* 51 (21):5794–5799.

Schadendorf, D., Heidel, J., Gawlik, C., Suter, L. and Czarnetzki (1995) Association with clinical outcome of expression of VLA-4 in primary cutaneous malignant melanoma as well as P-selectin and E-selectin on intratumoral vessels. *J. Natl. Cancer Inst.* 87: 366–371.

Schmitt, M., Jänicke, F. and Graeff, H. (1990) Tumour-associated fibrinolysis: the prognostic relevance of plasminogen activators uPA and tPA in human breast cancer. *Blood Coag. Fibrino.* 1:695–702.

Schultz-Hector, S. and Haghayegh, S. (1993) B-Fibroblast growth factor expression in human and murine squamous cell carcinomas and its relationship to regional endothelial cell proliferation. *Cancer Res.* 53:1444–1449.

Schweigerer, L., Malerstein, B. and Gospodarowicz, D. (1987) Tumor necrosis factor inhibits the proliferation of cultured capillary endothelial cells. *Biochem. Biophys. Res. Commun.* 143(3):997–1004.

Scott, P.A. and Harris, A.L. (1994) Current approaches to targeting cancer using antiangiogenesis therapies. *Cancer Treat. Rev.* 20(4):393 412.

Senger, D.R., Van, d. W.L., Brown, L.F., Nagy, J.A., Yeo, K.T., Yeo, T.K., Berse, B., Jackman, R.W., Dvorak, A.M. and Dvorak, H.F. (1993) Vascular permeability factor (VPF, VEGF) in tumor biology. *Cancer Metastasis Rev.* 12(3–4):303–324.

Shalaby, F., Rossant, J., Yamaguchi, T.P., Gertsenstein, M., Wu, X.F., Breitman, M.L. and Schuh, A.C. (1995) Failure of blood-island formation and vasculogenesis in Flk-1 deficient mice. *Nature* 376(6535):62–66.

Sheid, B. (1992) Angiogenic effects of macrophages isolated from ascitic fluid aspirated from women with advanced ovarian cancer. *Cancer Lett.* 62(2):153–158.

Shweiki, D., Itin, A., Soffer, D. and Keshet, E. (1992) Vascular endothelial growth factor induced by hypoxia may mediate hypoxia-initiated angiogenesis. *Nature* 359(6398):843–845.

Sightler, H., Borowsky, A., Dupont, W., Page, D. and Jensen, R. (1994) Evaluation of tumor angiogenesis as a prognostic marker in breast cancer. *Lab. Invest.* 70:22 A (abstract).

Simpson, J., Ahn, C., Battifora, H. and Esteban, J. (1994) Vascular surface area as a prognostic indicator in invasive breast carcinoma. *Lab. Invest.* 70:22 A (abstract)

Smith, H.S., Lu, Y., Deng, G., Martinez, O., Kramms, S., Ljung, B.M., Thor, A. and Lagios, M. (1993) Molecular aspects of early stages of breast cancer progression. *J. Cell. Biochem. Suppl.*

Smith-McCune, K. and Weidner, N. (1994) Demonstration and characterization of the angiogenic properties of cervical dysplasia. *Cancer Res.* 54:800–804.

Smolle, J., Soyer, H.P., Hofmann-Wellenhof, Smolle-Juettner, F.M. and Kerl, H. (1989) Vascular architecture of melanocytic skin tumors. *Path. Res. Pract.* 185:740–745.

Springer, T.A. (1990) Adhesion receptors of the immune system. *Nature* 346(6283):425–434.

Spyratos, F., Maudelonde, T., Brouillet, J.-P., Brunet, M., Defrenne, A., Andrieu, C., Hacene, K., Desplaces, A., Rouëssé, J. and Rochefort, H. (1989) Cathepsin D: an independent prognostic factor for metastasis of breast cancer. *Lancet* 8672:1115–1118.

Spyratos, F., Martin, P.M., Hacene, K., Romain, S., Andrieu, C., Ferrero, P.M., Deytieux, S., Le, D.V., Tubiana, H.M. and Brunet, M. (1992) Multiparametric prognostic evaluation of biological factors in primary breast cancer. *J. Natl. Cancer Inst.* 84(16):1266–1272.

Srivastava, A., Webster, D.J., Woodcock, J.P., Shrotria, S., Mansel, R.E. and Hughes, L.E. (1988) Role of Doppler ultrasound flowmetry in the diagnosis of breast lumps. *Br. J. Surg.* 75(9):851–853.

Stein, I., Neeman, M., Shweiki, D., Itin, A. and Keshet, E. (1995) Stabilization of vascular endothelial growth factor mRNA be hypoxia and hypoglycemia and coregulation with other ischemia-induced genes. *Mol. Cell. Biol.* 15(10):363–5368.

Steinhoff, G., Behrend, M. and Haverich, A. (1991) Signs of endothelial inflammation in human heart allografts. *Eur. Heart J.* 12 Suppl. D:141–143.

Strasser, A. Harris, A.W., Bath, M.L. and Cory, S. (1990) Novel primitive lymphoid tumours induced in transgenic mice by cooperation between myc and bcl-2. *Nature* 348(6299):331–333.

Stratton, M.R., Collins, N., Lakhani, S.R. and Sloane, J.P. (1995) Loss of heterozygosity in ductal carcinoma in situ of the breast. *J. Pathol.* 175(2):195–201.

Sumiyoshi, K., Baba, S., Sakaguchi, S., Urano, T., Takada, Y. and Takada, A. (1991) Increase in levels of plasminogen activator and type-1 plasminogen activator inhibitor in human breast cancer: possible roles in tumour progression and metastasis. *Thrombosis Res.* 63:59–71.

Sumizawa, T., Furikawa, T., Haraguchi, M., Yoshimura, A., Takeyasu, A., Ishizawa, M., Yamada, Y. and Akiyama, S. (1993) Thymidine phosphorylase activity associated with platelet-derived endothelial cell growth factor. *J. Biochem. Tokyo* 114(1):9–14.

Sunderkotter, C., Steinbrink, K., Goebeler, M., Bhardwaj, R. and Sorg, C. (1994) Macrophages and angiogenesis. *J. Leukoc. Biol.* 55(3):410–422.

Sutherland, R.M. (1988) Cell and environment interactions in tumor microregions: the multicell spheroid model. *Science* 240(4849):177–184.

Sutherland, R.M., McCredie, J.A. and Inch, W.R. (1971) Growth of multicell spheroids in tissue culture as a model of nodular carcinomas. *J. Natl. Cancer Inst.* 46(1):113–120.

Srivastava, A., Leidler, P., Davies, R., Horgan, K. and Hughes, L. (1988) The prognostic significance of tumor vascularity in intermediate-thickness (0.76–4.0 thick) skin melanoma. *Am. J. Pathol.* 133:419–423.

Takashashi, A., Sasaki, H., Kim, S.J., Tobisu, K., Kakizoe, T., Tsukamoto, T., Kumamoto, Y., Sugimura, T. and Terada, M. (1994) Markedly increased amounts of messenger RNAs for vascular endothelial growth factor and placenta growth factor in renal cell carcinoma associated with angiogenesis. *Cancer Res.* 54(15):4233–4237.

Takahashi, Y., Kitadai, Y., Bucana, C.D., Cleary, K.R. and Ellis, L.M. (1995) Expression of vascular endothelial growth factor and its receptor, KDR, correlates with vascularity, metastasis and proliferation of human colon cancer. *Cancer Res.* 55:3964–3968.

Tandon, A., Clark, G.M., Chamness, G.C., Chirgwin, J.M. and McGuire, W.L. (1990) Cathepsin D and prognosis in breast cancer. *N. Engl. J. Med.* 322:297–302.

Tang, D.G., Chen, Y.Q., Newman, P.J., Shi, L., Gao, X., Diglio, C.A. and Honn, K.V. (1993) Identification of PECAM-1 in solid tumor cells and its potential involvement in tumor cell adhesion to endothelium. *J. Biol. Chem.* 268(30):22883–22894.

Terman, B.I., Dougher, V.M., Carrion, M.E., Dimitrov, D., Armellino, D.C., Gospodarowicz, D. and Bohlen, P. (1992) Identification of the KDR tyrosine kinase as a receptor for vascular endothelial cell growth factor. *Biochem. Biophys. Res. Commun.* 187(3):1579–1586.

Tessler, S., Rockwell, P., Hicklin, D., Cohen, T., Levi, B.Z., Witte, L., Lemischka, I.R. and Neufeld, G. (1994) Heparin modulates the interaction of VEGF165 with soluble and cell associated flk-1 receptors. *J. Biol. Chem.* 269(17):12456–12461.

Thompson, T.C., Southgate, J., Kitchener, G. and Land, H. (1989) Multistage carcinogenesis induced by ras and myc oncogenes in a reconstituted organ. *Cell* 56(6):917–930.

Toi, M., Kashitani, J. and Tominaga, T. (1993) Tumor angiogenesis is an independent prognostic indicator in primary breast carcinoma. *Int. J. Cancer* 55(3):371–374.

Toi, M., Hoshina, S., Takayanagi, T. and Tominaga, T. (1994) Association of vascular endothelial growth factor expression with tumor angiogenesis and with early relapse in primary breast cancer. *Jpn. J. Cancer Res.* 85(10):1045–1049.

Toi, M., Hoshina, S., Taniguchi, T., Yamamoto, Y., Ishitsuka, H. and Tominaga, T. (1995) Expression of platelet-derived endothelial cell growth factor/thymidine phosphorylase in human breast cancer. *Int. J. Cancer* 64(2):79–82.

Tuszynski, G.P. and Nicosia, R.F. (1994) Localization of thrombospondin and its cysteine-serine-valine-threonine-cysteine-glycine-specific receptor in human breast carcinoma. *Lab. Invest.* 70(2):228–233.

Uria, J.A., Ferrando, A.A., Velasco, G., Freije, J.M. and Lopez, O.C. (1994) Structure and expression in breast tumors of human TIMP-3, a new member of the metalloproteinase inhibitor family. *Cancer Res.* 54(8):2091–2094.

Usuki, K., Saras, J., Waltenberger, J., Miyazono, K., Pierce, G., Thomason, A. and Heldin, C.H. (1992) Platelet-derived endothelial cell growth factor has thymidine phosphorylase activity. *Biochem. Biophys. Res. Commun.* 184(3):1311–1316.

van der Wal, A.C., Das, P.K., Tigges, A.J. and Becker, A.E. (1992) Adhesion molecules on the endothelium and mononuclear cells in human atherosclerotic lesion. *Am. J. Pathol.* 141(6):1427–1433.

Van Hoef, M.E., Knox, W.F., Dhesi, S.S., Howell, A. and Schor, A.M. (1993) Assessment of tumour vascularity as a prognostic factor in lymph node negative invasive breast cancer. *Eur. J. Cancer* 29A(8):1141–1145.

Van, M.E., Polverini, P.J., Chazin, V.R., Su, H.H., de, T.N. and Cavenee, W.K. (1994) Release of an inhibitor of angiogenesis upon induction of wild type p53 expression in glioblastoma cells. *Nat. Genet.* 8(2):171–176.

Vaporciyan, A.A., DeLisser, H.M., Yan, H.C., Mendiguren, I.I., Thom, S.R., Jones, M.L., Ward, P.A. and Albelda, S.M. (1993) Involvement of platelet-endothelial cell adhesion molecule-1 in neutrophil recruitment *in vivo*. *Science* 262(5139):1580–1582.

Vaupel, P., Kallinowski, F. and Okunieff, P. (1989) Blood flow, oxygen and nutrient supply and metabolis microenvironment of human tumors: a review. *Cancer Res.* 49(23): 6449–6465.

Vesalainen, S., Lipponen, P., Talja, M., Alhava, E. and Syrjanen, K. (1994) Tumor vascularity and basement membrane structure as prognostic factors in T1-2M0 prostatic adenocarcinoma. *Anticancer Res.* 14:709–714.

Vignon, F., Capony, F., Chambon, M., Freiss, G., Garcia, M. and Rochefort, H. (1986) Autocrine growth stimulation of the MCF7 breast cancer cells by the estrogen-regulated 52K protein. *Endocrinology* 118:1537–1545.

Visscher, D., Smilanetz, S., Drozdowicz, S. and Wykes, S. (1993) Prognostic significance of image morphometric microvessel enumeration in breast carcinoma. *Anal. Quant. Cytol.* 15:88–92.

Visscher, D., DeMattia, F. and Boman, S. (1994) Technical factors affecting image morphometric microvessel density counts in breast carcinomas. *Lab. Invest.* 70:168A (abstract)

Visscher, D.W., DeMattia, F., Ottosen, S., Sarkar, F.H. and Crissman, J.D. (1995) Biologic and clinical significance of basic fibroblast growth factor immunostaining in breast carcinoma. *Mod. Pathol.* 8:665–670.

Vlodavsky, I., Korner, G., Ishai, M.R., Bashkin, P., Bar, S.R. and Fuks, Z. (1990) Extracellular matrix-resident growth factors and enzymes: possible involvement in tumor metastasis and angiogenesis. *Cancer Metastasis Rev.* 9(3):203–226.

Volpert, O.V., Stellmach, V. and Bouck, N. (1995) The modulation of thrombospondin and other naturally occurring inhibitors of angiogenesis during tumor progression. *Breast Cancer Res. Treat.* 36:119–126.

Volpes, R., Van, D.O.J. and Desmet, V.J. (1992) Vascular adhesion molecules in acute and chronic liver inflammation. *Hepatology* 15(2):269–275.

Wakui, S., Furusato, M., Itoh, T., Sasaki, H., Akiyama, A., Kinoshita, I., Asano, K., Tokuda, T., Aizawa, S. and Ushigome, S. (1992) Tumour angiogenesis in prostatic carcinoma with and without bone marrow metastasis: a morphometric study. *J. Pathol.* 168(3): 257–262.

Waltenberger, J., Claesson, W.L., Siegbahn, A., Shibuya, M. and Heldin, C.H. (1994) Different signal transduction properties of KDR and Flt1, two receptors for vascular endothelial growth factor. *J. Biol. Chem.* 269(43):26988–26995.

Warren, B. (1979) The vascular morphology of tumors. *In:* H. Peterson (ed.): *Tumor Blood Circulation.* CRC Press, Boca Raton, 1–47.

Warren, B.A. and Shubik, P. (1966) The growth of the blood supply to melanoma transplants in the hamster cheek pouch. *Lab. Invest.* 15:464–478.

Weidner, N. and Gasparini, G. (1994) Determination of epidermal growth factor receptor provides additional prognostic information to measuring tumor angiogenesis in breast carcinoma patients. *Breast Cancer Res. Treat.* 29:97–108.

Weidner, N., Semple, J.P., Welch, W.R. and Folkman, J. (1991) Tumor angiogenesis and metastasis-correlation in invasive breast carcinoma. *N. Engl. J. Med.* 324(1): 1–8.

Weidner, N., Folkman, J., Pozza, F., Bevilacqua, P., Allred, E.N. Moore, D.H., Meli, S. and Gasparini, G. (1992) Tumor angiogenesis: a new significant and independent prognostic indicator in early-stage breast carcinoma [see comments]. *J. Natl. Cancer Inst.* 84(24): 1875–1887.

Weidner, N., Carroll, P.R., Flax, J., Blumenfeld, W. and Folkman, J. (1993) Tumor angiogenesis correlates with metastasis in invasive prostate carcinoma. *Am. J. Pathol.* 143(2): 401–409.

Weidner, N., Gasparini, G., Bevilacqua, P., Maluta, S., Palma, P., Caffo, O., Barbareschi, M., Borrachi, P., Marubini, E. and Pozza, F. (1994) Tumor microvessel density, p53 expression, and tumor size are relevant prognostic markers in node negative breast carcinomas. *Lab. Invest.* 70:A24 (abstract)

Weindel, K., Moringlane, J.R., Marme, D. and Weich, H.A. (1994) Detection and quantification of vascular endothelial growth factor/vascular permeability factor in brain tumor tissue and cyst fluid: the key to angiogenesis? *Neurosurgery* 35(3):439–448.

Weinstat, S.D. and Steeg, P.S. (1994) Angiogenesis and colonization in the tumor metastatic process: basic and applied advances. *FASEB J.* 8(6):401–407.

Wesseling, P., Vandersteenhoven, J.J., Downey, B.T., Ruiter, D.J. and Burger, P.C. (1993) Cellular components of microvascular proliferation inhuman glial and metastatic brain neoplasms. A light microscopic and immunohistochemical study of formalin-fixed, routinely processed material. *Acta Neuropathol. (Berl.)* 85(5):508–514.

Wiggins, D.L., Granai, C.O., Steinhoff, M.M. and Calabresi, P. (1995) Tumor angiogenesis as a prognostic factor in cervical carcinoma. *Gynecol. Oncol.* 56(3):353–356.

Williams, G.T. (1991) Programmed cell death: apoptosis and oncogenesis. *Cell* 65(7): 1097–1098.

Williams, J.K., Carlson, G.W., Cohen, C., Derose, P.B., Hunter, S. and Jurkiewicz, M.J. (1994) Tumor angiogenesis as a prognostic factor in oral cavity tumors. *Am. J. Surg.* 168(5): 373–380.

Winstanley, J., Leinster, S.J., Cooke, T.G., Westley, B.R., Platt-Higins, A.M. and Rudland, P.S. (1993) Prognostic significance of cathepsin-D in patients with breast cancer. *Br. J. Cancer.* 67:767–772.

Wong, S.Y., Purdie, A.T. and Han, P. (1992) Thrombospondin and other possible related matrix proteins in malignant and benign breast disease. An immunohistochemical study. *Am. J. Pathol.* 140(6):1473–1382.

Wu, C.C., Lee, C.N., Chen, T.M., Shyu, M.K., Hsieh, C.Y., Chen, H.Y. and Hsieh, F.J. (1994) Incremental angiogenesis assessed by color Doppler ultrasound in the tumorigenesis of ovarian neoplasms. *Cancer* 73(4):1251–1256.

Yamaguchi, T.P., Dumont, D.J., Conlon, R.A., Breitman, M.L. and Rossant, J. (1993) flk-1, an flt-related receptor tyrosine kinase is an early marker for endothelial cell precursors. *Development* 118(2):489–498.

Zabrenetzky, V., Harris, C.C., Steeg, P.S. and Roberts, D.D. (1994) Expression of the extracellular matrix molecule thrombospondin inversely correlates with malignant progression in melanoma, lung and breast carcinoma cell lines. *Int. J. Cancer* 59:191–195.

Zagzag, D., Brem, S. and Robert, F. (1988) Neovascularization and tumor growth in the rabbit brain. A model for experimental studies of angiogenesis and the blood-brain barrier. *Am. J. Pathol.* 131(2):361–372.

Zajchowski, D.A., Band, V., Trask, D.K., Kling, D., Connolly, J.L. and Sager, R. (1990) Suppression of tumor-forming ability and related traits in MCF-7 human breast cancer cells by fusion with immortal mammary epithelial cells. *Proc. Natl. Acad. Sci.* USA 87(6): 2314–2318.

Zarnegar, R. and DeFrances, M.C. (1993) Expression of HGF-SF in normal and malignant human tissues. EXS 65(181):181–199.

Zetter, B.R. (1993) Adhesion molecules in tumor metastasis. *Semin. Cancer Biol.* 4(4): 219–229.

Zhang, H.T., Craft, P., Scott, P.A., Ziche, M., Weich, H.A., Harris, A.L. and Bicknell, R. (1995) Enhancement of tumor growth and vascular density by transfection of vascular endothelial cell growth factor into MCF-7 human breast carcinoma cells. *J. Natl. Cancer Inst.* 7(3):213–219.

Ziche, M. and Gullino, P.M. (1982) Angiogenesis and neoplastic progression *in vitro*. *J. Natl. Cancer Inst.* 69(2):483–487.

Zocchi, M.R. and Poggi, A. (1993) Lymphocyte-endothelial cell adhesion molecules at the primary tumor site in human lung and renal cell carcinomas [letter]. *J. Natl. Cancer Inst.* 85(3):246–247.

Molecular mechanisms of angiogenesis regulation

Control of angiogenesis by cytokines and growth factors

Regulation of Angiogenesis
ed. by I.D. Goldberg & E.M. Rosen
© 1997 Birkhäuser Verlag Basel/Switzerland

Fibroblast growth factors as angiogenesis factors: New insights into their mechanism of action

S. Klein[1], M. Roghani[1] and D.B. Rifkin[1,2]

[1] *Department of Cell Biology and Kaplan Cancer Center, and*
[2] *The Raymond and Beverly Sackler Foundation Laboratory, New York University Medical Center, New York, New York 10016, USA*

Introduction

Angiogenesis is critical for physiological processes such as embryonic development and wound repair (D'Amore and Thompson, 1987; Folkman and Shing, 1992). Angiogenesis also contributes to several pathologies either directly, as in diabetic retinopathy, or indirectly by supporting the growth of pathological tissues, as in rheumatoid arthritis and tumor growth (Folkman, 1995). During angiogenesis, new capillaries arise as sprouts from preexisting capillaries or post-capillary venules. This phenomenon is accomplished by a series of sequential steps. In the initial phase of capillary sprouting, the basement membrane of endothelial cells in the parent blood vessel is degraded. Degradation of the basement membrane is mediated by endothelial cell proteinases, most notably plasminogen activators (PAs) and a variety of matrix-degrading metalloproteinases (MMPs). Once the basement membrane is degraded, endothelial cells bud out from the preexisting vessel and migrate into the perivascular space. Cells at the base of the sprout proliferate and replace migrated cells. Initially, the tubes or cords contain no lumen; a new basement membrane is formed, and after two contiguous sprouts have fused to form a loop, a lumen forms and blood begins to flow (Ausprunk and Folkman, 1977).

A number of growth factors and cytokines, such as vascular endothelial growth factor (VEGF), hepatocyte growth factor (HGF), epidermal growth factor (EGF), platelet-derived growth factor (PDGF), transforming growth factor-β (TGF-β), and fibroblast growth factor (FGF), are regulators of angiogenesis. The aim of this chapter is to describe the recent progress made in the understanding of the mechanisms mediating angiogenesis induced by the FGFs. The FGF family consists of nine structurally related polypeptides: the two prototypes, acidic FGF (FGF-1) (Jaye et al., 1986) and basic FGF (FGF-2) (Abraham et al., 1986a, b), and seven additional members, FGF-3 (int-2) (Dickson et al., 1984), FGF-4 (hst-1/kaposi-FGF) (Sakamoto et al., 1986), FGF-5 (Zhan et al., 1988), FGF-6 (hst-2) (Marics et al., 1989), FGF-7 (keratinocyte growth factor) (Rubin et al., 1989),

FGF-8 (androgen induced growth factor (Tanaka et al., 1992), and FGF-9 (glia-activating factor (Miyamoto et al., 1993). The cellular effects of the FGFs are mediated via specific binding to high-affinity tyrosine kinase receptors. Currently, the family of FGF receptors includes four members: FGF receptor-1 (the *flg* gene product), FGF receptor-2 (the *bek* gene product), FGF receptor-3 (FGFR-3), and FGF receptor-4 (FGFR-4).

In this review we have limited our discussion to FGF-1 and FGF-2 because these two molecules are the most extensively studied and have been shown to induce angiogenesis in various model systems. We will focus on the mechanisms of action of FGF-mediated proteolysis, migration, proliferation, and integrin expression.

Protein structure and tissue distribution of FGF-2 and FGF-1

FGF-2 was originally purified from the bovine pituitary gland as a 146-amino acid protein with a molecular mass (Mr) of 16,5 kDa and an isoelectric point of 9.6 (Esch et al., 1985a). Molecules with identical properties were subsequently isolated from bovine brain, retina, and adrenal, and from human brain (Gospodarowicz et al., 1984; Baird et al., 1985; Bohlen et al., 1985; Gimenez-Gallego et al., 1986; Gospodarowicz et al., 1986). FGF-2 molecules larger than 146 amino acids were isolated in the presence of protease inhibitors (Ueno et al., 1986; Story et al., 1987).

When both bovine and human FGF-2 cDNAs were cloned, an AUG codon was identified in the proper context to initiate translation of a protein of 155 amino acids. No additional 5' in-frame AUG codons were found (Abraham et al., 1986a, b). Therefore, translation was predicted to initiate at this AUG codon and to result in an 18 kDa protein. However, larger forms of FGF-2 were later purified from human placenta and from guinea pig brain (Moscatelli et al., 1987; Sommer et al., 1987). These higher Mr forms of FGF-2 were shown to arise from use of alternative translation initiation (CUG) codons upstream of the AUG codon. Thus, human FGF-2 is expressed in four forms: an 18 kDa form (155 amino acids) generated by initiation at the AUG codon, and 22, 22.5, and 24 kDa forms (196, 201, and 210 amino acids) arising from the CUG codons (Florkiewicz and Sommer, 1989; Prats et al., 1989). The high molecular weight (HMW) forms of FGF-2 contain the complete amino acid sequence of the 18 kDa form in addition to NH_2-terminal extensions of varying lengths.

FGF-2 has been found in all organs and tissues examined (Baird et al., 1986), and in the endothelial cells of some, but not all, blood vessels (Hanneken et al., 1989; Cordon-Cardo et al., 1990). FGF-2 is also synthesized by cultured fibroblasts, endothelial cells, glial cells, and smooth-muscle cells (Moscatelli et al., 1986a; Connolly et al., 1987; Schweigerer et al., 1987; Vlodavsky et al., 1987; Gospodarowicz et al., 1988; Hatten et al., 1988; Weich et al., 1990). Because these cell types are ubiquitous, the expression of FGF-2 may be widespread *in vivo*.

The relative amounts of the different Mr forms of FGF-2 vary among cell lines and in tissues during development, implying that the use of the alternative codons is highly regulated (Giordano et al., 1992; Liu et al., 1993; Dono and Zeller, 1994; Riese et al., 1995). It has been suggested that *cis*-acting elements in FGF-2 mRNA are involved in regulating the translation of different forms at the four initiation sites (Prats et al., 1992). In addition, translation initiation of FGF-2 mRNA does not involve the cap-dependent ribosome scanning mechanism but occurs through a process of internal ribosome entry mediated by the mRNA leader sequence. The presence of an internal ribosome entry site in FGF-2 mRNA may represent a mechanism to control FGF-2 expression during processes such as angiogenesis (Vagner et al., 1995).

FGF-1 was isolated originally as a 154-amino acid protein plus truncated forms of 140 and 134 amino acids (Esch et al., 1985b; Gimenez-Gallego et al., 1985; Burgess et al., 1986; Harper et al., 1986). The primary translation product of FGF-1 contains 155 amino acids (Jaye et al., 1986). There appear to be no NH$_2$-terminal extended forms of FGF-1, because a termination codon is found at position-1 to the AUG initiation codon. However, the existence of alternate 5' untranslated exons in FGF-1 mRNA has been described (Chiu et al., 1990; Crumley et al., 1990). The role of these untranslated sequences is unknown, but they may be involved in the differential regulation of translation of the molecule.

FGF-1 has 55% amino acid sequence identity with 18 kDa FGF-2. Homology extends over the entire sequence of the molecule, except for the 18 NH$_2$-terminal amino acids and a two-amino acid insert at positions 117 and 118. FGF-1 contains three cysteine residues, two of which are conserved among the four cysteines present in FGF-2. The cysteines do not seem to be required for biological activity, but modification of the cysteines may hinder the adaptation of biologically active conformations of the molecules (Crabb et al., 1986; Jaye et al., 1987; Fox et al., 1988; Linemeyer et al., 1990).

FGF-1 appears to have a more limited distribution than FGF-2. It has been found in neural tissue, kidney, prostate, and cardiac muscle (D'Amore and Klagsbrun, 1984; Thomas et al., 1984; Crabb et al., 1986; Gautschi-Sova et al., 1987; Quinkler et al., 1989; Casscells et al., 1990), and has also been identified in cultured vascular smooth muscle cells (Winkles et al., 1987; Weich et al., 1990).

Mechanism of FGF-1 and FGF-2 release from cells

An important question concerning the biology of FGF-1 ad FGF-2 is the mechanism of release of these growth factors from cells, as both molecules lack a signal sequence for secretion (Abraham et al., 1986b; Jaye et al.,

1986). In spite of the absence of a signal sequence, the growth factors are released and localized in the basement membrane and extracellular matrix (ECM) of numerous tissues. A number of other mammalian proteins, including interleukin-1α and interleukin-1β (March et al., 1985; Stevenson et al., 1992), platelet-derived endothelial cell growth factor (Ishikawa et al., 1989), ciliary neurotrophic factor (Lin et al., 1989), and sciatic nerve growth promoting activity (Leung et al., 1992), also lack conventional hydrophobic signal sequences required for protein secretion. Nevertheless, like the FGFs, these proteins exhibit important extracellular functions and are found in basement membrane and in culture medium, where they exert their biological activity through binding to specific cell surface receptors.

Originally, it was proposed that FGF-2 was released following cell death, disruption of plasma membrane integrity, or sublethal cell injury (McNeil et al., 1989; Muthukrishan et al., 1991). However, other studies have suggested that 18 kDa FGF-2 is exported from cells in a physiologically relevant manner without compromising cell integrity or requiring cell death. Mignatti et al. (1991a) showed that antibodies against FGF-2 inhibit migration of single, isolated cells. Thus, FGF-2 is released from viable cells and acts as an autocrine factor. Release of both FGF-2 and FGF-1 must be independent of the endoplasmic reticulum/Golgi complex as drugs that block secretion via the classical secretion pathway do not inhibit growth factor release from cells (Mignatti et al., 1992; Florkiewicz et al., 1995; Jackson et al., 1995).

The mechanism of FGF release is not understood. Exocytosis and/or an ATP-dependent pathway may be involved as drugs that affect either exocytosis (Mignatti et al., 1992) or energy-dependent steps (Florkiewicz et al., 1995) block FGF-release. Heat shock response may also be involved, as under conditions of temperature stress FGF-1 is released into the extracellular environment (Jackson et al., 1992). In response to heat shock FGF-1 is released as a latent homodimer that does not associate with heparin and has no biological activity. The latent extracellular FGF-1 that binds to heparin and is active (Jackson et al., 1992). Therefore, the extracellular redox potential of tissue microenvironments may ultimately be responsible for the activation of extracellular FGF-1.

Receptors

FGFs transduce signals through two classes of receptors: high-affinity tyrosine kinase receptors (Kd = $2-20 \times 10^{-11}$ M) and low-affinity heparan sulfate proteoglycans (Kd = $2 \times 10^{-9} - 2 \times 10^{-7}$ M). Four distinct genes, FGFR-1, FGFR-2, FGFR-3, and FGFR-4, encoding cell surface receptor tyrosine kinases, have been identified (Lee et al., 1989; Dionne et al., 1990; Keegan et al., 1991; Partanen et al., 1991). The FGFRs share a common structure consisting of a signal peptide, two or three immunoglobulin

(Ig)-like loops, and an acidic region between the first and second immuno-globulin loops. The intracellular domain includes the catalytic tyrosine kinase domain, which is split by a short kinase insert. FGF-1 binds with high affinity to all four tyrosine kinase receptors, whereas FGF-2 binds with higher affinity to FGFR-1 and FGFR-2. Multiple forms of the FGFRs are generated via alternative splicing (for review, see Johnson and Williams, 1993). cDNAs encoding soluble receptors, truncated receptors, and Ig loop variants have been isolated. Differential splicing in the extra-cellular region of FGFR-1 generates receptor variants with either two or three Ig loops that differ in their affinity for FGF-1 (Shi et al., 1993; Wang et al., 1995).

The second class of receptors to which FGFs bind, but with a lower affinity, are the heparan sulfate proteoglycans (HSPGs) present in abun-dance on the cell surface and in the ECM (Saksela et al., 1988; Vigny et al., 1988; Bashkin et al., 1989; Kiefer et al., 1990). Several roles have been proposed for FGF-1 and -2 binding to heparan sulfates. The interaction with heparan sulfates stabilizes and protects FGF-1 and -2 from thermal denaturation (Gospodarowicz and Cheng, 1986) and from the action of proteases (Saksela et al., 1988; Sommer and Rifkin, 1989), thus enhancing FGF-1 and -2 efficacy. FGF-1 and -2 bound to proteoglycans in the matrix or on the cell surface provide a reservoir that ensures a long-term response to a brief exposure to these growth factors (Flaumenhaft et al., 1989; Presta et al., 1989a). By binding to the low-affinity receptors, FGF-1 and -2 molecules diffuse less in the extracellular space and are concen-trated at the cell surface. This allows more frequent interactions of FGF-1 and -2 with the high-affinity signaling receptors, thereby promoting receptor activation (Schlessinger et al., 1995). FGF-1 and -2 can be releas-ed from the low-affinity binding sites by enzymatic digestion of the HSPG followed by release as a soluble complex. Enzymatic release can occur through degradation of the glycosaminoglycan chains with heparanases (Ishai-Michaeli et al., 1990), of the core protein with plasmin (Saksela and Rifkin, 1990), or in the case of phosphoinositol-linked HSPGs, of the phospholipid anchor by phospholipases (Brunner et al., 1991; Bashkin et al., 1992). These mechanisms generate soluble heparan sulfate-FGF-1 and -2 complexes that can diffuse and activate receptors at sites distant from the point of release (Flaumenhaft et al., 1990).

Another proposed role for HSPGs is that these molecules are necessary for FGF-2 binding to high-affinity receptors. This hypothesis derives from observations that FGF-2 binding to FGFR-1 is diminished in mutant Chinese hamster ovary cell lines defective in glycosaminoglycan meta-bolism (Yayon et al., 1991) and in myoblasts defective in their sulfation of HSPGs (Rapraeger et al., 1991). Subsequent analysis of the conditions required for high affinity binding of FGF-2 to the purified FGFR-1 extra-cellular domain confirmed the importance of the initial formation of specific FGF-2-heparin complexes for binding and activation of FGF receptors (Ornitz et al., 1992). Heparin was found to be essential for the mitogenic activity of FGF-2 and FGF-1 in IL-3-dependent cell lines made FGF-dependent by transfection with FGFR cDNA (Bernard et al., 1991; Mansukhani et al., 1992; Ornitz et al., 1992). Based on these findings, Klagsbrun and Baird (1991) proposed a dual receptor system that is composed first of a low-affinity HSPG receptor that binds FGF-2; this induces a conformational change in the molecule, and thus presents the ligand to its high-affinity receptor in a biologically active form. According to this model, a conformational change is necessary for the proper interaction of FGF-2 with its high affinity receptor. However, other studies with the purified FGFR-1 extracellular domain have demonstrated that heparin is not required for FGF-2 binding to high-affinity receptors (Kiefer et al., 1991; Bergonzoni et al., 1992; Pantoliano et al., 1994; Roghani et al., 1994). Cell surface HSPGs may modulate the action of FGF-2 by increasing its affinity for its receptor (Roghani et al., 1994; Pantoliano et al., 1994), thereby enhancing the FGF/FGFR interaction.

The FGFRs are activated by dimerization. Two mechanisms have been proposed to explain the role of FGF and heparin in receptor dimerization. One mechanism is based on the ability of heparin to provide a template for binding multiple FGFs when the growth factor is in excess (Ornitz et al., 1992; Mach et al., 1993; Spivak-Kroizman et al., 1994a; Thompson et al., 1994). Once bound to heparin, two molecules of FGF-1 have the potential to juxtapose two molecules of FGFR-1, facilitating receptor dimerization. This implies a mechanism in which a dimeric growth factor complex initiates receptor dimerization (stoichiometry of 2 FGFR: 2 FGF) and only one FGFR-1 binding site of FGF-1 would be required. A second mechanism for receptor dimerization suggests that monomeric FGF-2 binds two receptor molecules at two distinct FGFR-1 binding epitopes on opposite faces of the same FGF-2 molecule (stoichiometry of 2 FGFR: 1 FGF). This mechanism is analogous to the reported mechanism for human growth hormone interacting with its receptor; receptor dimerization is facilitated by monomeric human growth hormone containing two receptor binding domains (Cunningham, 1991). A primary role of heparin is to bind to FGF-2 and promote the secondary FGF-2/FGFR-1 interaction, thereby allowing receptor dimerization to occur (Springer et al., 1994).

Receptor dimerization induced by FGF binding, in turn, activates protein tyrosine kinase activity and autophosphorylation (for a review, see Schlessinger and Ullrich, 1992). Consequently, cellular target proteins, such as phospholipase $C\gamma$ (PLCγ) and Ras GTPase activating protein (GAP), bind to tyrosine autophosphorylation sites in the receptor cytoplasmic domain and become phosphorylated on tyrosine residues. Tyrosine autophosphorylation sites serve as binding sites for adaptor proteins such as Grb2, Shc, and Nck. Grb2 is bound to the Ras guanine nucleotide-releasing factor, Sos. The binding of Grb2/Sos complex to tyrosine autophosphorylated EGF receptor results in the translocation of Sos to the plasma membrane in the vicinity of Ras, and in the subsequent exchange of GDP for GTP and activation of Ras (for a review, see Schlessinger, 1993). This leads to activation of a kinase cascade composed of Raf, Map kinase kinase, and Map kinase (MAPK) (for a review, see Marshall, 1994). It is now well established that the Ras signaling pathway plays an important role in initiation of cell proliferation by many growth factors such as FGF.

One of the target molecules of FGFR-1 is PLCγ, which, upon ligand stimulation, binds to the receptor, becomes activated by tyrosine phosphorylation, and catalyzes the hydrolysis of phosphatidylinositol (PI) (for a review, see Rhee et al., 1991). Tyrosine 766 in the cytoplasmic tail of FGFR-1 is the binding site for PLCγ (Mohammadi et al., 1991). Replacement of tyrosine 766 with phenylalanine (Y766F) by site-directed mutagenesis prevents FGF-induced PI hydrolysis and Ca^{2+} release in FGFR-1-transfected myoblast cells (Mohammadi et al., 1992; Peters et al., 1992). However, FGF-1 still induces DNA synthesis in L6 cells or differentiation of PC12 cells expressing the mutant Y766F FGFR-1. This indicates that PI hydrolysis is not essential for FGF-induced mitogenesis of L6 myoblasts and neuronal differentiation of PC12 cells (Mohammadi et al., 1992; Peters et al., 1992; Spivak-Kroizman et al., 1994b). MAP kinase has been implicated as a critical component of the mitogenic signaling pathway. MAPK can be activated by both Ras-dependent and Ras-independent mechanisms (Burgering et al., 1993). In Ba/F3 hematopoietic cells expressing the mutant Y766F, FGFR-1, FGF-1 stimulates MAPK activation less than in cells expressing wild type FGFR-1. However, Ras is activated to the same extent in cells expressing either wild type or mutant receptor (Huang et al., 1995). A model suggesting that two pathways are utilized to transduce FGF-1 signals was proposed on the basis of these findings (Huang et al., 1995). One pathway depends upon Ras and a second pathway depends on PLCγ and PI hydrolysis. Raf-1 integrates signals from these two pathways to activate MAPK, ultimately leading to mitogenesis.

The mechanism regulating FGF-mediated cell migration is less understood. Sa and Fox (1994) proposed that the migratory response of vascular endothelial cells to FGF-2 is due to release of arachidonic acid, the primary fatty acid product of phospholipase A$_2$. The authors compared the role of signal-transducing G-proteins in migration and proliferation of bovine aortic endothelial cells. They found that pertussis toxin, which ADP-ribosylates susceptible G-proteins, reduces FGF-2-stimulated endothelial cell migration. In contrast, pertussis toxin does not inhibit FGF-2-stimulated endothelial cell proliferation. This suggests that distinct intracellular signaling mechanisms may be involved in the migratory and proliferative responses to FGF-2. Further studies will be necessary to elucidate these and other signal transduction pathways by which FGF-1 and -2 influence processes such as cell proliferation, migration, and differentiation.

Roles in development

Determination of prospective mesoderm cells in the embryo (mesoderm induction) starts during the earliest developmental stages and continues during primitive streak formation and subsequent gastrulation. Experiments with *Xenopus laevis* embryos have shown that members of the FGF family and their receptors play essential roles during mesoderm induction (reviewed by Kimelman, 1993; Smith, 1993). When added to animal caps (*Xenopus* blastular ectodermal explants), FGF-2 activates transcription of several mesoderm-specific genes, including *Xbra*, *1A11*, and *Xwnt8* (Christian et al., 1991; Smith et al., 1991; Taira et al., 1992).

Biochemical analysis has revealed that three FGF-2 forms of 18.5, 20.0, and 21.5 kDa are expressed during early avian embryogenesis (Dono and Zeller, 1994). The regulation of the subcellular distribution of FGF-2 appears to be a control mechanism during avian mesoderm induction and patterning (Shiurba et al., 1991; Godsave and Shiurba, 1992; Dono and Zeller, 1994; Riese et al., 1995). FGF-2 proteins are found to be predominantly nuclear in prestreak blastodisc cells committed to a mesodermal fate. The nuclear localization of FGF-2 during mesoderm induction appears to be evolutionarily conserved. Shiurba et al. (1991) reported that nuclear translocation of FGF-2 occurs during mesoderm induction in *Xenopus laevis* embryos. During primitive streak formation, however, FGF-2 starts to appear in the cytoplasm of epiblast cells but remains nuclear in the hypoblast. The FGF-2 proteins become predominantly cytoplasmic in all the cells during the subsequent developmental stages. High FGF-2 levels are also found in the extracellular basal lamina separating the epiblast from newly formed mesoderm. Extracellular FGF-2 may be required during the regression of Hensen's node and formation of mesodermal derivatives such as somites because blocking the binding of FGF to its receptors by heparin and/or suramin treatment of advanced stage gastrulating embryos interferes with these events (Riese et al., 1995). Thus, extracellular FGF-2 may be involved in mesoderm differentiation during the late stages of gastrulation, whereas nuclear FGF-2 may be important for mesoderm induction during the early phases preceding and at the onset of gastrulation.

In addition to the direct effects of FGFs on mesoderm induction, several studies have addressed the question of whether FGF-2 has indirect effects through the action of activin, a TGF-β family member (Green et al., 1992; Cornell and Kimelman, 1994; LaBonne and Whitman, 1994). Activin, which signals through a serine/threonine kinase receptor (Mathews and Vale, 1991), is a potent mesoderm inducer *in vivo*. Expression of a dominant negative activin receptor dramatically blocks mesoderm formation (Hemmati-Brivanlou and Melton, 1992). FGF-2 may be required for the effects of activin on mesoderm induction. This hypothesis is supported by experiments with dominant negative mutants either of the FGF receptor

(Cornell and Kimelman, 1994; LaBonne and Whitman, 1994) or of signaling molecules in the FGF receptor pathway, such as Ras and Raf (LaBonne and Whitman, 1994). Overexpression of any one of the dominant negative mutants inhibits the expression of mesoderm-specific genes induced by either FGF or activin. This finding suggests that FGF activation of a Ras-signaling pathway is required for FGF-induced and activin-induced mesoderm formation.

A role for HSPGs in FGF-2 mediated mesoderm induction has been proposed. Itoh and Sokol (1994) observed that the FGF-2 mediated stimulation of the expression of early (*Xbra, 1A11*) and late (cardiac actin) markers of mesoderm is inhibited by enzymatic removal of heparan sulfate from HSPGs by the heparinase treatment of explants of xenopus blastula ectoderm. In addition, heparinase treatment inhibits autonomous morphogenetic movements and mesodermal differentiation. These observations indicate that mesoderm induction and subsequent differentiation by FGF-2 depend on the presence of HSPGs. This finding is consistent with the proposed role of HSPGs in modulating FGF-2 binding to its high-affinity receptor (Rapraeger et al., 1991; Yayon et al., 1991; Nurcombe et al., 1993).

Progenitors of blood cells and endothelial cells differentiate from mesoderm. Endothelial progenitor cells, angioblasts, are defined as cells that have the potential to differentiate into endothelial cells but have not acquired all the characteristic markers (i.e. angiotensin converting enzyme, von Willebrand factor, VEGF receptors-1 and -2, vascular endothelial cadherin, platelet-endothelial cell adhesion molecule 1, P-selectin, and cd34) and not yet formed a lumen (for review, see Risau, 1995). In the yolk sac, angioblasts and hemopoietic precursor cells differentiate in close association with each other, forming blood islands. The formation of blood vessels from *in situ* differentiating angioblasts is called vasculogenesis. The subsequent processes of blood island fusion and formation of lumina by angioblasts lead to primordial vascular network. This vasculature is extended by sprouting of new capillaries from the preexisting network resulting in an elongated, branched vascular plexus. The specific role of FGF in vasculogenesis is suggested by experiments performed in an avian model system, in which endothelial cell differentiation is induced by incubation of epiblast cells with FGF-1 or FGF-2 (Flamme and Risau, 1992; Flamme et al., 1995). Other supportive findings show that dominant-negative mutants of FGFR-1 inhibit the induction of vasculogenic mesoderm and endothelial cell differentiation (Amaya et al., 1991). Thus, FGFs may have pleiotropic effects during vasculogenesis.

Regulation of the plasminogen activator system during angiogenesis

During the first phase of angiogenesis an increase in proteolytic activity is of pivotal importance as invasive endothelial cells must degrade the perivascular ECM in order to migrate into the tissue to be vascularized (Moscatelli and Rifkin, 1988; Pepper and Montesano, 1990). The FGF prototypes may be key modulators of the proteolytic response of endothelial cells during this process.

The best characterized protease system thought to be involved in angiogenesis is the plasminogen activator (PA) system (for reviews, see Vassalli et al., 1991; Mignatti and Rifkin, 1993). The PAs convert the zymogen plasminogen, a ubiquitous plasma protein, to plasmin. Plasmin has a broad trypsin-like substrate specificity and degrades several ECM components, such as fibronectin, laminin, and the protein core of proteoglycans (Werb et al., 1980). Plasmin does not degrade collagens and elastin but can degrade gelatins. In addition, plasmin also activates certain prometalloproteinases (Werb et al., 1977; Eaton et al., 1984), as well as latent elastase (Chapman and Stone, 1984).

There exist two types of PAs, the urokinase-type (uPA) and the tissue-type PA (tPA). tPA, whose activity is strongly enhanced by fibrin, is thought to be primarily involved in fibrinolysis. In contrast, uPA activity is not affected by fibrin and is thought to be primarily involved in processes such as cell invasion and tissue remodeling (for review, see Mignatti and Rifkin, 1993). uPA is secreted as a single chain proenzyme (pro-uPA) that is converted into the two chain polypeptide form by a single proteolytic cleavage. After secretion, pro-uPA interacts with a cell membrane binding site, the uPA receptor (uPAR), through its non-catalytic domain (Appella et al., 1987; Behrendt et al., 1990; Estreicher et al., 1990; Roldan et al., 1990), and is then activated by trace amounts of plasmin or by cathepsin B (Kobayashi et al., 1991). uPA bound to uPAR remains active on the cell membrane for several hours. The uPA/uPAR interaction accelerates plasminogen activation and stimulates several cell functions including proliferation, migration, and invasion (for a review, see Blasi, 1993). These cellular responses do not require uPA activity, and can be mediated by the non-catalytic amino terminal peptide of the enzyme (Fibbi et al., 1988; Odekon et al., 1992; Rabbani et al., 1992; Busso et al., 1994). Transfection of invasive cells with uPAR antisense cDNA inhibits their invasive potential, suggesting that interaction of uPA with uPAR is required for cell invasion (Kook et al., 1994). Unexpectedly, uPA and tPA knock-out mice show no abnormalities in vasculo- or angiogenesis (Carmeliet et al., 1994). It is possible that compensatory mechanisms mediate plasmin formation in the absence of PAs. However, abnormal endothelial cell functions may be observed after pharmacological or experimental manipulation of these mice.

Endothelial cells also produce plasminogen activator inhibitor-1 (PAI-1) (Pepper et al., 1992), a single chain molecule that binds the active two-chain forms of either uPA or tPA with a molar ratio of 1:1. PAI-1 bound to the extracellular matrix of endothelial cells can be released by proteases such as thrombin, which inactivates PAI-1 (Ehrlich et al., 1991), and by cathepsin G, which increases PAI-1 activity (Pintucci et al., 1993). Binding of PAI-1 to uPAR-bound uPA results in a rapid internalization and degradation of uPAR-uPA-PAI-1 complexes (for review, see Conese and Blasi, 1995). Thus, PAI-1 also has an important role in the turnover of uPA as it participates in the clearance of inactivated uPA from the cell surface.

Mobilization of FGF-2 from the ECM may contribute to the control of extracellular proteolysis mediated by vascular endothelial cells during angiogenesis. Plasmin appears to play a major role in this process (Saksela and Rifkin, 1990). Although other degradative enzymes such as heparitinases, gelatinases, stromelysins, and cathepsins can release FGF-2 from the ECM, the relative abundance of plasminogen in virtually all tissues, and the efficient amplification mechanism achieved by PA-mediated plasminogen activation, implicates plasmin as the most important proteinase in this process. FGF-2 released from the ECM stimulates uPA and collagenase expression in microvascular endothelial cells (Moscatelli et al., 1986b). This effect is independent of protein kinase C activation and requires Ca^{2+} influx (Presta et al., 1989b). Although the upregulation of uPA by FGF-2 appears to occur at the transcriptional level (Gualandris and Presta, 1995), a posttranscriptional mechanism has also been proposed. This hypothesis is based on the finding that, although the increase in uPA transcript level is detected early (4 h), the increase in uPA activity requires FGF-2 to be present in the extracellular environment for 12 h. It is speculated that internalized FGF-2 may exert its late effects on uPA expression by modulating the cytoplasmic translocation of mature nuclear uPA mRNA and/or the translation of the translocated message (Gualandris and Presta, 1995). However, further analysis is necessary to elucidate this mechanism.

FGF-2 also regulates uPAR expression in vascular endothelial cells (Mignatti et al., 1991b). Thus, FGF-2 exhibits a double regulatory effect on the PA system by enhancing the endothelial cell capacity both to synthesize and to bind uPA on the cell membrane.

FGF-2 signaling through its high affinity receptors increase PA levels (Moscatelli, 1987). A HSPG pathway may also mediate this effect as FGF-2 increases PA activity in L6 myoblasts lacking high-affinity receptors (Quarto and Amalric, 1994). Pretreatment of the cells with chlorate, which blocks HSPG sulfation (Rapraeger et al., 1991), blocks the expected FGF-induced increase in PA activity, suggesting that HSPGs are required for FGF-2 stimulation of PA levels (Quarto and Amalric, 1994).

The mobilization of FGF-2 from the ECM is one mechanism for mediating PA activity. A second mechanism involves the activation of latent TGF-β1. TGF-β1 is a potent inhibitor of cell proliferation, migration, and proteinase production in vascular endothelial cells (Mullins and Rohrlich, 1983; Hekman and Loskutoff, 1985; Rifkin et al., 1993). TGF-β1 is secreted constitutively by many tumor and normal cell types, including endothelial cells, as part of a HMW complex. The mature TGF-β (25 kDa) is associated non-covalently with its 75 kDa propeptide that is disulfide-linked to a 125–190 kDa protein, the latent TGF-β binding protein (LTBP). Mature TGF-β1 must be released from this complex to interact with its plasma membrane receptor and elicit a biological response. The active TGF-β1 formed in the extracellular milieu can counteract the stimulatory effects of FGF-2 on vascular endothelial cells by downregulating uPA and collagenase gene expression and stimulating the synthesis of PAI-1 and the tissue inhibitor of metalloproteinases (Pepper et al., 1990). As a consequence, plasmin formation and metalloproteinase activity are blocked. However, the activation of latent TGF-β1 requires the action of plasmin (Lyons et al., 1988, 1990; Sato and Rifkin, 1989; Sato et al., 1990). Therefore, FGF-2 can induce latent TGF-β1 activation by increasing plasmin production. The active TGF-β1 generated, in turn, dampens the response of endothelial cells to FGF-2 (Flaumenhaft et al., 1992). This represents a self-regulatory system at two levels: 1) alternate cycles of activation or inhibition of extracellular proteolysis, and 2) alternate cycles of release and/or activation of endothelial cell growth factors and growth inhibitors.

Roles in tumor angiogenesis

The growth of solid tumors, as well as their ability to metastasize, is dependent upon the process of angiogenesis (Folkman et al., 1989; Folkman, 1990; Weidner et al., 1991). Blood vessels penetrating into the tumor parenchyma provide nutrition and oxygen for the multiplying cells. In addition, blood vessels provide an entry route for infiltrating immune cells and an exit route into the systemic circulation for metastasizing tumor cells. A variety of experiments have demonstrated that tumors release angiogenic factors (Folkman and Klagsbrun, 1987). FGF-2 was one of the original angiogenic factors described in tumors (Folkman and Klagsbrun, 1987).

Consistent with a role for FGF-2 during tumor angiogenesis, the transition from a preneoplastic to a malignant vascularized fibrosarcoma is associated with release of FGF-2 from tumor cells (Kandel et al., 1991). Li et al. (1994) reported that the cerebrospinal fluid (CSF) of children and adults with brain tumors contained FGF-2, the level of which correlated with the extent of vascularization of the tumor. Both FGF-2 and VEGF are overexpressed in infantile hemangiomas during the proliferative phase, and urinary levels are concomitantly elevated during this period (Takahashi et al., 1994). High tissue levels of FGF-2 were observed in renal cell carcinomas (Nanus et al., 1993), and abnormally elevated levels of FGF-2 were also found in the serum and urine of patients with a variety of solid tumors, leukemias, and lymphomas (Nguyen et al., 1994).

The effect of neutralizing FGF-2 antibody on tumor growth and angiogenesis differs according to the tumor type: anti-FGF-2 antibodies inhibited the growth and angiogenesis of malignant human glial tumors (Stan et al., 1995), but had no effect on three different murine tumors (Dennis and

Rifkin, 1990). It is not surprising that administering anti-FGF-2 antibodies alone is not sufficient to block the growth of certain tumors *in vivo*. Other factors with growth and angiogenic properties, such as VEGF, TGF-β1, and HGF/SF, can also be synthesized by many neoplastic cells.

VEGF is a potent endothelial cell mitogen, as well as an angiogenic inducer *in vivo*. Several observations suggest that VEGF may play a crucial role during tumor angiogenesis. *In situ* hybridization studies have shown that high levels of VEGF mRNA are expressed by a variety of human tumors, including renal cell and colon carcinoma, glioblastoma multiforme, and capillary hemangioblastoma (Berse et al., 1992; Plate et al., 1992; Shweiki et al., 1992; Berkman et al., 1993). A strong correlation exists between the degree of vascularization of the tumor and VEGF mRNA expression (Berkman et al., 1993). In addition, monoclonal anti-VEGF antibodies inhibited both angiogenesis and tumor growth in mice injected with human rhabdomyosarcoma, glioblastoma multiforme, or leiomyosarcoma cell lines (Kim et al., 1993).

In addition to a direct effect on endothelial cells, another mechanism by which FGF-2 may act during tumor angiogenesis is by stimulating the synthesis of VEGF in tumor cells. FGF-2 has been shown to increase the secretion of VEGF in nine human malignant glioma cell lines and in a meningioma cell line (Tsai et al., 1995). This represents an additional positive feedback mechanism where FGF-2 released by tumor cells stimulates the secretion of a second potent angiogenesis factor, VEGF.

FGF-2 is released from tumor cells and tumor-associated macrophages (Lewis et al., 1995), or mobilized from the extracellular matrix by proteases, and acts by a paracrine mechanism on endothelial cells to promote angiogenesis. In addition, the FGF-2 produced by vascular endothelial cells can initiate tumor angiogenesis by an autocrine mechanism. This hypothesis is based on several findings. First, high levels of FGF-2 are present in the capillary endothelial cells of some tumors (Schulze-Osthoff et al., 1990). Second, certain tumor cell lines secrete a factor that rapidly upregulates both endothelial cell expression of FGF-2 and *in vitro* angiogenesis. Incubation of endothelial cells with the tumor cell-conditioned medium results in increased expression of uPA (Peverali et al., 1994), which may represent an essential feature for cell invasion during tumor angiogenesis (for review, see Mignatti and Rifkin, 1993). Although tumor cell-conditioned media may not contain detectable FGF-2 (Moscatelli et al., 1986a), addition of anti-FGF-2 antibody abolishes the uPA upregulation, and blocks *in vitro* angiogenesis, indicating that both the increase in uPA activity and angiogenesis are mediated by the endogenous FGF-2 of endothelial cells. VEGF, like FGF-2, may also act as an autocrine factor during tumor angiogenesis. Recent observations indicate that FGF-2 synthesized by endothelial cells induces VEGF expression in the same cells. This suggests that VEGF also acts as an autocrine angiogenic factor in response to FGF-2 synthesis (P. Mignatti, personal communication).

A recent, intriguing observation suggests that angiogenesis inhibitors may also be produced by tumor cells. O'Reilly and coworkers (1994) reported that the primary tumor produces angiogenic inhibitors that enter the circulation and prevent growth in the metastatic lesion. One important inhibitor may be angiostatin, a 32 kDa plasminogen fragment that has been identified in the urine of mice with Lewis lung carcinomas but not in urine from control mice (O'Reilly et al., 1994). Angiostatin specifically inhibits endothelial cell proliferation and blocks the neovascularization and growth of metastases *in vivo*. Although a mechanism for the action of angiostatin is not known, its binding to heparin suggests the possibility that it might displace FGF from HSPGs and inhibit the response of cells to this angiogenesis factor.

Control of integrin expression in endothelial cells

The interaction of microvascular endothelial cells with adjacent extracellular matrices changes dramatically during angiogenesis. Quiescent microvascular endothelial cells adhere to adjacent cells and to a basement membrane. Upon stimulation by angiogenic factors, endothelial cell sprouts hydrolyze and infiltrate through the basement membrane toward the source of the angiogenic factor. This degradative phase is followed by a differentiative phase in which endothelial cells synthesize new matrix components and reconstitute their basal lamina. Cell-cell interactions are re-established and the endothelial cells eventually return to a quiescent state. Endothelial cells, therefore, interact with different basement membrane and interstitial stroma components during the various phases of angiogenesis. Cell membrane molecules interact with the extracellular environment and transmit information across the membrane to the interior of the cell. One family of molecules important in such processes is the integrins, heterodimeric receptors composed of an α and β subunit. At present, eight different β and fifteen different α subunits have been identified that can combine to form twenty-one receptors with distinct ligand specificities. Endothelial cells express a variety of integrins that enable them to interact with extracellular matrix proteins such as laminin, fibronectin, collagens, and vitronectin (Albelda et al., 1989; Languino et al., 1989; Cheng et al., 1991).

Several studies have demonstrated the crucial role of integrins in mediating cell-matrix interactions during the growth and differentiation of endothelial cells. Gamble et al. (1993) found that endothelial tube formation is enhanced by decreasing the adhesion of human umbilical vein endothelial cells to either collagen or fibrin gels. An antibody against the $\alpha2\beta1$ integrin enhances tube formation in collagen gels by inhibiting $\alpha2\beta1$ binding to collagen; in contrast, an antibody against $\alpha v\beta3$ enhances tube formation in fibrin gels by blocking $\alpha v\beta3$ binding to fibrin. Thus, disruption of endothelial cell adhesion to matrix converts the cells from a proliferative to a differentiated state. These findings are consistent with those of Ingber and Folkman (1989) who showed that interactions of bovine capillary endothelial (BCE) cells with fibronectin, type I collagen, or gelatin can regulate cell behavior.

The effect of anti-integrin antibodies on *in vitro* tube formation has also been studied in other experimental systems. Addition of monoclonal antibodies against $\alpha6$ or $\beta1$ to a human endothelial cell line or to human umbilical vein endothelial cells grown on Matrigel markedly blocks cord formation (Bauer et al., 1992a; Davis and Camarillo, 1995). These results underscore the importance of specific integrins during angiogenesis.

Other studies have shown the role of specific integrins in vasculogenesis and angiogenesis *in vivo*. Mice with a disrupted $\alpha5$ gene fail to form a competent vascular system (Yang et al., 1993); antibodies against $\beta1$ or $\alpha v\beta3$ integrins injected into quail embryos arrest or severely disrupt vasculogenesis (Drake et al., 1992, 1995). Brooks et al. (1994) have shown that antibodies to $\alpha v\beta3$ severely perturb angiogenesis induced by a tumor fragment implanted onto the chorioallantoic membrane. These antibodies induce apoptosis in proliferative vascular cells; thus, ligation of $\alpha v\beta3$ may be required for the survival and maturation of newly forming blood vessels (Brooks et al., 1994).

Immunohistochemistry studies have shown that capillaries normally express relatively low levels of many of the integrins that bind the ECM components that are more abundant in areas where angiogenesis occurs (Albelda, 1991). Therefore, changes in the levels or functions of integrins may be necessary during angiogenesis. The observation that $\alpha2\beta1$, $\alpha6\beta4$, and $\alpha v\beta3$ integrins are specifically enriched at the growing tips of endothelial sprouts *in vivo* (Enenstein and Kramer, 1994) is consistent with this hypothesis. FGF-2 may play an important role in endothelial cell expression of integrins. This hypothesis is based on the findings that exposure of BCE cells to FGF-2 results in increased cell surface levels of $\alpha6\beta4$, $\alpha v\beta5$, and of various $\beta1$ integrins including $\alpha2\beta1$, $\alpha3\beta1$, $\alpha5\beta1$, and $\alpha6\beta1$; in contrast, the levels of $\alpha1\beta1$, $\alpha v\beta1$, and $\alpha v\beta3$ are decreased by FGF-2 (Klein et al., 1993). These effects result from changes in the synthesis and rate of processing of integrin subunits. A 48 h exposure to FGF-2 is required for maximal increase in $\beta1$ integrin levels. Therefore, the increase in $\beta1$ integrin levels may represent a late response of cultured endothelial cells to FGF-2, which could be related to the late differentiative phase that occurs *in vivo* during angiogenesis. FGF-2 induced increases in $\beta1$ integrins after shorter times may be necessary for cell migration and proliferation, which occur after 2–4 and 10–12 h exposure to FGF-2, respectively. In particular, FGF-2-mediated increases in $\alpha5\beta1$ levels may be important in endothelial cell migration and proliferation, as increases in this integrin have been demonstrated to increase migration of Chinese hamster ovary cells and neural crest-like cells (Bauer et al., 1992b; Beauvais et al., 1995), as well as the proliferation of carcinoma cells (Varner et al., 1995). These observations support the hypothesis that the observed FGF-2-mediated increase in the levels of some integrins and decrease in the levels of other occur during different stages of angiogenesis.

Mechanisms of action

Nuclear translocation of exogenous FGF-1 and FGF-2

The FGFs act through their interaction with a complex of high-affinity tyrosine kinase receptors and low-affinity HSPGs on the cell surface. It has traditionally been thought that the role of membrane receptors such as tyrosine kinase receptors is to transfer a signal, generated by ligand binding, from the external cell surface across the plasma membrane to within the cell. Other components of the signaling cascade such as second-messenger molecules and kinases then convey the signal from the cytoplasm to the nucleus to effect gene expression. Membrane receptor endocytosis is considered as part of the cellular downregulation and desensitization process rather than an active signaling function. However, following receptor-mediated internalization, certain growth factors, such as schwannoma-derived growth factor, PDGF, epidermal growth factor, nerve growth factor, and FGF-1 and -2, may be translocated to the nucleus, suggesting that they may have additional signaling roles.

Several lines of evidence suggest that nuclear translocation of exogenous FGF-2 is mediated by a specific mechanism(s), and that its presence in the nucleus serves important functions during angiogenesis. FGF-2 added to synchronized cultures of adult bovine aortic endothelial cells is translocated to the nucleus and nucleolus during the late G_1 phase of the cell cycle. In contrast, exogenously added FGF-2 accumulates in the cytoplasm continuously throughout the cell cycle (Baldin et al., 1990). In serum-starved coronary venular endothelial cells, exogenous FGF-2 accumulates in the nucleus more rapidly than in the cytoplasm (Hawker and Granger, 1992). Upon nuclear translocation, full-length FGF-2 persists in the nucleus for up to 24 h, with little evidence for proteolytic degradation or loss of biological activity (Hawker and Granger, 1992). In contrast, exogenous FGF-2 accumulated in the cytoplasm is quickly degraded, presumably by lysosomal proteinases (Hawker and Granger, 1992). These results suggest that cytoplasmic and nuclear uptake of exogenous FGF-2 are kinetically distinct events that are independently regulated. Nuclear FGF-2 may function within the nucleus throughout and well after the S and M phases. In a cell-free system, FGF-2 can interact with DNA and regulate transcription (Bouche et al., 1987; Nakanishi et al., 1992) suggesting a direct action of FGF-2 in the nucleus in addition to the signal transduction pathway originating at the cell surface.

These studies were performed with the 18 kDa [low molecular weight (LMW)] form of FGF-2. However, other MW forms of FGF-2 are also synthesized by endothelial cells and by most FGF-2 producing cell lines: these are the three high molecular weight forms of 22, 22.5, and 24 kDa. Both high and low molecular weight forms of exogenous FGF-2 are accumulated in the nucleus (Gualandris et al., 1994; Patry et al., 1994). Patry et al. (1994) found that more HMW FGF-2 compared with LMW FGF-2 localizes in the nucleus. Nuclear localization correlates with increased stimulation of PA synthesis. In contrast, Gualandris et al. (1994) found no differences. The reason for these discrepancies is unclear at this time.

Like FGF-2, exogenous FGF-1 can also be efficiently translocated to the nucleus where it accumulates in the nucleolus (Zhan et al., 1992, 1993; Imamura et al., 1994). Whereas incorporation of exogenous FGF-1 into the cytoplasm occurs continually, nuclear translocation occurs after initiation of the G_0 to G_1 transition (Imamura et al., 1994). Thus, both exogenous FGF-2 and FGF-1 accumulate in the nucleus, but the mechanism of their nuclearization is not understood. One possible mechanism for FGF-1 and -2 involves binding to the cell surface, endocytosis, transport of the protein across the endosomal membrane, and entrance of the factor into the nucleus through the nuclear pore.

FGF-2 can be internalized through two pathways: a high-affinity receptor-mediated and heparan sulfate-mediated pathway (Roghani and Moscatelli, 1992; Amalric et al., 1994; Gleizes et al., 1995). Internalized FGF-2 is considerably more stable than other growth factors that are rapidly degraded in lysosomes (Moscatelli, 1988; Moenner et al., 1989). This stability may be related to the association of FGF-2 with heparan sulfate molecules within the endosomes and lysosomes. This may protect FGF-2 from degradation by proteases (Gospodarowicz and Cheng, 1986; Saksela et al., 1988; Sommer and Rifkin, 1989). However, the acidic environment of these organelles would tend to make this unlikely.

A small subpopulation of the oligosaccharides generated by lysosomal degradation of cell surface heparan sulfates has been reported to accumulate in the nucleus (Fedarko and Conrad, 1986; Ishihara et al., 1986). This finding has generated the hypothesis that FGF-2 complexed with heparan sulfate may be co-translocated to the nucleus. Amalric et al. (1994) showed that HSPGs may be required for the transport of exogenous FGF-2 to the nucleus. In contrast with this mechanism, it was reported that heparin inhibits the ability of a fusion protein of FGF-1 with diphtheria toxin (a toxin that has the ability to cross the endosomal membrane) to enter the nucleus. Thus, heparin may rigidify the FGF-1 structure and prevent the necessary unfolding of the protein for transmembrane passage (Wiedlocha et al., 1992). These observations raise the question of whether the FGF-1/ HSPG complex, and by analogy the FGF-2/HSPG complex, translocates from endosomes to the cytosol. Fragmentation of the endocytic membrane could provide a means of releasing endogenous proteins into the cytosol.

However, this has yet to be described for FGFs or any other growth factor. In addition, the transport of heparan sulfate fragments to the nucleus has not yet been confirmed.

A requirement for nuclear translocation for the mitogenic activity of FGF-1 has been suggested by the finding that an FGF-1 mutant lacking the nuclear translocation sequence fails to induce mitogenesis (Imamura et al., 1990). However, further studies suggested that this effect may result from the structural instability of the deletion mutant (Friedman et al., 1994). Injection of a conjugate of FGF-1 with diphtheria toxin into the cytoplasm of FGF receptor-negative cells (another method used to examine the activity of exogenous FGF-1 in the nucleus) stimulates DNA synthesis only when the conjugate is translocated into the nucleus (Wiedlocha et al., 1994). This result implies that nuclearization of FGF-1 is required for DNA synthesis. However, before the nuclear activity of exogenous FGFs can be understood, a mechanism for translocation across an endosomal membrane to gain access into the cytosol and to nuclear pores needs to be elucidated.

Endogenous forms of FGF-2 in the nucleus

When considering the various mechanisms of action of FGFs, the intracellular pathways must be separated from the autocrine and paracrine pathways. The following is a discussion of the former.

The process of alternative initiation of translation has varying consequences for the ultimate fate of endogenous FGF-2 forms. Because all FGF-2 forms are cytoplasmic, the FGF-2 proteins are available in the cytoplasm for direct nuclear translocation. To study the fates of the various FGF-2 forms, their subcellular distribution was examined in cells transfected with various FGF-2 expression constructs as well as non-transfected cells using immunocytochemical and subcellular fractionation techniques (Renko et al., 1990; Brigstock et al., 1991; Bugler et al., 1991; Dell'Era et al., 1991; Florkiewicz et al., 1991; Powell and Klagsbrun, 1991). These studies indicated that endogenous HMW FGFs were primarily nuclear, whereas LMW FGF-2 appeared to be primarily cytosolic. Differences in the amounts of each form found in different fractions may relate to the absolute levels of synthesis. Cells producing more 18 kDa tended to have more of this form in the nucleus. This difference in the distribution of the various forms of FGF-2 suggested that there must be specific sequences in the NH_2-terminal extension that targeted the HMW FGF-2 molecules to the nucleus or retained them once they were in the nucleus. Nuclear proteins often shuttle back and forth between the nucleus and the cytoplasm, but a number have distribution restricted predominantly to the nucleus. A nuclear localization sequence (NLS) is thought to determine the topogenetic fate of this class of proteins (Garcia-Bustos et al., 1991). The possibility that the NH_2-terminal extension contained an NLS was examined by making fusion constructs in which the 5′ cDNA encoding the amino-terminal extension of FGF-2 was fused to a reporter gene cDNA encoding a non-nuclear protein (Bugler et al., 1991; Quarto et al., 1991a). The 5′ NH$_2$-terminal of HMW FGF-2 can drive normally cytosolic proteins into the nucleus. The sequence GRGRGR, which is duplicated in the NH$_2$-terminal extension of HMW FGF-2, may be important in the transport of the molecule to the nucleus (Rifkin et al., 1994).

An additional mechanism to regulate protein distribution and/or biological activity is post-translational modification (for a review, see Danpure, 1995). In recent years, two modifications have been described for FGF-2. The 18 kDa form of FGF-2 can undergo phosphorylation by a kinase associated with the cell surface (Vilgrain and Baird, 1991) or, alternatively, by a nuclear kinase activity detected in SK-Hep 1 cells (Vilgrain et al., 1993). A second post-translational modification occurs on high molecular weight forms of FGF-2. By pulse-chase experiments, Florkiewicz et al. (1991) observed that the 24 kDa, but not the 18 kDa form of FGF-2, undergoes an increase in its apparent Mr. Sommer et al. (1989) and Burgess et al. (1991) demonstrated that three specific arginine residues (residues -26, -24, and -22) of the HMW FGF-2 from guinea pig brain or transfected NIH 3T3 cells are mono- or di-methylated. The methyl-transferase inhibitor, 5′-deoxy-5′-methyl-thioadenosine, prevents the post-translational modification of HMW FGF-2 and markedly reduces its nuclear accumulation (Pintucci et al., 1996). Thus, methylation may be essential for the nuclear accumulation of the high molecular weight forms of FGF-2; in the absence of this post-translational modification, the differences observed in the intracellular distribution of 18 kDa versus HMW FGF-2 might be abolished.

In contrast to HMW FGF-2, the 18 kDa form of FGF-2 lacks an NLS. Nevertheless, endogenously synthesized 18 kDa FGF-2 has been detected in the nucleus (Renko et al., 1990; Florkiewicz et al., 1991), although the protein accumulates predominantly in the cytoplasm. Both endogenous 18 kDa and HMW FGF-2 bind to nuclear chromatin (Gualandris et al., 1993).

Like HMW FGF-2, FGF-1 contains an NLS which is able to direct a non-nuclear protein into the nucleus in transfected NIH 3T3 cells. Nevertheless, endogenous FGF-1 remains mostly cytoplasmic (Zhan et al., 1992). This finding suggests that the observed nuclear localization of FGF-1 occurs through an autocrine and not an intracellular pathway.

Functional differences among the endogenous forms of FGF-2

The finding that cells synthesize multiple forms of FGF-2 that have distinct subcellular distributions raises the question of whether these endogenous forms have specialized functions during angiogenesis. When added exogenously, both 18 kDa and HMW FGF-2 induce cell proliferation, migration, PA production, and compete for the binding to the same cell membrane high-affinity sites (Moscatelli et al., 1987; Gualandris et al., 1994; Bikfalvi et al., 1995). Furthermore, both high and low molecular weight forms induce angiogenesis *in vivo* when added as exogenous factors (Gualandris et al., 1994). The view that endogenous FGF-2 forms mediate endothelial cell migration is supported by an experiment in which the migration of BCE cells was blocked with anti-FGF-2 antibodies (Sato and Rifkin, 1988). Consistent with these results, NIH 3T3 cells expressing all FGF-2 forms have higher motility than control untransfected cells; this effect is abrogated by anti-FGF-2 antibodies (Mignatti et al., 1991 a).

In order to determine the specific roles of the individual forms of FGF-2, cell migration, proliferation, and modulation of integrin levels have been examined in transfected NIH 3T3 cell lines that synthesize either 18 kDa or HMW FGF-2 (Quarto et al., 1991 b; Bikfalvi et al., 1995, Klein et al., 1996). Cells synthesizing exclusively 18 kDa FGF-2 have high motility, $\beta 1$ integrin levels, and surface-associated 18 kDa FGF-2; cells synthesizing exclusively HMW FGF-2 have low motility and $\beta 1$ integrin levels, and virtually no surface-associated FGF-2. FGF receptors are downregulated in cells expressing 18 kDa FGF-2, but not in cells expressing HMW FGF-2. Cells expressing HMW FGF-2 have a reduced serum requirement for growth, whereas cells expressing 18 kDa FGF-2 proliferate poorly in low serum (Bikfalvi et al., 1995). These results show that 18 kDa and HMW FGF-2 have both unique and shared biological activities. Both forms increase cell proliferation; however, endogenous 18 kDa FGF-2 mediates cell migration and $\beta 1$ integrin levels, and endogenous HMW FGF-2 growth in low serum.

Expression of a dominant negative FGF receptor (FGFR-2 lacking the COOH-terminal domain) in cells synthesizing 18 kDa FGF-2 inhibits their migration, fails to modulate $\beta 1$ integrin levels, and suppresses their growth. In contrast, expression of a dominant negative receptor in cells expressing HMW FGF-2 has no effect on their growth (Bikfalvi et al., 1995). Therefore, 18 kDa and HMW FGF-2 may mediate certain functions through distinct mechanisms. 18 kDa FGF-2 modulates cell motility, proliferation, and $\beta 1$ integrin levels through its release and interaction with its cell surface receptors; HMW FGF-2 acts as a mitogen and an inducer of growth in low serum through an intracellular mechanism.

The discovery of a number of kinase activities located in the nucleus (Wang, 1994) has raised the possibility that specific phosphorylated intermediates can originate by nuclear pathways as well as surface-initiated mechanisms. The effects of the synthesis of different FGF-2 forms on the pattern of tyrosine phosphorylated proteins in both the nucleus and cytoplasm have been examined in transfected NIH 3T3 cells. Endogenous 18 kDa FGF-2 expression results in enhanced phosphorylation of both cytoplasmic and nuclear substrates. In contrast, HMW FGF-2 expression primarily induces the phosphorylation of nuclear substrates without inducing the phosphorylation of cytoplasmic substrates (G. Pintucci, unpublished observations). These results add further support to the hypothesis that two pathways exist for FGF action, a cell-surface initiated and a nuclear one.

Antibodies to FGF-2 also downregulate uPA levels in endothelial cells (Sato and Rifkin, 1988), suggesting that endogenous FGF-2 mediates uPA activity through an extracellular mechanism. As only 18 kDa FGF-2 is externalized by the cells, this form of endogenous FGF-2 appears to be the only one that mediates this activity. Suramin, a polysulfated aromatic compound that blocks FGF binding to high affinity receptors (Coffey et al., 1987), inhibits the effects of exogenous FGF-2 on uPA induction (unpublished observations). These findings support the hypothesis that endogenous 18 kDa FGF-2 upregulates uPA expression through its release and interaction with cell surface receptors.

Thus, during angiogenesis, 18 kDa FGF-2 synthesized by endothelial cells is released, interacts with high affinity and low-affinity receptors, which induces receptor phosphorylation and signal propagation, and ultimately triggers various biological responses. These include downregulation of FGF receptors, increased motility, proliferation and proteinase activity, and modulation of integrin levels. HMW FGF-2 may act on endothelial cell proliferation by an intracellular mechanism. This suggests that FGF-2 activities may be strictly regulated at the level of translation of the different forms (see above).

Concluding remarks

Historically, FGF-1 and FGF-2 have been studied as paracrine angiogenic factors but with the finding that endothelial cells synthesize FGF-2, they have also been viewed as autocrine factors. The discovery that multiple FGF-2 forms exist suggested that these factors may have distinct functions. The 18 kDa form, which is found on the cell surface and associated with the ECM, requires interaction with cell surface receptors to elicit its biological activities. This FGF-2 form may be the primary mediator of changes in endothelial cell behavior during angiogenesis as it increases cell proliferation, migration and protease production, and modulates integrin expression. In contrast, HMW FGF-2 forms, which are primarily nuclear and stimulate cell proliferation independently of cell surface receptors, may act as further mediators of endothelial cell proliferation during angiogenesis. Therefore, the 18 kDa and HMW FGF-2 forms most likely act in concert during angiogenesis. In addition to cell proliferation, other activities of HMW FGF-2 in the nucleus during angiogenesis remain to be identified.

The unique role(s) of FGF-1 and FGF-2 in physiological and pathological angiogenesis are unclear and difficult to establish, as other potent angiogenic factors, such as VEGF, and other FGF family members, have also been shown to participate in these processes. The interaction of these angiogenic factors is clearly complex as several factors may have redundant roles, act in different settings (e. g. normal vs. pathological), or influence another factor's behavior by modulating its expression and/or activity.

To elucidate further the mechanisms underlying FGF-1 and FGF-2-mediated angiogenesis, several important points must be addressed. How FGF-1 and 18 kDa FGF-2 are released from cells in the absence of a leader sequence, what mechanism mediates the nuclear translocation of exogenous FGF-1 and FGF-2, which forms of FGF act in specific biological phenomena, and which are the molecular components of the nuclear signaling pathway of HMW FGF-2, are questions that must be answered before the biological significance of FGFs in angiogenesis can be appreciated.

Acknowledgements
We would like to thank Dr. Paolo Mignatti for many helpful suggestions and for critical reading of the manuscript. We also thank Drs. Giuseppe Pintucci, Pierre-Emmanuel Gleizes, John Munger, and Irene Nunes for stimulating discussions. This work was supported by grants CA34282 and CA23753 (D. B. R.), 5T32GM07238-19 (S. K.), and CA42229 (M. R.) from the National Institutes of Health, and 194169 from the Juvenile Diabetes Foundation (D. B. R.). S. K. was supported by a Berlex student fellowship.

References

Abraham, J.A., Whang, J.L., Tumolo, A., Mergia, A. and Fiddes, J.C. (1986a) Human basic fibroblast growth factor: nucleotide sequence, genomic organization, and expression in mammalian cells. *Cold Spring Harbor Symp. Quant. Biol.* 51:657–668.

Abraham, J.A., Mergia, A., Whang, J.L., Tumolo, A., Friedman, J., Hjerrild, K.A., Gospodarowicz, D. and Fiddes, J.C. (1986b) Nucleotide sequence of a bovine clone encoding the angiogenic protein, basic fibroblast growth factor. *Science* 233:545–548.

Albelda, S.M., Daise, M., Levine, E.M. and Buck, C.A. (1989) Identification and characterization of cell-substratum adhesion receptors on cultured human endothelial cells. *J. Clin. Invest.* 83:1992–2002.

Albeda, S.M. (1991) Endothelial and epithelial cell adhesion molecules. *Am. J. Resp. Cell Mol. Biol.* 4:195–203.

Amalric, F., Bouche, G., Bonnet, H., Brethenou, P., Roman, A.M., Truchet, I. and Quarto, N. (1994) Fibroblast growth factor-2 (FGF-2) in the nucleus: translocation process and targets. *Biochem. Pharmacol.* 47:111–115.

Amaya, E., Musci, T.J. and Kirschner, M.W. (1991) Expression of a dominant negative mutant of the FGF receptor disrupts mesoderm formation in *Xenopus* embryos. *Cell* 66:257–270.

Appella, E., Robinson, E.A., Ullrich, S.J., Stoppelli, M.P., Corti, A., Cassani, G. and Blasi, F. (1987) The receptor-binding sequence of urokinase. A biological function for the growth-factor module of proteases. *J. Biol. Chem.* 262:4437–4440.

Ausprunk, D.H. and Folkman, J. (1977) Migration and proliferation of endothelial cells in preformed and newly formed blood vessels during tumor angiogenesis. *Microvascular Res.* 14:53–65.

Baird, A., Esch, F., Gospodarowicz, D. and Guillemin, R. (1985) Retina- and eye-derived endothelial cell growth factors: partial molecular characterization and identity with acidic and basic fibroblast growth factors. *Biochemistry.* 24:7855–7860.

Baird, A., Esch, F., Mormede, P., Ueno, N., Ling, N., Bohlen, P., Ying, S.Y., Wehrenberg, W.B. and Guillemin, R. (1986) Molecular characterization of fibroblast growth factor: distribution and biological activities in various tissues. *Rec. Prog. Horm. Res.* 42:143–205.

Baldin, V., Roman, A.M., Bosc-Bierne, I., Amalric, F. and Bouche, G. (1990) Translocation of bFGF to the nucleus is G1 phase cell cycle specific in bovine aortic endothelial cells. *EMBO J.* 9:1511–1517.

Bashkin, P., Doctrow, S., Klagsbrun, M., Svahn, C.M., Folkman, J. and Vlodavsky, I. (1989) Basic fibroblast growth factor binds to subendothelial extracellular matrix and is released by heparitinase and heparin-like molecules. *Biochemistry* 28:1737–1743.

Bashkin, P., Neufeld, G., Gitay-Goren, H. and Vlodavsky, I. (1992) Release of cell surface-associated basic fibroblast growth factor by glycosylphosphatidylinositol-specific phospholipase C. *J. Cell. Physiol.* 151:126–137.

Bauer, J., Margolis, M., Schreiner, C., Edgell, C.J., Azizkhan, J., Lazarowski, E. and Juliano, R.L. (1992a) In vitro model of angiogenesis using a human endothelium-derived permanent cell line: contributions of induced gene expression, G-proteins, and integrins. *J. Cell. Physiol.* 153:437–449.

Bauer, J.S., Schreiner, C.L., Giancotti, F.G., Ruoslahti, E. and Juliano, R.L. (1992b) Motility of fibronectin receptor-deficient cells on fibronectin and vitronectin: collaborative interactions among integrins. *J. Cell. Biol.* 116:477–487.

Beauvais, A., Erickson, C.A., Goins, T., Craig, S.E., Humphries, M.J., Thiery, J.P. and Dufour, S. (1995) Changes in the fibronectin-specific integrin expression pattern modify the migratory behavior of sarcoma S180 cells in vitro and in the embryonic environment. *J. Cell Biol.* 128:699–713.

Behrendt, N., Ronne, E., Ploug, M., Petri, T., Lober, D., Nielsen, L.S., Schleuning, W.D., Blasi, F., Appella, E. and Dano, K. (1990) The human receptor for urokinase plasminogen activator. NH_2-terminal amino acid sequence and glycosylation variants. *J. Biol. Chem.* 265:6453–6460.

Bergonzoni, L., Caccia, P., Cletini, O., Sarmientos, P. and Isacchi, A. (1992) Characterization of a biologically active extracellular domain of fibroblast growth factor receptor 1 expressed in Escherichia coli. *Eur. J. Biochem.* 210:823–829.

Berkman, R.A., Merrill, M.J., Reinhold, W.C., Monacci, W.T., Saxena, A., Clark, W.C., Robertson, J.T., Ali, I.U. and Oldfield, E.H. (1993) Expression of the vascular permeability factor/vascular endothelial growth factor gene in central nervous system neoplasms. *J. Clin. Invest.* 91:153–159.

Bernard, O., Li, M. and Reid, H.H. (1991) Expression of two different forms of fibroblast growth factor receptor 1 in different mouse tissues and cell lines. *Proc. Natl. Acad. Sci. USA* 88:7625–7629.

Berse, B., Brown, L.F., Van de Water, L., Dvorak, H.F. and Senger, D.R. (1992) Vascular permeability factor (vascular endothelial growth factor) gene is expressed differentially in normal tissues, macrophages, and tumors. *Mol. Biol. Cell* 3:211–220.

Bikfalvi, A., Klein, S., Pintucci, G., Quarto, N., Mignatti, P. and Rifkin, D.B. (1995) Differential modulation of cell phenotype by different molecular weight forms of basic fibroblast growth factor: possible intracellular signaling by the high molecular weight forms. *J. Cell Biol.* 129:233–243.

Blasi, F. (1993) Urokinase and urokinase receptor: a paracrine/autocrine system regulating cell migration and invasiveness. *Bioessays* 15:105–111.

Bohlen, P., Esch, F., Baird, A., Jones, K.L. and Gospodarowicz, D. (1985) Human brain fibroblast growth factor. Isolation and partial chemical characterization. *FEBS Lett.* 185:177–181.

Bouche, G., Gas, N., Prats, H., Baldin, V., Tauber, J.P., Teissie, J. and Amalric, F. (1987) Basic fibroblast growth factor enters the nucleolus and stimulates the transcription of ribosomal genes in ABAE cells undergoing G0–G1 transition. *Proc. Natl. Acad. Sci. USA* 84:6770–6774.

Brigstock, D.R., Sasse, J. and Klagsbrun, M. (1991) Subcellular distribution of basic fibroblast growth factor in human hepatoma cells. *Growth Factors* 4:189–196.

Brooks, P.C., Montgomery, A.M., Rosenfeld, M., Reisfeld, R.A., Hu, T., Klier, G. and Cheresh, D.A. (1994) Integrin $\alpha v \beta 3$ antagonists promote tumor regression by inducing apoptosis of angiogenic blood vessels. *Cell* 79:1157–1164.

Brunner, G., Gabrilove, J., Rifkin, D.B. and Wilson, E.L. (1991) Phospholipase C release of basic fibroblast growth factor from human bone marrow cultures as a biologically active complex with a phosphatidylinositol-anchored heparan sulfate proteoglycan. *J. Cell Biol.* 114:1275–1283.

Bugler, B., Amalric, F. and Prats, H. (1991) Alternative initiation of translation determines cytoplasmic or nuclear localization of basic fibroblast growth factor. *Mol. Cell. Biol.* 11: 573–577.

Burgering, B.M., de Vries-Smits, A.M., Medema, R.H., van Weeren, P.C., Tertoolen, L.G. and Bos, J.L. (1993) Epidermal growth factor induces phosphorylation of extracellular signal-regulated kinase 2 via multiple pathways. *Mol. Cell. Biol.* 13:7248–7256.

Burgess, W.H., Mehlman, T., Marshak, D.R., Fraser, B.A. and Maciag, T. (1986) Structural evidence that endothelial cell growth factor-β is the precursor of both endothelial cell growth factor-α and acidic fibroblast growth factor. *Proc. Natl. Acad. Sci. USA* 83:7216–7220.

Burgess, W.H., Bizik, J., Mehlman, T., Quarto, N. and Rifkin, D.B. (1991) Direct evidence for methylation of arginine residues in high molecular weight forms of basic fibroblast growth factor. *Cell Regul.* 2:87–93.

Busso, N., Masur, S.K., Lazega, D., Waxman, S. and Ossowski, L. (1994) Induction of cell migration by pro-urokinase binding to its receptor: possible mechanism for signal transduction in human epithelial cells. *J. Cell Biol.* 126:259–270.

Carmeliet, P., Schoonjans, L., Kieckens, L., Ream, B., Degen, J., Bronson, R., De Vos, R., van den Oord, J.J., Collen, D. and Mulligan, R.C. (1994) Physiological consequences of loss of plasminogen activator gene function in mice. *Nature* 368:419–424.

Casscells, W., Speir, E., Sasse, J., Klagsbrun, M., Allen, P., Lee, M., Calvo, B., Chiba, M., Haggroth, L., Folkman, J. and Epstein, S.E. (1990) Isolation, characterization, and localization of heparin-binding growth factors in the heart. *J. Clin. Invest.* 85:433–441.

Chapman, H.A.J. and Stone, O.L. (1984) Co-operation between plasmin and elastase in elastin degradation by intact murine macrophages. *Biochemical J.* 222:721–728.

Cheng, Y.F., Clyman, R.I., Enenstein, J., Waleh, N., Pytela, R. and Kramer, R.H. (1991) The integrin complex αvβ3 participates in the adhesion of microvascular endothelial cells to fibronectin. *Exp. Cell Res.* 194:69–77.

Chiu, I.M., Wang, W.P. and Lehtoma, K. (1990) Alternative splicing generates two forms of mRNA coding for human heparin-binding growth factor 1. *Oncogene* 5:755–762.

Christian, J.L., McMahon, J.A., McMahon, A.P. and Moon, R.T. (1991) *Xwnt-8*, a *Xenopus Wnt-1/int-1*-related gene responsive to mesoderm-inducing growth factors, may play a role in ventral mesodermal patterning during embryogenesis. *Development* 111: 1045–1055.

Coffey, R.J., Jr., Leof, E.B., Shipley, G.D. and Moses, H.L. (1987) Suramin inhibition of growth factor receptor binding and mitogenicity in AKR-2B cells. *J. Cell. Physiol.* 132: 143–148.

Conese, M. and Blasi, F. (1995) Urokinase/urokinase receptor system: internalization/degradation of urokinase-serpin complexes: mechanism and regulation. *Biol. Chem. Hoppe. Seyler* 376:143–155.

Connolly, D.T., Stoddard, B.L., Harakas, N.K. and Feder, J. (1987) Human fibroblast-derived growth factor is a mitogen and chemoattractant for endothelial cells. *Biochem. Biophys. Res. Commun.* 144:705–712.

Cordon-Cardo, C., Vlodavsky, I., Haimovitz-Friedman, A., Hicklin, D. and Fuks, Z. (1990) Expression of basic fibroblast growth factor in normal human tissues. *Lab. Invest.* 63: 832–840.

Cornell, R.A. and Kimelman, D. (1994) Activin-mediated mesoderm induction requires FGF. *Development* 120:453–462.

Crabb, J.W., Armes, L.G., Carr, S.A., Johnson, C.M., Roberts, G.D., Bordoli, R.S. and McKeehan, W.L. (1986) Complete primary structure of prostatropin, a prostate epithelial cell growth factor. *Biochemistry* 25:4988–4993.

Crumley, G., Dionne, C.A. and Jaye, M. (1990) The gene for human acidic fibroblast growth factor encodes two upstream exons alternatively spliced to the first coding exon. *Biochem. Biophys. Res. Commun.* 171:7–13.

Cunningham, B.C., Ultsch, M., De Vos, A.M., Mulkerrin, M.G., Clauser, K.R. and Wells, J.A. (1991) Dimerization of the extracellular domain of the human growth hormone receptor by a single hormone molecule. *Science* 254:821–825.

D'Amore, P.A. and Klagsbrun, M. (1984) Endothelial cell mitogens derived from retina and hypothalamus: biochemical and biological similarities. *J. Cell. Biol.* 99:1545–1549.

D'Amore, P.A. and Thompson, R.W. (1987) Mechanisms of angiogenesis. *Ann. Rev. Physiol.* 49:453–464.

Danpure, C.J. (1995) How can the products of a single gene be localized to more than one intra-cellular compartment? *Trends in Cell Biol.* 5:230–238.

Davis, G.E. and Camarillo, C.W. (1995) Regulation of endothelial cell morphogenesis by integrins, mechanical forces, and matrix guidance pathways. *Exp. Cell Res.* 216:113–123.

Dell'Era, P., Presta, M. and Ragnotti, G. (1991) Nuclear localization of endogenous basic fibro-blast growth factor in cultured endothelial cells. *Exp. Cell Res.* 192:505–510.

Dennis, P.A. and Rifkin, D.B. (1990) Studies on the role of basic fibroblast growth factor *in vivo*: inability of neutralizing antibodies to block growth. *J. Cell. Physiol.* 144:84–98.

Dickson, C., Smith, R., Brookes, S. and Peters, G. (1984) Tumorigenesis by mouse mammary tumor virus: proviral activation of a cellular gene in the common integration region int-2. *Cell* 37:529–536.

Dionne, C.A., Crumley, G., Bellot, F., Kaplow, J.M., Searfoss, G., Ruta, M., Burgess, W.H., Jaye, M. and Schlessinger, J. (1990) Cloning and expression of two distinct high-affinity receptors cross-reacting with acidic and basic fibroblast growth factors. *EMBO J.* 9:2685–2692.

Dono, R. and Zeller, R. (1994) Cell-type-specific nuclear translocation of fibroblast growth factor-2 isoforms during chicken kidney and limb morphogenesis. *Dev. Biol.* 163:316–330.

Drake, C.J., Davis, L.A. and Little, C.D. (1992) Antibodies to -β 1-integrins cause alterations of aortic vasculogenesis, *in vivo. Dev. Dynam.* 193:83–91.

Drake, C.J., Cheresh, D.A. and Little, C.D. (1995) An antagonist of integrin $\alpha v \beta 3$ prevents matu-ration of blood vessels during embryonic neovascularization. *J. Cell Science* 108:2655–2661.

Eaton, D.L., Scott, R.W. and Baker, J.B. (1984) Purification of human fibroblast urokinase proenzyme and analysis of its regulation by proteases and protease nexin. *J. Biol. Chem.* 259:6241–6247.

Ehrlich, H.J., Gebbink, R.K., Preissner, K.T., Keijer, J., Esmon, N.L., Mertens, K. and Pannekoek, H. (1991) Thrombin neutralizes plasminogen activator inhibitor 1 (PAI-1) that is complexed with vitronectin in the endothelial cell matrix. *J. Cell Biol.* 115:1773–1781.

Enenstein, J. and Kramer, R.H. (1994) Confocal microscopic analysis of integrin expression on the microvasculature and its sprouts in the neonatal foreskin. *J. Invest. Dermatol.* 103:381–386.

Esch, F., Baird, A., Ling, N., Ueno, N., Hill, F., Denoroy, L., Klepper, R., Gospodarowicz, D., Bohlen, P. and Guillemin, R. (1985a) Primary structure of bovine pituitary basic fibroblast growth factor (FGF) and comparison with the amino-terminal sequence of bovine brain acidic FGF. *Proc. Natl. Acad. Sci. USA* 82:6507–6511.

Esch, F., Ueno, N., Baird, A., Hill, F., Denoroy, L., Ling, N., Gospodarowicz, D. and Guillemin, R. (1985b) Primary structure of bovine brain acidic fibroblast growth factor (FGF). *Biochem. Biophys. Res. Commun.* 133:554–562.

Estreicher, A., Muhlhauser, J., Carpentier, J.L., Orci, L. and Vassalli, J.D. (1990) The receptor for urokinase type plasminogen activator polarizes expression of the protease to the leading edge of migrating monocytes and promotes degradation of enzyme-inhibitor complexes. *J. Cell Biol.* 111:783–792.

Fedarko, N.S. and Conrad, H.E. (1986) A unique heparan sulfate in the nuclei of hepatocytes: structural changes with the growth state of the cells. *J. Cell Biol.* 102:587–599.

Fibbi, G., Ziche, M., Morbidelli, L., Magnelli, L. and Del Rosso, M. (1988) Interaction of urokinase with specific receptors stimulates mobilization of bovine adrenal capillary endo-thelial cells. *Exp. Cell Res.* 179:385–395.

Flamme, I. and Risau, W. (1992) Induction of vasculogenesis and hematopoiesis *in vitro. Development* 116:435–439.

Flamme, I., Breier, G. and Risau, W. (1995) Vascular endothelial growth factor (VEGF) and VEGF receptor 2 (flk-1) are expressed during vasculogenesis and vascular differentiation in the quail embryo. *Dev. Biol.* 169:699–712.

Flaumenhaft, R., Moscatelli, D., Saksela, O. and Rifkin, D.B. (1989) Role of extracellular matrix in the action of basic fibroblast growth factor: matrix as a source of growth factor for long-term stimulation of plasminogen activator production and DNA synthesis. *J. Cell. Physiol.* 140:75–81.

Flaumenhaft, R., Moscatelli, D. and Rifkin, D.B. (1990) Heparin and heparan sulfate increase the radius of diffusion and action of basic fibroblast growth factor. *J. Cell Biol.* 111:1651–1659.

Flaumenhaft, R., Abe, M., Mignatti, P. and Rifkin, D.B. (1992) Basic fibroblast growth factor-induced activation of latent transforming growth factor -β in endothelial cells: regulation of plasminogen activator activity. *J. Cell Biol.* 118:901–909.

Florkiewicz, R.Z. and Sommer, A. (1989) Human basic fibroblast growth factor gene encodes four polypeptides: three initiate translation from non-AUG codons. *Proc. Natl. Acad. Sci. USA* 86:3978–3981.

Florkiewicz, R.Z., Baird, A. and Gonzalez, A.M. (1991) Multiple forms of bFGF: differential nuclear and cell surface localization. *Growth Factors* 4:265–275.

Florkiewicz, R.Z., Majack, R.A., Buechler, R.D. and Florkiewicz, E. (1995) Quantitative export of FGF-2 occurs through an alternative, energy-dependent, non-ER/Golgi pathway. *J. Cell. Physiol.* 162:388–399.

Folkman, J. and Klagsbrun, M. (1987) Angiogenic factors. *Science* 235:442–447.

Folkman, J., Watson, K., Ingber, D. and Hanahan, D. (1989) Induction of angiogenesis during the transition from hyperplasia to neoplasia. *Nature* 339:58–61.

Folkman, J. (1990) What is the evidence that tumors are angiogenesis dependent? *J. Natl. Cancer Inst.* 82:4–6.

Folkman, J. and Shing, Y. (1992) Angiogenesis. *J. Biol. Chem.* 267:10931–10934.

Folkman, J. (1995) Angiogenesis in cancer, vascular, rheumatoid and other disease. *Nature Med.* 1:27–31.

Fox, G.M., Schiffer, S.G., Rohde, M.F., Tsai, L.B., Banks, A.R. and Arakawa, T. (1988) Production, biological activity, and structure of recombinant basic fibroblast growth factor and an analog with cysteine replaced by serine. *J. Biol. Chem.* 263:18452–18458.

Friedman, S., Zhan, X. and Maciag, T. (1994) Mutagenesis of the nuclear localization sequence in FGF-1 alters protein stability but not mitogenic activity. *Biochem. Biophys. Res. Commun.* 198:1203–1208.

Gamble, J.R., Matthias, L.J., Meyer, G., Kaur, P., Russ, G., Faull, R., Berndt, M.C. and Vadas, M.A. (1993) Regulation of *in vitro* capillary tube formation by anti-integrin antibodies. *J. Cell Biol.* 121:931–943.

Garcia-Bustos, J., Heitman, J. and Hall, M.N. (1991) Nuclear protein localization. *Biochim. Biophys. Acta* 1071:83–101.

Gautschi-Sova, P., Jiang, Z.P., Frater-Schroder, M. and Bohlen, P. (1987) Acidic fibroblast growth factor is present in nonneural tissue: isolation and chemical characterization from bovine kidney. *Biochemistry* 26:5844–5847.

Gimenez-Gallego, G., Rodkey, J., Bennett, C., Rios-Candelore, M., DiSalvo, J. and Thomas, K. (1985) Brain-derived acidic fibroblast growth factor: complete amino acid sequence and homologies. *Science* 230:1385–1388.

Gimenez-Gallego, G., Conn, G., Hatcher, V.B. and Thomas, K.A. (1986) Human brain-derived acidic and basic fibroblast growth factors: amino terminal sequences and specific mitogenic activities. *Biochem. Biophys. Res. Commun.* 135:541–548.

Giordano, S., Sherman, L., Lyman, W. and Morrison, R. (1992) Multiple molecular weight forms of basic fibroblast growth factor are developmentally regulated in the central nervous system. *Dev. Biol.* 152:293–303.

Gleizes, P.-E., Noaillac-Depeyre, J., Amalric, F. and Gas, N. (1995) Basic fibroblast growth factor (FGF-2) internalization through the heparan sulfate proteoglycans-mediated pathway: an ultrastructural approach. *Eur. J. Cell Biol.* 66:47–59.

Godsave, S.F. and Shiurba, R.A. (1992) *Xenopus* blastulae show regional differences in competence for mesoderm induction: correlation with endogenous basic fibroblast growth factor levels. *Dev. Biol.* 151:506–515.

Gospodarowicz, D., Cheng, J., Lui, G.M., Baird, A. and Bohlen, P. (1984) Isolation of brain fibroblast growth factor by heparin-Sepharose affinity chromatography: identity with pituitary fibroblast growth factor. *Proc. Natl. Acad. Sci. USA* 81:6963–6967.

Gospodarowicz, D. and Cheng, J. (1986) Heparin protects basic and acidic FGF inactivation. *J. Cell. Physiol.* 128:475–484.

Gospodarowicz, D., Baird, A., Cheng, J., Lui, G.M., Esch, F. and Bohlen, P. (1986) Isolation of fibroblast growth factor from bovine adrenal gland: physicochemical and biological characterization. *Endocrinology* 118:82–90.

Gospodarowicz, D., Ferrara, N., Haaparanta, T. and Neufeld, G. (1988) Basic fibroblast growth factor: expression in cultured bovine vascular smooth muscle cells. *Eur. J. Cell Biol.* 46:144–151.

Green, J.B., New, H.V. and Smith, J.C. (1992) Responses of embryonic *Xenopus* cells to activin and FGF are separated by multiple dose thresholds and correspond to distinct axes of the mesoderm. *Cell* 71:731–739.

Gualandris, A., Coltrini, D., Bergonzoni, L., Isacchi, A., Tenca, S., Ginelli, B. and Presta, M. (1993) The NH$_2$-terminal extension of high molecular weight forms of basic fibroblast growth factor (bFGF) is not essential for the binding of bFGF to nuclear chromatin in transfected NIH 3T3 cells. *Growth Factors* 8:49–60.

Gualandris, A., Urbinati, C., Rusnati, M., Ziche, M. and Presta, M. (1994) Interaction of high-molecular-weight basic fibroblast growth factor with endothelium: biological activity and intracellular fate of human recombinant M(r) 24,000 bFGF. *J. Cell. Physiol.* 161:149–159.

Gualandris, A. and Presta, M. (1995) Transcriptional and posttranscriptional regulation of urokinase-type plasminogen activator expression in endothelial cells by basic fibroblast growth factor. *J. Cell. Physiol.* 162:400–409.

Hanneken, A., Lutty, G.A., McLeod, D.S., Robey, F., Harvey, A.K. and Hjelmeland, L.M. (1989) Localization of basic fibroblast growth factor to the developing capillaries of the bovine retina. *J. Cell. Physiol.* 138:115–120.

Harper, J.W., Strydom, D.J. and Lobb, R.R. (1986) Human class 1 heparin-binding growth factor: structure and homology to bovine acidic brain fibroblast growth factor. *Biochemistry* 25:4097–4103.

Hatten, M.E., Lynch, M., Rydel, R.E., Sanchez, J., Joseph-Silverstein, J., Moscatelli, D. and Rifkin, D.B. (1988) *In vitro* neurite extension by granule neurons is dependent upon astrolial-derived fibroblast growth factor. *Dev. Biol.* 125:280–289.

Hawker, J.R., Jr. and Granger, H.J. (1992) Internalized basic fibroblast growth factor translocates to nuclei of venular endothelial cells. *Am. J. Physiol.* 262:1525–1537.

Hekman, C.M. and Loskutoff, D.J. (1985) Endothelial cells produce a latent inhibitor of plasminogen activators that can be activated by denaturants. *J. Biol. Chem.* 260:11581–11587.

Hemmati-Brivanlou, A. and Melton, D.A. (1992) A truncated activin receptor inhibits mesoderm induction and formation of axial structures in *Xenopus* embryos. *Nature* 359:609–614.

Huang, J., Mohammadi, M., Rodrigues, G.A. and Schlessinger, J. (1995) Reduced activation of RAF-1 and MAP kinase by a fibroblast growth factor receptor mutant deficient in stimulation of phosphatidylinositol hydrolysis. *J. Biol. Chem.* 270:5065–5072.

Imamura, T., Engleka, K., Zhan, X., Tokita, Y., Forough, R., Roeder, D., Jackson, A., Maier, J.A., Hla, T. and Maciag, T. (1990) Recovery of mitogenic activity of a growth factor mutant with a nuclear translocation sequence. *Science* 249:1567–1570.

Imamura, T., Oka, S., Tanahashi, T. and Okita, Y. (1994) Cell cycle-dependent nuclear localization of exogenously added fibroblast growth factor-1 in BALB/c 3T3 and human vascular endothelial cells. *Exp. Cell. Res.* 215:363–372.

Ingber, D.E. and Folkman, J. (1989) Mechanochemical switching between growth and differentiation during fibroblast growth factor-stimulated angiogenesis *in vitro*: role of extracellular matrix. *J. Cell Biol.* 109:317–330.

Ishai-Michaeli, R., Eldor, A. and Vlodavski, I. (1990) Heparanase activity expressed by platelets, neutrophils, and lymphoma cells releases active fibroblast growth factor from extracellular matrix. *Cell Regul.* 1:833–842.

Ishihara, M., Fedarko, N.S. and Conrad, H.E. (1986) Transport of heparan sulfate into the nuclei of hepatocytes. *J. Biol. Chem.* 261:13575–13580.

Ishikawa, F., Miyazono, K., Hellman, U., Drexler, H., Wernstedt, C., Hagiwara, K., Usuki, K., Takaku, F., Risau, W. and Heldin, C.H. (1989) Identification of angiogenic activity and the cloning and expression of platelet-derived endothelial cell growth factor. *Nature* 338:557–562.

Itoh, K. and Sokol, S.Y. (1994) Heparan sulfate proteoglycans are required for mesoderm formation in *Xenopus* embryos. *Development* 120:2703–2711.

Jackson, A., Friedman, S., Zhan, Y., Engleka, K.A., Forough, R. and Maciag, T. (1992) Heat shock induces the release of fibroblast growth factor 1 from NIH 3T3 cells. *Proc. Natl. Acad. Sci. USA* 89:10691–10695.

Jackson, A., Tarantini, F., Gamble, S., Friedman, S. and Maciag, T. (1995) The release of fibroblast growth factor-1 from NIH 3T3 cells in response to temperature involves the function of cysteine residues. *J. Biol. Chem.* 270:33–36.

Jaye, M., Howk, R., Burgess, W., Ricca, G.A., Chiu, I.-M., Ravera, M.W., O'Brien, S.J., Modi, W.S., Maciag, T. and Drohan, W.N. (1986) Human endothelial cell growth factor: cloning, nucleotide sequence, and chromosome localization. *Science* 233:541–545.

Jaye, M., Burgess, W.H., Shaw, A.B. and Drohan, W.N. (1987) Biological equivalence of natural bovine and recombinant human α-endothelial cell growth factors. *J. Biol. Chem.* 262: 16612–16617.

Johnson, D.E. and Williams, L.T. (1993) Structural and functional diversity in the FGF receptor multigene family. *Adv. Cancer Res.* 60:1–41.

Kandel, J., Bossy-Wetzel, E., Radvanyi, F., Klagsbrun, M., Folkman, J. and Hanahan, D. (1991) Neovascularization is associated with a switch to the export of bFGF in the multistep development of fibrosarcoma. *Cell* 66:1095–1104.

Keegan, K., Johnson, D.E., Williams, L.T. and Hayman, M.J. (1991) Isolation of an additional member of the fibroblast growth factor receptor family, FGFR-3. *Proc. Natl. Acad. Sci. USA* 88:1095–1099.

Kiefer, M.C., Stephans, J.C., Crawford, K., Okino, K. and Barr, P.J. (1990) Ligand-affinity cloning and structure of a cell surface heparan sulfate proteoglycan that binds basic fibroblast growth factor. *Proc. Natl. Acad. Sci. USA* 87:6985–6989.

Kiefer, M.C., Baird, A., Nguyen, T., George-Nascimento, C., Mason, O.B., Boley, L.J., Valenzuela, P. and Barr, P.J. (1991) Molecular cloning of a human basic fibroblast growth factor receptor cDNA and expression of a biologically active extracellular domain in a baculovirus system. *Growth Factors* 5:115–127.

Kim, K.J., Li, B., Winer, J., Armanini, M., Gillett, N., Phillips, H.S. and Ferrara, N. (1993) Inhibition of vascular endothelial growth factor-induced angiogenesis suppresses tumour growth *in vivo*. *Nature* 362:841–844.

Kimelman, D. (1993) Peptide growth factors and the regulation of early amphibian development. *Biochim. Biophys. Acta* 1155:227–237.

Klagsbrun, M. and Baird, A. (1991) A dual receptor system is required for basic fibroblast growth factor activity. *Cell* 67:229–231.

Klein, S., Giancotti, F.G., Presta, M., Albelda, S.M., Buck, C.A. and Rifkin, D.B. (1993) Basic fibroblast growth factor modulates integrin expression in microvascular endothelial cells. *Mol. Biol. Cell* 4:973–982.

Klein, S., Bikfalvi, A., Birkenmeier, T.M., Giancotti, F.G. and Rifkin, D.B. (1996) Integrin regulation by endogenous expression of 18 kDa fibroblast growth factor-2. *Mol. Biol. Cell,* in press.

Kobayashi, H., Schmitt, M., Goretzki, L., Chucholowski, N., Calvete, J., Kramer, M., Gunzler, W.A., Janicke, F. and Graeff, H. (1991) Cathepsin B efficiently activates the soluble and the tumor cell receptor-bound form of the proenzyme urokinase-type plasminogen activator (Pro-uPA). *J. Biol. Chem.* 266:5147–5152.

Kook, Y., H., Adamski, J., Zelent, A. and Ossowski, L. (1994) The effect of antisense inhibition of urokinase receptor in human squamous cell carcinoma on malignancy. *EMBO J.* 13:3983–3991.

LaBonne, C. and Whitman, M. (1994) Mesoderm induction by activin requires FGF-mediated intracellular signals. *Development* 120:463–472.

Languino, L.R., Gehlsen, K.R., Wayner, E., Carter, W.G., Engvall, E. and Ruoslahti, E. (1989) Endothelial cells use αvβ1 integrin as a laminin receptor. *J. Cell Biol.* 109:2455–2462.

Lee, P.L., Johnson, D.E., Cousens, L.S., Fried, V.A. and Williams, L.T. (1989) Purification and complementary DNA cloning of a receptor for basic fibroblast growth factor. *Science* 245:57–60.

Leung, D.W., Parent, A.S., Cachianes, G., Esch, F., Coulombe, J.N., Nikolics, K., Eckenstein, F.P. and Nishi, R. (1992) Cloning, expression during development, and evidence for release of a trophic factor for ciliary ganglion neurons. *Neuron* 8:1045–1053.

Lewis, C.E., Leek, R., Harris, A. and McGee, J.O. (1995) Cytokine regulation of angiogenesis in breast cancer: the role of tumor-associated macrophages. *J. Leuk. Biol.* 57:747–751.

Li, V.W., Folkerth, R.D., Watanabe, H., Yu, C., Rupnick, M., Barnes, P., Scott, R.M., Black, P.M., Sallan, S.E. and Folkman, J. (1994) Microvessel count and cerebrospinal fluid basic fibroblast growth factor in children with brain tumours. *Lancet* 344:82–86.

Lin, L.-F.H., Mismer, D., Lile, J.D., Armes, L.G., Butler, E.T. III., Vannice, J.L. and Collins, F. (1989) Purification, cloning, and expression of ciliary neurotrophic factor (CNTF). *Science* 246:1023–1025.

Linemeyer, D.L., Menke, J.G., Kelly, L.J., DiSalvo, J., Soderman, D., Schaeffer, M.T., Ortega, S., Gimenez-Gallego, G. and Thomas, K.A. (1990) Disulfide bonds are neither required, present, nor compatible with full activity of human recombinant acidic fibroblast growth factor. *Growth Factors* 3:287–298.

Liu, L., Doble, B.W. and Kardami, E. (1993) Perinatal phenotype and hypothyroidism are associated with elevated levels of 21.5- to 22-kDa basic fibroblast growth factor in cardiac ventricles. *Dev. Biol.* 157:507−516.

Lyons, R.M., Keski-Oja, J. and Moses, H.L. (1988) Proteolytic activation of latent transforming growth factor-β from fibroblast-conditioned medium. *J. Cell Biol.* 106:1659−1665.

Lyons, R.M., Gentry, L.E., Purchio, A.F. and Moses, H.L. (1990) Mechanism of activation of latent recombinant transforming growth factor β1 by plasmin. *J. Cell Biol.* 110:1361−1367.

Mach, H., Volkin, D.B., Burke, C.J., Middaugh, C.R., Linhardt, R.J., Fromm, J.R., Loganathan, D. and Mattsson, L. (1993) Nature of the interaction of heparin with acidic fibroblast growth factor. *Biochemistry* 32:5480−5489.

Mansukhani, A., Dell'Era, P., Moscatelli, D., Kornbluth, S., Hanafusa, H. and Basilico, C. (1992) Characterization of the murine BEK fibroblast growth factor (FGF) receptor: activation by three members of the FGF family and requirement for heparin. *Proc. Natl. Acad. Sci. USA* 89:3305−3309.

March, C.J., Mosley, B., Larsen, A., Cerretti, D.P., Braedt, G., Price, V., Gillis, S., Henney, C.S., Kronheim, S.R., Grabstein, K., Conlon, P.J., Hopp, T.P. and Cosman, D. (1985) Cloning, sequence and expression of two distinct human interleukin-1 complementary DNAs. *Nature* 315:641−647.

Marics, I., Adelaide, J., Raybaud, F., Mattei, M.G., Coulier, F., Planche, J., de Lapeyriere, O. and Birnbaum, D. (1989) Characterization of the HST-related FGF-6 gene, a new member of the fibroblast growth factor gene family. *Oncogene* 4:335−340.

Marshall, C.J. (1994) MAP kinase kinase kinase, MAP kinase kinase and MAP kinase. *Curr. Opin. Genet. Dev.* 4:82−89.

Mathews, L.S. and Vale, W.W. (1991) Expression of an activin receptor, a predicted transmembrane serine kinase. *Cell* 65:973−982.

McNeil, P.L., Muthukrishnan, L., Warder, E. and D'Amore, P.A. (1989) Growth factors are released by mechanically wounded endothelial cells. *J. Cell Biol.* 109:811−822.

Mignatti, P., Morimoto, T. and Rifkin, D.B. (1991a) Basic fibroblast growth factor released by single, isolated cells stimulates their migration in an autocrine manner. *Proc. Natl. Acad. Sci. USA* 88:11007−11011.

Mignatti, P., Mazzieri, R. and Rifkin, D.B. (1991b) Expression of the urokinase receptor in vascular endothelial cells is stimulated by basic fibroblast growth factor. *J. Cell Biol.* 113:1193−1201.

Mignatti, P., Morimoto, T. and Rifkin, D.B. (1992) Basic fibroblast growth factor, a protein devoid of secretory signal sequence, is released by cells via a pathway independent of the endoplasmic reticulum-Golgi complex. *J. Cell. Physiol.* 151:81−93.

Mignatti, P. and Rifkin, D.B. (1993) Biology and biochemistry of proteinases in tumor invasion. *Physiol. Rev.* 73:161−195.

Miyamoto, M., Naruo, K., Seko, C., Matsumoto, S., Kondo, T. and Kurokawa, T. (1993) Molecular cloning of a novel cytokine cDNA encoding the ninth member of the fibroblast growth factor family, which has a unique secretion property. *Mol. Cell. Biol.* 13:4251−4259.

Moenner, M., Gannoun-Zaki, L., Badet, J. and Barritault, D. (1989) Internalization and limited processing of basic fibroblast growth factor on Chinese hamster lung fibroblasts. *Growth Factors* 1:115−123.

Mohammadi, M., Honegger, A.M., Rotin, D., Fischer, R., Bellot, F., Li, W., Dionne, C.A., Jaye, M., Rubinstein, M. and Schlessinger, J. (1991) A tyrosine-phosphorylated carboxy-terminal peptide of the fibroblast growth factor receptor (Flg) is a binding site for the SH2 domain of phospholipase C-γ1. *Mol. Cell. Biol.* 11:5068−5078.

Mohammadi, M., Dionne, C.A., Li, W., Li, N., Spivak, T., Honegger, A.M., Jaye, M. and Schlessinger, J. (1992) Point mutation in FGF receptor eliminates phosphatidylinositol hydrolysis without affecting mitogenesis. *Nature* 358:681−684.

Moscatelli, D., Presta, M., Joseph-Silverstein, J. and Rifkin, D.B. (1986a) Both normal and tumor cells produce basic fibroblast growth factor. *J. Cell. Physiol.* 129:273−276.

Moscatelli, D., Presta, M. and Rifkin, D.B. (1986b) Purification of a factor from human placenta that stimulates capillary endothelial cell protease production, DNA synthesis, and migration. *Proc. Natl. Acad. Sci. USA* 83:2091−2095.

Moscatelli, D. (1987) High and low affinity binding sites for basic fibroblast growth factor on cultured cells: absence of a role for low affinity binding in the stimulation of plasminogen activator production by bovine capillary endothelial cells. *J. Cell. Physiol.* 131:123–130.

Moscatelli, D., Joseph-Silverstein, J., Manejias, R. and Rifkin, D.B. (1987) Mr 25,000 heparin-binding protein from guinea pig brain is a high molecular weight form of basic fibroblast growth factor. *Proc. Natl. Acad. Sci. USA* 84:5778–5782.

Moscatelli, D. and Rifkin, D.B. (1988) Membrane and matrix localization of proteinases: a common theme in tumor cell invasion and angiogenesis. *Biochim. Biophys. Acta* 948:67–85.

Moscatelli, D. (1988) Metabolism of receptor-bound and matrix-bound basic fibroblast growth factor by bovine capillary endothelial cells. *J. Cell Biol.* 107:753–759.

Mullins, D.E. and Rohrlich, S.T. (1983) The role of proteinases in cellular invasiveness. *Biochim. Biophys. Acta* 695:177–214.

Muthukrishnan, L., Warder, E. and McNeil, P. (1991) Basic fibroblast growth factor is efficiently released from a cytosolic storage site through plasma membrane disruptions of endothelial cells. *J. Cell. Physiol.* 148:1–16.

Nakanishi, Y., Kihara, K., Mizuno, K., Masamune, Y., Yoshitake, Y. and Nishikawa, K. (1992) Direct effect of basic fibroblast growth factor on gene transcription in a cell-free system. *Proc. Natl. Acad. Sci. USA* 89:5216–5220.

Nanus, D.M., Schmitz-Drager, B.J., Motzer, R.J., Lee, A.C., Vlamis, V., Cordon-Cardo, C., Albino, A.P. and Reuter, V.E. (1993) Expression of basic fibroblast growth factor in primary human renal tumors: correlation with poor survival. *J. Natl. Cancer Inst.* 85:1597–1599.

Nguyen, M., Watanabe, H., Budson, A.E., Richie, J.P., Hayes, D.F. and Folkman, J. (1994) Elevated levels of an angiogenic peptide, basic fibroblast growth factor, in the urine of patients with a wide spectrum of cancers. *J. Natl. Cancer Inst.* 86:356–361.

Nurcombe, V., Ford, M.D., Wildschut, J.A. and Bartlett, P.F. (1993) Developmental regulation of neural response to FGF-1 and FGF-2 by heparan sulfate proteoglycan. *Science* 260:103–106.

O'Reilly, M.S., Holmgren, L., Shing, Y., Chen, C., Rosenthal, R.A., Moses, M., Lane, W.S., Cao, Y., Sage, E.H. and Folkman, J. (1994) Angiostatin: a novel angiogenesis inhibitor that mediates the suppression of metastases by a Lewis lung carcinoma. *Cell* 79:315–328.

Odekon, L.E., Sato, Y. and Rifkin, D.B. (1992) Urokinase-type plasminogen activator mediates basic fibroblast growth factor-induced bovine endothelial cell migration independent of its proteolytic activity. *J. Cell. Physiol.* 150:258–263.

Ornitz, D.M., Yayon, A., Flanagan, J.G., Svahn, C.M., Levi, E. and Leder, P. (1992) Heparin is required for cell-free binding of basic fibroblast growth factor to a soluble receptor and for mitogenesis in whole cells. *Mol. Cell. Biol.* 12:240–247.

Pantoliano, M.W., Horlick, R.A., Springer, B.A., Van Dyk, D.E., Tobery, T., Wetmore, D.R., Lear, J.D., Nahapetian, A.T., Bradley, J.D. and Sisk, W.P. (1994) Multivalent ligand-receptor binding interactions in the fibroblast growth factor system produce a cooperative growth factor and heparin mechanism for receptor dimerization. *Biochemistry* 33:10229–10248.

Partanen, J., Makela, T.P., Eerola, E., Korhonen, J., Hirvonen, H., Claesson-Welsh, L. and Alitalo, K. (1991) FGFR-4, a novel acidic fibroblast growth factor receptor with a distinct expression pattern. *EMBO J.* 10:1347–1354.

Patry, V., Arnaud, E., Amalric, F. and Prats, H. (1994) Involvement of basic fibroblast growth factor NH$_2$ terminus in nuclear accumulation. *Growth Factors* 11:163–174.

Pepper, M.S., Belin, D., Montesano, R., Orci, L. and Vassalli, J.D. (1990) Transforming growth factor-induced proteolytic and angiogenic properties of endothelial cells *in vitro*. *J. Cell. Biol.* 111:743–755.

Pepper, M.S. and Montesano, R. (1990) Protolytic balance and capillary morphogenesis. *Cell. Differ. Dev.* 32:319–328.

Pepper, M.S., Sappino, A.P., Montesano, R., Orci, L. and Vassalli, J.D. (1992) Plasminogen activator inhibitor-1 is induced in migrating endothelial cells. *J. Cell. Physiol.* 153:129–139.

Peters, K.G., Marie, J., Wilson, E., Ives, H.E., Escobedo, J., Del Rosario, M., Mirda, D. and Williams, L.T. (1992) Point mutation of an FGF receptor abolishes phosphatidylinositol turnover and Ca^{2+} flux but not mitogenesis. *Nature* 358:678–681.

Peverali, F.A., Mandriota, S.J., Ciana, P., Marelli, R., Quax, P., Rifkin, D.B., Della Valle, G. and Mignatti, P. (1994) Tumor cells secrete an angiogenic factor that stimulates basic fibroblast growth factor and urokinase expression in vascular endothelial cells. *J. Cell. Physiol.* 161:1–14.

Pintucci, G., Iacoviello, L., Castelli, M.P., Amore, C., Evangelista, V., Cerletti, C. and Donati, M.B. (1993) Cathepsin G-induced release of PAI-1 in the culture medium of endothelial cells: a new thrombogenic role for polymorphonuclear leukocytes? *J. Lab. Clin. Med.* 122:69–79.

Pintucci, G., Quarto, N. and Rifkin, D.B. (1996) Methylation of high molecular weight FG-2 determines post-translational increases in molecular weight and affects its intracellular distribution. *Mol. Biol. Cell*, in press.

Plate, K.H., Breier, G., Weich, H.A. and Risau, W. (1992) Vascular endothelial growth factor is a potential tumour angiogenesis factor in human gliomas *in vivo. Nature* 359:845–848.

Powell, P.P., and Klagsbrun, M. (1991) Three forms of rat basic fibroblast growth factor are made from a single mRNA and localize to the nucleus. *J. Cell. Physiol.* 148:202–210.

Prats, H., Kaghad, M., Prats, A.C., Klagsbrun, M., Lelias, J.M., Liauzun, P., Chalon, P., Tauber, J.P., Amalric, F., Smith, J.A. and Caput, D. (1989) High molecular mass forms of basic fibroblast growth factor are initiated by alternative CUG codons. *Proc. Natl. Acad. Sci. USA* 86:1836–1840.

Prats, A.C., Vagner, S., Prats, H. and Amalric, F. (1992) cis-acting elements involved in the alternative translation initiation process of human basic fibroblast growth factor mRNA. *Mol. Cell. Biol.* 12:4796–4805.

Presta, M., Maier, J.A., Rusnati, M. and Ragnotti, G. (1989a) Basic fibroblast growth factor is released from endothelial extracellular matrix in a biologically active form. *J. Cell. Physiol.* 140:68–74.

Presta, M., Maier, J.A., Ragnotti, G. (1989b) The mitogenic signaling pathway but not the plasminogen activator-inducing pathway of basic fibroblast growth factor is mediated through protein kinase C in fetal bovine aortic endothelial cells. *J. Cell Biol.* 109:1877–1884.

Quarto, N., Finger, F.P. and Rifkin, D.B. (1991a) The NH_2-terminal extension of high molecular weight bFGF is a nuclear targeting signal. *J. Cell. Physiol.* 147:311–318.

Quarto, N., Talarico, D., Florkiewicz, R. and Rifkin, D.B. (1991b) Selective expression of high molecular weight basic fibroblast growth factor confers a unique phenotype of NIH 3T3 cells. *Cell Regul.* 2:699–708.

Quarto, N. and Amalric, F. (1994) Heparan sulfate proteoglycans as transducers of FGF-2 signaling. *J. Cell Sci.* 107:3201–3212.

Quinkler, W., Maasberg, M., Bernotat-Danielowski, S., Luthe, N., Sharma, H.S. and Schaper, W. (1989) Isolation of heparin-binding growth factors from bovine, porcine and canine hearts. *Eur. J. Biochem.* 181:67–73.

Rabbani, S.A., Mazar, A.P., Bernier, S.M., Haq, M., Bolivar, I., Henkin, J. and Goltzman, D. (1992) Structural requirements for the growth factor activity of the amino-terminal domain of urokinase. *J. Biol. Chem.* 267:14151–14156.

Rapraeger, A.C., Krufka, A. and Olwin, B.B. (1991) Requirement of heparan sulfate for bFGF-mediated fibroblast growth and myoblast differentiation. *Science* 252:1705–1708.

Renko, M., Quarto, N., Morimoto, T. and Rifkin, D.B. (1990) Nuclear and cytoplasmic localization of different basic fibroblast growth factor species. *J. Cell. Physiol.* 144:108–114.

Rhee, S.G., Kim, H., Suh, P.G. and Choi, W.C. (1991) Multiple forms of phosphoinositide-specific phospholipase C and different modes of activation. *Biochem. Soc. Trans.* 19:337–341.

Riese, J., Zeller, R. and Dono, R. (1995) Nucleo-cytoplasmic translocation and secretion of fibroblast growth factor-2 during avian gastrulation. *Mech. Dev.* 49:13–22.

Rifkin, D.B., Kojima, S., Abe, M. and Harpel, J.G. (1993) TGF-β: structure, function, and formation. *Thromb. Haemost.* 70:177–179.

Rifkin, D.B., Moscatelli, D., Roghani, M., Nagano, Y., Quarto, N., Klein, S. and Bikfalvi, A. (1994) Studies on FGF-2: nuclear localization and function of high molecular weight forms and receptor binding in the absence of heparin. *Mol. Reprod. Dev.* 39:102–105.

Risau, W. (1995) Differentiation of endothelium. *FASEB J.* 9:926–933.

Roghani, M. and Moscatelli, D. (1992) Basic fibroblast growth factor is internalized through both receptor-mediated and heparan sulfate-mediated mechanisms. *J. Biol. Chem.* 267:22156–22162.

Roghani, M., Mansukhani, A., Dell'Era, P., Bellosta, P., Basilico, C., Rifkin, D.B. and Moscatelli, D. (1994) Heparin increases the affinity of basic fibroblast growth factor for its receptor but is not required for binding. *J. Biol. Chem.* 269:3976–3984.

Roldan, A.L., Cubellis, M.V., Masucci, M.T., Behrendt, N., Lund, L.R., Dano, K., Appella, E. and Blasi, F. (1990) Cloning and expression of the receptor for human urokinase plasminogen activator, a central molecule in cell surface, plasmin dependent proteolysis. *EMBO J.* 9:467–474.

Rubin, J.S., Osada, H., Finch, P.W., Taylor, W.G., Rudikoff, S. and Aaronson, S.A. (1989) Purification and characterization of a newly identified growth factor specific for epithelial cells. *Proc. Natl. Acad. Sci. USA* 86:802–806.

Sa, G. and Fox, P.L. (1994) Basic fibroblast growth factor-stimulated endothelial cell movement is mediated by a pertussis toxin-sensitive pathway regulating phospholipase A2 activity. *J. Biol. Chem.* 269:3219–3225.

Sakamoto, H., Mori, M., Taira, M., Yoshida, T., Matsukawa, S., Shimizu, K., Sekiguchi, M., Terada, M. and Sugimura, T. (1986) Transforming gene from human stomach cancers and a noncancerous portion of stomach mucosa. *Proc. Natl. Acad. Sci. USA* 83:3997–4001.

Saksela, O., Moscatelli, D., Sommer, A. and Rifkin, D.B. (1988) Endothelial cell-derived heparan sulfate binds basic fibroblast growth factor and protects it form proteolytic degradation. *J. Cell Biol.* 107:743–751.

Saksela, O. an Rifkin, D.B. (1990) Release of basic fibroblast growth factor-heparan sulfate complexes from endothelial cells by plasminogen activator-mediated proteolytic activity. *J. Cell. Biol.* 110:767–775.

Sato, Y. and Rifkin, D.B. (1988) Autocrine activities of basic fibroblast growth factor: regulation of endothelial cell movement, plasminogen activator synthesis, and DNA synthesis. *J. Cell Biol.* 107:1199–1205.

Sato, Y. and Rifkin, D.B. (1989) Inhibition of endothelial cell movement by pericytes and smooth muscle cells: activation of a latent transforming growth factor-β 1-like molecule by plasmin during co-culture. *J. Cell Biol.* 109:309–315.

Sato, Y., Tsuboi, R., Lyons, R., Moses, H. and Rifkin, D.B. (1990) Characterization of the activation of latent TGF-β by co-cultures of endothelial cells and pericytes or smooth muscle cells: a self-regulating system. *J. Cell Biol.* 111:757–763.

Schlessinger, J. and Ullrich, A. (1992) Growth factor signaling by receptor tyrosine kinases. *Neuron* 9:383–391.

Schlessinger, J. (1993) How receptor tyrosine kinases activate Ras. *Trends Biochem. Sci.* 18:273–275.

Schlessinger, J., Lax, I. and Lemmon, M. (1995) Regulation of growth factor activation by proteoglycans; what is the role of the low affinity receptors? *Cell,* 83:357–360.

Schulze-Osthoff, K., Risau, W., Vollmer, E. and Sorg, C. (1990) In situ detection of basic fibroblast growth factor by highly specific antibodies. *Am. J. Pathol.* 137:85–92.

Schweigerer, L., Neufeld, G., Friedman, J., Abraham, J.A., Fiddes, J.C. and Gospodarowicz, D. (1987) Capillary endothelial cells express basic fibroblast growth factor, a mitogen that promotes their own growth. *Nature* 325:257–259.

Shi, E., Kan, M., Xu, J., Wang, F., Hou, J. and McKeehan, W.L. (1993) Control of fibroblast growth factor receptor kinase signal transduction by heterodimerization of combinatorial splice variants. *Mol. Cell. Biol.* 13:3907–3918.

Shiurba, R.A., Jing, N., Sakakura, T. and Godsave, S.F. (1991) Nuclear translocation of fibroblast growth factor during *Xenopus* mesoderm induction. *Development* 113:487–493.

Shweiki, D., Itin, A., Soffer, D. and Keshet, E. (1992) Vascular endothelial growth factor induced by hypoxia may mediate hypoxia-initiated angiogenesis. *Nature* 359:843–845.

Smith, J.C., Price, B.M., Green, J.B., Weigel, D. and Herrmann, B.G. (1991) Expression of a *Xenopus* homolog of *Brachyury (T)* is an immediate-early response to mesoderm induction. *Cell* 67:79–87.

Smith, J.C. (1993) Mesoderm-inducing factors in early vertebrate development. *EMBO J.* 12:4463–4470.

Sommer, A., Brewer, M.T., Thompson, R.C., Moscatelli, D., Presta, M. and Rifkin, D.B (1987) A form of human basic fibroblast growth factor with an extended amino terminus. *Biochem. Biophys. Res. Commun.* 144:543–550.

Sommer, A., Moscatelli, D. and Rifkin, D.B. (1989) An amino-terminally extended and post-translationally modified form of a 25 kD basic fibroblast growth factor. *Biochem. Biophys. Res. Commun.* 160:1267–1274.

Sommer, A. and Rifkin, D.B. (1989) Interaction of heparin with human basic fibroblast growth factor: protection of the angiogenic protein from proteolytic degradation by a glycosaminoglycan. *J. Cell. Physiol.* 138:215–220.

Spivak-Kroizman, T., Lemmon, M.A., Dikic, I., Ladbury, J.E., Pinchasi, D., Huang, J., Jaye, M., Crumley, G., Schlessinger, J. and Lax I. (1994a) Heparin-induced oligomerization of FGF molecules is responsible for FGF receptor dimerization, activation, and cell proliferation. *Cell* 79:1015–1024.

Spivak-Kroizman, T., Mohammadi, M., Hu, P., Jaye, M., Schlessinger, J. and Lax, I. (1994b) Point mutation in the fibroblast growth factor receptor eliminates phosphatidylinositol hydrolysis without affecting neuronal differentiation of PC12 cells. *J. Biol. Chem.* 269:14419–14423.

Springer, B.A., Pentoliano, M.W., Barbera, F.A., Gunyuzlu, P.L., Thompson, L.D., Herblin, W.F., Rosenfeld, S.A. and Book, G.W. (1994) Identification and concerted function of two receptor binding surfaces on basic fibroblast growth factor required for mitogenesis. *J. Biol. Chem.* 269:26879–26884.

Stan, A.C., Nemati, M.N., Pietsch, T., Walter, G.F. and Dietz, H. (1995) *In vivo* inhibition of angiogenesis and growth of the human U-87 malignant glial tumor by treatment with an antibody against basic fibroblast growth factor. *J. Neurosurg.* 82:1044–1052.

Stevenson, F.T., Torrano, F., Locksley, R.M. and Lovett, D.H. (1992) Interleukin 1: the patterns of translation and intracellular distribution support alternative secretory mechanisms. *J. Cell. Physiol.* 152:223–231.

Story, M.T., Esch, F., Shimasaki, S., Sasse, J., Jacobs, S.C. and Lawson, R.K. (1987) Amino-terminal sequence of a large form of basic fibroblast growth factor isolated from human benign prostatic hyperplastic tissue. *Biochem. Biophys. Res. Commun.* 142:702–709.

Taira, M., Jamrich, M., Good, P.J. and Dawid, I.B. (1992) The LIM domain-containing homeo box gene Xlim-1 is expressed specifically in the organizer region of *Xenopus* gastrula embryos. *Genes Dev.* 6:356–366.

Takahashi, K., Mulliken, J.B., Kozakewich, H.P., Rogers, R.A., Folkman, J. and Ezekowitz, R.A. (1994) Cellular markers that distinguish the phases of hemangioma during infancy and childhood. *J. Clin. Invest.* 93:2357–2364.

Tanaka, A., Miyamoto, K., Minamino, N., Takeda, M., Sato, B., Matsuo, H. and Matsumoto, K. (1992) Cloning and characterization of an androgen-induced growth factor essential for the androgen-dependent growth of mouse mammary carcinoma cells. *Proc. Natl. Acad. Sci. USA* 89:8928–8932.

Thomas, K.A., Rios-Candelore, M. and Fitzpatrick, S. (1984) Purification and characterization of acidic fibroblast growth factor from bovine brain. *Proc. Natl. Acad. Sci. USA* 81:357–361.

Thompson, L.D., Pantoliano, M.W. and Springer, B.A. (1994) Energetic characterization of the basic fibroblast growth factor-heparin interaction: identification of the heparin binding domain. *Biochemistry* 33:3831–3840.

Tsai, J.-C., Goldman, C.K. and Gillespie, G.Y. (1995) Vascular endothelial growth factor in human glioma cell lines: induced secretion by EGF, PDGF-BB, and bFGF. *J. Neurosurg.* 82:864–873.

Ueno, N., Baird, A., Esch, F., Ling, N. and Guillemin, R. (1986) Isolation of an amino terminal extended form of basic fibroblast growth factor. *Biochem. Biophys. Res. Commun.* 138:580–588.

Vagner, S., Gensac, M.C., Maret, A., Bayard, F., Amalric, F., Prats, H. and Prats, A.-C. (1995) Alternative translation of human fibroblast growth factor 2 mRNA occurs by internal entry of ribosomes. *Mol. Cell. Biol.* 15:35–44.

Varner, J.A., Emerson, D.A. and Juliano, R.L. (1995) Integin $\alpha 5\beta 1$ expression negatively regulates cell growth: reversal by attachment to fibronectin. *Mol. Biol. Cell* 6:725–740.

Vassalli, J.D., Sappino, A.P. and Belin, D. (1991) The plasminogen activator/plasmin system. *J. Clin. Invest.* 88:1067–1072.

Vigny, M., Ollier-Hartmann, M.P., Lavigne, M., Fayein, N., Jeanny, J.C., Laurent, M. and Courtois, Y. (1988) Specific binding of basic fibroblast growth factor to basement membrane-like structures and to purified heparan sulfate proteoglycan of the EHS tumor. *J. Cell. Physiol.* 137:321–328.

Vilgrain, I. and Baird, A. (1991) Phosphorylation of basic fibroblast growth factor by a protein kinase associated with the outer surface of a target cell. *Mol. Endocrinol.* 5:1003–1012.

Vilgrain, I., Gonzalez, A.M. and Baird, A. (1993) Phosphorylation of basic fibroblast growth factor (FGF-2) in the nuclei of SK-Hep-1 cells. *FEBS Lett.* 331:228–232.

Vlodavsky, I., Folkman, J., Sullivan, R., Fridman, R., Ishai-Michaeli, R., Sasse, J. and Klagsbrun, M. (1987) Endothelial cell-derived basic fibroblast growth factor: synthesis and deposition into subendothelial extracellular matrix. *Proc. Natl. Acad. Sci. USA* 84:2292–2296.

Wang, J.Y. (1994) Nuclear protein tyrosine kinases. *Trends Biochem. Sci.* 19:373–376.

Wang, F., Kan, M., Yan, G., Xu, J. and McKeehan, W.L. (1995) Alternately spliced NH_2 terminal immunoglobulin-like Loop I in the ectodomain of the fibroblast growth factor (FGF) receptor 1 lowers affinity for both heparin and FGF-1. *J. Biol. Chem.* 270:10231–10235.

Weich, H.A., Iberg, N., Klagsbrun, M. and Folkman, J. (1990) Expression of acidic and basic fibroblast growth factors in human and bovine vascular smooth muscle cells. *Growth Factors* 2:313–320.

Weidner, N., Semple, J.P., Welch, W.R. and Folkman, J. (1991) Tumor angiogenesis and metastasis – correlation in invasive breast carcinoma. *N. Engl. J. Med.* 324:1–8.

Werb, Z., Mainardi, C.L., Vater, C.A. and Harris, E.D.J. (1977) Endogenous activation of latent collagenase by rheumatoid synovial cells. Evidence for a role of plasminogen activator. *N. Engl. J. Med.* 296:1017–1023.

Werb, Z., Banda, M.J. and Jones, P.A. (1980) Degradation of connective tissue matrices by macrophages. Proteolysis of elastin, glycoproteins, and collagen by proteinases isolated from macrophages. *J. Exp. Med.* 152:1340–1357.

Wiedlocha, A., Madshus, I.H., Mach, H., Middaugh, C.R. and Olsnes, S. (1992) Tight folding of acidic fibroblast growth factor prevents its translocation to the cytosol with diphtheria toxin as vector. *EMBO J.* 11:4835–4842.

Wiedlocha, A., Falnes, P.O., Madshus, I.H., Sandvig, K. and Olsnes, S. (1994) Dual mode of signal transduction by externally added acidic fibroblast growth factor. *Cell* 76:1039–1051.

Winkles, J.A., Friesel, R., Burgess, W.H., Howk, R., Mehlman, T., Weinstein, R. and Maciag, T. (1987) Human vascular smooth muscle cells both express and respond to heparin-binding growth factor I (endothelial cell growth factor). *Proc. Natl. Acad. Sci. USA* 84:7124–7128.

Yang, J.T., Rayburn, H. and Hynes, R.O. (1993) Embryonic mesodermal defects in α-5 integrin-deficient mice. *Development* 119:1093–1105.

Yayon, A., Klagsbrun, M., Esko, J.D., Leder, P. and Ornitz, D.M. (1991) Cell surface, heparin-like molecules are required for binding of basic fibroblast growth factor to its high affinity receptor. *Cell* 64:841–848.

Zhan, X., Bates, B., Hu, X.G. and Goldfarb, M. (1988) The human FGF-5 oncogene encodes a novel protein related to fibroblast growth factors. *Mol. Cell. Biol.* 8:3487–3495.

Zhan, X., Hu, X., Friedman, S. and Maciag, T. (1992) Analysis of endogenous and exogenous nuclear translocation of fibroblast growth factor-1 in NIH 3T3 cells. *Biochem. Biophys. Res. Commun.* 188:982–991.

Zhan, X., Hu, X., Friesel, R. and Maciag, T. (1993) Long term growth factor exposure and differential tyrosine phosphorylation are required for DNA synthesis in BALB/c 3T3 cells. *J. Biol. Chem.* 268:9611–9620.

Regulation of Angiogenesis
ed. by I.D. Goldberg & E.M. Rosen
© 1997 Birkhäuser Verlag Basel/Switzerland

Regulation of angiogenesis by scatter factor

E.M. Rosen and I.D. Goldberg

Long Island Jewish Medical Center, The Long Island Campus for Albert Einstein College of Medicine, Department of Radiation Oncology, New Hyde Park, New York, New York 11040, USA

Summary. Scatter factor (SF, hepatocyte growth factor) is a cytokine that stimulates motility, proliferation, and morphogenesis of epithelia. These responses are transduced through a tyrosine kinase receptor that is encoded by a proto-oncogene (c-*met*). SF is a potent angiogenic molecule, and its angiogenic activity is mediated, in part, through direct actions on endothelial cells. These include stimulation of cell motility, proliferation, protease production, invasion, and organization into capillary-like tubes. SF also stimulates the proliferation of smooth muscle cells and pericytes, cell types that also participate in the formation of capillaries and other microvessels. SF is chronically overexpressed in tumors, and it is postulated SF may function as a tumor angiogenesis factor. SF production in tumors may be due, in part, to an abnormal tumor: stroma interaction in which tumor cells secrete soluble proteins (SF-inducing factors) that stimulate stromal cell SF production and in part to autocrine production by the tumor cells themselves. Recent studies suggest a linkage between tumor suppressors (anti-oncogenes) and inhibition of angiogenesis. We hypothesize that tumor suppressor gene mutations may contribute to activation of an SF → c-*met* signalling pathway, leading to an invasive and angiogenic tumor phenotype. Modulation of this pathway may provide clinically useful methods of enhancing or inhibiting angiogenesis.

Introduction

Scatter Factor (SF)

SF was originally characterized as a soluble mesenchymal cell-derived protein that induces dissociation (scattering) of colonies of epithelial cells (Stoker et al., 1987; Rosen et al., 1989). SF was subsequently found to be identical to hepatocyte growth factor (HGF) (Weidner et al., 1991a; Bhargava et al., 1992), a serum-derived growth factor for primary cultures of hepatocytes that is thought to be an hepatotrophic factor for liver repair (Miyazawa et al., 1989; Nakamura et al., 1989). SF is a 90 kDa heparin-binding glycoprotein consisting of a 60 kDa α-chain and a 30 kDa β-chain linked by an interchain disulfide bond (Gherardi et al., 1989; Rosen et al., 1990a; Weidner et al., 1990). The α-chain contains an N-terminal hairpin loop and four "kringle" domains (triple disulfide looped structures that mediate protein:protein interactions). The β-chain is homologous to serine proteases, but lacks protease activity (Rosen et al., 1990b) due to replacement of two essential amino acids at the catalytic triad of the β-chain (Nakamura et al., 1989). SF has similar subunit structure and about 38% amino acid sequence identity to the serum zymogen plasminogen

(Nakamura et al., 1989). In addition, SF is about 50% identical to macro-phage-stimulating protein (MSP), a serum-derived cytokine that induces macrophages to become competent to respond to chemotactic and phago-cytic stimuli (Yoshimura et al., 1993).

SF is synthesized as a 728 amino acid precursor (preproSF) (Nakamura et al., 1989). Intracellular cleavage of a 31 amino acid signal peptide results in its secreted single-chain form (proSF), which is biologically inactive (Lokker et al., 1992). Extracellular cleavage of proSF at [494]arg-[495]val yields active two-chain SF, similar to the arg-val cleavage required for conversion of the plasminogen to plasmin and the conversion of pro-MSP to active MSP. While plasminogen activators (uPA and tPA) can also cleave and activate proSF, they do so only at supraphysiologic concentrations (Naldini et al., 1992; Mars et al., 1993). HGF activator, a 34 kDa two-chain novel serine protease homologous to coagulation factor XII (Hagemann factor), may be a physiologic cleavage enzyme for SF (Miyazawa et al., 1993). HGF activator is produced in pro-enzyme form, and it may be activated by a proteolytic cascade initiated by tissue injury (Miyazawa et al., 1994). It appears that physiologic activation of the structurally similar kringle proteins SF, plasminogen, and MSP is accomplished via different enzymes.

c-Met receptor

The SF receptor has been identified as the protein product of the c-*met* proto-oncogene (Bottaro et al., 1991), which encodes a transmembrane tyrosine kinase expressed predominantly by epithelia (Gonzatti-Haces et al., 1988). The c-met receptor is a 190 kDa glycoprotein consisting of a 145 kDa membrane-spanning β-chain and a 50 kDa α-chain expressed on the cell surface. The extracellular binding, transmembrane, intracellular kinase, and non-catalytic phosphate acceptor domains are located on the β-chain. A rearranged and truncated form of c-*met* called *tpr-met* encodes a tyrosine kinase that lacks the extracellular ligand-binding domain, is constitutively activated, and induces cell transformation (Gonzatti-Haces et al., 1988). *Tpr-met* results from a translocation involving a powerful promoter on chromosome 1 (called *t*ransposed *p*romoter *r*egion) and the 3′ portion of c-*met* on chromosome 7 (Park et al., 1986).

Recent studies suggest that much of the signal transduction from the SF-activated c-met receptor occurs through a unique signalling site charac-terized by a tandem YV(H/N)V. Phosphorylation of one or both tyrosine residues in this site allows interaction with the *src* homology-2 (SH2) domains of various intracellular signalling molecules, including phosphat-idylinositol-3′-kinase, protein tyrosine phosphatase 2, phospholipase C-gamma, pp60[c-*src*], and the *grb2/hSos1* complex (Ponzetto et al., 1994). Two receptor tyrosine kinases related to c-*met*, *c-sea* and *Ron*, have been described (Ronsin et al., 1993; Huff et al., 1994). The ligand for the *Ron* receptor was recently identified as MSP (see above) (Wang et al., 1994a); Yoshimura et al., 1993), while the ligand for *c-sea* has not yet been identified.

SF biologic actions

SF induces three major classes of cellular responses *in vitro*: motility, proliferation, and morphogenesis. Studies employing chimeric receptor constructs indicate that each of these responses is transduced by the c-*met* tyrosine kinase (Weidner et al., 1993; Zhu et al., 1994). In addition to cell dissociation, SF induces random movement of isolated epithelial and carcinoma cells, chemotactic (gradient-directed) migration, migration from carrier beads to flat surfaces, and invasion through extracellular matrix proteins (Rosen et al., 1990a, b, c, 1991a; Weidner et al., 1990; Bhargava et al., 1992; Li et al., 1994). SF stimulates the expression of booth urokinase (uPA) and uPA receptor (uPAR) (Pepper et al., 1992; Grant et al., 1993; Rosen et al., 1994b). The net effect is to increase the amount of uPA bound to uPAR on the cell surface. Receptor bound uPA on the cell surface is catalytically active and is thought to mediate focal degradation of the extracellular matrix necessary to clear a path for invading cells (Saksela and Rifkin, 1988). Thus, SF appears to turn on a whole program of cellular activities required for invasion.

SF is mitogenic for various normal cell types, including epithelial cells, vascular endothelial cells, and melanocytes (Kan et al., 1991; Rubin et al., 1991; Halaban et al., 1992). SF is also a potent morphogen. SF induces MDCK epithelial cells incubated in collagen I gels to organize into a network of branching tubules with the proper apical-basolateral polarity (Montesano et al., 1991; Santos et al., 1993). Similarly, SF induces mammary epithelial cells to form duct-like structures (Tsarfaty et al., 1992). Thus, SF can active specific programs of cell differentiation depending upon the cell type and environment.

SF biologic actions on vascular cell types

Endothelial cells (ECs)

During the early stages of angiogenesis *in vivo*, ECs from pre-existing small vessels (usually venules that lack a smooth muscle covering) focally degrade the subendothelial basement membrane, migrate out into the interstitium toward an angiogenic stimulus, and form capillary sprouts (Folkman, 1985). Sprouting ECs proximal to the migrating tip proliferate; and subsequently, the EC sprouts organize into an anastomosing network of capillary tubes. Finally, these ECs synthesize new basement membrane.

Adhesion of SMCs and pericytes and formation of new basement membrane are processes associated with the termination of the angiogenic response (Folkman, 1985; Antonelli-Orlidge et al., 1989). Stimulation of EC motility, proliferation, and capillary-like tube formation *in vitro* are generally thought to correlate with the ability to induce angiogenesis *in vivo*, since each of these processes occurs during the formation of new blood vessels (Folkman, 1985).

Large vessel- and microvessel-derived ECs express the c-*met* receptor and respond to SF in various *in vitro* assay systems (Rosen et al., 1990b, c, 1991b, c; Bussolino et al., 1992; Grant et al., 1993; Naidu et al., 1994). SF is chemotactic to ECs and stimulates random motility, as demonstrated in assays using microwell modified Boyden chambers (Rosen et al., 1990b, 1991b). In addition, SF induces ECs attached to microcarrier beads to migrate from the carrier beads to flat culture surfaces (Rosen et al., 1990b, c). SF also induces invasion of ECs through porous filters coated with Matrigel, a reconstituted matrix of basement membrane (Rosen et al., 1991b). Maximal chemotaxis, bead migration, and invasion of human umbilical vein ECs (HUVEC), calf pulmonary artery ECs (CPAE), bovine aortic EC (BAEC), and bovine brain ECs (BBEC) are typically observed at SF concentrations of 2–20 ng/ml. In studies utilizing the bead migration assay, we found that BBEC migration was stimulated 5-fold by SF but was not affected by basic fibroblast growth factor (FGF) or by epidermal growth factor (EGF) (Rosen et al., 1991c). On the other hand, transforming growth factor-β (TGF-β) blocked both basal and SF-stimulated migration of BBEC. SF-induced migration from carrier beads was blocked by inhibitors of protein synthesis (cycloheximide) but not by inhibitors of DNA synthesis (hydroxyurea), evidence that the motile and proliferative phenotypes of endothelial cells are regulated independently (Rosen et al., 1991c).

In addition to motility, SF stimulates DNA synthesis and proliferation of some EC types, including HUVEC and human omental microvessel ECs (Rubin et al., 1991; Morimoto et al., 1991). Capillary tube formation appears to be an independent property of ECs, not directly related to motility or proliferation (Grant et al., 1989). When ECs are seeded onto basement membrane (Matrigel), they stop proliferating, extend long cytoplasmic processes, and begin to organize into a network of capillary-like tubes. SF stimulates the formation of these capillary-like tubes in HUVEC and BBEC cultures by up to (5–10)-fold, as measured via computerized image analysis of stained cultures (Rosen et al., 1991b; Grant et al., 1993).

SF also stimulates large increases in the production of uPA activity by EC cultures (Rosen et al., 1991b; Grant et al., 1993). Most of the SF-induced uPA activity is cell-associated rather than secreted. The majority of cell-associated uPA is bound to uPA receptor on the cell surface, where it is well-positioned to mediate focal degradation of extracellular matrix proteins, a prerequisite for invasion (Saksela and Rifkin, 1988). Taken together, these findings indicate the SF can induce most or all of the phenotypic characteristics expressed by ECs during the process of angiogenesis.

Smooth muscle cells (SMCs) and pericytes

In vitro, bovine aortic, human iliac artery, and rat arterial SMCs produce SF at rates comparable to those of high producer human fibroblast cell lines (e.g. MRC5, WI38) (64–128 scatter units/10^6 cells/48 h) (Rosen et al., 1989, 1990a). The biologic and chemical properties of SMC-derived SF are very similar to those of fibroblast-derived SF, and it is likely that these molecules are identical (Rosen et al., 1990a, b). Thus, vascular smooth muscle appear to be a significant SF producer cell type.

In addition to producing SF, SMCs and pericytes, which are thought to be microvascular SMCs, express the c-met receptor and are capable of responding to SF. SMCs and pericytes of the microvessel wall in psoriatic skin and in the microvasculature of tumors such as bladder carcinomas immunostain positively for c-met protein (Grant et al., 1993; Joseph et al., 1995), suggesting that these cell types are potential target cells for SF. Recruitment of SMCs and pericytes is an essential component of angiogenesis. These cells are thought to stabilize newly formed vessels, thus contributing to termination of the angiogenic process (Antonelli-Orlidge et al., 1989). Therefore, it seems logical that SF itself might induce the influx and/or proliferation of these cells at the appropriate time during the angiogenic response.

Cultured pericytes from bovine retina express c-*met* mRNA, as demonstrated by reverse-transcriptase PCR analysis of pericyte RNA (unpublished results), consistent with the finding that pericytes express immunoreactive c-*met* protein *in vivo*. Moreover, we have found that SF stimulates the proliferation of bovine aortic SMC and bovine retinal pericytes *in vitro* (Rosen and Goldberg, 1995). The maximum degrees of stimulation of proliferation in medium containing 1% and 5% calf serum were 2.4- and 1.8-fold for SMCs, and 1.7- and 2.2-fold for pericytes, respectively. These maximal values were observed at 20–100 ng/ml of SF. These findings are consistent with the putative role of SMCs in angiogenesis and the presence of SMCs in new microvessels induced by SF. The role of SF as a paracrine and/ or autocrine modulator of smooth muscle and pericyte function during neovascularization remains to be further elucidated.

In vivo angiogenic activity of SF

We used two different assays, the recently described mouse Matrigel angiogenesis assay (Kibbey et al., 1992) and the conventional rat corneal neovascularization assay (Polverini and Leibovich, 1984), to demonstrate the ability of SF to induce new blood vessel formation *in vivo* (Grant et al., 1993; Naidu et al., 1994). In the former assay, different doses of SF were mixed with 0.5 ml of Matrigel, a preparation of reconstituted basement membrane that remains a liquid in the cold (4 °C). The Matrigel was injected subcutaneously into athymic nude mice or C57/BL mice. At body temperature, Matrigel rapidly forms a solid gel which retains the SF bound to extracellular matrix proteins and allows prolonged exposure of the surrounding tissues to it. Animals were sacrificed after ten days, and the Matrigel plugs were excised. The ingrowth of blood vessels into the plugs was quantitated by computerized image analysis of histologic sections stained with Masson's trichrome, which stains endothelial cells reddish-purple. Analysis revealed that SF-containing plugs contained increased numbers of migrated endothelial cells, which was conformed by immunostaining for Factor VIII-related antigen. Angiogenesis increased in a dose-dependent manner from 2–200 ng/ml of SF, up to four to five times control values. At the higher doses, well-formed blood vessels were observed. Responses were quantitatively similar in nude mice and C57/BL mice. Inflammatory responses were not observed in athymic mice at any SF dose and were found only at supramaximal SF doses (2000 ng/ml) in C57/BL mice.

In the second assay, SF was dissolved in Hydron polymer, which was air-dried to form a solid pellet. The Hydron pellets were placed in surgically created pockets about 1–2 mm from the limbus of the avascular rat cornea. Animals were perfused with colloidal carbon and sacrificed after seven days; and the growth of new vessels from the limbus toward the pellet was assessed. SF induced dose-dependent corneal neovascularization similar to that observed in the mouse Matrigel assay (Grant et al., 1993). Purified native mouse SF and recombinant human SF elicited equal angiogeneic responses; and the maximal responses induced by SF were similar in intensity to that induced by basic FGF (Grant et al., 1993). Two different antibody preparations against SF blocked SF-induced angiogenesis but did not affect FGF-induced angiogenesis. Inflammatory responses, assessed by F4/80 immunostaining to detect infiltration of monocyte/macrophage-like cells, were observed only at supramaximal doses of SF. Therefore, SF appears to be as potent an inducer of angiogenesis as is basic FGF.

Our findings indicate that recruitment of inflammatory cells does not play a major role in SF-induced angiogenesis. However, in the mouse Matrigel assay, histologic sections of Matrigel plugs prepared at early times (days 2–3) revealed the presence of many SMC/pericyte-like cells, which stained positively for α-actin. Moreover, at higher doses of SF, histologic sections prepared on day 10 revealed SMCs in some of the newly formed vessels in the Matrigel (Grant et al., 1993). Thus, SF-induced angiogenesis appears to be mediated by direct effects on ECs and, in addition, by direct and/or indirect effects on SMCs. This conclusion is supported by the *in vitro* studies described above.

Tumor angiogenesis

Angiogenesis in human cancers

Recent clinical studies suggest that tumor angiogenesis, as indicated by increased numbers of microvessels in the tumor stroma, is an independent indicator of aggressive biologic behavior in breast cancer (Weidner et al., 1991 b, 1992; Bosari et al., 1992; Toi et al., 1993; Guidi et al., 1994) and in other tumor types (Bochner et al., 1995). Experimental studies of human and animal tissues suggest that an angiogenic phenotype may be observed in earlier lesions (e.g. hyperplasia or dysplasia) of breast or other tissues (Brem et al., 1978; Folkman et al., 1989). It has not been established whether angiogenesis is required for malignant progression or merely reflects an underlying aggressive biology However, a large body of evidence suggests that angiogenesis is a critical requirement for continued local growth as well as metastasis of most solid tumors (Folkman, 1992).

While physiologic angiogenesis in adult tissues (e.g. as occurs during wound healing, corpus luteum formation, and placental implantation) is tightly regulated spatially and temporally, tumor angiogenesis is characterized by *persistent* neovascularization. This situation may occur because of: (a) overproduction and accumulation of pro-angiogenic factors within the tumor microenvironment; (b) underproduction of anti-angiogenic factors; and/or (c) failure of other homeostatic mechanisms that normally terminate the angiogenic process or induce vascular regression. A modest number of growth factors and cytokines are capable of inducing angiogenesis in various *in vivo* and *in vitro* assay systems [e.g. angiogenin, various forms of FGF, EGF, TGF-α, interleukin-8 (IL-8), platelet-derived endothelial cell growth factor (PDECGF), SF, tumor necrosis factor-α (TNF-α), TGF-β, and vascular endothelial cell growth factor (VEGF)]. These angiogenic factors may be produced by tumor cells, host stromal cells (e.g. fibroblasts, SMCs), or infiltrating leukocytes (e.g., lymphocytes, macrophages, mast cells) (Polverini, 1989; Leek et al., 1994).

A smaller number of naturally occurring anti-angiogenic proteins have been identified, including thrombospondin-1, platelet factor-4, α-interferon, and interleukin-12 (Good et al., 1990; Folkman, 1992; Maione et al., 1990; Majewski et al., 1995; Ricketts et al., 1994; Voest et al., 1995). The regulation of production of these pro-angiogenic and anti-angiogenic factors, their precise contributions to angiogenesis in specific human cancers, and the cell types that produce them, are not well delineated.

SF as a potential tumor angiogen

In Kaposi's sarcoma (KS)

KS is a complex multicellular neoplasm characterized by populations of interstitial spindle-shaped tumor cells, macrophage-like dendritic cells, and a major component of proliferating ECs and angiogenesis. The epidemic form of KS occurs frequently in homosexual males with Acquired Immunodeficiency Syndrome (AIDS). AIDS-KS is regarded as a multifocal cytokine-dependent tumor rather than a malignant metastasizing cancer (Mallery et al., 1994). In a recent study, SF was found to mediate several functions that may be related to the pathogenesis of AIDS KS: (1) *in vitro* "transdifferentiation" of normal human ECs into a KS tumor cell-like phenotype; and (2) autocrine stimulation of KS tumor cell proliferation. Moreover, SF and its receptor, c-*met*, were found to be expressed in various cell types present in AIDS KS lesions. In addition, SF was found to be the major angiogenic molecule in conditioned medium from human T lymphotrophic virus type II-infected T lymphocytes (HTLV-II CM) (Naidu et al., 1994). HTLV-II CM is required for the long-term cultivation of KS tumor cells. These findings suggest that SF may contribute to generation of the KS tumor cell phenotype, proliferation of ECs, and angiogenesis in KS tumors.

In human carcinomas

Both SF and c-*met* appear to be upregulated and downregulated in precisely coordinated patterns during normal developmental and reparative processes (Sonnenberg et al., 1993; Matsumoto and Nakamura, 1993; Joannidis et al., 1994). In a study of the developing mouse embryo, a pattern characterized by expression of c-*met* mRNA in epithelium and SF mRNA in adjacent mesenchyme was observed (Sonnenberg et al., 1993). Thus, it appeared that the SF:c-*met* ligand:receptor pair might be involved in a *paracrine* signalling mechanism related to normal development. On the other hand, SF is chronically overexpressed in tumors (Rosen et al., 1994b; Yamashita et al., 1994; Joseph et al., 1995). Several studies indicate that cultured carcinoma cells secrete soluble protein factors ("scatter factor-inducing factors") that stimulate fibroblasts and other stromal cell types (e.g. endothelium, smooth muscle cells) to produces SF. In a recent study of human breast tissue sections, it was found that in addition to stromal cell types, the carcinoma cells themselves express SF mRNA transcripts (Wang et al., 1994b). Thus, overproduction of SF in tumors appears to be due both to a paracrine tumor:stroma interaction and to autocrine production of SF by tumor cells, which also express the c-*met* receptor.

In a study of 258 primary invasive breast carcinomas, a high titer of SF in extracted tumor tissue was found to be a powerful *independent* predictor of relapse and death (Yamashita et al., 1994). In patients with transitional cell bladder cancers, higher titers of SF were found in high grade, muscle invasive cancers than in low grade non-invasive or superficially invasive cancers (Joseph et al., 1995). Patients in the former category usually fare poorly in comparison with patients in the latter category.

Since both SF content and tumor angiogenesis appear to be prognostic indicators for breast and bladder carcinoma, it is reasonable to speculate that SF may function as a tumor angiogen in these settings. If SF is indeed a tumor angiogen, it is expected that the tumoral SF content would be correlated with the quantitative extent of tumor angiogenesis. Studies to determine the relationship between SF content and tumor angiogenesis are currently in progress in our laboratory. Such a correlation may no be exact since, as described above, a variety of other factors may function as positive or negative regulators of angiogenesis. Since many or most of these factors are found in tumors, the net angiogenic phenotype of the tumor may be determined by the balance of pro-angiogenic and anti-angiogenic factors present.

New directions for angiogenesis and SF research

Tumor suppressors, anti-angiogenesis, and SF

Recent studies have established a linkage between certain tumor suppressor gene products and inhibition of angiogenesis (Bouck, 1990). Thrombospondin-1 (TSP-1) is a complex multidomain adhesive glycoprotein of the extracellular matrix (ECM) that mediates various cell:cell and cell:matrix interactions via its modular structure (Lawler, 1993; Bornstein, 1995). A portion of the TSP-1 molecule with angio-inhibitory activity (GP140) is encoded by a putative tumor suppressor gene that is inactivated when BHK hamster cells are chemically transformed into a tumorigenic phenotype (Rastinjad et al., 1989; Good et al., 1990). Both GP140 and native TSP-1 inhibit EC proliferation, EC migration, and angiogenesis induced by basic FGF and macrophage conditioned medium (Good et al., 1990; Taraboletti et al., 1990). Preliminary findings from our laboratory indicate that TSP-1 also blocks chemotaxis of capillary ECs toward SF and SF-induced neovascularization in the rat cornea assay (Polverini PJ and Rosen EM, unpublished results).

The mechanism of TSP-1's anti-angiogenic activity has not been elucidated. However, a recent study suggests most of this activity is concentrated in several 15–20 amino acid regions within the procollagen homology domain and type I repeats of the central stalk of the molecule (Tolsma et al., 1993). The central stalk of TSP-1 has been implicated in the binding of TSP-1 to multiple different ECM proteins, raising the possibility that TSP-1 acts by interfering with critical cell:matrix interactions required for angiogenesis. Recent studies from our laboratory indicate that SF binds to TSP-1 with high affinity, probably via a site(s) on TSP-1's central stalk, and that TSP-1 is a major component of the breast cancer

ECM responsible for binding SF (Lamszus et al., 1996). Thus, molecular interactions between SF and TSP-1 may modulate the angiogenic phenotype of breast cancers.

The p53 tumor suppressor gene encodes a nuclear phosphoprotein that generally regulates cell growth by transactivating or repressing transcription (Tominaga et al., 1992; Mack et al., 1993). Los of p53 function in Li-Fraumeni cancer syndrome fibroblasts results in decreased expression of TSP-1 and acquisition of an angiogenic phenotype; whereas transfection of a wild-type p53 expression construct into fibroblasts lacking a functional p53 allele results in upregulation of TSP-1 and loss of the angiogenic phenotype (Dameron et al., 1994). Wild type p53 also suppresses expression of the angiogenic peptide basic FGF in glioblastoma and hepatoma cells (Ueba et al., 1994). A mutant p53 construct was found to potentiate protein kinase C-mediated stimulation of VEGF mRNA expression (Kieser et al., 1994). These findings raise the possibility that mutations resulting in loss of p53 function in tumors may contribute to the angiogenic phenotype by enhancing expression of pro-angiogenic factors and/or inhibiting expression of anti-angiogenic factors.

In addition to tumor suppressor genes, steroid hormone- and vitamin-related compounds have recently been implicated in the regulation of angiogenesis. For example, recent studies suggest that partial estrogen antagonists such as tamoxifen have significant ability to inhibit endothelial cell growth and angiogenesis (Gagliari and Collins, 1993; Haran et al., 1994). The mechanism of this anti-angiogenic action is not certain, but does not necessarily involve inhibition of the estrogen effect. Thus, inhibition of angiogenesis may contribute to the clinical anti-tumor activity of tamoxifen in human breast cancer. Interestingly, steroid hormone-response elements have been identified on both the SF and c-*met* promoters (Liu et al., 1994; Moghul et al., 1994), raising the possibility these hormones and their antagonists may modulate SF-mediated angiogenesis. Similarly, retinoic acid and various retinoids, compounds which are active in chemoprevention of tumors, exert potent anti-angiogenic activity in *in vitro* and *in vivo* angiogenesis assay systems (Bollag et al., 1994; Majewski et al., 1995). The vitamin D analog 1,25-dihydroxyvitamin D3 inhibits tumor angiogenesis *in vivo* (Bollag et al., 1994) and inhibits SF production by myeloid cell lines *in vitro* (Inaba et al., 1993). A potential vitamin D-responsive regulatory element has also been identified on the SF gene promoter (Liu et al., 1994). The relationship between retinoid- and vitamin D-related inhibition of angiogenesis and the SF : c-met signalling pathway remains to be elucidated.

Future directions and potential clinical applications

Based on the foregoing discussion, we postulate that: (1) mutations of tumor suppressor genes (e.g. p53) induce activation of the SF → c-met signalling pathway: (2) the resulting overexpression of SF and/or c-*met* contributes to persistent neovascularization in tumors; and (3) expression of angio-inhibitory molecules (e.g. TSP-1) in this setting is insufficient to neutralize the pro-angiogenic activity of SF and other tumor angiogens in growing tumors. To delineate the contribution of SF and c-*met* to tumor angiogenesis further, we propose the following as potentially important areas for further study: (1) the mechanisms by which SF and c-*met* expression are regulated in tumors; (2) the linkage of these regulatory mechanisms to tumor suppressor systems; (3) interactions between SF and tumor suppressor gene-encoded angiogenesis inhibitors (e.g. TSP-1); and (4) interactions between anti-angiogenic compounds with anti-tumor activity (e.g. tamoxifen, retinoids, vitamin D analogs) with the SFF:c-*met* mediated cellular signalling pathway.

If SF was found to be an important human angiogenesis factor, several types of clinical applications may be envisaged. First, additional SF may be supplied in circumstances in which increased angiogenesis would be bene-ficial (e.g. to promote healing of burns, viability of transplanted tissues or organs, regeneration of ischemic myocardium). Second, inhibition of SF might be used to treat selected angiogenesis dependent diseases. The latter include chronic inflammatory disorders (e.g. rheumatoid arthritis and psoriasis) and most malignant solid tumors (Polverini, 1989; Folkman, 1992). A truncated SF α-chain peptide (designated NK2) is produced by some human fibroblast-types via an alternatively spliced SF mRNA trans-cript (Chan et al., 1991). NK2 binds with high affinity to cellular c-*met* receptors and blocks SF mitogenic activity for epithelial cells. A smaller recombinant fragment of SF (NK1) has similar properties (Lokker et al., 1992). Preliminary studies from our laboratories indicate that NK and NK2 block SF-induced EC migration and SF-induced angiogenesis (unpublish-ed results). A chimeric antibody has been produced, which consists of bivalent extracellular binding domains of the c-met receptor fused to the constant region of the human IgG heavy chain (Mark et al., 1992). This antibody inhibits the binding of SF to cellular c-met receptors, and thus may also have potential use as an SF antagonist. Additional avenues for SF inhibition include development of agents that block synthesis of SF [e.g. by targetting SF-inducing factors (Rosen et al., 1994a)] or conversion of proSF to SF (e.g. by targetting HGF activator). Thus, a variety of different approaches might be used to promote or inhibit SF-mediated angiogenesis in various clinical conditions.

Acknowledgements
Supported in part by the United States Public Health Service (grants CA50516 and CA64869) and the American Heart Association (Established Investigator Award 90–195). We thank Genentech, Inc., South San Francisco, CA, for providing recombinant human SF and sheep antiserum to human SF. We thank Dr George Vande Woude, National Cancer Institute (Frederick, MD) for providing rabbit antiserum to human c-met peptide. We are grateful to Dr. James Kinselly. National Institute on Aging, Baltimore, MD, for performing the capillary tube formation assays. We thank Drs. Patricia D'Amore and Andrea Dodge, Department of Surgery, Children's Hospital, Boston, MA, for performing proliferation assays of smooth

muscle cells, pericytes, and capillary endothelium, and for providing pericyte cultures. PCR analysis of pericyte cDNA was performed by Dr Morag Park and Monica Naujokas, Royal Victoria Hospital and McGill University, Montreal, Quebec.

References

Antonelli-Orlidge, A., Smith, S.R. and D'Amore, P.A. (1989) Influence of pericytes on capillary endothelial cell growth. *Am. Rev. Resp. Dis.* 140:1129–1131.

Bhargava, M., Joseph, A., Knesel, J., Halaban, R., Li, Y., Pang, S., Goldberg, I., Setter, E., Donovan, M.A., Zarnegar, R., Michalopoulos, G.A., Nakamura, T., Faletto, D. and Rosen, E.M. (1992) Scatter factor and hepatocyte growth factor: Activities, properties, and mechanism. *Cell Growth & Differen.* 3:11–20.

Bochner, B.H., Cote, R.J., Weidner, N., Groshen, S., Chen, S.C., Skinner, D.G. and Nichols, P.W. (1995) Angiogenesis in bladder cancer: relationship between microvessel density and tumor prognosis. *J. Natl. Cancer Inst.* 87:1603–1612.

Bollag, W., Majewski, S. and Jablonska, S. (1994) Cancer combination chemotherapy with retinoids: experimental rationale. *Leukemia* 8:1453–1457.

Bornstein, P. (1995) Diversity of function is inherent in matricellular proteins: An appraisal of thrombospondin-1. *J. Cell Biol.* 130:503–506.

Bosari, S., Lee, A.K., DeLellis, R.A., Wiley, B.D., Heatley, G.J. and Silverman, M.L. (1992) Microvessel quantitation and prognosis in invasive breast carcinoma. *Hum. Pathol.* 23: 755–761.

Bottaro, D.P., Rubin, J.S., Faletto, D.L. Chan, A.M.-L., Kmiecik, T.E., Vande Woude, G.F. and Aaronson, S.A. (1991) Identification of the hepatocyte growth factor receptor as the c-*met* proto-oncogene product. *Science* 251:802–804.

Bouck, N. 81990) Tumor angiogenesis: the role of oncogenes and tumor suppressor genes. *Cancer Cells* 2:179–185.

Brem, S.S., Jensen, H.M. and Gullino, P.M. (1978) Angiogenesis as a marker of neoplastic lesions of the human breast. *Cancer* 41:239–244.

Bussolino, F., DiRenzo, M.F., Ziche, M., Bocchieto, E., Olivero, M., Naldini, L., Gaudino, G., Tamagnone, L., Coffer, A. and Comoglio, P.M. (1992) Hepatocyte growth factor is a potent angiogenic factor which stimulates endothelial cell motility and growth. *J. Cell Biol.* 119: 629–641.

Chan, A.M.-L., Rubin, J.S., Bottaro, D.P., Hirschfield, D.W., Chedid, M. and Aaronson, S.A. (1991) Identification of a competitive antagonist encoded by an alternative transcript. *Science* 254:1382–1385.

Dameron, K.M., Volpert, O.V., Tainsky, M.A. and Bouck, N. (1994) Control of angiogenesis in fibroblasts by p53 regulation of thrombospondin-1. *Science* 265:1582–1590.

Folkman, J. (1985) Tumor angiogenesis. *Adv. Cancer Res.* 43:175–203.

Folkman, J. (1992) The role of angiogenesis in tumor growth. *Semin. Cancer Biol.* 3:65–71.

Folkman, J., Watson, K., Ingber, D. and Hanahan, D. (1989) Induction of angiogenesis during the transition from hyperplasia to neoplasia. *Nature* 339:58–61.

Gagliardi, A. and Collins, D.C. (1993) Inhibition of angiogenesis by antiestrogens. *Cancer Res.* 53:533–535.

Gherardi, E., Gray, J., Stoker, M., Perryman, M. and Furlong, R. (1989) Purification of scatter factor, a fibroblast-derived basic protein which modulates epithelial interactions and movement. *Proc. Natl. Acad. Sci. USA* 86:5844–5848.

Gonzatti-Haces, M., Seth, A., Park, M., Copeland, T., Oroszlan, S. and Vande Woude, G.F. (1988) Characterization of the TPR-MET oncogene p65 and the MET protooncogene p140 protein tyrosine kinases. *Proc. Natl. Acad. Sci. USA* 85:21–25.

Good, D.J., Polverini, P.J., Rastinejad, F., Le Beau, M.M., Lemons, R.S., Frazier, W.A. and Bouck, N.P. (1990) A tumor supressor-dependent inhibitor of angiogenesis is immunologically and functionally indistinguishable from a fragment of thrombospondin. *Proc. Natl. Acad. Sci. USA* 87:6624–6628.

Grant, D.S., Tashiro, K.I., Segui-Real, B., Yamada, Y., Martin, G.R. and Kleinman, H.K. (1989) Two different laminin domains mediate the differentiation of human endothelial cells into capillary-like structures *in vitro. Cell* 58:933–943.

Grant, D.S., Kleinman, H.K., Goldberg, I.D., Bhargava, M., Nickoloff, B.J., Polverini, P. and Rosen, E.M. (1993) Scatter factor induces blood vessel formation *in vivo. Proc. Natl. Acad. Sci. USA* 90:1937–1941.

Guidi, A.J., Fischer, L., Harris, J.R. and Schnitt, S.J. (1994) Microvessel density and distribution in ductal carcinoma *in situ* of the breast. *J. Natl. Cancer Inst.* 86:356–361.

Halaban, R., Rubin, J., Funusaka, Y., Cobb, M., Boulton, T., Faletto, D., Rosen, E., Chan, A., Yoko, K., White, W., Cook, C. and Moellmann, G. (1992) Met and hepatocyte growth factor/scatter factor signal transduction in normal melanocytes and melanoma cells. *Oncogene* 7:2195–2206.

Haran, E.F., Maretzek, A.F., Goldberg, I., Horowitz, A and Degani, H. (1994) Tamoxifen enhances cell death in implanted MCF7 breast cancer by inhibiting endothelium growth. *Cancer Res.* 54:5511–5514.

Huff, J.L., Jelinek, M.A., Borgman, C.A. and Lansing, T.J. (1994) The protooncogene c-sea encodes a transmembrane protein-tyrosine kinase related to the Met/HGF/SF receptor. *Proc. Natl. Acad. Sci. USA* 90:6140–6144.

Inaba, M., Koyama, H., Hino, M., Okuno, S., Terada, M., Nishizawa, Y., Nishino, T. and Morii, H. (1993) Regulation of release of hepatocyte growth factor from human promyelocytic leukemia cells, HL-60, by 1,25-dihydroxyvitamin D3, 12-O-tetradecanoylphorbol 13-acetate, and dibutyryl cyclic adenosine monophosphate. *Blood* 82:53–59.

Joannidis, M., Spokes, K., Nakamura, T., Faletto, D. and Cantley, L.G. (1994) Regional expression of hepatocyte growth factor/c-met in experimental renal hypertrophy and hyperplasia. *Am. J. Physiol.* 267:F231–F236.

Joseph, A., Weiss, G.H., Jin, L., Fuchs, A., Chowdhury, S., O'Saughnessy, P., Goldberg, I.D. and Rosen, E.M. (1995) Expression of scatter factor in human bladder carcinoma. *J. Natl. Cancer Inst.* 87:372–377.

Kan, M., Zhang, G.H., Zarnegar, R., Michalapoulos, G., Myoken, Y., McKeehan, W.L. and Stevens, J.L. (1991) Hepatocyte growth factor-hepatopoietin A stimulates the growth of rat proximal tubule epithelial cells (rpte), rat non-parenchymal liver cells, human melanoma cells, mouse keratinocytes, and stimulates anchorage-independent growth of SV-40-transformed rpte. *Biochem. Biophys. Res. Commun.* 174:331–337.

Kibbey, M.C., Grant, D.S. and Kleinman, H.K. (1992) The role of the SIKVAV site of laminin in promotion of angiogenesis and tumor growth: an *in vivo* Matrigel Model. *J. Natl. Cancer Inst.* 84:1633–1638.

Kieser, A., Weich, H.A., Brandner, G., Marme, D. and Kolch, W. (1994) Mutant p53 potentiates protein kinase C induction of vascular endothelial growth factor expression. *Oncogene* 9:963–969.

Lamszus, K., Joseph, A., Jin, L., Yao, Y., Chowdhury, S., Fuchs, A., Goldberg, I.D. and Rosen, E.M. (1996) Scatter factor binds to thrombospondin and other extracellular matrix proteins. *Am. J. Pathol.* In press.

Lawler, J. (1993) Thrombospondin. *In:* Kreis, T. and Vale, R. (eds): *Guidebook to the extracellular matrix and adhesion proteins.* Oxford University Press, New York, pp 95–96.

Leek, R.D., Harris, A.L. and Lewis, C.E. (1994) Cytokine networks in solid human tumors: regulation of angiogenesis. *J. Leuk. Biol.* 56:423–435.

Li, Y., Bhargava, M.M., Joseph, A., Jin, L., Rosen, E.M. and Goldberg, I.D. (1994) The effect of scatter factor and hepatocyte growth factor on motility and morphology of non-tumorigenic and tumor cells. *In Vitro Cell Dev. Biol.* 30A:105–110.

Liu, Y., Michalopoulos, G.K. and Zarnegar, R. (1994) Structural and functional characterization of the mouse hepatocyte growth factor gene promoter. *J. Biol. Chem.* 269:4152–4160.

Lokker, N.A., Mark, M.R., Luis, E.A., Bennett, G.L., Robbins, K.A., Baker, J.B. and Godowski, P.J. (1992) Structure-function analysis of hepatocyte growth factor: Identification of variants that lack mitogenic activity yet retain high affinity receptor binding. *EMBO J.* 11:2403–2410.

Mack, D.H., Vartikar, J., Pipas, J.M. and Laimins, L.A. (1993) Specific repression of TATA-mediated but not initiator-mediated transcription. *Nature* 363:281–283.

Maione, T.E., Gray, G.S., Petro, J., Hunt, A.J., Donner, A.L., Bauer, S.J., Carson, H.F. and Sharpe, R.J. (1990) Inhibition of angiogenesis by recombinant human platelet factor-4 and related peptides. *Science* 247:77–79.

Majewski, S., Marczak, M., Szmurlo, A., Jablonska, S. and Bollag, W. (1995) Retinoids, interferon alpha, 1,25-dihydroxyvitamin D3 and their combination inhibit angiogenesis induced by non-HPV-harboring tumor cell lines. RAR alpha mediates the antiangiogenic effect of retinoids. *Cancer Lett.* 89:117–124.

Mallery, S.R., Lantry, L.E., Hegtvedt, A., Lazo, A., Titterington, L.C., Hout, B.L., Brierley, G.P. and Stephens, R.E. (1994) Origin of spindle-shaped cells in Kaposi sarcoma. *Lymphology* 27:45–48.

Mark, M.R., Lokker, N.A., Zioncheck, T.F., Luis, E.A. and Godowski, P.J. (1992) Expression and characterization of hepatocyte growth factor receptor-IgG fusion proteins. Effects of mutations in the potential proteolytic cleavage sites on processing and ligand binding. *J. Biol. Chem.* 267:26166–26171.

Mars, W.M., Zarnegar, R. and Michalopoulos, G.K. (1993) Activation of hepatocyte growth factor by the plasminogen activators uPA and tPA. *Am. J. Pathol.* 143:949–958.

Matsumoto, K. and Nakamura, T. (1993) Roles of HGF as a pleiotropic factor in organ regeneration. *In:* Goldberg, I.D. and Rosen, E.M. (eds): *Hepatocyte Growth Factor – Scatter Factor and the c-Met Receptor.* Birkhäuser Verlag, Basel, pp 225–250.

Miyazawa, K., Tsubouchi, H., Naka, D., Takahashi, K., Okigaki, M., Arakaki, N., Nakayama, S., Hirono, S., Sakiyama, O., Gohda, E., Daikuhara, Y. and Kitamura, N. (1989) Molecular cloning and sequence analysis of cDNA for human hepatocyte growth factor. *Biochem. Biophys. Res. Commun.* 163:967–973.

Miyazawa, K., Shimomura, T., Kitamura, A., Kondo, J., Morimoto, Y. and Kitamura, N. (1993) Molecular cloning and sequence analysis of the cDNA for a human serine protease responsible for activation of hepatocyte growth factor. Structural similarity of protease precursor to blood coagulation factor XII. *J. Biol. Chem.* 268:10024–10028.

Miyazawa, K., Shimomura, T., Naka, D. and Kitamura, N. (1994) Proteolytic activation of hepatocyte growth factor in response to tissue injury. *J. Biol. Chem.* 269:8966–8970.

Moghul, A., Lin, L., Beedle, A., Kanbour-Shakir, A., DeFrances, M.C., Liu, Y. and Zarnegar, R. (1994) Modulation of c-MET proto-oncogene (HGF receptor) mRNA abundance by cytokines and hormones: evidence for rapid decay of the 8 kb c-MET transcript. *Oncogene* 9:2045–2052.

Montesano, R., Matsumoto, K., Nakamura, T. and Orci, L. (1991) Identification of a fibroblast-derived epithelial morphogen as hepatocyte growth factor. *Cell* 67:901–908.

Morimoto, A., Okamura, K., Hamanaka, R., Sato, Y., Shima, N., Higashio, K. and Kuwano, M. (1991) Hepatocyte growth factor modulates migration and proliferation of microvascular endothelial cells in culture. *Biochem. Biophys. Res. Commun.* 179:1042–1049.

Naidu, Y.M., Rosen, E.M., Zitnik, R., Goldberg, I., Park, M., Naujokas, M., Polverini, P.J. and Nickoloff, B.J. (1994) Role of scatter factor in the pathogenesis of AIDS-related Kaposi's sarcoma. *Proc. Natl. Acad. Sci. USA* 91:5281–5285.

Naldini, L., Tamagnone, L., Vigna, E., Sachs, M., Hartmann, I., Birchmeier, W., Daikuhara, Y., Tsubouchi, H., Blasi, F. and Comoglio, P.M. (1992) Extracellular proteolytic cleavage by urokinase is required for activation off hepatocyte growth factor/scatter factor. *EMBO J.* 11:4825–4833.

Nakamura, T., Nishizawa, T., Hagiya, M., Seki, T., Shimonishi, M., Sugimura, A. and Shimizu, S. (1989) Molecular cloning and expression of human hepatocyte growth factor. *Nature* 342:440–443.

Park, M., Dean, M., Cooper, C.S., Schmidt, M., O'Brien, S.J., Blair, D.G. and Vande Woude, G.F. (1986) Mechanism of met oncogene activation. *Cell* 45:895–904.

Pepper, M.S., Matsumoto, K., Nakamura, T., Orci, L. and Montesano, R. (1992) Hepatocyte growth factor increases urokinase-type plasminogen activator (u-PA) and u-PA receptor expression in Madin-Darby canine kidney epithelial cells. *J. Biol. Chem.* 267:20493–20496.

Polverini, P.J. (1989) Macrophage-induced angiogenesis: A review. *Cytokines* Vol 1:54–73.

Polverini, P.J. and Leibovich, S.J. (1984) Induction of neovascularization *in vivo* and endothelial cell proliferation *in vivo* by tumor-associated macrophages. *Lab. Invest.* 51:635–642.

Ponzetto, C., Bardelli, A., Zhen, Z., Maina, F., Zonca, P., Giordano, S., Graziani, A., Panayoyou, G. and Comoglio, P.M. (1994) A multifunctional docking site mediates signalling and transformation by the HGF/SF receptor family. *Cell* 77:261–271.

Rastinejad, F., Polverini, P.J. and Bouck, N.P. (1989) Regulation of the activity of a new inhibitor of angiogenesis by a cancer suppressor gene. *Cell* 56:345–355.

Ricketts, R.R., Hatley, R.M., Corden, B.J., Sabio, H. and Howell, C.G. (1994) Interferon-alpha-2a for the treatment of complex hemangiomas of infancy and childhood. *Ann. Surg.* 219:605–612.

Ronsin, C., Muscatelli, F., Mattei, M.-G. and Breathnach, R. (1993) A novel putative receptor tyrosine kinase of the met family. *Oncogene* 8:1195–1202.

Rosen, E.M. and Goldberg, I.D. (1995) Scatter factor and angiogenesis. *Adv. Cancer. Res.* 67:257–279.

Rosen, E.M., Goldberg, I.D., Kacinski, B.M., Buckholz, T. and Vinter, D.W. (1989) Smooth muscle releases an epithelial cell scatter factor which binds to heparin. *In Vitro Cell Dev. Biol.* 25:163–173.

Rosen, E.M., Meromsky, L., Setter, E., Vinter, D.W. and Goldberg, I.D. (1990a) Smooth muscle-derived factor stimulates mobility of human tumor cells. *Invasion Metast.* 10:49–64.

Rosen, E.M., Meromsky, L., Setter, E., Vinter, D.W. and Goldberg, I.D. (1990b) Purification and migration-stimulating activities of scatter factor. *Proc. Soc. Exp. Biol. Med.* 195:34–43.

Rosen, E.M., Meromsky, L., Setter, E., Vinter, D.W. and Goldberg, I.D. (1990c) Quantitation of cytokine-stimulated migration of endothelium and epithelium by a new assay using microcarrier beads. *Exp. Cell Res.* 1886:22–31.

Rosen, E.M., Goldberg, I.D., Liu, D., Setter, E., Donovan, M.A., Bhargava, M., Reiss, M. and Kacinski, B.M. (1991a) Tumor necrosis factor stimulates epithelial tumor cell motility. *Cancer Res.* 57:5315–5321.

Rosen, E.M., Grant, D., Kleinman, H., Jaken, S., Donovan, M.A., Setter, E., Luckett, P.M. and Carley, W. (1991b) Scatter factor stimulates migration of vascular endothelium and capillary-like tube formation. *In:* Goldberg, I.D. and Rosen, E.M. (eds): *Cell Motility Factors.* Birkhäuser Verlag, Basel, pp 76–88.

Rosen, E.M., Jaken, S., Carley, W., Setter, E., Bhargava, M. and Goldberg, J.D. (1991c) Regulation of motility in bovine brain endothelial cells. *J. Cell. Physiol.* 146:325–335.

Rosen, E.M., Joseph, A., Jin, L., Rockwell, S., Elias, J.A., Knesel, J., Wines, J., McClellan, J., Kluger, M.J., Goldberg, I.D. and Zitnik, R. (1994a) Regulation of scatter factor production via a soluble inducing factor. *J. Cell. Biol.* 127:225–234.

Rosen, E.M., Knesel, J., Goldberg, I.D., Bhargava, M., Joseph, A., Zitnik, R., Wines, J., Kelley, M. and Rockwell, S. (1994b) Scatter factor modulates the metastatic phenotype of the EMT6 mouse mammary tumor. *Int. J. Cancer* 57:706–714.

Rubin, J.S., Chan, A.M.-L. Bottaro, D.P., Burgess, W.H., Taylor, W.G., Cech, A.C., Hirschfield, D.W., Wong, J., Miki, T., Finch, P.W. and Aaronson, S.A. (1991) A broad spectrum human lung fibroblast-derived mitogen is a variant of hepatocyte growth factor. *Proc. Natl. Acad. Sci. USA* 88:415–419.

Saksela, O., and Rifkin, D.B. (1988) Cell-associated plasminogen activation: regulation and physiologic functions. *Ann. Rev. Cell. Biol.* 4:93–126.

Santos, O.F.P., Moira, L.A., Rosen, E.M. and Nigam, S.K. (1993) Modulation of HGF-induced tubulogenesis and branching by multiple phosphorylation mechanisms. *Dev. Biol.* 159:538–545.

Sonnenberg, E., Meyer, D., Weidner, K.M. and Birchmeier, C. (1993) Scatter factor/hepatocyte growth factor and its receptor, the c-met tyrosine kinase, can mediate a signal exchange between mesenchyme and epithelia during mouse development. *J. Cell Biol.* 123:223–235.

Stoker, M., Gherardi, E., Perryman, M. and Gray, J. (1987) Scatter factor is a fibroblast-derived modulator of epithelial cell mobility. *Nature* 327:238–242.

Taraboletti, G., Roberts, D., Liotta, L.A. and Giavazzi, R. (1990) Platelet thrombospondin modulates endothelial cell adhesion, motility, and growth: a potential angiogenesis regulatory factor. *J. Cell Biol.* 111:765–772.

Toi, M., Kashitani, J. and Tominaga, T. (1993) Tumor angiogenesis is an independent prognostic indicator in primary breast carcinoma. *Int. J. Cancer* 55:371–374.

Tolsma, S.S., Volpert, O.V., Good, D.J., Frazier, W.A., Polverini, P.J. and Bouck, N. 81993) Peptides derived from two separate domains of the matrix protein thrombospondin-1 have anti-angiogenic activity. *J. Cell Biol.* 122:497–511.

Tominaga, O., Hamelin, R., Remvikos, Y., Salmon, R.J. and Thomas, G. (1992) P53 from basic research to clinical applications. *Crit. Rev. Oncogenesis* 3:257–282.

Tsarfaty, I., Resau, J.H., Rulong, S., Keydar, I., Faletto, D. and Vande Woude, G.F. (1992) The *met* proto-oncogene receptor and lumen formation. *Science* 257:1258–1261.

Ueba, T., Nosaka, T., Takahashi, J.A., Shibata, F., Florkiewicz, R.Z., Vogelstein, B., Oda, Y., Kikuchi, H. and Hatanaka, M. 81994) Transcriptional regulation of basic fibroblast growth factor by p53 in human glioblastoma and hepatocellular carcinoma cells. *Proc. Natl. Acad. Sci. USA* 91:9009–9013.

Voest, E.E., Kenyon, B.M., O'Reilly, M.S., Truitt, G., D'Amato, R.J. and Folkman, J. (1995) Inhibition of angiogenesis *in vivo* by interleukin 12. *J. Natl. Cancer Inst.* 87:581–586.

Wang, M.H., Ronsin, C., Gesnel, M.-C., Coupey, L., Skeel, A., Leonard, E.J. and Breathnach, R. (1994a) Identification of the ron gene product as the receptor for the human macrophage stimulating protein. *Science* 266:117–119.

Wang, Y., Selden, A.C., Morgan, N., Stamp, G.W. and Hodgson, H.J. (1994b) Hepatocyte growth factor/scatter factor expression in human mammary epithelium. *Am. J. Pathol.* 144: 675–682.

Weidner, K.M., Behrens, J., Vandekerckhove, J. and Birchmeier, W. (1990) Scatter factor: Molecular characteristics and effect on invasiveness of epithelial cells. *J. Cell Biol.* 111: 2097–2108.

Weidner, K.M., Arakaki, N., Vandekerckhove, J., J. Weingart, S., Hartmann, G., Rieder, H., Fonatsch, C., Tsubouchi, H., Hishida, T., Daikuhara, Y. and Birchmeier, W. (1991a) Evidence for the identity of human scatter factor and human hepatocyte growth factor. *Proc. Natl. Acad. Sci. USA* 88:7001–7005.

Weidner, N., Semple, J.P., Welch, W.R. and Folkman, J. (1991b) Tumor angiogenesis and metastasis – correlation in invasive breast carcinoma. *N. Engl. J. Med.* 324:1–8.

Weidner, N., Folkman, J., Pozza, F., Bevilacqua, P., Allred, E.N., Moore, D.H., Meli, S. and Gasparini, G. (1992) Tumor angiogenesis: a new significant and independent prognostic indicator in early stage breast carcinoma. *J. Natl. Cancer Inst.* 84:1875–1887.

Weidner, K.M., Sachs, M. and Birchmeier, W (1993) The met receptor tyrosine kinase transduces motility, proliferation, and morphogenic signals of scatter factor/hepatocyte growth factor in epithelial cells. *J. Cell Biol.* 121:145–154.

Yamashita, J., Ogawa, M., Yamashita, S., Nomura, K., Kuramoto, M., Saishoji, T. and Sadahito, S. (1994) Immunoreactive hepatocyte growth factor is a strong and independent predictor of recurrence and survival in human breast cancer. *Cancer Res.* 54:1630–1633.

Yoshimura, T., Yuhki, N., Wang, M.H., Skeel, A. and Leonard, E.J. (1993) Cloning, sequencing, and expression of human macrophage stimulating protein (MSP, MST1) confirms MSP as a member of the family of kringle proteins and locates the MSP gene on chromosome 3. *J. Biol. Chem.* 268:15461–15468.

Zhu, H., Naujokas, M.A. and Park, M. (1994) Receptor chimeras indicate that the met tyrosine kinase mediates the motility and morphogeneic responses of hepatocyte growth factor/scatter factor. *Cell Growth Differen.* 5:359–366.

Regulation of Angiogenesis
ed. by I.D. Goldberg & E.M. Rosen
© 1997 Birkhäuser Verlag Basel/Switzerland

Vascular endothelial growth factor: Basic biology and clinical implications

N. Ferrara and B. Keyt

Department of Cardiovascular Research, Genentech, Inc., South San Francisco, California 94080, USA

Introduction

The development of a vascular supply is a fundamental requirement for organ development and differentiation in multicellular organisms (Hamilton et al., 1962) and is also necessary for tissue repair and reproductive functions in the adult (Klagsbrun and D'Amore, 1991; Folkman and Shing, 1992). Angiogenesis is also implicated in the pathogenesis of a variety of disorders. These include: proliferative retinopathies, age-related macular degeneration, tumors, rheumatoid arthritis, and psoriasis (Klagsbrun and D'Amore, 1991; Folkman, 1991). In the case of proliferative retinopathies and age-related macular degeneration, the new blood vessels are directly responsible for many of the destructive events characteristic of these conditions (Garner, 1994). Conversely, in neoplasms the neovascularization provides nourishment to the growing tumor, thus allowing the tumor cells to express their critical growth advantage, and also permits the establishment of continuity with the vasculature of the host (Folkman, 1991). A strong correlation has been noted between density of microvessels in primary breast carcinoma sections, nodal metastases and survival (Weidner et al., 1991, 1992; Horak et al., 1992; Vartanian and Weidner, 1994). Similarly, a correlation has been reported between vascularity and invasive behavior in bladder (Chodak et al., 1980), prostate (Wakui et al., 1992; Bigler et al., 1993), non-small cell lung (Macchiarini et al., 1992), and uterine cervix (Sillman et al., 1981) carcinomas and in cutaneous melanomas (Srivastava et al., 1988). Furthermore, recent studies have shown a statistically significant increase in microvessel count in severe uterine cervix dysplasia (CIN III) compared to low grade lesions (CIN I) (Smith-McCune et al., 1994).

Several factors have previously been identified as potential positive regulators of angiogenesis: aFGF, bFGF, EGF, TGF-α, TGF-β, PGE_2, monobutyrin, HGF, TNF-α, PD-ECGF, angiogenin and interleukin-8 (Klagsbrun and D'Amore, 1991; Folkman and Shing, 1992).

This chapter will review the molecular and biological properties of vascular endothelial growth factor (VEGF), and endothelial cell mitogen and angiogenesis inducer. The potential clinical applications of VEGF and VEGF antagonists will also be discussed. The pivotal role played by VEGF in physiological angiogenesis is emphasized by the recent finding that heterozygous mutations inactivating the VEGF gene result in embryonic lethality (Ferrara et al., 1996). Also, VEGF administration has a therapeutic effect in animal models of coronary or limb ischemia (Ferrara, 1993). Furthermore, recent studies point to VEGF as a crucial mediator of neo-vascularization associated with tumors and proliferative retinopathies (Kim et al., 1993; Aiello et al., 1994).

Biological activities of VEGF

VEGF is a potent mitogen (ED_{50} 2–10 pM) for vascular endothelial cells but it is devoid of consistent and appreciable mitogenic activity for other cell types (Ferrara and Henzel, 1989; Ploüet et al., 1989; Conn et al., 1990). VEGF is also able to induce a marked angiogenic response in a variety of *in vivo* models including the chick chorioallentoic membrane (Leung et al., 1989; Ploüet et al., 1989), the rabbit cornea (Phillips et al., 1995), the primate iris (Tolentino et al., in press), the rabbit bone (Connolly et al., 1989a), etc. VEGF has been shown to promote angiogenesis in a tri-dimensional *in vitro* model, inducing confluent microvascular endothelial cells to invade a collagen gel and form tube-like structures (Pepper et al., 1992). Also, VEGF induces sprouting from rat aortic rings embedded in a collagen gel (Nicosia et al., 1995). Furthermore, VEGF induces expression of the serine proteases urokinase-type and tissue-type plasmi-nogen activators (PA) and also PA inhibitor-1 (PAI-1) in cultured bovine microvascular endothelial cells (Pepper et al., 1991). VEGF also induces expression of the metalloproteinase interstitial collagenase in human umbilical vein endothelial cells but not in dermal fibroblasts (Unemori et al., 1992). Interstitial collagenase is able to degrade types I and III collagen (Gross and Nagai, 1965). The co-induction of plasminogen activators and collagenase by VEGF is expected to promote a pro-degradative environ-ment that facilitates migration of endothelial cells. The expression of PAI-1 may serve to regulate and balance the process (Pepper and Montesano, 1990). Very recent studies have shown that VEGF induces expression of urokinase receptor in vascular endothelial cells (Mandriota et al., 1995). Considering that the PA-plasmin system, and in particular the interaction of uPA with uPAR, is an important element in the chain of cellular processes that mediate cellular invasion and tissue remodeling (Mignatti et al., 1989; Rifkin et al., 1990), these findings are clearly consistent with the known pro-angiogenic activities of VEGF.

VEGF has been independently purified and clones as a vascular per-meability factor (VPF) based on its ability to induce vascular leakage in the guinea pig skin (Connolly et al., 1989b; Keck et al., 1989). It has been proposed that an increase in microvascular permeability is a crucial step in angiogenesis associated with tumors and wounds (Senger et al., 1983; Dvorak, 1986). According to this hypothesis, a major function of VPF/VEGF in the angiogenic process would be inducing plasma protein leakage. This would result in the formation of an extravascular fibrin gel, a substrate for endothelial and tumor cell growth (Dvorak et al., 1987; Dvorak et al., 1995).

An additional effect of VEGF on the vascular endothelium is the stimulation of hexose transport (Pekala et al., 1990). Exposure of bovine aortic endothelial cells to VEGF or TNF-α resulted in a significant increase in the rat of hexose transport. The combination of factors had an additive effect. It is tempting to speculate that such effect is important for increased energy demands during endothelial cell proliferation or inflammation.

VEGF has also been shown to induce vasodilatation *in vitro* in a dose-dependent fashion (Ku et al., 1993), and to produce a transient hypotension *in vivo* (Yang et al., in press). Such effects appear to be mediated primarily by endothelial cell-derived NO, as assessed by the requirement for an intact endothelium and the prevention of the effect by N-methyl-arginine (Ku et al., 1993; Yang et al., 1996).

Recent studies (Park et al., 1994) indicate that the mitogenic and the permeability-enhancing activity of VEGF can be potentiated by placenta growth factor (PlGF), a molecule having a significant degree of structural homology with VEGF (Maglione et al., 1991; Hauser and Weich, 1993). While PlGF has little or no direct mitogenic or permeability-enhancing activity, it is able to potentiate significantly the activity of low, marginally efficacious, concentrations of VEGF (Park et al., 1994).

The VEGF gene

cDNA sequence analysis of a variety of human VEGF clones indicated that VEGF may exist as one of four different molecular species, having respectively 121, 165, 189 and 206 amino acids (VEGF$_{121}$, VEGF$_{165}$, VEGF$_{206}$) (Leung et al., 1989; Houck et al., 1991; Tisher et al., 1991). VEGF$_{165}$ is the predominant isoform secreted by a variety of normal and transformed cells. Transcripts encoding VEGF$_{121}$ and VEGF$_{189}$ are detected in the majority of cells and tissues expressing the VEGF gene (Houck et al., 1991). In contrast, VEGF$_{206}$ is a very rare form, so far identified only in a human fetal liver cDNA library (Houck et al., 1991). Compared to VEGF$_{165}$, VEGF$_{121}$ lacks 44 amino acids; VEGF$_{189}$ has an insertion of 24 amino acids highly enriched in basic residues and VEGF$_{206}$ has an additional insertion of 17 amino acids. Rodent and bovine VEGF isoforms are shorter by one amino acid (Leung et al., 1989; Conn et al., 1990). The organization of the human VEGF gene has been elucidated (Houck et al., 1991; Tisher et al., 1991). It is known that alternative splicing of RNA, rather than transcription of separate genes, is the basis for the molecular heterogeneity evidenced by cDNA sequence analysis (Houck et al., 1991; Tisher et al., 1991). The human VEGF gene is organized in eight exons and the size of its coding region has been estimated to be approximately 14 kb (Tisher et al., 1991). $VEGF_{165}$ lacks the residues encoded by exon 6, while $VEGF_{121}$ lacks the residues encoded by exons 6 and 7. Interestingly, there is no intron between the coding sequence of the 24 amino acid insertion in VEGF and the additional 17 amino acid insertion found in $VEGF_{206}$. The 5′ end of the 51 base pair insertion of $VEGF_{206}$ begins with GT, the consensus sequence for the 5′-splice donor necessary for mRNA processing. Therefore, the definition of the 5′-splice donor site for removal of a 1 kb intron sequence is variable (Houck et al., 1991). Analysis of the VEGF gene promoter region reveals a single major transcription start which lies near a cluster of potential Sp1 factor binding sites. Several potential binding sites for the transcription factors AP-1 and AP-2 are also present in the promoter region (Tisher et al., 1991).

Regulation of VEGF expression

Several mechanisms have been shown to be involved in the regulation of VEGF gene expression. Oxygen tension appears to play a pivotal regulatory role, both *in vitro* and *in vivo*. VEGF mRNA expression is rapidly and reversibly induced by exposure to low pO_2 in a variety of cultured cells including retinal pigmented epithelium cells (Shima et al., 1995), myoblasts (Shweiki et al., 1992), cardiomyocytes (Banai et al., 1994b) and tumor cells (Shweiki et al., 1992). In sections of glioblastoma multiforme, VEGF mRNA is highly expressed in ischemic tumor cells that are juxtaposed to areas of necrosis (Plate et al., 1992; Shweiki et al., 1992). Furthermore, occlusion of the left anterior descending coronary artery results in a dramatic increase in VEGF RNA levels in the pig myocardium (Banai et al., 1994b).

It has been shown recently that similarities exist between the mechanisms leading to hypoxic regulation of VEGF and erythropoietin (Epo) (Goldberg and Schneider, 1994). Furthermore, hypoxia-inducibility is conferred to both genes by homologous sequences. By deletion and mutation analysis, a 28 base sequence has been identified in the 5′ promoter of the rat VEGF gene which mediated hypoxia-induced transcription in transient assays (Levy et al., 1995). This sequence has a high degree of homology and similar protein binding characteristics as the hypoxia-inducible factor-1 (HIF-1) binding site within the Epo gene, which behaves like a classic 3′ transcriptional enhancer (Madan and Curtin, 1993). HIF-1 has been purified and cloned as a mediator of transcriptional responses to hypoxia (Wang et al., 1995).

Very recent studies have provided evidence that activation of c-Src also participates in the hypoxic up-regulation of VEGF gene (Mukhopadhyay et al., 1995). Hypoxia increases the kinase activity of pp60[c-src] and its phosphorylation on tyrosine 416.

Consistent with the presence of AP-1 and AP-2 sites in the VEGF gene promoter, phorbol esters and forskolin, a potent activator of adenylate cyclase, induce VEGF mRNA expression (Garrido et al., 1993). Accordingly, luteotrophic hormone, a known activator of adenylate cyclase, has been shown to induce expression of VEGF mRNA in cultured bovine ovarian granulosa cells (Garrido et al., 1993).

Several cytokines or growth factors can upregulate VEGF mRNA expression and/or induce release of VEGF protein. For example, exposure of quiescent human keratinocytes to serum, EGF, TGF-β or KGF resulted in a marked induction of VEGF mRNA expression (Frank et al., 1995). In addition, treatment of quiescent cultures of several epithelial and fibroblastic cell lines with TGF-β resulted in induction of VEGF mRNA and release of VEGF protein in the medium (Pertovaara et al., 1994). Based on these findings, it has been proposed that VEGF may function as a paracrine mediator for indirect-acting angiogenic agents such as TGF-β (Pertovaara et al., 1994). Furthermore, IL-1 beta induces VEGF expression in aortic smooth muscle cells (Li et al., 1995).

Differentiation also plays a major role in the regulation of VEGF gene expression, at least in some models of cellular differentiation (Claffey et al., 1992). The VEGF mRNA was markedly upregulated during the conversion of 3T3 preadipocytes into adipocytes or during the myogenic differentiation of C2C12 cells. Conversely, VEGF gene expression was dramatically suppressed during the differentiation of the pheochromocytoma cell line PC12 into non-malignant, neuron-like, cells.

Specific transforming events also result in induction of VEGF gene expression. For example, a mutated form of the murine p53 tumor suppressor gene (Ala^{135} → Val) has been shown to induce VEGF mRNA expression and potentiate phorbol ester-stimulated VEGF mRNA expression in NIH 3T3 cells in transient transfection assays (Kieser et al., 1994). Likewise, oncogenic mutations or amplification of *ras* lead to VEGF upregulation (Rak et al., 1995). This effect is blocked by treatment with inhibitors of *ras* farnesyl transferase. Therefore, it is tempting to speculate that VEGF-induced angiogenesis is a final common pathway for multiple and apparently unrelated alterations in cell growth regulatory pathways, leading to uncontrolled proliferation and tumorigenesis.

The VEGF isoforms

VEGF purified from a variety of species and sources is a basic, heparin-binding, homodimeric glycoprotein of 45,000 daltons (Ferrara et al., 1992).

VEGF is inactivated by reducing agents, but it is heat-stable and acid-stable. These properties correspond to those of $VEGF_{165}$, the predominant isoform. $VEGF_{121}$ is a weakly acidic polypeptide that fails to bind to heparin (Houck et al., 1992). $VEGF_{189}$ and $VEGF_{206}$ are more basic and bind to heparin with greater affinity than $VEGF_{165}$ (Houck et al., 1992). Such differences in the isoelectric point and in affinity for heparin profoundly affect the bioavailability of the VEGF isoforms (Houck et al., 1992; Park et al., 1993). $VEGF_{121}$ is secreted as a freely soluble protein in the conditioned medium of transfected cells. $VEGF_{165}$ is also secreted although a significant fraction remains bound to the cell surface or the extracellular matrix (ECM). In contrast, $VEGF_{189}$ and $VEGF_{206}$ are almost completely sequestered in the ECM (Park et al., 1993). However, they may be released from the bound state by heparin or heparinase, suggesting that their binding sites is represented by proteoglycans containing heparin-like moieties. Furthermore, the long forms may be released by plasmin (Houck et al., 1992; Park et al., 1993) following cleavage at the COOH terminus. This action generates a bioactive proteolytic fragment having molecular weight of ~34,000 daltons (Houck et al., 1992; Park et al., 1993). Recent studies have shown that the bioactive product of plasmin action is comprised of the first 110 NH_2-terminal amino acids of VEGF (Keyt et al., 1996a). Plasminogen activation and generation of plasmin have been shown to play an important role in the angiogenesis cascade (Mignatti et al., 1989). However, it is possible that this property is not confined to plasmin. It may be that cleavage of VEGF can be brought about by several inflammation-associated proteases. Thus the VEGF proteins may become available to endothelial cells by at least two different mechanisms: as freely diffusible proteins ($VEGF_{121}$, $VEGF_{165}$) or following protease activation and cleavage of the longer isoforms. Generation of bioactive VEGF by proteolytic cleavage may be especially important in the microenvironment of a tumor where increased expression of proteases, including plasminogen activators, is well documented (Rifkin et al., 1990; Sreenath et al., 1992). However, loss of heparin binding, whether due to alternative splicing or plasmin cleavage, results in a substantial loss of mitogenic activity for vascular endothelial cells (Keyt et al., 1996a).

Interestingly, naturally occurring heterodimers between $VEGF_{164}$ and PlGF have been recently identified in the conditioned medium of a rat glioma cell line (DiSalvo et al., 1995). The VEGF:PlGF heterodimer was approximately 7-fold less potent than the VEGF homodimer in promoting endothelial cell growth.

The VEGF receptors

Ligand autoradiography studies on fetal and adult rat tissue sections revealed that high affinity VEGF binding sites are localized to the vascular endothelium of large or small vessels but not to other cell types (Jakeman et al., 1992, 1993). Specific binding co-localizes with factor VIII-like immuno-reactivity and is apparent on both proliferating and quiescent endothelial cells (Jakeman et al., 1992, 1993).

Two tyrosine kinases have been identified as VEGF receptors (deVries et al., 1992; Terman et al., 1992; Millauer et al., 1993; Quinn et al., 1993). The Flt-1 (fms-like-tyrosine kinase) (Shibuya et al., 1990) and KDR (kinase domain region) (Terman et al., 1991) proteins have been shown to bind VEGF with high affinity. Flk-1 (fetal liver kinase-1), the murine homologue of KDR, shares 85% sequence identity with human KDR (Matthews et al., 1991). Both Flt-1 and KDR/Flk-1 have seven immuno-globulin (Ig)-like domains in the extracellular domain (ECD), a single transmembrane region and a consensus tyrosine kinase sequence which is interrupted by a kinase-insert domain (Shibuya et al., 1990; Matthews et al., 1991; Terman et al., 1991). An additional member of the family of tyrosine kinases with seven Ig-like domains in the ECD is Flt-4, which, however, is not a receptor for VEGF (Mustonen and Alitalo, 1995). Flt-1 has the highest affinity for rhVEGF$_{165}$, with a K_d of approximately 10–20 pM (de Vries et al., 1992). KDR has a somewhat lower affinity for VEGF: the K_d has been estimated to be approximately 75–125 pM (Terman et al., 1992).

In situ hybridization studies have revealed that the flk-1 mRNA is expressed in the vascular endothelium in the mouse embryo (Millauer et al., 1993; Quinn et al., 1993). There is evidence that the *flk-1* mRNA is downregulated in adult endothelial cells as compared to fetal endothelial cells (Millauer et al., 1993; Quinn et al., 1993). The *flt-1* mRNA is selec-tively expressed in vascular endothelial cells, both in fetal and adult mouse tissues (Peters et al., 1993). Similarly to the high affinity VEGF binding (Jakeman et al., 1992, 1993), the *flt-1* mRNA is expressed in both prolife-rating and quiescent endothelial cells (Peters et al., 1993).

The Flt-1 and KDR proteins have been shown to have different signal transduction properties (Waltenberger et al., 1994). Porcine aortic endo-thelial cells lacking endogenous VEGF receptors display chemotaxis and mitogenesis in response to VEGF when transfected with an expression vector coding for KDR. In contrast, transfected cells expressing Flt-1 lack such responses (Waltenberger et al., 1994; Seetharam et al., 1995). While Flk-1/KDR undergoes strong ligand-dependent tyrosine phosphorylation in intact cells (Millauer et al., 1993; Quinn et al., 1993). Flt-1 reveals at very weak or undetectable response (de Vries et al., 1992; Waltenberger et al., 1994; Seetharam et al., 1995). Transfection of Flt-1 cDNA in NIH 3T3 led to a weak VEGF-dependent tyrosine phosphorylation but did not generate any mitogenic signal (Seetharam et al., 1995). These findings are in agreement with other studies showing that PlGF, which binds with high affinity to Flt-1 but not to Flk-1/KDR, lacks direct mitogenic or permeabil-ity-enhancing properties or the ability to stimulate effectively tyrosine phosphorylation in endothelial cells (Park et al., 1994). Therefore, it appears that interaction with Flk-1/KDR is a critical requirement to induce the full spectrum of VEGF biologic responses. In agreement with this hypothesis. VEGF mutants which bind selectively to Flk-1/KDR are fully active endothelial cell mitogens (see below) (Keyt et al., 1996b). Whether VEGF induces formation of heterodimers between Flt-1 and Flk-1/KDR, which could confer new properties or ligand specificity upon these receptors, remains to be established. Further studies are clearly required to characterize the molecular events involved in VEGF signal transduction.

Very recent studies have demonstrated that both Flt-1 and Flk-1/KDR are essential for normal development of embryonic vasculature, although their respective roles in endothelial cell proliferation and differentiation appear to be distinct (Fong et al., 1995; Shalaby et al., 1995). Mouse embryos homozygous for a targeted mutation in the *flt-1* locus died *in utero* at day 8.5 (Fong et al., 1995). Endothelial cells developed in both embryonic and extraembryonic sites but failed to organize in normal vascular channels. Mice where the *flk-1* gene had been inactivated revealed a more profound deficit, as they not only lacked vasculogenesis but also failed to develop blood islands (Shalaby et al., 1995). Hematopoietic precursors were severely disrupted and organized blood vessels failed to develop throughout the embryo or the yolk sac resulting in death *in utero* between day 8.5 and 9.5.

Structural requirements for VEGF binding to Flt-1 and Flk-1/KDR

As noted above, the VEGF proteins may be expressed as multiple homodimeric forms with progressively higher affinity for heparin. The higher molecular weight forms of VEGF can be cleaved by plasmin to release a diffusible form(s) of VEGF (Houck et al., 1992). We isolated plasmin fragments of VEGF and evaluated their relative interaction with soluble VEGF receptors (Keyt et al., 1996a). The receptor affinity of homodimeric $VEGF_{165}$ was compared to that of heterodimeric $VEGF_{165/110}$ and homodimeric $VEGF_{110/110}$ and $VEGF_{121/121}$. The carboxy terminal polypeptide (111–165) displayed no affinity for KDR or Flt-1 receptors. The various isoforms of VEGF (165, 165/110, 110, and 121) bound soluble KDR receptor with similar affinity (approx. 30 pM). In contrast, soluble Flt-1 differentiated VEGF isoforms (165, 165/110, 110, and 121) with apparent affinities of 10, 30, 120 and 200 pM, respectively. These findings indicate that the amino-terminal domain (residues 1–110) predominantly contains the receptor binding determinants for KDR and Flt-1 receptors, although the carboxy-terminal domain (residues 111–165) also contributes significantly to the interaction with Flt-1 (Keyt et al., 1996a).

Mutational analysis was used to localize further the determinants on VEGF which mediate binding to Flk-1/KDR and Flt-1. Alanine-scanning mutagenesis was performed to identify a positively charged surface in

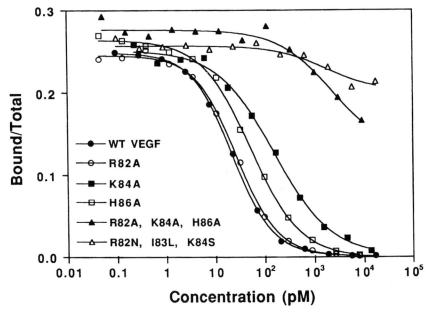

Figure 1. Competitive displacement of ^{125}I-labeled VEGF$_{165}$ from KDR with single and multiple alanine scan or glycosylation mutants. Displacement curves with KDR-IgG binding labeled VEGF in competition with wildtype VEGF (filled circles), R82A VEGF (open circles), K84A VEGF (filled boxes), H86A VEGF (open boxes), R82A, K84A, H86A VEGF (filled triangles), RIK(84–84)NLS VEGF (open triangles). These experiments were performed in duplicate.

VEGF that mediates receptor binding (Keyt et al., 1996b). Arg[82], Lys[84] and His[86], located in a hairpin loop, were found to be critical for binding KDR/Flk-1, while negatively charged residues, Asp[63], Glu[64] and Glu[67], were associated with Flt-1 binding. The single mutations, R82A, K84A and H86A were found to display modestly decreased KDR binding (Fig. 1). The triple mutants involving alanine replacement or neo-glycosylation sites, R82A, K84A, H86A VEGF and R82N, I83L, K84S VEGF, exhibited minimal binding to KDR receptor.

Mitogenic activities of VEGF and mutants of VEGF were determined using bovine adrenal cortical capillary endothelial cells. The half-maximally effective concentrations (EC$_{50}$) for most of the VEGF mutants were similar to those observed for wildtype VEGF (Fig. 2). The most significant effect on endothelial cell proliferation was observed with mutations in the 82–86 region. The EC$_{50}$ or R82A, K84A, H86A VEGF increased 20-fold, such that mitogenic potency of this mutant was decreased to 5% of wildtype VEGF. The neo-glycosylation site mutant, RIK(82–84)NLS VEGF, exhibited 60-fold reduced potency in endothelial cell proliferation. The mutational analysis of VEGF by alanine scanning and extra-glycosylation provides strong evidence that binding to Flk-1/KDR on endothelial cells is critical for the induction of proliferation observed with VEGF.

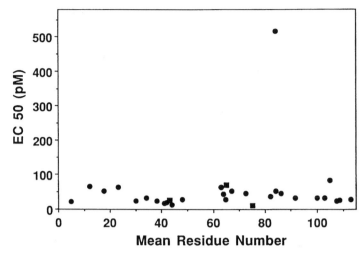

Figure 2. Activity of VEGF mutants as endothelial cell mitogens. The VEGF mutants were expressed in 293 cells; the conditioned media were used to stimulate mitogenesis of bovine adrenal cortical capillary endothelial cells. The mean residue number indicates the location of the mutations. The values are expressed as the concentration required to half-maximally stimulate endothelial cell proliferation (EC_{50}). Alanine mutants of VEGF are indicated as filled circles and potential extra-glycosylated VEGF mutants as filled boxes. These experiments were done in triplicate.

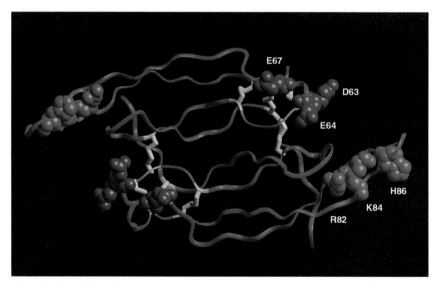

Figure 3. Three dimensional model of the (1–110) domain of VEGF dimer and the positions of two types of mutations. The model is based on the crystal structure of PDGFb dimer (Oefner et al., 1992). Shown as an oval ribbon diagram, the polypeptide backbone is in green, and disulfides in yellow. The sidechains of acidic amino acids involved in Flt-1 receptor binding are shown in red for Asp[63], Glu[64], and Glu[67]. The sidechains of basic amino acids involved in KDR receptor binding are shown in blue for Arg[82], Lys[84], and His[86].

A VEGF model based on PDGFb indicated these positively and negatively charged regions are distal in the monomer, but are spatially close in the dimer (Fig. 3). Mutations within the KDR site had minimal effect on Flt-1 binding and mutants deficient in Flt-1 binding did not affect KDR binding. Endothelial cell mitogenesis was abolished in mutants lacking KDR affinity, but Flt-1-deficient mutants induced normal proliferation. These results suggest dual sets of determinants in the VEGF dimer that cross-link cell surface receptors, triggering endothelial cell growth and angiogenesis. Furthermore, this mutational analysis implicates KDR, but not Flt-1, in VEGF-induced endothelial cell proliferation (Keyt et al., 1996b).

Role of VEGF in physiological angiogenesis

The proliferation of blood vessels is crucial for a wide variety of physiological processes such as embryonic development, normal growth and differentiation, wound healing and reproductive functions. The VEGF mRNA is expressed within the first few days following implantation in the giant cells of the trophoblast (Breier et al., 1992; Jakeman et al., 1993), suggesting a role for VEGF in the induction of vascular growth in the decidua, placenta and vascular membranes. At later developmental stages in mouse or rat embryos, the VEGF mRNA is expressed in several organs, including heart, vertebral column, kidney, and along the surface of the spinal cord and brain (Breier et al., 1992; Jakeman et al., 1993). These studies indicate that a variety of cells express the VEGF mRNA. However, there is no evidence that endothelial cells express the VEGF mRNA, suggesting that VEGF is a purely paracrine mediator. In the developing mouse brain, the highest levels of mRNA expression are associated with the choroid plexus and the ventricular epithelium (Breier et al., 1992).

Very recent studies (Ferrara et al., 1996) emphasize the key role played by VEGF in embryonic development. Heterozygous mutations inactivating the VEGF gene in mice result in embryonic lethality between day 11.5 and 12.5. Mutant embryos revealed dramatic deficits in angiogenesis and hematopoiesis as well as in the development of the cardiovascular and other systems. These findings indicate that VEGF is a most-critical factor in vasculogenesis and angiogenesis, as other factors cannot compensate for even reduced levels of VEGF. This may be the first example of embryonic lethality following loss of a single allele of a gene.

In the human fetus (16–22 weeks), VEGF mRNA expression is detectable in virtually all tissues and is most abundant in lung, kidney and spleen. VEGF protein, as assessed by immunocytochemistry, is expressed in epithelial cells and myocytes, but not vascular endothelial cells (Shifren et al., 1994). Interestingly, VEGF expression is also detectable, both in fetus and in the adult, around microvessels in areas where endothelial cells are quiescent, such as kidney glomerulus, pituitary, heart, lung, and brain (Ferrara et al., 1992; Monacci et al., 1993; Shifren et al., 1994). These findings raise the possibility that VEGF is required not only to induce active vascular proliferation but also for the maintenance of the differentiated state of blood vessels, at least in some circumstances (Ferrara et al., 1992).

Role of VEGF in pathologic angiogenesis

Tumor angiogenesis

Numerous tumor cell lines express the VEGF mRNA and secrete a VEGF-like protein in the medium (Senger et al., 1986; Rosenthal et al., 1990). Furthermore, *in situ* hybridization studies have shown that the VEGF mRNA is markedly upregulated in most human tumors so far examined. These include: kidney, bladder (Brown et al., 1993a), breast (Brown et al., 1995a) ovarian (Olson et al., 1994) and gastrointestinal tract (Brown et al., 1993b) carcinomas and several intracranial tumors including glioblastoma multiforme (Plate et al., 1992; Shweiki et al., 1992) and sporadic, as well as von Hippel-Lindau syndrome-associated, capillary hemangioblastoma (Berkman et al., 1993; Wizigmann-Voss, 1994). Only sections of lobular carcinoma of the breast and papillary carcinoma of the bladder failed to reveal significant VEGF mRNA expression (Dvorak et al., 1995). In all of these circumstances, the VEGF mRNA is expressed in tumor cells but not in endothelial cells. This is consistent with the hypothesis that VEGF is a purely paracrine mediator (Ferrara et al., 1992). However, immunohistochemical studies have localized the VEGF protein not only to the tumor cells but also to the vasculature (Plate et al., 1992; Brown et al., 1993). This discrepancy indicates that tumor-secreted VEGF accumulates in the target cells. In addition, the mRNA for the VEGF receptors, Flt-1 and KDR, is upregulated in the tumor-associated endothelial cells in comparison with the vasculature of the surrounding tumor-free tissue (Plate et al., 1992; Brown et al., 1993; Warren et al., 1995). Interestingly, a correlation has been observed between VEGF expression, as assessed by immunohistochemistry, and microvessel density in primary breast cancer sections (Toi et al., 1994). Postoperative survey indicated that the relapse-free survival rate of VEGF-rich tumors was significantly worse than that of VEGF-poor, suggesting that expression of VEGF is associated with stimulation of angiogenesis and with early relapse in primary breast cancer (Toi et al., 1994). In tumors with a significant component of necrosis such as glioblastoma multiforme, VEGF mRNA expression is not uniform but occurs primarily in clusters of tumor cells at the border between viable tumor and necrotic areas (Plate et al., 1992; Shweiki et al., 1992). This localization is consistent with local hypoxia being a major inducer of VEGF gene expression and suggests that a VEGF gradient is responsible for angiogenesis and tumor expension toward ischemic areas.

More direct evidence for a role of VEGF in tumorigenesis has been made possible by the availability of specific monoclonal antibodies capable of inhibiting VEGF-induced angiogenesis *in vivo* and *in vitro* (Kim et al., 1992). Such antibodies exert a dramatic inhibitory effect on the growth of a variety of human tumor cell lines injected subcutaneously into nude mice, including glioblastoma multiforme, rhabdomyosarcoma, leiomyosarcoma, and colon carcinoma (Kim et al., 1993; Warren et al., 1995). However, neither the antibodies nor VEGF have any effect on the *in vitro* growth of the tumor cells. In agreement with the hypothesis that inhibition of angiogenesis is the mechanism of tumor suppression, the density of microvessels was significantly lower in sections of tumors from antibody-treated animals as compared with controls (Kim et al., 1993; Warren et al., 1995). Intravital microscopy techniques have recently provided a direct demonstration that anti-VEGF antibodies indeed block tumor angiogenesis (Börgstrom et al., in press). Tumor spheroids of A673 rhabdomyosarcoma cells were implanted in dorsal skinfold chambers inserted in nude mice. Non-invasive visualization of the vasculature, three, seven or fourteen days following tumor spheroid implant, revealed a dramatic suppression of tumor angiogenesis in anti-VEGF treated animals as compared with controls, at all time points.

Warren et al. (1995) have demonstrated that VEGF is a major mediator of the *in vivo* growth of human colon carcinoma cells in a nude mouse model of liver metastasis where the tumor cells are injected into the spleen. Similarly to human tumors, in this murine model the expression of Flk-1 mRNA was markedly upregulated in the vasculature associated with liver metastases. Treatment with anti-VEGF monoclonal antibodies resulted in a dramatic decrease in the number and size of metastases. Most of the tumors in the treated group were under 1 mm in diameter and all were under 3 mm. Also, neither blood vessels nor Flk-1 mRNA expression could be demonstrated in such metastases.

An independent verification of the hypothesis that VEGF action is necessary for tumorigenesis has been provided by the finding that retrovirus-mediated expression of a negative dominant Flk-1 mutant suppresses the growth of glioblastoma cells *in vivo* (Millauer et al., 1994).

Angiogenesis associated with other disorders

Diabetes mellitus, occlusion of central retinal vein, or prematurity with subsequent exposure to oxygen, can all be associated with intraocular vascular proliferation (Patz, 1980). The new blood vessels may lead to vitreous hemorrhage, retinal detachment, neovascular glaucoma, and eventual blindness (Garner, 1994). Diabetic retinopathy is the leading cause of blindness in the working population (Olk and Lee, 1993). All of these conditions are known to be associated with retinal ischemia (Patz, 1980). As early as in 1948, Michaelson proposed that the ischemic retina is able to release into the vitreous diffusible angiogenic factor(s) responsible for retinal and iris neovascularization. Even though IGF-1 and bFGF have been implicated in this process, these factors do not show a consistent increase as would be expected if they played a significant causative role (Hannehan et al., 1991; Meyer-Schwickerath, 1993). VEGF, by virtue of its diffusible nature and hypoxia-inducibility, is an attractive candidate as a retina-derived mediator of intraocular neovascularization. Recently, elevations of VEGF levels in the aqueous and vitreous of eyes with proliferative retino-pathy have been reported (Adamis et al., 1994; Aiello et al., 1994; Malecaze et al., 1994). In a large series, a strong correlation was found between levels of immunoreactive VEGF in the aqueous and vitreous humors and active proliferative retinopathy (Aiello et al., 1994). VEGF levels were undetec-table or very low (less than 0.5 ng/ml) in the eyes of patients affected by non-neovascular disorders or diabetes without proliferative retinopathy. In contrast, the VEGF levels were in the range of $3-10$ ng/ml in the presence of active proliferative retinopathy associated with diabetes, occlusion of central retinal vein or prematurity. Remarkably, the VEGF levels were again very low in the eyes of patients with quiescent proliferative retino-pathy, a phase of vascular regression that follows the period of active vascular proliferation in diabetic and other retinopathies (Aiello et al., 1994). Thus, although the involvement of other factors cannot be rules out, VEGF is the molecule that correlates best with ocular angiogenesis (Ferrara, 1995).

In agreement with these findings, *in situ* hybridization studies have demonstrated upregulation of VEGF mRNA in the retina of patients with proliferative retinopathies secondary to diabetes, central retinal vein occlusion, retinal detachment or intraocular tumors (Pe'er et al., 1995). Remarkably, VEGF mRNA expression was confined to the specific retinal layer(s) expected to be ischemic.

More direct evidence for the hypothesis that VEGF is a mediator of intra-ocular neovascularization has recently been provided in a primate model of iris neovascularization and in a murine model of retinopathy of prematurity. In the former, intraocular administration of anti-VEGF antibodies drama-tically inhibits the neovascularization that follows occlusion of central retinal veins (Adamis et al., in press). Likewise, soluble Flt-1 or Flk-1 fused to an IgG suppresses retinal angiogenesis in the mouse model (Aiello et al., 1995).

It has also been proposed that VEGF is involved in the pathogenesis of another important disease where angiogenesis plays a significant role, rheumatoid arthritis (RA) (Fava et al., 1994; Koch et al., 1994). The RA synovium is characterized by the formation of pannus, an extensive-ly vascularized tissue that invades and destroys the articular cartilage

(Fassbender et al., 1983). Because of its vascularity and rapid proliferation rate, the RA synovium has been likened to a tumor (Hamilton et al., 1976). Levels of immunoreactive VEGF were high in the synovial fluid of RA patients while they were very low or undetectable in the synovial fluid of patients affected by other forms of arthritis or by degenerative joint disease. Furthermore, anti-VEGF antibodies significantly reduced the endothelial cell chemotactic activity of the RA synovial fluid (Koch et al., 1994).

It has been shown that VEGF expression is increased in psoriatic skin (Brown et al., 1995 b. Increased vascularity and permeability are characteristic of psoriasis. Also VEGF mRNA expression has recently been examined in three bullous disorders with subepidermal blister formation, bullous pemphigoid, erythema multiforme and dermatitis herpetiformis (Brown et al., 1995). Inn all of these conditions, VEGF mRNA was markedly upregulated not only in the epidermis over blisters, but also at a distance from blisters, in areas adjacent to dermal inflammatory infiltrates.

Intriguingly, at least two sequences having a significant homology to VEGF have been identified in the genome of two different strains of *orf* virus, a parapoxvirus that affects goats, sheeps and occasionally humans (Lyttle et al., 1994). Interestingly, the lesions of goats and humans following *orf* virus infection are characterized by extensive microvascular proliferation in the skin, raising the possibility that the product of the viral VEGF-like gene is responsible for such lesions.

VEGF as a potential therapeutic agent

The availability of agents able to promote the growth of new collateral vessels would potentially be of major therapeutic value for disorders characterized by inadequate tissue perfusion and might constitute an alternative to surgical reconstruction procedures. For example, chronic limb ischemia, most frequently caused by obstructive atherosclerosis affecting the superficial femoral artery, is associated with a high rate of morbidity and mortality, and treatment is currently limited to surgical revascularization or endovascular interventional therapy (Topol, 1990; Thompson and D'Amore, 1990). No pharmacological therapy has been shown to be effective for this condition. It has recently been shown that intra-arterial or intra-muscular administration of rhVEGF$_{165}$ may significantly augment perfusion and development of collateral vessels in a rabbit model where chronic hindlimb ischemia was created by surgical removal of the femoral artery (Takeshita et al., 1994 a; 1994 b). These studies provided angiographic evidence of neovascularization in the ischemic limbs. Arterial gene transfer with a cDNA encoding VEGF$_{165}$ also led to revascularization of rabbit ischemic limbs to an extent comparable to that achieved with the recombinant protein (Takeshita et al., in press). In addition, the hypothesis that the angiogenesis initiated by the administration of VEGF improved muscle function in ischemic limbs was recently tested (Walder et al., in press). A single intra-arterial injection of rhVEGF$_{165}$ augmented muscle function in this rabbit model of peripheral limb ischemia. This exercise-induced hyperemia was significantly improved in ischemic limbs treated with rhVEGF$_{165}$ (Walder et al., 1996). Such improvement in perfusion was, however, not seen in other non-ischemic tissues including the contralateral limb. Similarly, Bauters et al. (1994) have shown that both maximal flow velocity and maximal blood flow, as assessed by doppler, are significantly increased in ischemic limbs following VEGF administration. Recent studies have shown that VEGF administration also leads to a recovery of normal endothelial reactivity in dysfunctional endothelium (Bauters et al., 1995). Following obstruction of a large artery and development of collateral vessels, the increase in blood flow which normally follows acetylcholine infusion is severely blunted; serotonin paradoxically leads to a decrease in blood flow (Sellke et al., 1992). Thirty days after a single intra-arterial bolus of VEGF$_{165}$, restoration of normal increase in blood flow in ischemic rabbit hindlimb following acetylcholine or serotonin infusion was demonstrated (Bauters et al., 1995).

Furthermore, it has been shown that VEGF administration results in increased coronary blood flow in a dog model of coronary insufficiency (Banai et al., 1994a). Following occlusion of the left circumflex coronary artery, daily intraluminal injections of rhVEGF distal to the occlusion resulted in a significant enhancement in collateral blood flow over a four week period. In addition, Harada et al. (1996) have recently demonstrated that extraluminal administration of as little as 20µg of rhVEGF by an osmotic pump results in a significant increase in coronary blood flow in a pig model of chronic myocardial ischemia created by ameroid occlusion of proximal circumflex artery. Remarkably, VEGF treatment led to 2.6-fold decrease in the size of left ventricular infarct in this model (Harada et al., 1996).

Another potential therapeutic application of VEGF is the prevention of restenosis following percutaneous transluminal angioplasty (PTA). Between 15% and 75% of patients undergoing PTA for occlusive coronary or peripheral arterial disease develop restenosis within six months. The frequency of clinical stenosis depends on the size and location of the artery and the definition of stenosis (Graor and Gray, 1991). It has been proposed that damage to the endothelium is the crucial event triggering fibrocellular intimal proliferation (Essed et al., 1983). Recent studies have shown that VEGF also accelerates reendothelialization and attenuates intimal hyperplasia in balloon-injured rat carotid artery (Asahara et al., 1995). Therefore, it is tempting to speculate that rapid reendothelialization promoted by VEGF may prove effective at preventing the cascade of events leading to neointima formation and restenosis in patients.

Conclusions

The recent findings that heterozygous mutations inactivating the VEGF gene and homozygous mutations inactivating the Flt-1 or Flk-1/KDR genes result in profound deficits in vasculogenesis and blood island formation, leading to early intrauterine death, emphasize the pivotal role played by the VEGF/VEGF-receptor system in the development of the vascular system.

An attractive possibility is that recombinant VEGF or gene therapy with the VEGF gene may be used to promote endothelial cell growth and collateral vessel formation. This would represent a novel therapeutic modality for conditions that are frequently refractory to conservative measures and unresponsive to pharmacological therapy.

The high expression of VEGF mRNA in human tumors, and the presence of the VEGF protein in ocular fluids of individuals with proliferative retinopathies and in the synovial fluid of RA patients, strongly support the hypothesis that VEGF is a key mediator of angiogenesis associated with various pathological conditions. Therefore, anti-VEGF antibodies or VEGF antagonists have the potential to be of therapeutic value for a variety of highly vascularized and aggressive malignancies as well as for other angiogenic disorders. An anti-VEGF therapy may have low toxicity, perhaps limited to inhibition of wound healing and ovarian and endometrial function, since endothelial cells are essentially quiescent in most adult tissues.

In conclusion, recent evidence strongly suggests that, in spite of the plurality of factors potentially involved in pathological angiogenesis, strategies aimed at antagonizing one specific endothelial cell mitogen, VEGF, may form the basis for an effective treatment of a variety of tumors and proliferative retinopathies.

References

Adamis, A.P., Miller, J.W., Bernal, M.T., D'Amico, D., Folkman, J., Yeo, T.-K. and Yeo, K.-T. (1994) Increased vascular endothelial growth factor in the vitreous of eyes with proliferative diabetic retinopathy. *Am. J. Ophthalmol.* 118:445–450.

Adamis, A.P., Shima, D.T., Tolentino, M., Gragoudas, E., Ferrara, N., Folkman, J., D'Amore, P.A. and Miller, J.W. (1996) Inhibition of VEGF prevents iris neovascularization in a non-human primate. *Arch. Ophthalmol,* 114:66–71.

Aiello, L.P., Avery, R., Arrigg, R., Keyt, B., Jampel, H., Shah, S., Pasquale, L., Thieme, H., Iwamoto, M., Park, J.E., Nguyen, H., Aiello, L.M., Ferrara, N. and King, G.L. (1994) Vascular endothelial growth factor in ocular fluid of patients with diabetic retinopathy and other retinal disorders. *N. Engl. J. Med.* 331:1480–1487.

Aiello, L.P., Pierce, E.A., Foley, E.D., Takagi, H., Riddle, L., Chen, H., Ferrara, N., King, G.L. and Smith, L.E. (1995) Suppression of retinal neovascularization in vivo by inhibition of vascular endothelial growth factor (VEGF) using soluble VEGF-receptor chimeric proteins. *Proc. Natl. Acad. Sci. USA* 92:10457–10461.

Asahara, T., Bauters, C., Pastore, C., Bunting, S., Ferrara, N., Symes, J.F. and Isner, J.M. (1995) Local delivery of vascular endothelial growth factor accelerates reendothelialization and attenuates intimal hyperplasia in balloon-injured rat carotid artery. *Circulation* 91:2802–2809.

Banai, S., Jaktlish, M.T., Shou, M., Lazarous, D.F., Scheinowitz, M., Biro, S., Epstein, S. and Unger, E. (1994a) Angiogenic-induced enhancement of collateral blood flow to ischemic myocardium by vascular endothelial growth factor in dogs. *Circulation* 89:2183–2189.

Banai, S., Shweiki, D., Pinson, A., Chandra, M., Lazarovici, G. and Keshet, E. (1994b) Upregulation of vascular endothelial growth factor expression induced by myocardial ischemia: implications for coronary angiogenesis. *Cardiovasc. Res.* 28:1176–1179.

Bauters, C., Asahara, T., Zheng, L.P., Takeshita, S., Bunting, S., Ferrara, N., Symes, J.F. and Isner, J.M. (1994) Physiologic assessment of augmented vascularity induced by VEGF in a rabbit ischemic hindlimb model. *Am. J. Physiol.* 267:H1263–H1271.

Bauters, C., Asahara, T., Zheng, L.P., Takeshita, S., Bunting, S., Ferrara, N., Symes, J.F. and Isner, J.M. (1995) Recovery of disturbed endothelium-dependent flow in collateral-perfused rabbit ischemic hindlimb following administration of VEGF. *Circulation* 91:2793–2801.

Berkman, R.A., Merrill, M.J., Reinhold, W.C., Monacci, W.T., Saxena, A., Clark, W.C., Robertson, J.T., Ali, I.U. and Oldfield, E.H. (1993) Expression of the vascular permeability/vascular endothelial growth factor gene in central nervous system neoplasms. *J. Clin. Invest.* 91:153–159.

Bigler, S.A., Deering, R.E. and Brawer, M.K. (1993) Comparison of microscopic vascularity in benign and malignant prostatic tissue. *Hum. Pathol.* 24:220–226.

Borgström, P., Hillan, K.J., Sriramoa, S. and Ferrara, N. (1996) In vivo imaging of tumor vasculature by intravital microscopy: Inhibition of angiogenesis by anti-VEGF antibodies. *Cancer Res.* In press.

Breier, G., Albrecht, U., Sterrer, S. and Risau, W. (1992) Expression of vascular endothelial growth factor during embryonic angiogenesis and endothelial cell differentiation. *Development* 114:521–532.

Brown, L.F., Berse, B., Jackman, R.W., Tognazzi, K., Manseau, E.J., Dvorak, H.F., Senger, D.R. (1993a) Increased expression of vascular permeability factor (vascular endothelial growth factor) and its receptors in kidney and bladder carcinomas. *Am. J. Pathol.* 143:1255–1262.

Brown, L.F., Berse, B., Jackman, R.W., Tognazzi, K., Manseau, E.J., Senger, D.R. and Dvorak, H.F. (1993b) Expression of vascular permeability factor (vascular endothelial growth factor) and its receptors in adenocarcinomas of the gastrointestinal tract. *Cancer Res.* 53:4727–4735.

Brown, L.F., Berse, B., Jackman, R.W., Guidi, A.J., Dvorak, H.F., Senger, D.R., Connolly, J.L. and Schnitt, S.J. (1995a) Expression of vascular permeability factor (vascular endothelial growth factor) and its receptors in breast cancer. *Hum. Pathol.* 26:86–91.

Brown, L.F., Harris, T.J., Yeo, K.T., Stahle-Backdahl, M., Jackman, R.W., Berse, B., Tognazzi, K., Dvorak, H.F. and Detmar, M. (1995b) Increased expression of vascular permeability factor (vascular endothelial growth factor) in bullous pemphigoid, dermatitis herpetiformis and herythema multiforme. *J. Invest. Dermatol.* 104:744–749.

Chodak, G.W., Haudenschild, C., Gittes, R.F. and Folkman, J. (1980) Angiogenic activity as a marker of neoplastic and preneoplastic lesions of the human bladder. *Ann. Surg.* 192:762–771.

Claffey, K.P., Wilkinson, W.O. and Spiegelman, B.M. (1992) Vascular endothelial growth factor. Regulation by cell differentiation and activated second messenger pathways. *J. Biol. Chem.* 267:16317–16322.

Conn, G., Bayne, M., Soderman, L., Kwok, P.W., Sullivan, K.A., Palisi, T.M., Hope, D.A. and Thomas, K.A. (1990) Amino acid and cDNA sequence of a vascular endothelial cell mitogen homologous to platelet-derived growth factor. *Proc. Natl. Acad. Sci USA* 87:2628–2632.

Connolly, D.T., Heuvelman, D.M., Nelson, R., Olander, J.V., Eppley, B.L., Delfino, J.J., Siegel, N.R., Leimgruber, R.M. and Feder, J. (1989a) Tumor vascular permeability factor stimulates endothelial cell growth and angiogenesis. *J. Clin. Invest.* 84:1470–1478.

Connolly, D.T., Olander, J.V., Heuvelman, D., Nelson, R., Monsell, R., Siegel, N., Haymore, B.L., Leingruber, R. and Feder, J. (1989b) Human vascular permeability factor. Isolation from U937 cells. *J. Biol. Chem.* 254:20017–20024.

deVries, C., Escobedo, J.A., Ueno, H., Houck, K.A., Ferrara, N. and Williams, L.T. (1992) The fms-like tyrosine kinase, a receptor for vascular endothelial growth factor. *Science* 255:989–991.

DiSalvo, J., Bayne, M.L., Conn, G., Kwok, P.W., Trivedi, P.G., Soderman, D.D., Palisi, T.M., Sullivan, K.A. and Thomas, K.A. (1995) Purification and characterization of a naturally occurring vascular endothelial growth factor placenta growth factor heterodimer. *J. Biol. Chem.* 270:7717–7723.

Dvorak, H.F. (1986) Tumors: wound that do not heal. Similarity between tumor stroma generation and wound healing. *N. Engl. J. Med.* 315:1650–1658.

Dvorak, H.F., Harvey, V.S., Estrella, P., Brown, L.F., McDonagh, J. and Dvorak, A.M. (1987) Fibrin containing gels induce angiogenesis: implications for tumor stroma generation and wound healing. *Lab. Invest.* 57:673–686.

Dvorak, H.F., Brown, L.F., Detmar, M. and Dvorak, A.M. (1995) Vascular permeability factor/vascular endothelial growth factor, microvascular permeability and angiogenesis. *Am. J. Pathol.* 146:1029 1039.

Essed, C.D., Brand, M.V.D. and Becker, A.E. (1983) Transluminal coronary angioplasty and early restenosis. *Br. Heart J.* 49:393–402.

Fassbender, H.J. and Simling-Annefeld, M. (1983) The potential aggressiveness of synovial tissue in rheumatoid arthritis. *J. Pathol.* 139:399–406.

Fava, R.A., Olsen, N.J., Spencer-Green, G., Yeo, T.-K., Yeo, K.-T., Berse, B., Jackman, R.W., Senger, D.R., Dvorak, H.F. and Brown, L.F. (1994) Vascular permeability factor/vascular endothelial growth factor (VPF/VEGF): accumulation and expression in human synovial fluids and rheumatoid arthritis. *J. Exp. Med.* 180:340–346.

Ferrara, N. and Henzel, W.J. (1989) Pituitary follicular cells secrete a novel heparin-binding growth factor specific for vascular endothelial cells. *Biochem. Biophys. Res. Commun.* 161:851–859.

Ferrara, N., Houck, K., Jakeman, L. and Leung, D.W. (1992) Molecular and biological properties of the vascular endothelial growth factor family of proteins. *Endocr. Rev.* 13:18–32.

Ferrara, N. (1993) Vascular endothelial growth factor. *Trends Cardiovasc. Med.* 3:244–250.

Ferrara, N. (1995) Vascular endothelial growth factor – The trigger for neovascularization in the eye. (Editorial) *Lab. Invest.* 72:615–618.

Ferrara, N., Carver-Moore, K., Chen, H., Dowd, M., Hillan, K.J., Lu, L., O'Shea, S., Powell-Braxton, L. and Moore, M. (1996) Heterozygous embryonic lethality induced by targeted inactivation of the VEGF gene. *Nature* 380:439–442.

Folkman, J. (1991) What is the evidence that tumors are angiogenesis-dependent? *J. Natl. Cancer Inst.* 82:4–6.

Folkman, J. and Shing, Y. Angiogenesis. *J. Biol. Chem.* 267:10931–10934.

Fong, G.-H., Rassant, J., Gertenstein, M. and Breitman, M. (1995) Role of Flt-1 receptor tyrosine kinase in regulation of assembly of vascular endothelium. *Nature* 376:6670.

Frank, S., Hubner, G., Breier, G., Longaker, M.T., Greenhalgh, D.G. and Werner, S. (1995) Regulation of VEGF expression in cultured keratinocytes. Implications for normal and impaired wound healing. *J. Biol. Chem.* 20:12607–12613.

Garner, A. (1994) Vascular diseases. *In:* A. Garner and G.K. Klintworth (eds): *Pathobiology of ocular disease. A dynamic approach.* 2nd Edition, Marcel Dekker, New York, pp 1625–1710.

Garrido, C., Saule, S. and Gospodarowicz, D. (1993) Transcriptional regulation of vascular endothelial growth factor gene expression in ovarian bovine granulosa cells. *Growth Factors* 8:109–117.

Goldberg, M.A. and Schneider, T.J. (1994) Similarities between the oxygen-sensing mechanisms regulating the expression of vascular endothelial growth factor and erythropoietin. *J. Biol. Chem.* 269:4355–4361.

Graor, R.A. and Gray, B.H. (1991) Interventional treatment of peripheral vascular disease. *In:* J.R. Young, R.A. Graor, J.W. Olin and J.R. Bartholomew (eds): *Peripheral Vascular Diseases*, Mosby, St. Louis, pp 111–133.

Gross, J. and Nagai, Y. (1965) Specific degradation of the collagen molecule by tadpole collagenolytic enzyme. *Proc. Natl. Acad. Sci. USA* 54:1197–1204.

Hamilton, J. (1976) Hypothesis: In vitro evidence for the invasive and tumor-like properties of the rheumatoid pannus. *J. Rheumatol.* 10:845–851.

Hamilton, W.J., Boyd, J.D. and Mossman, H.W. (1962) *Human Embryology*, William & Wilkins, Baltimore.

Hannehan, A., deJuan, E., Lutti, G.A., Fox, G.M., Schiffer, S. and Hjelmeland, L.M. (1991) Altered distribution of basic fibroblast growth factor in diabetic retinopathy. *Arch. Ophthalmol.* 109:1005–1011.

Harada, K., Friedman, M., Lopez, J., Prasad, P.V., Hibberd, M., Pearlman, J.D., Sellke, F.W. and Simons, M. (1996) Vascular endothelial growth factor improves coronary flow and myocardial function in chronically ischemic porcine hearts. *Am. J. Physiol.* 270:1791–1802.

Hauser, S. and Weich, H.A. (1993) A heparin-binding form of placenta growth factor (PlGF-2) is expressed in human umbilical vein endothelial cells and in placenta. *Growth Factors* 9:259–268.

Horak, E.R., Leek, R., Klenk, N., Lejeune, S., Smith, K., Stuart, M., Greenall, M. and Harris, A.L. (1992) Quantitative angiogenesis assessed by anti-PECAM antibodies: correlation with node metastasis and survival in breast cancer. *Lancet* 340:1120–1124.

Houck, K.A., Ferrara, N., Winer, J., Cachianes, G., Li, B. and Leung, D.W. (1991) The vascular endothelial growth factor family: Identification of a fourth molecular species and characterization of alternative splicing of RNA. *Mol. Endocrinol.* 5:1806–1814.

Houck, K.A., Leung, D.W., Rowland, A.M., Winer, J. and Ferrara, N. (1992) Dual regulation of vascular endothelial growth factor bioavailability by genetic and proteolytic mechanisms. *J. Biol. Chem.* 267:26031–26037.

Jakeman, L.B., Winer, J., Bennett, G.L., Altar, C.A. and Ferrara, N. (1992) Binding sites for vascular endothelial growth factor rare localized on endothelial cells in adult rat tissues. *J. Clin. Invest.* 89:244–253.

Jakeman, L.B., Armanini, M., Phillips, H.S. and Ferrara, N. (1993) Developmental expression of binding sites and mRNA for vascular endothelial growth factor suggests a role or this protein in vasculogenesis and angiogenesis. *Endocrinology* 133:848–859.

Keck, P.J., Hauser, S.D., Krivi, G., Sanzo, K., Warren, T., Feder, J. and Connolly, D.T. (1989) Vascular permeability factor, an endothelial cell mitogen related to platelet derived growth factor. *Science* 246:1309–1312.

Keyt, B., Berleau, L., Nguyen, H., Chen, H., Heinsohn, H., Vandler, R. and Ferrara, N. (1996a) The Carboxy-Terminal Domain (111–165) of Vascular Endothelial Growth Factor is Critical for its Mitogenic Potency. *J. Biol. Chem.* 271:7788–7795.

Keyt, B., Nguyen, H., Berleau, L., Duarte, C., Park, J., Chen, H. and Ferrara, N. (1996b) Identification of VEGF Determinants for Binding KDR and FLT-1 Receptors. *J. Biol. Chem.* 271(9): 5638–5646.

Kieser, A., Weich, H., Brandner, G., Marme, D. and Kolch, W. (1994) Mutant p53 potentiates protein kinase C induction of vascular endothelial growth factor expression. *Oncogene* 9:963–969.

Kim, K.J., Li, B., Houck, K., Winer, J. and Ferrara, N. (1992) The vascular endothelial growth factor proteins: Identification of biologically relevant regions by neutralizing monoclonal antibodies. *Growth Factors* 7:53–64.

Kim, K.J., Li, B., Winer, J., Armanini, M., Gillett, N., Phillips, H.S. and Ferrara, N. (1993) Inhibition of vascular endothelial growth factor-induced angiogenesis suppresses tumour growth *in vivo. Nature* 362:841–844.

Klagsbrun, M. and D'Amore, P.A. (1991) Regulators of angiogenesis. *Ann. Rev. Physiol.* 53:217–239.

Koch, E., Harlow, I, Haines, G.K., Amento, E.P., Unemori, E.N., Wong, W.-L., Pope, R.M. and Ferrara, N. (1994) Vascular endothelial growth factor: a cytokine modulating endothelial function in rheumatoid arthritis. *J. Immunol.* 152:4149–4155.

Ku, D.D., Zaleski, J.K., Liu, S. and Brock, T. (1993) Vascular endothelial growth factor induces EDRF-dependent relaxation of coronary arteries. *Am. J. Physiol.* 265, h586–592.

Leung, D.W., Cachianes, G., Kuang, W.-J., Goeddel, D.V. and Ferrara, N. (1989) Vascular endothelial growth factor is a secreted angiogenic mitogen. *Science* 246:1306–1309.

Levy, A.P., Levy, N.S., Wegner, S. and Goldberg, M.A. (1995) Transcriptional regulation of the rat vascular endothelial growth factor gene by hypoxia. *J. Biol. Chem.* 270:13333–13340.

Li, J., Perrella, M.A., Tsai, J.C., Yet, S.F., Hsieh, C.M., Yoshizumi, M., Patterson, C., Endego, W.O., Zhou, F. and Lee, M. (1995) Induction of vascular endothelial growth factor gene expression by interleukin-1 beta in rat aortic smooth muscle cells. *J. Biol. Chem.* 270:308–312.

Lyttle, D.J., Fraser, K.M., Flemings, S.B., Mercer, A.A. and Robinson, A.J. (1994) Homology of vascular endothelial growth factor are encoded by the poxvirus orf virus. *J. Virol.* 68:84–92.

Macchiarini, P., Fontanini, G., Hardin, M.J., Squartini, F. and Angeletti, C.A. (1992) Relation of neovascularization to metastasis of non-small cell lung carcinoma. *Lancet* 340:145–146.

Madan, A. and Curtin, P.T. (1993) A 24-base pair sequence 3' to the human erythropoietin contains a hypoxia-responsive transcriptional enhancer. *Proc. Natl. Acad. Sci. USA* 90:3928–3932.

Maglione, D., Guerriero, V., Viglietto, G., Delli-Bovi, P. and Persico, M.G. (1991) Isolation of a human placenta cDNA coding for a protein related to the vascular permeability factor. *Proc. Natl. Acad. Sci. USA* 88:9267–9271.

Malecaze, F., Clemens, S., Simorer-Pinotel, V., Mathis, A., Chollet, P., Favard, P., Bayard, F. and Plöüet, J. (1994) Detection of vascular endothelial growth factor mRNA and vascular endothelial growth factor-like activity in proliferative diabetic retinopathy. *Arch. Ophthalmol.* 112:1476–1482.

Mandriota, S., Montesano, R., Orci, L., Seghezzi, G., Vassalli, J.-D., Ferrara, N., Mignatti, P. and Pepper, M.S. (1995) Vascular endothelial growth factor increases urokinase receptor expression in vascular endothelial cells. *J. Biol. Chem.* 270:9709–9716.

Matthews, W., Jordan, C.T., Gavin, M., Jenkins, N.A., Copeland, N.G. and Lemischka, I.R. (1991) A receptor tyrosine kinase cDNA isolated from a population of enriched primitive hematopoietic cells and exhibiting close genetic linkage to c-kit. *Proc. Natl. Acad. Sci. USA* 88:9026–9030.

Meyer-Schwickerath, R., Pfeiffer, A., Blum, W.F., Freyberger, H., Klein, M., Losche, C., Rollman, R. and Schatz, H. (1993) Vitreous levels of the insulin-like growth factors I and II, and the insulin-like growth factor binding proteins 2 and 3 increase in neovascular disease. *J. Clin. Invest.* 92:2620–2625.

Michaelson, I.C. (1948) The mode of development of the vascular system of the retina with some observations on its significance for certain retinal disorders. *Trans. Ophthalmol. Soc. U.K.* 68:137–180.

Mignatti, P., Tsuboi, R., Robbins, E. and Rifkin, D.B. (1989) In vitro angiogenesis on the human amniotic membrane: requirement for basic fibroblast growth factor-induced proteinases. *J. Cell Biol.* 108:671–682.

Millauer, B., Wizigmann-Voos, S., Schnurch, H., Martinez, R., Moller, N.P., Risau, W. and Ullrich, A. (1993) High affinity binding and developmental expression suggest Flk-1 as a major regulator of vasculogenesis and angiogenesis. *Cell* 72:8358–846.

Millauer, B., Shawver, L.K., Plate, K.H., Risau, W. and Ullrich, A. (1994) Glioblastoma growth is inhibited in vivo by a negative dominant Flk-1 mutant. *Nature* 367:576–579.

Monacci, W., Merrill, M. and Oldfield, E. (1993) Expression of vascular permeability factor/vascular endothelial growth factor in normal rat tissues. *Am. J. Physiol.* 264:c995–c1002.

Mukhopadhyay, D., Tsilokas, L., Zhou, X.M., Foster, D., Brugge, J.S. and Sukhatme, Y.P. (1995) Hypoxic induction of human vascular endothelial growth factor expression through c-Src activation. *Nature* 375:577–581.

Mustonen, T. and Alitalo, K. (1995) Endothelial receptor tyrosine kinases involved in angiogenesis. *J. Cell Biol.* 129:895–898.

Nicosia, R.F., Nicosia, S.V. and Smith, M. (1995) Vascular endothelial growth factor, platelet-derived growth factor and insulin-like growth factor-1 promote rat aortic angiogenesis in vitro. *Am. J. Pathol.* 145:1023–1029.

Oefner, C., D'Arcy, A., Winkler, F.K., Eggimann, B. and Hosang, M. (1992) Crystal Structure of human platelet-derived growth factor BB. *EMBO J.* 11:3921–3926.

Olk, R.J. and Lee, C.M. (1993) *Diabetic retinopathy: practical management,* Lippincott Co. Philadelphia, PA.

Olson, T.A., Mohanraj, D., Carson, L.F. and Ramakrishnan, S. (1994) Vascular permeability factor gene expression in normal and neoplastic human ovaries. *Cancer Res.* 54:276–280.

Park, J.E., Keller, G.-A. and Ferrara, N. (1993) The vascular endothelial growth factor (VEGF) isoforms: Differential deposition into the subepithelial extracellular matrix and bioactivity of ECM-bound VEGF. *Mol. Biol. Cell* 4:1317–1326.

Park, J.E., Chen, H., Winer, J., Houck, K. and Ferrara, N. (1994) Placenta growth factor. Potentiation of vascular endothelial growth factor bioactivity, in vitro and in vivo, and high affinity binding to Flt-1 but not to Flk-1/KDR. *J. Biol. Chem.* 269:25646–25654.

Patz, A. (1980) Studies on retinal neovascularization. *Invest. Ophthalmol. Vis. Sci.* 19:1133–1138.

Pe'er, J., Shweiki, D., Itin, A., Hemo, I., Gnessi, H. and Keshet, E. (1995) Hypoxia-induced expression of vascular endothelial growth factor (VEGF) by retinal cells in a common factor in neovascularization. *Lab. Invest.* 72:638–645.

Pekala, P., Marlow, M., Heuvelman, D. and Connolly, D. (1990) Regulation of hexose transport in aortic endothelial cells by vascular permeability factor and tumor necrosis factor-alpha, but not by insulin. *J. Biol. Chem.* 265:18051–18054.

Pepper, M.S. and Montesano, R. (1990) Proteolytic balance and capillary morphogenesis. *Cell Diff. Dev.* 32:319–331.

Pepper, M.S., Ferrara, N., Orci, L. and Montesano, R. (1991) Vascular endothelial growth factor (VEGF) induces plasminogen activators and plasminogen activator inhibitor type 1 in microvascular endothelial cells. *Biochem. Biophys. Res. Commun.* 181:902–908.

Pepper, M.S., Ferrara, N., Orci, K. and Montesano, R. (1992) Potent synergism between vascular endothelial growth factor and basic fibroblast growth factor in the induction of angiogenesis in vitro. *Biochem. Biophys. Res. Commun.* 189:824–831.

Pertovaara, L., Kaipainen, A., Mustonen, T., Orpana, A., Ferrara, N., Saksela, O. and Alitalo, K. (1994) Vascular endothelial growth factor is induced in response to transforming growth factor-β in fibroblastic and epithelial cells. *J. Biol. Chem.* 269:6271–6274.

Peters, K.G., DeVries, C. and Williams, L.T. (1993) Vascular endothelial growth factor receptor expression during embryogenesis and tissue repair suggests a role in endothelial differentiation and blood vessel growth. *Proc. Natl. Acad. Sci. USA* 90:8915–8919.

Phillips, H.S., Hains, J., Leung, D.W. and Ferrara, N. (1990) Vascular endothelial growth factor is expressed in rat corpus luteum. *Endocrinology* 127:965–968.

Phillips, G.D., Stone, A.M., Jones, B.D., Schultz, J.C., Whitehead, R.A. and Knighton, D.R. (1993) Vascular endothelial growth factor (rhVEGF165) stimulates direct angiogenesis in the rabbit cornea. *In Vivo* 8:961–965.

Plate, K.H., Breier, G., Weich, H.A. and Risau, W. (1992) Vascular endothelial growth is a potential tumour angiogenesis factor in vivo. *Nature* 359:845–847.

Plöuet, J., Schilling, J. and Gospodarowicz, D. (1989) Isolation and characterization of a newly identified endothelial cell mitogen produced by AtT20 cells. *EMBO J.* 8:3801–3807.

Quinn, T., Peters, K.G., de Vries, C., Ferrara, N. and Williams, L.T. (1993) Fetal liver kinase 1 is a receptor for vascular endothelial growth factor and is selectively expressed in vascular endothelium. *Proc. Natl. Acad. Sci. USA* 90:7533–7537.

Rak, J. Mitsuhashi, Y., Bayko, I., Filmus, J., Shirasawa, S., Sasazuki, T. and Kerbel, R.S. (1995) Mutant ras oncogenes upregulate VEGF/VPF expression: implications for induction and inhibition of tumor angiogenesis. *Cancer Res.* 55:4575–4580.

Ravindranath, N., Little-Ihrig, L., Phillips, H.S., Ferrara, N. and Zeleznick, A.J. (1992) Vascular endothelial growth factor mRNA expression in the primate ovary. *Endocrinology* 131:254–260.

Rifkin, D.B., Moscatelli, D., Bizik, J., Quarto, N., Blei, F., Dennis, P., Flaumenhaft, R. and Mignatti, P. (1990) Growth factor control of extracellular proteolysis. *Cell Differ. Dev.* 32:313–318.

Rosenthal, R., Megyesi, J.F., Henzel, W.J., Ferrara, N. and Folkman, J. (1990) Conditioned medium from mouse sarcoma 180 cells contains vascular endothelial growth factor. *Growth Factors* 4:53–59.

Seetharam, L., Gotoh, N., Maru, Y., Neufeld, G., Yamaguchi, S. and Shibuya, M. (1995) A unique signal transduction pathway for the FLT tyrosine kinase, a receptor for vascular endothelial growth factor. *Oncogene* 10:135–137.

Sellke, F.W., Kagaya, Y., Johnson, R.G., Shafique, T., Schoen, F.J., Grossman, W. and Weintraub, R.M. (1993) Endothelial modulation of porcine corinary microcirculation perfused via immature vessels. *Am. J. Physiol.* 262:h1669–1695.

Senger, D.R., Galli, S.J., Dvorak, A.M., Perruzzi, C.A., Harvey, V.S. and Dvorak, H.F. (1983) Tumor cells secrete a vascular permeability factor that promotes accumulation of ascites fluid. *Science* 219:983–985.

Senger, D., Perruzzi, C.A., Feder, J. and Dvorak, H.F. (1986) A highly conserved vascular permeability factor secreted by a variety of human and rodent tumor cell lines. *Cancer Res.* 36:5269–5275.

Shalaby, F., Rossant, J., Yamaguchi, T.P., Gertenstein, M., Wu, X.-F., Breitman, M.L. and Schuh, A.C. (1995) Failure of blood island formation and vasculogenesis in Flk-1 deficient mice. *Nature* 376:62–66.

Shibuya, M., Yamaguchi, S., Yamane, A., Ikada, T., Tojo, T., Matsushima, H. and Sato, M. (1990) Nucleotide sequence and expression of a novel human receptor-type tyrosine kinase (*flt*) closely related to the *fms* family. *Oncogene* 8:519–527.

Shifren, J.L., Doldi, N., Ferrara, N., Mesiano, S. and Jaffe, R.B. (1994) In the human fetus, vascular endothelial growth factor (VEGF) is expressed in epithelial cells and myocytes, but not vascular endothelium: implications for mode of action. *J. Clin. Endocrinol. Metab.* 79:316–322.

Shima, D.T., Adamis, A.P., Ferrara, N., Yeo, K.-T., Yeo, T.K., Folkman, J. and D'Amore, P.A. (1995) Hypoxic induction of vascular endothelial cell growth factors in the retina: Identification and characterization of vascular endothelial growth factor (VEGF) as the sole mitogen. *Molec. Med.* 2:64–71.

Shweiki, D., Itin, A., Soffer, D. and Keshet, E. (1992) Vascular endothelial growth factor induced by hypoxia may mediate hypoxia-initiated angiogenesis. *Nature* 359:843–845.

Shweiki, D., Itin, A., Neufeld, G., Gitay-Goren, H. and Keshet, E. (1993) Patterns of expression of vascular endothelial growth factor (VEGF) and VEGF receptors in mice suggest a role in hormonally regulated angiogenesis. *J. Clin. Invest.* 91:2235–2243.

Sillman, F., Boyce, J. and Fruchter, R. (1981) The significance of atypical vessels and neovascularization in cervical neoplasia. *Am. J. Obstet. Gynecol.* 139:154–157.

Smith-McCune, K.S., Weidner, N. (1994) Demonstration and characterization of the angiogenic properties of cervical dysplasia. *Cancer Res.* 54:804–808.

Srenath, T., Matrisian, L.M., Stetler-Stevenson, W., Gattoni-Celli, S. and Pozzatti, R.O. (1992) Expression of matrix metalloproteinases in transformed rat cell lines of high and low metastatic potential. *Cancer Res.* 52:4942–4947.

Srivastava, A., Laidler, P., Davies, R., Horgan, K. and Hughes, L.E. (1988) The prognostic significance of tumor vascularity in intermediate-thickness (0.76–4.00 mm thick) skin melanoma. *Am. J. Pathol.* 133:419–423.

Takeshita, S., Zhung, L., Brogi, E., Kearney, M., Pu, L.-Q., Bunting, S., Ferrara, N., Symes, J.F. and Isner, J.M. (1994a) Therapeutic angiogenesis: A single intra-arterial bolus of vascular endothelial growth factor augments collateral vessel formation in a rabbit ischemic hindlimb model. *J. Clin. Invest.* 93:662–670.

Takeshita, S., Pu, L.-Q., Stein, L.A., Sniderman, A.D., Bunting, S., Ferrara, N., Isner, J.M. and Symes, J.F. (1994b) Intramuscular administration of vascular endothelial growth factor induces dose-dependent collateral artery augmentation in a rabbit model of chronic limb ischemia. *Circulation* 90:II228–234.

Takeshita, S., Zheng, L.P., Cheng, D., Riessen, R., Weir, L., Symes, J.F., Ferrara, N. and Isner, J.M. (1996) Therapeutic angiogenesis following arterial gene transfer of vascular endothelial in a rabbit model of hind-limb ischemia. *Lab. Invest.* In press.

Terman, B.I., Carrion, M.E., Kovacs, E., Rasmussen, B.A., Eddy, R.L. and Shows, T.B. (1991) Identification of a new endothelial cell growth factor receptor tyrosine kinase. *Oncogene* 6:519–524.

Terman, B.I., Vermazen, M.D., Carrion, M.E., Dimitrov, D., Armellino, D.C., Gospodarowicz, D. and Bohlen, P. (1992) Identification of the KDR tyrosine kinase as a receptor for vascular endothelial growth factor. *Biochem. Biophys. Res. Commun.* 34:1578–1586.

Tisher, E., Mitchell, R., Hartmann, T., Silva, M., Gospodarowicz, D., Fiddes, J. and Abraham, J. (1991) The human gene for vascular endothelial growth factor. *J. Biol. Chem.* 266:11947–11954.

Thompson, R.W. and D'Amore, P.A. (1990) Recruitment of growth and collateral circulation. *In*: G.B. Zelenock, L.G. D'Alecy, J.C. Fantone III, M. Shlafer, J.C. Stanley (eds): *Clinical ischemic syndromes: mechanisms and consequences of tissue injury.* C.V. Mosby, St. Louis, pp 117 137.

Toi, M., Hoshima, S., Takayanagi, T. and Tominaga, T. (1994) Association of vascular endothelial growth factor expression with tumor angiogenesis and with early relapse in primary breast cancer. *Jpn. J. Cancer Res.* 85:1045–1049.

Tolentino, M.J., Miller, J.W., Gragoudas, E.S., Chatzistefanou, K., Ferrara, N. and Adamis, A.P. (1996) VEGF is sufficient to produce iris neovascularization and neovascular glaucoma in an non-human primate, *Ophthalmology.* In press.

Topol, E.J. (1996) *Textbook of Interventional Cardiology*, W.B. Saunders Co., Philadelphia.

Unemori, E., Ferrara, N., Bauer, E.A. and Amento, E.P. (1992) Vascular endothelial growth factor induces interstitial collagenase expression in human endothelial cells. *J. Cell. Physiol.* 153:557–562.

Vaisman, N., Gospodarowicz, D. and Neufeld, G. (1990) Characterization of the receptors for vascular endothelial growth factor. *J. Biol. Chem.* 265:19461–19469.

Vartanian, R.K. and Weidner, N. (1994) Correlation of intramural endothelial cell proliferation with microvessel density (tumor angiogenesis) and tumor cell proliferation. *Am. J. Pathol.* 144:1188–1194.

Wakui, S., Furusato, M., Sasaki, H., Akiyama, A., Kinoshito, I., Asano, K., Tokuda, T., Aizawa, S. and Ushigome, S. (1992) Tumor angiogenesis in prostatic carcinoma with and without bone metastasis: a morphometric study. *J. Pathol.* 168:257–262.

Walder, C.E., Errett, C.J., Ogez, J., Heinshon, H., Bunting, S., Lindquist, P., Ferrara, N. and Thomas, G.R. (1996) Vascular endothelial growth factor (VEGF) improves blood flow and function in a chronic ischemic hind-limb model. *J. Cardiovasc. Pharmacol.* 27:91–98.

Waltenberger, J., Claesson-Welsh, L., Siegbahn, A., Shibuya, M. and Heldin, C.H. (1994) Different signal transduction properties of KDR and Flt1, two receptors for vascular endothelial growth factor. *J. Biol. Chem.* 269:26988–26995.

Wand, G.L. and Semenza, G.L. (1995) Purification and characterization of hypoxia-inducible factor-1. *J. Biol. Chem.* 270:1230–1237.

Wang, G.L., Jiang, B.H., Rue, E.A. and Semenza, G.L. (1995) Hypoxia-inducible factor-1 is a basic helix-loop helix-PAS heterodimer regulated by cellular O2 tension. *Proc. Natl. Acad. Sci. USA* 92:5510–5514.

Warren, R.S., Yuan, H., Matli, M.R., Gillett, N.A. and Ferrara, N. (1995) Regulation by vascular endothelial growth factor of human colon cancer tumorigenesis in a mouse model of experimental liver metastasis. *J. Clin. Invest.* 95:1789–1797.

Weidner, N., Semple, P., Welch, W. and Folkman, J. (1991) Tumor angiogenesis and metastasis. Correlation in invasive breast carcinoma. *N. Engl. J. Med.* 324:1–6.

Weidner, N., Folkman, J., Pozza, F., Bevilacqua, P., Allred, E.N., Moore, D.H., Meli, S. and Gasparini, G. (1992) Tumor angiogenesis: A new significant and independent prognostic indicator in early-stage breast carcinoma. *J. Natl. Cancer Inst.* 84:1875–1888.

Wizigmann-Voss, S., Breier, G., Risau, W. and Plate, K. (1994) Up-regulation of vascular endothelial growth factor and its receptors in von Hippel-Lindau disease-associated and sporadic hemangioblastoma. *Cancer Res.* 55:1358–1364.

Yang, R., Thomas, G.R., Bunting, S., Ko, A., Keyt, B., Ferrara, N., Ross, J. and Jin, H. (1996) Effects of hemodynamics and cardiac performance. *J. Cardiovasc. Pharmacol.,* 27:838–844.

Regulation of Angiogenesis
ed. by I.D. Goldberg & E.M. Rosen
© 1997 Birkhäuser Verlag Basel/Switzerland

Vascular permeability factor/vascular endothelial growth factor: A multifunctional angiogenic cytokine

L. F. Brown, M. Detmar, K. Claffey, J. A. Nagy, D. Feng, A. M. Dvorak and H. F. Dvorak

Departments of Pathology, Beth Israel Hospital and Harvard Medical School, Boston, Massachusetts 02159, USA

Introduction

Vascular permeability factor (VPF), variously known as a vascular endothelial growth factor (VEGF) and vasculotropin (VAS), is a multifunctional cytokine with important roles in vasculogenesis and both pathological and physiological angiogenesis. Originally described in the late 1970s as a tumor-secreted protein that potently increased microvascular permeability to plasma proteins, VPF/VEGF (as we will designate the molecule in this review) exerts a variety of effects on vascular endothelial cells which together promote the formation and growth of new blood vessels. Thus, in addition to rendering microvessels hyperpermeable, VPF/VEGF stimulates endothelial cells to migrate and divide and profoundly alters their pattern of gene expression.

Because of its central role in neoplasia, in many non-neoplastic disorders, and in normal adult physiology, angiogenesis (along with the vasculogenesis of development) is at present a "hot" scientific topic. Therefore, it is not surprising that VPF/VEGF has excited considerable interest and, since 1990, has been the subject of more than 300 publications. However, there are many angiogenic factors (some would say too many) and the current excitement over VPF/VEGF is attributable in large part to the selectivity of its action. Thus, whereas many cytokines and growth factors promote endothelial cell division, some (e. g., bFGF) with greater potency than VPF/VEGF, VPF/VEGF alone acts selectively on endothelial cells; this specificity is attributable to the fact that both of the high affinity receptor tyrosine kinases by which VPF/VEGF induces signal transduction are expressed primarily on vascular endothelial cells. As will be seen, however, more recent experiments have shown that VPF/VEGF receptors are expressed by cells other than vascular endothelium, suggesting that VPF/VEGF's activities may be less specific to endothelium than had originally been thought.

Given the explosion of recent interest in VPF/VEGF, any review article is likely to be dated by the time of its publication. This chapter reviews the literature published through mid-1995.

VPF/VEGF and its receptors: Structure and function

VPF/VEGF structure

Discovery of VPF/VEGF resulted from the immunohistochemical observation that tumor stroma was rich in fibrin (Dvorak et al., 1979 a, b). Fibrin is generated by the clotting and crosslinking of plasma fibrinogen, an elongated, cigar-shaped protein of large mass that normally is confined almost entirely within the blood vasculature. Therefore, the presence of fibrin *outside* of blood vessels in tumor stroma indicated that tumor microvessels were hyperpermeable to fibrinogen; based on this finding, we postulated that tumor cells secreted a factor that rendered microvessels leaky. In fact, this postulate proved to be correct; serum-free conditioned medium from a variety of cultured tumor cells contained an activity that rendered normal blood vessels hyperpermeable to albumin and other plasma proteins as could be demonstrated in the Miles assay (Dvorak et al., 1979b; Senger et al., 1983, 1986). Soon after it was established that the tumor-secreted vascular permeability factor was non-dialyseable and not affected by inhibitors of histamine or serotonin, common inflammatory mediators that also permeabilize microvessels (Dvorak et al., 1979b; Senger et al., 1983). Moreover, it is now agreed that tumor blood vessels are generally hyperpermeable to circulating macromolecules such as plasma proteins (reviewed in Dvorak et al., 1995).

VPF/VEGF was originally described in guinea pig tumor cells (Dvorak et al., 1979b; Senger et al., 1983). Based on reports of complete sequences of cDNAs encoding human, bovine, rat, guinea pig and mouse VPF/VEGF, it is a highly conserved, disulfide-bonded dimeric glycoprotein of Mr 34–45kD (Keck et al., 1989; Leung et al., 1989; Conn et al., 1990a; Berse et al., 1992; Claffey et al., 1992). Upon reduction, VPF/VEGF separates into major bands of 17–23 kD (Connolly et al., 1989b; Ferrara and Henzel, 1989; Gospodarowicz et al., 1989; Conn et al., 1990b; Senger et al., 1990; Myoken et al., 1991; Yeo et al., 1991b) and loses all of its biological activity (Sioussat et al., 1993; Potgens et al., 1994; Claffey et al., 1995). VPF/VEGF possesses a single N-glycosylation site but full biological activity is retained in the absence of glycosylation (Yeo et al., 1991b; Peretz et al., 1992; Claffey et al., 1995).

At least four different VPF/VEGF transcripts have been described in human cells, encoding polypeptides of 206, 189, 165, and 121 amino acids (corresponding murine proteins are one amino acid shorter) (Leung et al., 1989; Claffey et al., 1992). The several isoforms arise from the alternative splicing of a single gene (Tischer et al., 1991; Park et al., 1993). The human

VPF/VEGF gene has a coding region comprised of eight exons, the first of which encodes a hydrophobic leader sequence, typical of secreted proteins (Tischer et al., 1991). The 189 amino acid VPF/VEGF isomer is encoded by all eight exons. The shorter isoforms all include the first five exons as well as exon eight (Leung et al., 1989; Houck et al., 1991; Tischer et al., 1991). VPF/VEGF$_{165}$ lacks the residues encoded by exon 6 and VPF/VEGF$_{121}$ lacks those encoded by both exons 6 and 7. A recently described 145 amino acid form is predicted to lack the amino acids coded by exon 7 (Charnock-Jones et al., 1993). VPF/VEGF$_{206}$, the least common isoform, includes 17 additional codons beyond the 24 amino acid insertion present in VPF/VEGF$_{189}$ (Houck et al., 1991). Mapping of the 5' end of the human VPF/VEGF gene revealed a single major transcriptional start site 1038 bp upstream from the ATG initiation codon (Tischer et al., 1991). No TATA box promoter was identified but potential binding sites were found for Sp1, AP-1 and AP-2 transcriptional regulatory proteins.

As might be anticipated, the proteins encoded by the several different VPF/VEGF mRNA isoforms have distinct physical properties. The 189 and 165 amino acid forms contain 16 cysteines; 8 of these cysteines are present in exons 3 and 4 and are common to all VPF/VEGF isoforms, as is the single cysteine encoded by exon 8. Seven additional cysteines are encoded by exon 7 which is absent from VPF/VEGF$_{121}$. VPF/VEGF$_{121}$, and to an intermediate extent VPF/VEGF$_{165}$, are secreted in soluble form whereas VPF/VEGF$_{189}$ remains largely cell- or matrix-associated. This likely reflects the strong basic charge characteristic of the longer isoforms (Tischer et al., 1991; Houck et al., 1992; Park et al., 1993). Thus, bound VPF/VEGF$_{189}$ can be released in soluble form into the culture medium by heparin or heparan sulfate (Houck et al., 1992). VPF/VEGF$_{165}$, the predominant isoform in most systems that have been studied thus far, is partly cell bound and partly secreted and that portion which is cell bound can be released by heparin (Houck et al., 1992). The 189 and 165 amino acid forms of VPF/VEGF can also be released from their bound states as biologically active 34 kD dimers by plasmin.

VPF/VEGF$_{165}$ was originally purified to homogeneity on the basis of its affinity for heparin (Senger et al., 1983, 1990); however, the affinity of VPF/VEGF$_{165}$ for heparin is substantially lower than that of other typical heparin-binding growth factors such as basic fibroblast growth factor (Senger et al., 1983). VPF/VEGF$_{121}$ does not bind to heparin (Houck et al., 1992). Finally antibodies to peptides representing defined segments of VPF/VEGF have revealed additional features of VPF/VEGF structure (Sioussat et al., 1993). In particular, antibodies to peptides representing the N- and C-termini have proven more effective than antibodies to internal sequences at binding native VPF/VEGF in solution and in blocking its actions as an endothelial cell mitogen and vascular permeabilizer.

Despite significant physical differences, the several VPF/VEGF isoforms have been found to express identical biological activities, i.e., so far as is known, all isoforms render microvessels hyperpermeable and serve as endothelial cell mitogens.

VPF/VEGF receptors and signal transduction

Selective binding of VPF/VEGF to endothelial cells has been demonstrated both *in vitro* (Olander et al., 1991) and *in vivo* (Jakeman et al., 1992). Two high-affinity transmembrane tyrosine kinase receptors for VPF/VEGF are selectively expressed by vascular endothelium, flt-1 (fms-like tyrosine kinase) and KDR/flk-1 (fetal liver kinase 1); both have seven immunoglobulin-like extracellular domains and a kinase insert sequence (Matthews et al., 1991; Terman et al., 1991; De Vries et al., 1992; Terman et al., 1992; Millauer et al., 1993; Quinn et al., 1993). In various cultured endothelial cells, Scatchard analysis revealed that one of these (flt-1) has a K_D of $1-20$ pM and is present at a frequency of ~ 3000 copies per cells; the other receptor, KDR/flk-1, has a K_D of $50-770$ pM and is present at a frequency of $\sim 40\,000$ copies per cell (Vaisman et al., 1990; de Vries et al., 1992; Terman et al., 1992; Millauer et al., 1993; Waltenberger et al., 1994; Detmar et al., 1995b; Mustonen and Alitalo, 1995). On the other hand, Thieme et al. (1995) report that bovine retinal endothelial cells have a single class of high affinity ($K_D \sim 49$ pM) receptors for VPF/VEGF that is present in higher copy number (1×10^5/cell) than the VPF/VEGF receptors reported on other endothelial cells. Cell surface-associated, heparin-related proteoglycans may contribute to VPF/VEGF binding to endothelial cells and exogenous heparin increases such binding (Gitay-Goren et al., 1992). On the other hand, the serum protease inhibitor α_2 macroglobulin binds to and inactivates VPF/VEGF as it does several other growth factors (Soker et al., 1993). Platelet factor 4 also blocks VPF/VEGF mitogenic activity (Gengrinovitch et al., 1995).

A number of proteins are phosphorylated as a consequence of VPF/VEGF interaction with its endothelial cell receptors. In addition to autophosphorylation of both high affinity receptors, other phosphorylated proteins include: phospholipase Cγ, phosphatidylinositol 3-kinase; GAP, the Ras GTPase-activating protein; two GAP-associated proteins, p190 and p62; and the oncogenic adaptor protein NcK (Dougher-Vermazen et al., 1994; Waltenberger et al., 1994; Guo et al., 1995; Seetharam et al., 1995). Recent studies have shown that VPF/VEGF activates the mitogen-activated protein kinase (MAPK), a step blocked by the 16-kDa N-terminal fragment of human prolactin (D'Angelo et al., 1995).

In several instances, one or the other VPF/VEGF receptor or its mRNA has been found in cells other than endothelium; e.g. malignant melanoma and choriocarcinoma cell lines, human cytotrophoblasts (Gitay-Goren et al., 1993; Charnock-Jones et al., 1994; Cohen et al., 1995). Both flt-1 and KDR/flk-1 are expressed at increased levels on endothelial cells during development and in a variety of pathological and physiological circumstances associated with angiogenesis (see below).

VPF/VEGF homologues

VPF/VEGF shares low but significant sequence homology with platelet derived growth factor (PDGF), conserving all eight of the cysteines found in the PDGF B chain (Tischer et al., 1991). As in PDGF, the disulfide-linked monomer chains are likely aligned in anti-parallel fashion (Sioussat et al., 1993; Potgens et al., 1994).

VPF/VEGF shares more extensive homology (53%) with a dimeric, N-glycosylated protein that has been isolated from placenta, placenta growth factor (PlGF) (Maglione et al., 1991); interestingly, PlGF, unlike VPF/VEGF, is also expressed prominently by vascular endothelium (Barleon et al., 1994). Two PlGF isoforms have been identified as the result of alternative RNA splicing and only the larger of these ($PlGF_{152}$) binds to heparin (Park et al., 1994). Both PlGF isoforms bound with high affinity to one of the tyrosine kinase VPF/VEGF receptors (flt-1) but neither bound to the second VPF/VEGF receptor, KDR/flk-1. Not itself a vascular perme-abilizing factor, PlGF nonetheless does potentiate the vascular permeabiliz-ing activity of VPF/VEGF. Since PlGF bound avidly to flt-1 ($K_d \sim 200$ pM), the authors postulated that flt-1 was either not involved in the transduction of VPF/VEGF's signal for increasing vessel permeability or actually suppressed such activity; therefore, occupation of flt-1 by PlGF would presumably free up more VPF/VEGF to react with KDR/flk-1, the putative transducing receptor, or, at the very least, would not send an inhibitory signal by way of flt-1.

Recently, DiSalvo et al. (1995) have reported that cultured glioma cells secrete not only VPF/VEGF and PlGF *homodimers* but, in addition, VPF/VEGF-PlGF *heterodimers*. These heterodimers act as endothelial cell mitogens that are nearly as potent as VPF/VEGF homodimers; in contrast, PlGF homodimers were only mitogenic for endothelial cells at high, possibly non-physiological concentrations. Discovery of VPF/VEGF-PlGF heterodimers thus extends the similarity between VPF/VEGF and PDGF whose A and B chains form heterodimers as well as homodimers. On the other hand, these results leave PlGF homodimers without an obvious function.

Finally, it has recently been shown that a poxvirus *(orf)* encodes a poly-peptide with low level (16–27%) homology to VPF/VEGF (Lyttle et al., 1994). The *orf* virus infects sheep and goats and occasionally humans, resulting in dermal lesions that exhibit angio-proliferation, possibly attri-butable to expression of the VPF/VEGF homologue. Taken together, these results indicate that VPF/VEGF belongs to a larger family of proteins, other members of which likely remain to be discovered.

Biological activities of VPF/VEGF

VPF/VEGF is a multifunctional cytokine that exerts a number of direct effects on vascular endothelial cells. As already noted, VPF/VEGF induces a sequence of protein phosphorylations, beginning with the autophosphorylation of both of its receptor tyrosine kinases. $[Ca^{2+}]_i$ begins to increase after a lag phase of ~ 15 seconds following exposure of cultured endothelial cells to sub-pM concentrations of VPF/VEGF, making this measurement the most sensitive assay currently available for assessing VPF/VEGF activity (Brock et al., 1991). VPF/VEGF's effect on endothelial cell $[Ca^{2+}]_i$, resulting in up to a 4-fold increase, is roughly similar to that of other endothelial cell agonists such as thrombin and histamine; like these, VPF/VEGF also induces increased release of von Willebrand factor from endothelial cells. However, as noted above, VPF/VEGF acts by way of distinct endothelial cell receptors and its action is unaffected by inhibitors of thrombin and histamine (Brock et al., 1991). Like other agonists which increase $[Ca^{2+}]_i$, VPF/VEGF stimulates IP_3 accumulation (Brock et al., 1991) and is thought to act through a phosphoinositide-specific phospholipase C (Berridge, 1993).

VPF/VEGF was originally discovered because of its ability to increase the permeability of microvessels, primarily post-capillary venules and small veins, to circulating macromolecules. It is one of the most potent vascular permeabilizing agents known, acting at concentrations below 1 nM, and, as measured in the Miles assay, with a potency some 50000 times that of histamine (Dvorak et al., 1979b; Senger et al., 1983, 1993; Collins et al., 1993; Senger and Dvorak, 1993). VPF/VEGF acts to permeabilize a number of vascular beds, including those of the skin, subcutaneous tissues, peritoneal wall, mesentery and diaphragm (Dvorak et al., 1979b; Collins et al., 1993; Nagy et al., 1995b). The vascular permeabilizing effect occurs rapidly, becoming evident within several minutes of local VPF/VEGF injection (Dvorak et al., 1979b; Senger et al., 1983). Like its effects on endothelial cell $[Ca^{2+}]_i$, the vascular hyperpermeability resulting from a single pulse of VPF/VEGF is transient and reversible, persisting for less than 30 min and is not associated with detectable cell injury. Also, VPF/VEGF does not provoke mast cell degranulation or induce a significant inflammatory cell infiltrate. VPF/VEGF's vascular permeabilizing action is not blocked by antihistamines (Dvorak et al., 1979b; Senger et al., 1983) or by any of a variety of inhibitors of inflammation including those that inhibit platelet activating factor (DA Senger, unpublished data; Collins et al., 1993).

Recent studies have shown that VPF/VEGF increases microvascular permeability in tumor vessels primarily by enhancing the functional activity of a recently described organelle, the vesicular-vacuolar organelle or VVO (Kohn et al., 1992; Feng et al., 1995; Qu-Hong et al., 1995; Dvorak et al., 1996). VVOs are grape-like clusters of uncoated vesicles and vacuoles deployed at intervals in the cytoplasm of endothelial cells lining venules and small veins. The individual vesicles and vacuoles comprising VVOs are bounded by trilaminar unit membranes and are reminiscent of the plasmalemmal vesicles or caveolae that have been described in capillary endothelium (Palade, 1988). In contrast to these, however, VVOs are elaborate structures each of which, in a single section plane, is comprised of an average of 12 (up to as many as 20) individual vesicles or vacuoles which exhibit considerably larger average size and much greater size heterogeneity than caveolae. In serial sections, VVOs span the entire thickness of endothelial cell cytoplasm from the luminal to the abluminal plasma membranes (Feng et al., 1995). The individual vesicles and vacuoles that comprise VVOs interconnect with each other and with the endothelial cell plasma membranes by means of fenestrae or somata that may be open or are closed by thin diaphragms. When these stomata are open, macromolecular tracers are able to pass between interconnecting vesicles and vacuoles and tracer flow can be followed sequentially across the endothelial cell cytoplasm from the vascular lumen to the abluminal basal lamina; thus, VVOs provide a pathway whereby plasma and plasma proteins may exit the circulation and enter the tissues.

In normal adult tissues, extravasion of circulating macromolecules from venules and small veins is quite limited but, to the extent that it occurs, takes place largely (e.g. horseradish peroxidase) or entirely (ferritin, dextrans) by way of VVOs (Kohn et al., 1992; Dvorak et al., 1996). In tumor vessels, or in normal skin following local injection of VPF/VEGF, VVO function is substantially upregulated (Feng et al., 1995, 1996). The mechanisms by which VPF/VEGF upregulates VVO function remain to be determined but VPF/VEGF has been localized by immunocytochemistry to the plasma membrane and to VVOs of tumor-supplying vascular endothelium (Qu-Hong et al., 1995); together these data suggest that VPF/VEGF may act to open stomata, thereby facilitating tracer extravasation. VPF/VEGF may also increase the extravasation of macromolecular tracers by inducing endothelial cell gaps, especially with particulate tracers such as colloidal carbon (D Feng et al., unpublished data; Senger et al., 1993; Roberts and Palade, 1995).

VPF/VEGF-induced increases in $[Ca^{2+}]_i$ commonly serve as a second messenger in signal transduction; however, such increases may also modulate the effect of Ca^{2+}-sensitive enzymes such as nitric oxide synthase, leading to increased synthesis of NO by endothelial cells and resultant relaxation of muscular arteries (Ku et al., 1993). VPF/VEGF also causes endothelial cells from several different sources to assume an elongated shape, migrate and divide (Connolly et al., 1989a; Ferrara and Henzel, 1989; Gospodarowicz et al., 1989; Favard et al., 1991; Senger and Dvorak, 1993; Koch et al., 1994). More generally, VPF/VEGF alters the pattern of endothelial cell gene activation, upregulating the expression of both plasminogen activators, uPA and tPA, as well as the plasminogen activator inhibitor PAI-1 and the urokinase receptor (Pepper et al., 1991; Mandriota et al., 1995). VPF/VEGF also acts on endothelial cells to induce the expression of interstitial collagenase (Unemori et al., 1992) and tissue factor (Clauss et al., 1990) as well as to stimulate increased hexose transport, apparently by stimulating the expression of the GLUT-1 glucose transporter (Pekala et al., 1990). Finally, VPF/VEGF promotes angiogenesis in a variety of *in vitro* and *in vivo* assay systems (Connolly et al., 1989a; Leung et al., Wilting et al., 1992; Phillips et al., 1994). Interestingly, potent synergism has been found between VPF/VEGF and bFGF in inducing endothelial cells to form tubes in collagen gels in culture, a process widely regarded as an *in vitro* surrogate for angiogenesis (Pepper et al., 1992; Goto et al., 1993).

While VPF/VEGF is primarily noted for its activation of endothelial cells, a number of reports have indicated that it also acts on other cellular targets, consistent with the finding that one or both VPF/VEGF receptors are, at least under some circumstances, represented on cells other than vascular endothelium. Thus, VPF/VEGF has been shown to stimulate mouse and human mononuclear phagocytes to migrate and express tissue factor, though at VPF/VEGF concentrations approximately 10-fold higher than are required to induce endothelial cell proliferation (Clauss et al., 1990; Shen et al., 1993). A single, as yet unconfirmed report indicates that VPF/VEGF is mitogenic for IL-2-stimulated lymphocytes (Praloran et al., 1991). Other reports indicate that VPF/VEGF stimulates both the migration and the differentiation of fetal bovine osteoblasts (Midy and Plouet, 1994) and of human retinal pigment epithelial cells (Guerrin et al., 1995). However, in at least some non-endothelial cells that express VPF/VEGF receptors, VPF/VEGF has been found to induce no measurable response (Gitay-Goren et al., 1993).

Regulation of VPF/VEGF and VPF/VEGF receptor expression

Though constitutively expressed by many tumor cells and transformed cell lines, VPF/VEGF expression is also subject to regulation by both protein kinase C and cAMP-dependent kinase pathways (Claffey et al., 1992; Finkenzeller et al., 1992; Garrido et al., 1993; Fischer et al., 1995; Sato et al., 1995) and by mechanisms that involve alterations in both mRNA transcription and stability (Finkenzeller et al., 1992; Goldberg and Schneider, 1994; Harada et al., 1994; Ladoux and Frelin, 1994; Minchenko et al., 1994b; Li et al., 1995b). Other factors that can regulate VPF/VEGF expression include the degree of cell differentiation; local concentrations of oxygen, glucose, and serum; cytokines; hormones; prostaglandins; modulators of protein kinase C; calcium influx; the electron transport chain; depolarizing agents; angiotensin II; stimulators of adenylate cyclase; nitric oxide; and expression levels of certain oncogenes (Frank et al., 1995; Levy et al., 1995a; Sato et al., 1995; Shweiki et al., 1995; Williams et al., 1995a, b).

Oxygen concentration

Many cell lines express increased amounts of VPF/VEGF when subjected to hypoxic culture (Koos and Olson, 1991; Shweiki et al., 1992; Ladoux and Frelin, 1993; Goldberg and Schneider, 1994; Minchenko et al., 1994a; Hata et al., 1995). Further upregulation of VPF/VEGF mRNA expression has also been demonstrated *in situ* in hypoxic zones of tumors immediately adjacent to areas of tumor necrosis (Plate et al., 1992; Brown et al., 1993a, b) and in a variety of normal tissues exposed to hypobaric hypoxia, functional anemia or localized ischemia as in the heart following coronary artery occlusion (Shweiki et al., 1992; Ladoux and Frelin, 1993; Banai et al., 1994b; Goldberg and Schneider, 1994; Iizuka et al., 1994; Ladoux and Frelin, 1994; Minchenko et al., 1994a; Li et al., 1995; Shweiki et al., 1995). Both transcriptional and post-transcriptional mechanisms are apparently involved (Goldberg and Schneider, 1994; Ladoux and Frelin, 1994; Finkenzeller et al., 1995; Levy et al., 1995b). Like another oxygen-sensitive protein, erythropoietin, VPF/VEGF expression is significantly upregulated by cobalt chloride, presumably because cobalt replaces iron in the protoporphyrin ring of a heme-containing protein oxygen sensor (Goldberg and Schneider, 1994; Ladoux and Frelin, 1994; Minchenko et al., 1994a). Transfection assays have been used to identify hypoxia/cobalt responsive enhancer elements in both the 5′ and 3′ regions flanking the coding portion of the VPF/VEGF gene. The 3′ enhancer includes a 12 base pair sequence with homology to sequences in the erythropoietin hypoxia-responsive enhancer (Minchenko et al., 1994b). A 5′ flanking enhancer is contained in a 100 bp fragment located about 800 bp upstream from the start site and does not contain significant homology with the 5′ erythropoietin enhancer (Minchenko et al., 1994b). Recently, a 28 base pair element in the 5′ promoter has been identified that mediates hypoxia-induced transcription of VPF/VEGF; this element has sequence and protein binding similarities to the hypoxia-inducible factor 1 binding site within the erythropoietin 3′ enhancer (Levy et al., 1995b; Liu et al., 1995).

Hypoxia may also be acting to induce VPF/VEGF expression by a second, indirect mechanism, that of inducing increased concentrations of adenosine as the result of enforced anaerobic metabolism. Adenosine and adenosine agonists up-regulate VPF/VEGF expression in cultured porcine brain microvascular endothelium and U937 cells and adenosine receptor antagonists abolish this effect (Hashimoto et al., 1994a; Fischer et al., 1995).

Hormonal regulation

VPF/VEGF is expressed by many cells that make steroid hormones (adrenal cortex, corpus luteum, Leydig cells) and that are themselves under hormonal regulation (Berse et al., 1992; Cullinan-Bove and Koos, 1993; Shweiki et al., 1993). VPF/VEGF expression is clearly subject to hormonal regulation in the cycling uterus and ovary. In the former, both estrogen and progesterone upregulate VPF/VEGF mRNA (Cullinan-Bove and Koos, 1993). Luteotrophin, a known activator of adenylyl cyclase, upregulates VPF/VEGF transcription in bovine ovarian granulosa cells (Garrido et al., 1993). Recently, Koos (1995) has shown that follicular maturation induced in rats by gonadotropin stimulation led to a markedly increased expression of VPF/VEGF mRNA immediately prior to the large increase in the permeability of peri-follicular capillaries that is characteristic of ovulation. This study also postulated a role for VPF/VEGF in follicle rupture.

Of additional interest is the fact that VPF/VEGF is expressed by the peptide hormone-producing cells of the thyroid follicle and its production in culture is upregulated by a number of agents including insulin, dibutyryl cAMP, thyroid stimulating hormone, and the IgG of Graves' disease (Sato et al., 1995).

Regulation by cytokines and other agents

Different cell types are stimulated to upregulate VPF/VEGF expression by different cytokines *in vitro* but it is difficult to determine which of these has physiological or pathological significance. Moreover, cytokines that increase VPF/VEGF expression in one cell or tissue may not do so in others. Thus, PDGF-BB has been reported to stimulate VPF/VEGF mRNA production in smooth muscle cells (Brogi et al., 1994; Stavri et al., 1995a; Williams et al., 1995b) and NIH 3T3 cells (Finkenzeller et al., 1992) but not in monocyte-like U937 cells (Dolecki and Connolly, 1991). TGF-β has been reported to upregulate VPF/VEGF mRNA in smooth muscle cells (Brogi et al., 1994) and to a small extent in U937 cells (Dolecki and Connolly, 1991) and both VPF/VEGF mRNA and protein in mouse AKR-2B fibroblastic cells and A549 lung adenocarcinoma cells (Pertovaara et al., 1994). IL-1β also upregulates VPF/VEGF mRNA and protein in rat aortic smooth muscle cells (Li et al., 1995b) and bFGF upregulates VPF/VEGF in rabbit vascular smooth muscle cells, an effect that is synergistic with hypoxia (Stavri et al., 1995b).

TGF-α, and less potently EGF, upregulate VPF/VEGF mRNA and protein production in early passage skin keratinocytes in a dose-dependent manner; this effect is mediated through EGF receptors that are also expressed by these cells, in that antibodies to the EGF receptor blocked VPF/VEGF induction (Detmar et al., 1994). This finding is of particular interest since keratinocytes themselves express TGF-α, suggesting an autocrine mechanism for stimulating VPF/VEGF production in the epidermis. This finding probably also has more general relevance to cancer in that many epithelial tumors that express VPF/VEGF also secrete TGF-α

and possess EGF surface receptors (Derynck et al., 1987). Also of interest, cytokines likely to be expressed in healing wounds and in inflammation such as keratinocyte growth factor (KGF), TNF-α, and in some hands TGF-β, may also upregulate VPF/VEGF mRNA expression in keratinocytes (Frank et al., 1995).

Prostaglandins E1 and E2 both induce VPF/VEGF mRNA and protein in cultured osteoblasts (Harada et al., 1994). Both of these prostaglandins are potent stimulators of bone production, and recent studies have localized increased VPF/VEGF production, vascular hyperpermeability and angiogenesis to sites of mineralization in the long bones of newborn rodents (R. Fava, personal communication). Conversely, dexamethasone blocked the VPF/VEGF-inducing effect of PGE 2 (Harada et al., 1994); this finding may be or practical interest in that steroids inhibit bone formation and can induce avascular necrosis of bone. Others have also found that dexamethasone downregulates or prevents cytokine (but not hypoxic) upregulation of VPF/VEGF mRNA expression (Finkenzeller et al., 1995; Rak et al., 1995 a). Finally, there is evidence to suggest that nitric oxide acts to decrease VPF/VEGF production in the lung (Tuder et al., 1995).

Regulation by oncogenes

VPF/VEGF itself is not an oncogene (Ferrara et al., 1993) but at least two oncogenes *(src, ras)* and a tumor suppressor gene *(p53)* have been implicated in VPF/VEGF regulation.

Mukhopadhyay et al. (1995) have shown that hypoxia of the type leading to upregulated expression of VPF/VEGF mRNA is accompanied by pp60[c-src] autophosphorylation and increased kinase activity. Moreover, transfection of cells with mutant v-*src* leads to a constitutive rise in VPF/VEGF expression without hypoxia, whereas transfection with a dominant-negative *src* mutant strongly inhibits hypoxia-induced stimulation of VPF/VEGF expression. In addition, hypoxic upregulation of VPF/VEGF expression was strikingly decreased in c-*src* (−) cells and disruption of *raf-1* a downstream step in the Src pathway, also interfered with hypoxia- or Src-induced VPF/VEGF expression.

Very recently, two groups have shown that cells transfected with mutant *ras* oncogenes express increased levels of VPF/VEGF mRNA and protein (Grugel et al., 1995; Rak et al., 1995 a, b). Moreover, genetic disruption of mutant K-*ras* in human colon carcinoma cells led to decreased VPF/VEGF expression; pharmacological disruption of Ras activity by a protein farnesyltransferase inhibitor had a similar effect (Rak et al., 1995 a, b). Finally, recent studies have shown that cells transfected with mutant forms of *p53* induce increased expression of VPF/VEGF mRNA and potentiate that induced by stimulation with phorbol esters (Kieser et al., 1994). The inescapable conclusion from studies such as these is that oncogenes and tumor suppressor genes act not only by increasing tumor cell proliferation but also by increasing VPF/VEGF expression and therefore by potentiating angiogenesis.

Expression of VPF/VEGF and its receptors in neoplasia

Although not itself an oncogene, VPF/VEGF is substantially overexpressed at both the mRNA and protein levels in a high percentage of malignant animal and human tumors and in many immortalized and transformed cell lines as determined by Northern analysis, *in situ* hybridization, Western blotting, "sandwich" or other immunoassay, and immunohistochemistry (Dvorak et al., 1979a, b; Senger et al., 1983; 1986; Berse et al., 1992; Plate et al., 1992; Brown et al., 1993a, b, 1995; Charnock-Jones et al., 1993; Sugihara et al., 1994). Expression of VPF/VEGF and its receptors by various autochthonous human tumors is listed in Table 1. Clearly, VPF/VEGF and both of its receptors are overexpressed in the great majority of clinically important human cancers. However, important exceptions exist even among tumors that are clearly able to induce new blood vessels, suggesting that at least some tumors induce angiogenesis by VPF/VEGF-independent mechanisms. Interestingly, at least one type of cancer, malignant melanoma, overexpresses VPF/VEGF when it arises in the eye (Vinores et al., 1995), but apparently not when it occurs in the skin (Brown et al., unpublished data). Finally, at least some malignant tumors overexpress PlGF as well as VPF/VEGF (Weindel et al., 1994).

Overexpression of VPF/VEGF by cancer cells is likely to have clinical significance. Thus, VPF/VEGF expression generally correlates with vascular hyperpermeability to macromolecules, resulting in alterations to the extracellular matrix that favor angiogenesis and the generation of new stroma (see below). Moreover, VPF/VEGF mRNA levels correlate with vascular density in both cervical and breast carcinomas (Toi et al., 1994; Guidi et al., 1995). At least one report has also linked high levels of VPF/VEGF expression to early relapse in primary breast cancer (Toi et al., 1994). Although not as yet exhaustively investigated, it is clear that tumor metastases also overexpress VPF/VEGF much as do the primary tumors from which they arose (Strugar et al., 1994; Brown et al., 1995a).

As noted earlier, many cell lines and tissues can be induced to upregulate VPF/VEGF expression by hypoxia. Further upregulation can also be observed in tumors in hypoxic zones immediately adjacent to areas of tumor necrosis and may involve stromal cells as well as tumor cells (Plate et al., 1992; Shweiki et al., 1992; Brown et al., 1993a, b, 1995a, b; Hlatky et al., 1994). It is important to remember, however, that malignant cells are generally constitutive overproducers of VPF/VEGF under normoxic conditions and that hypoxia serves only to upregulate further an already high level of VPF/VEGF expression (Dvorak et al., 1979b; Senger et al., 1983, 1986).

Table 1. Expression of VPF/VEGF and its receptors in various autochthonous malignant human tumors

Malignant tumors that overexpress VPF/VEGF and its receptors

Tumor	Reference
Adenocarcinomas arising in the gastrointestinal tract (stomach, duodenum, colon)	(Brown et al., 1993b)
Adenocarcinomas of the pancreas	(Brown et al., 1993b)
Most carcinomas of the breast, including DCIS	(Brown et al., 1995b)
Carcinoma of the cervix, including preinvasive lesions	(Guidi et al., 1995)
Carcinoma of the bladder	(Brown et al., 1993a)
Renal cell carcinoma of the kidney	(Brown et al., 1993a; Sato et al., 1994; Takahashi et al., 1994)
Ovarian carcinoma	(Olson et al., 1994)
Hemangioblastoma, sporadic and associated with von Hippel-Lindau disease	(Berkman et al., 1993; Morii et al.,1993; Brown et al., 1995b; Wizigmann-Voos et al., 1995)
Astrocytoma, particularly glioblastoma multiforme	(Plate et al., 1992, 1993, 1994; Senger and Dvorak, 1993; Weindel et al., 1994; Samoto et al., 1995)
Meningioma	(Berkman et al., 1993)
Choriocarcinoma	(Charnock-Jones et al., 1994)
Malignant melanomas arising in the eye	(Vinores et al., 1995)

Malignant tumors that do NOT overexpress VPF/VEGF and its receptors

Papillary carcinoma of the kidney	(Brown et al., 1993a)
Lobular carcinoma of the breast	(Brown et al., 1995a)
Non-astrocytic glioma	(Berkman et al., 1993)
Adenocarcinoma of the prostate	(Brown et al., unpublished data)
Malignant melanomas arising in the skin	(Brown et al., unpublished data)

AIDS-associated Kaposi's sarcoma and angiosarcomas represent something of an anomaly. A number of reports indicate that cells derived from such lesions overexpress VPF/VEGF mRNA in culture (Weindel et al., 1992). Recent experience from our laboratory, however, has demonstrated detectable but relatively weak expression of VPF/VEGF mRNA and protein in most Kaposi's lesions and in angiosarcomas as they occur *in vivo*, though expression and immunostaining levels may be quite high in foci adjacent to areas of ulceration or necrosis, i.e. in areas that might be expected to be hypoxic (Brown et al., 1995b). Paradoxically, endothelial cells in the stromal vessels supplying these tumors strongly overexpress both KDR and flt-1 mRNAs, suggesting that, though not abundantly expressed, VPF/VEGF may nonetheless be an important regulator of the edema and angiogenesis seen in these tumors. Moreover, the tumor cells of both

Kaposi's and angiosarcomas themselves strongly expressed KDR (but not flt-1), suggesting that VPF/VEGF could also have an autocrine effect on tumor cells. Like the more malignant vascular tumors, benign capillary hemangiomas expressed little VPF/VEGF mRNA; however, a majority strongly expressed both KDR and flt-1 mRNAs, in contrast to the high level expression of KDR alone that was observed in Kaposi's sarcoma and angiosarcoma (Brown et al., 1995 b).

VPF/VEGF protein may be detected by immunohistochemistry in tumors that overexpress VPF/VEGF mRNA. More surprising, however, is our finding (Dvorak et al., 1991 b), now confirmed by several other laboratories (Plate et al., 1992, 1994), that large amounts of immunoreactive VPF/VEGF protein are recognized on tumor microvascular endothelium by antibodies with specificity against the VPF/VEGF N-terminus. While this observation makes sense in that these blood vessels are though to be the target of tumor-secreted VPF/VEGF, it is nonetheless a surprising finding in that immunohistochemistry is a comparatively insensitive method for detecting the minute concentrations of VPF/VEGF (pM range) that are sufficient to activate microvascular endothelium. Therefore, relatively large amounts of VPF/VEGF, in excess of those required to stimulate endothelium, are apparently associated with tumor microvascular endothelium, and microvessel-associated VPF/VEGF has been proposed as a possible target for tumor imaging and/or therapy (Dvorak et al., 1991 a, b). Of interest is the observation that tumor microvessel-associated VPF/VEGF has been localized to the endothelial cell plasma membrane as well as to intra-endothelial cell VVOs by electron microscopic immunocyto-chemistry (Qu- Hong et al., 1995).

Accumulation of protein-rich fluid in the extravascular space (i.e. an exudate) and angiogenesis are important consequences of VPF/VEGF overexpression by tumor cells. Fluid accumulation results from VPF/VEGF-induced leakage of plasma from hyperpermeable microvessels but is also favored by the fact that tumors in general lack lymphatic vessels and hence are unable to drain extravasated proteinaceous fluid effectively. When, as is commonly the case, tumors arise in confined tissue spaces, fluid accumulation leads to increased interstitial pressure and many human and animal tumors exhibit internal pressures in excess of 100 mm of Hg (Boucher et al., 1991). Fluid accumulation and increased interstitial pressures are particularly devastating in brain tumors because the bony skull permits little room for edematous expansion; as a result, increased intracranial pressure may rapidly cause death from brain herniation (Berkman et al., 1993; Criscuolo, 1993; Weindel et al., 1994). In tumors arising in or metastasizing to body cavities, on the other hand, substantial amounts of fluid (many liters in man, many milliliters in mice) can accumulate before causing the patient or mouse symptoms such as impaired breathing or electrolyte imbalance due to loss of plasma proteins (Nagy et al., 1989, 1991, 1993, 1995 a, b, c). Immunoreactive and bioactive

VPF/VEGF protein is commonly found in these malignant fluid accumulations, sometimes in very high concentrations (Nagy et al., 1993, 1995b; Yeo et al., 1993; Olson et al., 1994; Weindel et al., 1994).

Benign tumors have been less carefully studied than their malignant counterparts for VPF/VEGF expression. Non-dysplastic adenomatous polyps arising sporadically in the colon or as the result of familial adenomatous polyposis do not express VPF/VEGF above levels found in normal colonic mucosa (Brown et al., 1993b). At least in the colon, therefore, overexpression of VPF/VEGF is not an early feature of malignant progression. Islet cell tumors of the pancreas were found to express only modestly greater levels of VPF/VEGF mRNA than normal islet cells and expression of neither VPF/VEGF receptor was detected in the course of tumorigenesis (Christofori et al., 1995; Kuroda et al., 1995). Pituitary adenomas (Berkman et al., 1993), hemangiomas (Brown et al., 1995b) also rarely overexpress VPF/VEGF whereas leiomyomas of the uterus are reported to do so (Harrison-Woolrych et al., 1995). In contrast, VPF/VEGF is commonly overexpressed by ductal *in situ* carcinomas arising in the breast and in pre-malignant cervical neoplasia; such tumors exhibit both increased VPF/VEGF expression and angiogenesis that is evident even prior to malignant invasion (Brown et al., 1995a; Guidi et al., 1995).

Of considerably interest is the fact that tumors that overexpress VPF/VEGF are, so far without exception, supplied by microvessels whose lining endothelial cells overexpress both VPF/VEGF receptors, flt-1 ad KDR/flk-1 (Plate et al., 1992, 1993, 1994; Brown et al., 1993a, b, 1995a; Guidi et al., 1995; Hatva et al., 1995; Wizigmann-Voos et al., 1995). The mechanisms relating VPF/VEGF receptor overexpression to that of their ligand are largely unknown. However, there is evidence that chronic exposure to high levels of VPF/VEGF enhances the expression of both VPF/VEGF receptors on microvascular endothelial cells in vivo, as demonstrated in transgenic mice that overexpress VPF/VEGF in the skin under the control of a keratin promoter (Detmar et al., 1995a).

Expression of VPF/VEGF and its receptors in the angiogenesis associated with non-neoplastic pathology

VPF/VEGF and one or both of its endothelial cell receptors are also overexpressed in a number of pathological entities that involve angiogenesis but that are not associated with neoplasia. These include wound healing, delayed hypersensitivity reactions, rheumatoid arthritis, psoriasis, diabetic and other retinopathies and various examples of tissue ischemia and infarction. In all of these examples, overexpression of VPF/VEGF and its receptors has been accompanied by increased microvascular hyperpermeability, leakage of plasma proteins, and deposition of an extravascular fibrin gel.

Taken together, these findings suggest that the angiogenesis that develops in these entities shares a common pathogenesis with tumor angiogenesis (Dvorak, 1986; Dvorak et al., 1995). However, there is at least one basic difference that distinguishes non-neoplastic from tumor angiogenesis; namely, that in the latter VPF/VEGF overexpression is a constitutive property of neoplastic cells (though admittedly subject to further potential regulation by environmental factors).

The healing of split-thickness skin wounds provides a good illustration of regulated overexpression of VPF/VEGF (Brown et al., 1992). Within 24 h of wounding rodent skin, VPF/VEGF mRNA expression increases dramatically in epidermal keratinocytes at the wound edge and in residual hair follicles in the wound base. VPF/VEGF overexpression reaches a peak at 2–3 days and persists at an elevated level for about 1 week, the time required for granulation tissue to form and migrating keratinocytes to cover the wound defect. Certain mononuclear cells infiltrating the dermis, most likely a subpopulation of macrophages, also overexpresses VPF/VEGF. Increased expression of at least one VPF/VEGF receptor mRNA, flt-1 (flk-1 was not investigated), has been documented in the endothelial cells lining the new blood vessels that form as part of the developing granulation tissue that accompanies wound healing (Peters et al., 1993).

In contrast to normal mice, congenitally diabetic *db/db* mice have elevated endogenous levels of VPF/VEGF mRNA in their flank skin which increases transiently after wounding. However, the rise in VPF/VEGF is not sustained and, as granulation tissue forms, VPF/VEGF expression plummets to barely detectable levels, thus associating decreased VPF/VEGF expression with defective wound healing in diabetes (Frank et al., 1995).

VPF/VEGF expression is also strikingly upregulated in cardiac myocytes following ischemia or infarction. Adult cardiac myocytes normally express detectable levels of VPF/VEGF mRNA and these levels increase dramatically within 30 min of transient (5–10 min) ischemic perfusion; overexpression of VPF/VEGF mRNA can also be induced by hypoxia, both *in vivo* and in culture (Ladoux and Frelin, 1993; Banai et al., 1994b; Hashimoto et al., 1994b; Minchenko et al., 1994a). VPF/VEGF, flt-1, and flk-1 mRNAs were all substantially elevated as early as 1 h after mouse coronary artery ligation and remained elevated for at least one week thereafter, returning to normal only by 6 weeks (Li et al. 1995a). This is not surprising in that healing of myocardial infarctions is an example of organ-specific wound healing.

VPF/VEGF and VPF/VEGF receptor mRNAs are also increased in the lung following *ex vivo* perfusion under hypoxic conditions or following chronic exposure to simulated high altitudes (Tuder et al., 1995). Hypoxia has been found to stimulate VPF/VEGF mRNA expression in a variety of cultured cells (see above) and therefore may be anticipated in any tissue subjected to ischemia.

Overexpression of VPF/VEGF and its receptors has also been demonstrated in the retinas of patients or animals subjected to various forms of ocular ischemia, e.g., diabetes mellitus, central vein occlusion, prematurity, etc. (Adamis et al., 1993, 1994; Aiello et al., 1994; Miller et al., 1994; Pe'er et al., 1995; Pierce et al., 1995). Several types and layers of retinal cells have been found capable of producing VPF/VEGF, and, in various disease states, upregulated VPF/VEGF mRNA expression is localized to those layers of retina expected to be ischemic, e.g., the outer retinal layer in chronic retinal detachment (Pe'er et al., 1995). Immunoassays of aqueous or vitreous humour samples have provided a strong correlation between VPF/VEGF content and proliferative retinopathy in patients with diabetes, central retinal vein occlusion or prematurity; on the other hand, VPF/VEGF levels were very low in the eyes of patients affected by disorders that did not involve vascular proliferation (e.g., diabetes without retinopathy) and in the eyes of patients with quiescent vascular proliferations. Moreover, the retinas of diabetic rats have been found to display greatly increased amounts of immunohistochemically reactive VPF/VEGF in all layers, and retinal vessels were hyperpermeable and stained with VPF/VEGF (Murata et al., 1995), much as has been described for tumor blood vessels (Dvorak et al., 1991 b).

Conversely, shut-off of VPF/VEGF production as occurs in the neuroglia of newborn rats exposed to high oxygen results in apoptotic loss of retinal vessels, suggesting that, among other functions, VPF/VEGF behaves as a survival factor for newly formed blood vessels (Alon et al., 1995). These findings have obvious implications for an understanding of the retinopathy of prematurity.

Overexpression of VPF/VEGF mRNA by keratinocytes and infiltrating mononuclear cells is also observed in delayed hypersensitivity reactions of both the tuberculin and contact allergy type as elicited in rodent and human skin (Brown et al., 1995 e). In addition, both flt-1 and KDR mRNAs were overexpressed in the microvascular endothelium supplying human contact reactions. Similar findings have been observed in rheumatoid arthritis (Fava et al., 1994; Koch et al., 1994), often classified as an example of cell-mediated hypersensitivity of the delayed type. Both VPF/VEGF mRNA and protein are overexpressed by the synovial macrophages lining joints with active inflammation, and both flt-1 and KDR are overexpressed in microvessels supplying the pannus. Consistent with these findings, elicited (but not native) peritoneal macrophages express readily detectable levels of VPF/VEGF mRNA. In addition, substantial amounts of immunoreacive and bioactive VPF/VEGF are present in rheumatoid joint fluids; these amounts are comparable to those found in tumor ascites (Yeo et al., 1993; Fava et al., 1994; Koch et al., 1994). By way of comparison, joint fluids taken from patients with other forms of arthritis (e.g. osteoarthritis) contain no or significantly lower amounts of VPF/VEGF.

Overexpression of VPF/VEGF and its receptors has also been found in a variety of inflammatory skin disorders. Prominent among these are such important entities as psoriasis, bullous pemphigoid, dermatitis herpetiformis and erythema multiforme (Detmar et al., 1994; Brown et al., 1995 d). In all of these instances, VPF/VEGF mRNA expression is strikingly elevated in epidermal keratinocytes and both flt-1 and KDR are overexpressed in dermal microvessels. Also, papillary dermal edema is characteristic of these disorders and high levels of immunoreactive and bioactive VPF/VEGF may be detected in blister fluid.

It is likely that further work will implicate VPF/VEGF in the pathogenesis of still other disease states. Recently, for example, high levels of VPF/VEGF protein have been found in the ascites fluid of ovarian hyper-stimulation syndrome (McClure et al., 1994). Studies cited earlier have implicated VPF/VEGF, as well as its receptors, in the angiogenesis of Graves' disease (Sato et al., 1995). Also, a vascular permeabilizing activity has been described in supernatants of lymphocytes isolated from patients with minimal change nephrotic syndrome (Tomizawa et al., 1990; Heslan et al., 1991; Maruyama et al., 1994); however, it remains to be determined whether this factor is responsible for the observed kidney disease and/or is related to VPF/VEGF.

Expression of VPF/VEGF and its receptors in vasculogenesis and in the physiological angiogenesis of adulthood

VPF/VEGF is widely and abundantly expressed in many tissues (with the notable exception of vascular endothelium) during fetal development and has been convincingly implicated in primary vasculogenesis of both birds and mammals, including humans (Breier et al., 1992; Jakeman et al., 1993; Shifren et al., 1994; Flamme et al., 1995). During murine development, the entire 7.5 day endoderm expresses large amounts of VPF/VEGF (Breier et al., 1992). During mouse brain development, VPF/VEGF transcripts are abundantly expressed in the ventricular neuroectoderm of embryonic and postnatal brain when vasculogenesis is active and fall to low levels in the adult brain when vascular endothelial cells are no longer proliferating rapidly (Breier et al., 1992). Other studies have called attention to a close relationship between the expression of VPF/VEGF and its receptors in renal ontogenesis (Simon et al., 1995). That expression of VPF/VEGF in an appropriate temporal and spatial sequence is critical for normal development has been demonstrated by studies in which recom-binant VPF/VEGF was injected into early stage quail embryos, inducing a variety of vascular malformations (Drake and Little, 1995). Similarly, mRNAs encoding VPF/VEGF have been readily detected in all human midgestation (16–22 week) fetuses and were particularly abundant in lung, kidney and spleen (Shifren et al., 1994). VPF/VEGF, like PlGF, is abundantly expressed in human placental trophoblasts and stromal cells (Sharkey et al., 1993; Jackson et al., 1994).

The mRNAs of both VPF/VEGF receptors are also expressed strongly in fetal vascular endothelium and its precursors in vasculogenesis and at later stages of fetal development (Eichmann et al., 1993; Millauer et al., 1993; Dumont et al., 1995; Mustonen and Alitalo, 1995; Shalaby et al., 1995). flk-1, the mouse equivalent of human KDR, is expressed in mesodermal yolk-sac blood island progenitors as early as day 7.0 postcoitum in the mouse and may also be a marker for the hemangioblast, the common precursor to both endothelial cell and hematopoetic lineages (Shalaby et al., 1995). Mice rendered null for *flk-1* by homologous recombination die *in utero* between days 8.5 and 9.5 and exhibit defects in the development of both blood and endothelial cells, e.g. failure to form yolk-sac blood islands or organized blood vessels (Shalaby et al., 1995). flt-1 mRNA is also expressed strongly during vasculogenesis (Peters et al., 1993; Fong et al., 1995) and, while not essential for endothelial cell development, is required for organization of the embryonic vasculature. In *flt-1* null mice, blood islands form abnormally and, although endothelial cells differentiate in both embryonic and extra-embryonic regions, their assembly into vascular channels is abnormal and embryos die *in utero* at the mid-somite stage (Fong et al., 1995).

Large amounts of VPF/VEGF mRNA and protein are also found in adult female reproductive tissues in association with the hormonally-regulated angiogenesis that takes place in the ovary and endometrium at specific stages of the menstrual cycle and in pregnancy. Thus, VPF/VEGF mRNA and protein are both strongly expressed in the ovary by granulosa and theca lutein cells late in follicle development, and, subsequent to ovulation, by granulosa and theca lutein cells of the developing corpus luteum (Phillips et al., 1990; Ravindranath et al., 1992; Shweiki et al., 1993; Kamat et al., 1995). Subsequently, strong VPF/VEGF expression persists in the corpus luteum of pregnancy but declines rapidly in the corpus luteum of menstruation and in the subsequent corpus albicans (Phillips et al., 1990; Ravindranath et al., 1992; Kamat et al., 1995).

VPF/VEGF (predominantly the two smaller isoforms) is also expressed in the uterus in a cycle-dependent manner (Charnock-Jones et al., 1993; Cullinan-Bove and Koos, 1993; Shweiki et al., 1993; Li et al., 1994; Harrison-Woolrych et al., 1995). During the proliferative phase, strong VPF/VEGF mRNA expression was found in a subset of stromal cells whereas glandular cells exhibited a lower level of expression (Charnock-Jones et al., 1993). During the secretory phase, this pattern became altered as expression by stromal cells disappeared while expression of message by glandular epithelium increased, culminating in very intense glandular hybridization at the time of menstruation. A somewhat different distribution was found when immunohistochemistry was used to detect VPF/VEGF protein (Li et al., 1994). Thus, staining of glandular cells was detected throughout the cycle whereas stromal cell staining was observed only during the proliferative phase; in addition, staining of vascular endothelial cells, reminiscent of that described in tumors (Dvorak et al., 1991b; Plate et al., 1992) and rheumatoid arthritis (Fava et al., 1994), was observed in late proliferative phase. VPF/VEGF is also strongly expressed by decidual cells in the pregnant uterus (Shweiki et al., 1993).

Expression of VPF/VEGF in normal adult tissues without angiogenesis

In the adult, strong expression of VPF/VEGF mRNA can be detected in several tissues in the absence of angiogenesis, particularly kidney, lung, adrenal gland, and heart (Berse et al., 1992). Thus, VPF/VEGF transcripts are readily detected by *in situ* hybridization in the visceral glomerular epithelium (podocytes) of normal adult kidneys, in epithelial cells of lung alveoli and adrenal cortex, and in cardiac myocytes (Berse et al., 1992; Breier et al., 1992; Monacci et al., 1993; Li et al., 1995a). Of possible relevance, two of these tissues (renal glomeruli, adrenal cortex) are associated with fenestrated endothelium, and a relation between VPF/VEGF expression and endothelial fenestration has been noted (Berse et al., 1992; Breier et al., 1992; Sioussat et al., 1993; Roberts and Palade, 1995; Simon et al., 1995).

Lower levels of VPF/VEGF mRNA are also detected in several other normal adult tissues including spleen and the epithelium of liver, gastric mucosa and breast (Berse et al., 1992). In addition, VPF/VEGF protein can be detected by immunohistochemistry in the smooth muscle cells of normal arteries and of uterus. Angiogenesis is not evident in any of these tissues. Therefore, it is assumed that in these situations VPF/VEGF plays a maintenance role, perhaps regulating vascular permeability (especially important in the kidney) and may have other as yet poorly understood roles in vascular homeostasis. One explanation for the absence of angiogenesis in association with abundant VPF/VEGF expression in normal adult tissues is the fact that neither VPF/VEGF receptor may be overexpressed by endothelial cells. Indeed, regulation of the expression of VPF/VEGF receptor mRNAs by vascular endothelium is likely to be as important for inducing angiogenesis as upregulation of VPF/VEGF itself.

As noted above, recent studies have shown that VPF/VEGF mRNA is expressed in readily detectable amounts by a variety of normal adult cells that secrete both peptide and steroid hormones. In addition to examples discussed earlier in the context of hormonally regulated physiological angiogenesis, a number of hormone-secreting cells express significant amounts of VPF/VEGF without inducing local angiogenesis; these include islet cells of the pancreas (Christofori et al., 1995; Kuroda et al., 1995), follicular cells of the ovary (Ravindranath et al., 1992; Kamat et al., 1995), Leydig cells of the testis (Shweiki et al., 1993), and the glandular epithelia of normal adult prostate and seminal vesicles (Brown et al., 1995c). The function of the VPF/VEGF secreted by these cells and its target (whether local and/or systemic) are as yet undetermined. However, in the case of prostate and seminal vesicles VPF/VEGF is likely synthesized for export as high levels of immunoreactive VPF/VEGF protein (greater than 200 pM) are present in semen. The role of VPF/VEGF in semen is unclear but it could have a role in sperm development or maintenance or a role in the female genital tract, as, for example, in implantation.

Potential diagnostic and therapeutic applications of VPF/VEGF

The significance of VPF/VEGF in a number of important disease states and physiological processes suggests that reagents that react with this cytokine with high specificity and affinity could have practical use as diagnostic, imaging or therapeutic agents. For example, VPF/VEGF has been found to be present in high concentrations in malignant effusions, in the joint fluids of patients with rheumatoid arthritis, and in skin blisters, semen and ocular fluids. Therefore measurements of this cytokine in body fluids might be useful in diagnosis as well as in an assessment of prognosis. Yeo et al. (1993) developed a sensitive and specific sandwich-type immunoassay for VPF/VEGF that made use of anti-peptide antibodies to the molecule's N- and C-termini (Sioussat et al., 1993). This assay detected elevated levels of VPF/VEGF in animal and human malignant effusions and in several instances improved upon cytologic diagnosis; lower but still elevated levels of VPF/VEGF were detected in inflammatory effusions. However, VPF/VEGF was not detected in the plasma of patients bearing large tumor loads. Japanese workers recently developed a similar assay that used a single antibody to recombinant VPF/VEGF for both capture and detection steps (Kondo et al., 1994); they found significantly higher levels of VPF/VEGF in the sera of tumor-bearing mice and cancer patients and suggested that their success was attributable to the use of antibody that was able to detect circulating VPF/VEGF that had undergone partial proteolysis or had been complexed with α_2 macroglobulin. Recently, nuclease-resistant oligonucleotide probes have been developed which recognize VPF/VEGF with high affinity and specificity; these have potential as possible alternatives to antibodies in both diagnostic and therapeutic applications (Green et al., 1995).

The importance of VPF/VEGF in initiating tumor and other forms of angiogenesis had led to attempts at therapeutic interventions that might inhibit or stimulate the angiogenic process. In the case of tumors, inhibition of VPF/VEGF activity would be expected to reduce angiogenesis and therefore tumor growth. Indeed, antibodies that neutralize VPF/VEGF reduce tumor growth, presumably by interfering with tumor angiogenesis (Kim et al., 1992; Kondo et al., 1993). Moreover, use of a dominant-negative VPF/VEGF receptor mutant (Millauer et al., 1994) has resulted in inhibition of tumor growth in nude mice, apparently by binding

VPF/VEGF in an ineffectual complex and thereby preventing tumor angio-genesis.

In other circumstances, such as those associated with ischemia of the heart, limbs or skin flaps, increased local levels of VPF/VEGF might be expected to have use in increasing new blood vessel development. In fact, a number of studies have suggested that systemically or locally administered VPF/VEGF will be useful for this purpose (Hom and Assefa, 1992; Banai et al., 1994a, b; Bauters et al., 1994; Takashita et al., 1994a, b; Stepnick et al., 1995). VPF/VEGF may also be useful in stimulating endothelial cell growth following vascular injury (Callow et al., 1994; Asahara et al., 1995; Burke et al., 1995). Finally, VPF/VEGF shows promise as a therapeutic agent for stimulating the development of a collateral circulation at sites of vascular insufficiency, e.g. ischemic limbs (Bauters et al., 1994, 1995a, b; Takeshita et al., 1994b).

VPF/VEGF, microvascular hyperpermeability and the mechanisms of angiogenesis

The work reviewed here has established a clear correlation between the increased expression of VPF/VEGF and its receptors with vasculogenesis as it occurs in development and with angiogenesis as it occurs in tumors, in a variety of non-neoplastic pathologies, and in certain physiological states such as ovarian and uterine cycling and pregnancy. Although the permea-bility of the developing vasculature to circulating macromolecules has not been carefully investigated, microvascular hyperpermeability to plasma proteins and other circulating macromolecules has been described in all of the states of angiogenesis where it has been studied, i.e. solid and ascites tumors, wound healing, various types of inflammation including delayed hypersensitivity, rheumatoid arthritis, psoriasis, etc., and in physiological circumstances such as corpus luteum formation in the ovary, implantation of the fertilized ovum in the uterus, etc. This consistent association between overexpression of VPF/VEGF and its receptors, microvascular hyperper-meability and angiogenesis suggests a causal relationship. Therefore an additional question must be considered. How could VPF/VEGF-induced microvascular hyperpermeability lead to angiogenesis?

A likely hypothesis continues to be that which we put forth more than a decade ago with reference to tumor angiogenesis, that plasma proteins extravasate from leaky tumor blood vessels to form a new, provisional extravascular matrix which permits and indeed favors the inward migration of endothelial cells and fibroblasts (Dvorak et al., 1983a; Dvorak, 1986). Acting in concert, immigrating endothelial cells form new blood vessels and fibroblasts synthesize and secrete the matrix proteins, proteoglycans and glycosaminoglycans that comprise mature tumor stroma (Dvorak, 1986; 1979a, 1983a, 1984a; Yeo et al., 1991a; Yeo and Dvorak, 1995). It is noteworthy that this mature stroma differs significantly in composition from that found in the normal adult, resembling more closely, in some respects, the stroma of embryonic development. In fact, it is reasonable to postulate that the extracellular matrix of normal adult tissues is *anti-angiogenic*, i.e. the normal adult matrix limits or even prohibits mesenchymal cell immigration and, for angiogenesis to occur, must be replaced by an altered, embryonic-like matrix. This is not an unreasonable premise in that division and migration of endothelial cells or fibroblasts are inappropriate in a mature tissue in which new blood vessel formation is contraindicated. Therefore, if new blood vessels and supporting connective tissue are to develop in adult tissues, as, for example, in response to the increased metabolic needs of a growing tumor, native tissue stroma must first be replaced with a matrix that, like that of the developing organism, supports mesenchymal cell migration.

There is now considerable evidence that supports or is at least consistent with this hypothesis. Plasma proteins leaking from hyperpermeable blood vessels that supply animal tumors do in fact alter dramatically the extracellular matrix of normal adult tissues and provide a new matrix that favors mesenchymal cell migration. Within a few hours of tumor cell transplant, preexisting microvessels become measurably hyperpermeable to macromolecules and tumor-supplying microvessels eventually become 4 – 10-fold more permeable than their normal tissue counterparts (Dvorak et al., 1979b, 1984b, 1988). Moreover, plasma fibrinogen that extravasates at tumor sites rapidly clots to form crosslinked fibrin (Dvorak et al., 1979 a, b, 1984b, 1985; Brown et al., 1988a, b). This last finding indicates, moreover, that the plasma protein extravasation that tumors induce is relatively non-discriminating in that, to form an extravascular clot, a variety of plasma clotting proteins are required and must extravasate along with fibrinogen; these include at least clotting factors V, VII, X, XIII and prothrombin (Dvorak et al., 1981, 1983b). The fibrin gel that is deposited in tumor stroma is itself modulated by plasmin which is generated locally from another leaked plasma protein, plasminogen, by the action of tumor- (and perhaps endothelial cell-) secreted plasminogen activators (Dvorak et al., 1979b, 1983a, 1984b, Danø et al., 1985; Dvorak, 1986; Pepper et al., 1991). Plasmin also activates matrix metalloproteases which are capable of digesting other matrix elements. The extent of fibrin deposition and its persistence over time vary extensively among different tumors, as does the amount of mature stroma generated by different tumors. Apparently these differences depend on quantitative balances among vascular hyperpermeability, clotting and fibrinolysis that are unique to individual tumors (Dvorak et al., 1984b). These balances are further complicated because, as noted earlier, VPF/VEGF acts to stimulate the synthesis of proteins that affect both coagulation (tissue factor) (Clauss et al., 1990) and fibrinolysis (uPA, tPA, and PAI-1) (Pepper et al., 1991).

Supportive evidence for our hypothesis also comes from *in vitro* studies which have provided direct evidence that crosslinked fibrin of the type deposited in tumors provides a matrix that supports and favors cell migration (Lanir et al., 1988; Brown et al., 1993c). Furthermore, implantation of crosslinked fibrin in guinea pigs in the absence of tumor cells, induces the progressive ingrowth of new blood vessels and fibroblasts which together generate vascularized stroma of the type found in many tumors (Dvorak et al., 1979a, 1983a, 1984a, 1987; Dvorak, 1986). Fibrin exerts its pro-angiogenic effect at least in part by providing a favorable surface for cell adhesion and migration; presumably this is mediated, again at least in part, by the arg-gly-asp (RGD) sequences present in both its alpha and gamma chains. Other circulating RGD-containing plasma proteins such as fibronectin and vitronectin probably also extravasate from leaky blood vessels at sites of tumor growth and there contribute to the generation of an extracellular matrix that favors angiogenesis and new stroma formation.

Evidence that this hypothesis, originally proposed for tumor angiogenesis, has more general applicability comes from the analysis of angiogenesis as it occurs in non-neoplastic settings. We (Dvorak, 1986; Brown et al., 1992; Dvorak et al., 1995) and others (Haddow, 1972) have called attention to the many similarities between tumor stroma generation and wound healing. Other examples in which overexpression of VPF/VEGF and its receptors, plasma protein extravasation, and deposition of extravascular fibrin accompany angiogenesis were discussed earlier and include rheumatoid arthritis, psoriasis and delayed hypersensitivity. Together these findings suggest that angiogenesis, when it occurs in adult tissues in response to diverse stimuli, is mediated by the triggering of a common pathway, initiated by over-expression of both VPF/VEGF and its receptors, by heightened vascular permeability, and by an altered extracellular matrix that favors endothelial cell and fibroblast migration.

Summary and conclusions

VPF/VEGF is a multifunctional cytokine that contributes to angiogenesis by both direct and indirect mechanisms. On the one hand, VPF/VEGF stimulates the endothelial cells lining nearby microvessels to proliferate, to migrate and to alter their pattern of gene expression. On the other hand, VPF/VEGF renders these same microvascular endothelial cells hyper-permeable so that they spill plasma proteins into the extravascular space, leading to profound alterations in the extracellular matrix that favor angio-genesis. These same principles apply in tumors, in several examples of non-neoplastic pathology, and in physiological processes that involve angio-genesis and new stroma generation. In all of these examples, microvascular hyperpermeability and the introduction of a provisional, plasma-derived matrix precede and accompany the onset of endothelial cell division and new blood vessel formation. It would seem, therefore, that tumors have made use of fundamental pathways that developed in multicellular organisms for purposes of tissue defense, renewal and repair.

VPF/VEGF, therefore, has taught us something new about angiogenesis; namely, that vascular hyperpermeability and consequent plasma protein extravasation are important – perhaps essential – elements in its generation. However, this finding raises a paradox. While VPF/VEGF induces vascular hyperpermeability, other potent angiogenic factors apparently do not, at least in sub-toxic concentrations that are more than sufficient to induce angiogenesis (Connolly et al., 1989a). Nonetheless, wherever angiogenesis has been studied, the newly generated vessels have been found to be hyperpermeable. How, therefore, do angiogenic factors other than VPF/VEGF lead to the formation of new and leaky blood vessels? We do not as yet have a complete answer to this question. One possibility is that at least some angiogenic factors mediate their effect by inducing or stimulating VPF/VEGF expression. In fact, there are already clear example of this. A number of putative angiogenic factors including small molecules (e.g. prostaglandins, adenosine) as well as many cytokines (e.g. TGF-α, bFGF, TGF-β, TNF-α, KGF, PDGF) have all been shown to upregulate VPF/VEGF expression. Further studies that elucidate the crosstalk among various angiogenic factors are likely to contribute significantly to a better understanding of the mechanisms by which new blood vessels are formed in health and in disease.

Acknowledgements

This work was supported by USPHS NIH grants CA-50453, CA-58845, HL-54519 and AI-33372, by the BIH Pathology Foundation, Inc., and under terms of a contract from the National Foundation for Cancer Research.

References

Adamis, A., Shima, D., Yeo, K.-T., Yeo, T.-K., Brown, K., Berse, B., D'Amore, P. and Folkman, J. (1993) Synthesis and secretion of vascular permeability factor/vascular endothelial growth factor by human retinal pigment epithelial cells. *Biochem. Biophys. Res. Comm.* 193:631–638.

Adamis, A.P., Miller, J.W., Bernal, M.T., D'Amico, D.J., Folkman, J., Yeo, T.K. and Yeo, K.T. (1994) Increased vascular endothelial growth factor levels in the vitreous of eyes with proliferative diabetic retinopathy. *Am. J. Ophthalmol.* 118:445–450.

Aiello, L.P., Avery, R.L., Arrigg, P.G., Keyt, B.A., Jampel, H.D., Shah, S.T., Pasquale, L.R., Thieme, H., Iwamoto, M.A., Park, J.E., Nguyen, M.S., Aiello, L.M., Ferrara, N. (1994) Vascular endothelial growth factor in ocular fluid of patients with diabetic retinopathy and other retinal disorders. *N. Engl. J. Med.* 331:1480–1487.

Alon, T., Hemo, I., Itin, A., Pe'er, J., Stone, J. and Keshet, E. (1995) Vascular endothelial growth factor acts as a survival factor for newly formed retinal vessels and has implications for retinopathy of prematurity. *Nature Medicine* 1:1024–1028.

Asahara, T., Bauters, C., Pastore, C., Kearney, M., Rossow, S., Bunting, S., Ferrara, N., Symes, J.F. and Isner, J.M. (1995) Local delivery of vascular endothelial growth factor accelerates reendothelialization and attenuates intimal hyperplasia in balloon-injured rat carotid artery. *Circulation* 91:2793–2801.

Banai, S., Jaklitsch, M.T., Shou, M., Lazarous, D.F., Scheinowitz, M., Biro, S., Epstein, S.E. and Unger, E.F. (1994a) Angiogenic-induced enhancement of collateral blood flow to ischemic myocardium by vascular endothelial growth factor in dogs. *Circulation* 89:2183–2189.

Banai, S.D., Shweiki, A., Pinson, M., Chandra, G., Lazarovici, G. and Keshet, E. (1994b) Upregulation of vascular endothelial growth factor expression induced by myocardial ischaemia: implications for coronary angiogenesis. *Cardiovasc. Res.* 28:1176–1179.

Barleon, B., Hauser, S., Schollmann,, C., Weindel, K., Marme, D., Yayon, A. and Weich, H.A. (1994) Differential expression of the two VEGF receptors flt and KDR in placenta and vascular endothelial cells. *J. Cell Biochem.* 54:56–66.

Bauters, C., Asahara, T., Zheng, L.P., Takeshita, S., Bunting, S., Ferrara, N., Symes, J.F. and Isner, J.M. (1994) Physiological assessment of augmented vascularity induced by VEGF in ischemic rabbit hindlimb. *Am. J. Physiol.* 267:h1263–1271.

Bauters, C., Asahara, T., Zheng, L.P., Takeshita, S., Bunting, S., Ferrara, N., Symes, J.F. and Isner, J.M. (1995a) Recovery of disturbed endothelium-dependent flow in the collateral-perfused rabbit ischemic hindlimb after administration of vascular endothelial growth factor. *Circulation* 91:2802–2809.

Bauters, C., Asahara, T., Zheng, L.P., Takeshita, S., Bunting, S., Ferrara, N., Symes, J.F. and Isner, J.M. (1995b) Site-specific therapeutic angiogenesis after systemic administration of vascular endothelial growth factor. *J. Vasc. Surg.* 21:314–324.

Berkman, R.A., Merrill, M.J., Reinhold, W.C., Monacci, W.T., Saxena, A., Clark, W.C., Robertson, J.T., Ali, I.U. and Oldfield, E.H. (1993) Expression of the vascular permeability factor/vascular endothelial growth factor gene in central nervous system neoplasms. *J. Clin. Invest.* 91:153–159.

Berridge, M.J. (1993) Inositol trisphosphate and calcium signalling. *Nature* 361:315–325.

Berse, B., Brown, L.F., Van De Water, L., Dvorak, H.F. and Senger, D.R. (1992) Vascular permeability factor (vascular endothelial growth factor) gene is expressed differentially in normal tissues, macrophages, and tumors. *Mol. Biol. Cell* 3:211–220.

Boucher, Y., Kirkwood, J.M., Opacic, D., Desantis, M. and Jain, R.K. (1991) Interstitial hypertension in superficial metastatic melanomas in humans. *Cancer Res.* 51:6691–6694.

Breier, G., Albrecht, U., Sterrer, S. and Risau, W. (1992) Expression of vascular endothelial growth factor during embryonic angiogenesis and endothelial cell differentiation. *Development* 114:521–532.

Brock, T.A., Dvorak, H.F., Senger, D.R. (1991) Tumor-secreted vascular permeability factor increases cytosolic Ca^{2+} and von Willebrand factor release in human endothelial cells. *Am. J. Pathol.* 138:213–221.

Brogi, E., Wu, T., Namiki, A. and Isner, J.M. (1994) Indirect angiogenic cytokines upregulate VEGF and bFGF gene expression in vascular smooth muscle cells, whereas hypoxia upregulates VEGF expression only. *Circulation* 90:649–652.

Brown, L.F., Asch, B., Harvey, V.S., Buchinsky, B. and Dvorak, H.F. (1988$20a) Fibrinogen influx and accumulation of cross-linked fibrin in mouse carcinomas. *Cancer Res.* 48:1920–1925.

Brown, L.F., Van De Water, L., Harvey, V.S. and Dvorak, H.F. (1988$20b) Fibrinogen influx and accumulation of cross-linked fibrin in healing wounds and in tumor stroma. *Am. J. Pathol.* 130:455–465.

Brown, L.F., Yeo, K.-T., Berse, B., Yeo, T.-K., Senger, D.R., Dvorak, H.F. and Van De Water, L. (1992) Expression of vascular permeability factor (vascular endothelial growth factor) by epidermal keratinocytes during wound healing. *J. Exp. Med.* 176:1375–1379.

Brown, L.F., Berse, B., Jackman, R.W., Tognazzi, K., Manseau, E.J., Dvorak, H.F. and Senger, D.R. (1993a). Increased expression of vascular permeability factor (vascular endothelial growth factor) and its receptors in kidney and bladder carcinomas. *Am. J. Pathol.* 143:1255–1262.

Brown, L.F., Berse, B., Jackman, R.W., Tognazzi, K., Manseau, E.J., Senger, D.R. and Dvorak, H.F. (1993b) Expression of vascular permeability factor (vascular endothelial growth factor) and its receptors in adenocarcinomas of the gastrointestinal tract. *Cancer Res.* 53:4727–4735.

Brown, L.F., Lanir, N., McDonagh, J., Czarnecki, K., Estrella, P., Dvorak, A.M. and Dvorak, H.F. (1993c) Fibroblast migration in fibrin gel matrices. *Am. J. Pathol.* 142:273–283.

Brown, L., Berse, B., Jackman, R., Tognazzi, K., Guidi, A., Dvorak, H., Senger, D., Connolly, J. and Schnitt, S. (1995) Expression of vascular permeability factor (vascular endothelial growth factor) and its receptors in breast cancer. *Hum. Pathol.* 26:86–91.

Brown, L., Tognazzi, K., Dvorak, H. and Harrist, T. (1996) Strong expression of KDR, a vascular permeability factor/vascular endothelial growth factor receptor in AIDS-associated Kaposi's sarcoma and cutaneous angiosarcoma. *Am. J. Pathol.* 148:1065–1074.

Brown, L., Yeo, K.-T., Berse, B., Morgentaler, A., Dvorak, H. and Rosen, S. (1995c) Vascular permeability factor (vascular endothelial growth factor) is strongly expressed in the normal male genital tract and is present in substantial quantities in semen. *J. Urology* 154:576–579.

Brown, L.F., Harrist, T.J., Yeo, K.T., Stahle, B.M., Jackman, R.W., Berse, B., Tognazzi, K., Dvorak, H.F. and Detmar, M. (1995d) Increased expression of vascular permeability factor (vascular endothelial growth factor) in bullous pemphigoid, dermatitis herpetiformis, and erythema multiforme. *J. Invest. Dermatol.* 104:744–749.

Brown, L.F., Olbricht, S.M., Berse, B., Jackman, R.W., Matsueda, G., Tognazzi, K.A., Manseau, E.J., Dvorak, H.F. and Van De Water, L. (1995e) Overexpression of vascular permeability factor (VPF/VEGF) and its endothelial cell receptors in delayed hypersensitivity skin reactions. *J. Immunol.* 154:2801–2807.

Burke, P.A., Lehmann, B.K. and Powell, J.S. (1995) Vascular endothelial growth factor causes endothelial proliferation after vascular injury. *Biochem. Biophys. Res. Commun.* 207:348–354.

Callow, A.D., Choi, E.T., Trachtenberg, J.D., Stevens, S.L., Connolly, D.T., Rodi, C. and Ryan, U.S. (1994) Vascular permeability factor accelerates endothelial regrowth following balloon angioplasty. *Growth Factors* 10:223–228.

Charnock-Jones, D., Sharkey, A., Rajput, W.J., Burch, D., Schofield, J.P., Fountain, S.A., Boocock, C.A. and Smith, S.K. (1993) Identification and localization of alternately spliced mRNAs for vascular endothelial growth factor in human uterus and estrogen regulation in endometrial carcinoma cell lines. *Biol. Reprod.* 48:1120–1128.

Charnock-Jones, D., Sharkey, A., Boocock, C., Ahmed, A., Plevin, R., Ferrara, N. and Smith, S. (1994) Vascular endothelial growth factor receptor localization and activation in human trophoblast and choriocarcinoma cells. *Biol. Reprod.* 51:524–530.

Christofori, G., Naik, P. and Hanahan, D. (1995) Vascular endothelial growth factor and its receptors, *flt-1* and *flk-1*, are expressed in normal pancreatic islets and throughout islet cell tumorigenesis. *Mol. Endocrinol.* 9:1760–1770.

Claffey, K.P., Wilkison, W.O. and Spiegelman, B.M. (1992) Vascular endothelial growth factor. Regulation by cell differentiation and activated second messenger pathways. *J. Biol. Chem.* 267:16317–16322.

Claffey, K.P., Senger, D.R. and Spiegelman, B.M. (1995) Structural requirements for dimerization, glycosylation, secretion, and biological function of VPF/VEGF. *Biochim. Biophys. Acta* 1246:1–9.

Clauss, M., Gerlach, M., Gerlach, H., Brett, J., Wang, F., Familletti, P.C., Pan, Y.-C.E., Olander, J.V., Connolly, D.T. and Stern, D. (1990) Vascular permeability factor: A tumor-derived polypeptide that induces endothelial cell and monocyte procoagulant activity, and promotes monocyte migration. *J. Exp. Med.* 172:1535–1545.

Cohen, T., Gitay, G.H., Sharon, R., Shibuya, M., Halaban, R., Levi, B.Z. and Neufeld, G. (1995) VEGF121, a vascular endothelial growth factor (VEGF) isoform lacking heparin binding ability, requires cell-surface heparan sulfates for efficient binding to the VEGF receptors of human melanoma cells. *J. Biol. Chem.* 270:11322–11326.

Collins, P.D., Connolly, D.T. and Williams, T.J. (1993) Characterization of the increase in vascular permeability induced by vascular permeability factor in vivo. *Br. J. Pharmacol.* 109:195–199.

Conn, G., Bayne, M.L., Soderman, D.D., Kwok, P.W., Sullivan, K.W., Palisi, T.M., Hope, D.A. and Thomas, K.A. (1990a) Amino acid and cDNA sequences of a vascular endothelial cell mitogen that is homologous to platelet-derived growth factor. *Proc. Natl. Acad. Sci. USA* 87:2628–2632.

Conn, G., Soderman, D.D., Schaeffer, M.-T., Wile, M., Hatcher, V.B. and Thomas, K.A. (1990b) Purification of a glycoprotein vascular endothelial cell mitogen from a rat glioma-derived cell line. *Proc. Natl. Acad. Sci. USA* 87:1323–1327.

Connolly, D.T., Heuvelman, D.M., Nelson, R., Olander, J.V., Eppley, B.L., Delfino, J.J., Siegel, N.R., Leimgruber, R.M. and Feder, J. (1989a) Tumor vascular permeability factor stimulates endothelial cell growth and angiogenesis. *J. Clin. Invest.* 84:1470–1478.

Connolly, D.T., Olander, J.V., Heuvelman, D., Nelson, R., Monsell, R., Siegel, N., Haymore, B.L., Leimgruber, R. and Feder, J. 1989b) Human vascular permeability factor. Isolation from U937 cells. *J. Biol. Chem.* 264:20017–20024.

Criscuolo, G.R. (1993) The genesis of peritumoral vasogenic brain edema and tumor cysts: a hypothetical role for tumor-derived vascular permeability factor. *Yale J. Biol. Med.* 66:277–314.

Cullinan-Bove, K. and Koos, R. (1993) Vascular endothelial growth factor/vascular permeability factor expression in the rat uterus: Rapid stimulation by estrogen correlates with estrogen-induced increases in uterine capillary permeability and growth. *Endocrinology* 133: 829–837.

D'Angelo, G., Struman, I., Martial, J. and Weiner, R.I. (1995) Activation of mitogen-activated protein kinases by vascular endothelial growth factor and basic fibroblast growth factor in capillary endothelial cells is inhibited by the antiangiogenic factor 16-kDa N-terminal fragment of prolactin. *Proc. Natl. Acad. Sci. USA* 92:6374–6378.

Danø, K., Andreasen, P.A., Grøndahl-Hansen, J., Kristensen, P., Nielsen, L.S. and Skriver, L. (1985) Plasminogen activators, tissue degradation, and cancer. *Adv. Cancer Res.* 44: 139–266.

de Vries, C., Escobedo, J.A., Ueno, H., Houck, K., Ferrara, N. and Williams, L.T. (1992) The *fms*-like tyrosine kinase, a receptor for vascular endothelial growth factor. *Science* 255:989–991.

Derynck, R., Goeddel, D.V., Ullrich, A., Gutterman, J.U., Williams, R.D., Bringman, T.S. and Berger, W.H. (1987) Synthesis of messenger RNAs for transforming growth factors alpha and beta and the epidermal growth factor receptor by human tumors. *Cancer Res.* 47:707–712.

Detmar, M., Brown, L., Claffey, K., Yeo, K.-T., Kocher, O., Jackman, R., Berse, B. and Dvorak, H. (1994) Overexpression of vascular permeability factor and its receptors *in psoriasis*. *J. Exp. Med.* 180:1141–1146.

Detmar, M., Brown, L., Dvorak, H. and Claffey, K. (1995a) Overexpression of VPF/VEGF in the skin of transgenic mice leads to increased skin vascularization. *J. Invest. Derm.* 105:448 (abstract).

Detmar, M., Yeo, K.T., Nagy, J.A., Van De Water, L., Brown, L.F., Berse, B., Elicker, B.M., Ledbetter, S. and Dvorak, H.F. (1995b) Keratinocyte-derived vascular permeability factor (vascular endothelial growth factor) is a potent mitogen for dermal microvascular endothelial cells. *J. Invest. Dermatol.* 105:44–50.

DiSalvo, J., Bayne, M.L., Conn, G., Kwok, P.W., Trivedi, P.G., Soderman, D.D., Palisi, T.M., Sullivan, K.A. and Thomas, K.A. (1995) Purification and characterization of a naturally occurring vascular endothelial growth factor placenta growth factor heterodimer. *J. Biol. Chem.* 270:7717–7723.

Dolecki, G.J. and Connolly, D.T. (1991) Effects of a variety of cytokines and inducing agents on vascular permeability factor mRNA levels in U937 cells. *Biochem. Biophys. Res. Commun.* 180:572–578.

Dougher-Vermazen, M., Hulmes, J.D., Bohlen, P. and Terman, B.I. (1994) Biological activity and phosphorylation sites of the bacterially expressed cytosolic domain of the KDR VEGF-receptor. *Biochem. Biophys. Res. Commun.* 205:728–738.

Drake, C. and Little, C. (1995) Exogenous vascular endothelial growth factor induces malformed and hyperfused vessels during embryonic neovascularization. *Proc. Natl. Acad. Sci. USA* 92:7657–7661.

Dumont, D.J., Fong, G.H., Puri, M.C., Gradwohl, G., Alitalo, K. and Breitman, M.L. (1995) Vascularization of the mouse embryo: a study of flk-1, tek, tie, and vascular endothelial growth factor expression during development. *Dev. Dyn.* 203:80–92.

Dvorak, H.F. (1986) Tumors: Wounds that do not heal. Similarities between tumor stroma generation and wound healing. *N. Engl. J. Med.* 315:1650–1659.

Dvorak, H.F., Dvorak, A.M., Manseau, E.J., Wiberg, L. and Churchill, W.H. (1989a) Fibrin gel investment associated with line 1 and line 10 solid tumor growth, angiogenesis, and fibroplasia in guinea pigs. Role of cellular immunity, myofibroblasts, microvascular damage, and infarction in line 1 tumor regression. *J. Natl. Cancer Inst.* 62:1459–1472.

Dvorak, H.F., Orenstein, N.S., Carvalho, A.C., Churchill, W.H., Dvorak, A.M., Galli, S.J., Feder, J., Bitzer, A.M., Rypysc, J. and Giovinco, P. (1979b) Induction of a fibrin-gel investment: An early event in line 10 hepatocarcinoma growth mediated by tumor-secreted products. *J. Immunol.* 122:166–174.

Dvorak, H.F., Quay, S.C., Orenstein, N.S., Dvorak, A.M., Hahn, P., Bitzer, A.M. and Carvalho, A.C. (1981) Tumor shedding and coagulation. *Science* 212:923–924.

Dvorak, H.F., Senger, D.R. and Dvorak, A.M. (1983a) Fibrin as a component of the tumor stroma: Origins and biological significance. *Cancer Metastasis Rev.* 2:41–73.

Dvorak H.F., Van De Water, L., Bitzer, A.M., Dvorak, A.M., Anderson, D., Harvey, V.S., Bach, R., Davis, G.L., De Wolf, W. and Carvalho, A.C.A. (1983b) Procoagulant activity associated with plasma membrane vesicles shed by cultured tumor cells. *Cancer Res.* 43:4434–4442.

Dvorak, H.F., Form, D.M., Manseau, E.J. and Smith, B.D. (1984a) Pathogenesis of desmo-
plasia. I. Immunofluorescence identification and localization of some structural proteins of
line 1 and line 10 guinea pig tumors and of healing wounds. *J. Natl. Cancer Inst.* 73:
1195–1205.

Dvorak, H.F., Harvey, V.S. and McDonagh, J. (1984b) Quantitation of fibrinogen influx and
fibrin deposition and turnover in line 1 and line 10 guinea pig carcinomas. *Cancer Res.*
44:3348–3354.

Dvorak, H.F., Senger, D.R., Dvorak, A.M., Harvey, V.S. and Donagh, J. (1985) Regulation of
extravascular coagulation by microvascular permeability. *Science* 277:1059–1061.

Dvorak, H.F., Harvey, V.S., Estrella, P., Brown, L.F., McDonagh, J. and Dvorak, A.M. (1987)
Fibrin containing gels induce angiogenesis. Implications for tumor stroma generation and
wound healing. *Lab. Invest.* 57:673–686.

Dvorak, H.F., Nagy, J.A., Dvorak, J.T. and Dvorak, A.M. (1988) Identification and characterization of the blood vessels of solid tumors that are leaky to circulating macromolecules.
Am. J. Pathol. 133:95–109.

Dvorak, H.F., Nagy, J.A. and Dvorak, A.M. (1991a) Structure of solid tumors and their
vasculature: Implications for therapy with monoclonal antibodies. *Cancer Cells* 3:77–85.

Dvorak, H.F., Sioussat, T.M., Brown, L.F., Berse, B., Nagy, J.A., Sotrel, A., Manseau, E.J., Van
De Water, L. and Senger, D.R. (1991b) Distribution of vascular permeability factor (vascular
endothelial growth factor) in tumors. Concentration in tumor blood vessels. *J. Exp. Med.*
174:1275–1278.

Dvorak, H.F., Brown, L.F., Detmar, M. and Dvorak, A.M. (1995) Vascular permeability
factor/vascular endothelial growth factor, microvascular hyperpermeability, and angiogenesis. *Am. J. Pathol.* 146:1029–1039.

Dvorak, A., Kohn, S., Morgan, E., Fox, P., Nagy, J. and Dvorak, H. (1996) The vesiculo-vacuolar
organelle (VVO): a distinct endothelial cell structure that provides a transcellular pathway
for macromolecular extravasation. *J. Leukocyte Biol.* 59:100–115.

Eichmann, A., Marcelle, C., Breant, C. and Le Douarin, N.M. (1993) Two molecules related to
the VEGF receptor are expressed in early endothelial cells during avian embryonic development. *Mech. Dev.* 42:33–48.

Fava, R., Olsen, N., Spencer-Green, G., Yeo, K.-T., Yeo, T.-K., Berse, B., Jackman, R., Senger,
D., Dvorak, H. and Brown, L. (1994) Vascular permeability factor/endothelial growth factor
(VPF/VEGF): Accumulation and expression in human synovial fluids and rheumatoid
synovial tissue. *J. Exp. Med.* 180:341–346.

Favard, C., Moukadiri, H., Dorey, C., Praloran, V. and Plouet, J. (1991) Purification and biological properties of vasculotropin, a new angiogenic cytokine. *Biol. Cell* 73:1–6.

Feng, D., Nagy, J., Hipp, J., Dvorak, H. and Dvorak, A. (1995) Vascular permeability factor/vas-
cular endothelial growth factor (VPF/VEGF), vesiculo-vacuolar organelles (VVOs), and
regulation of vascular permeability. *Mol. Biol. Cell* 6 (suppl):335a.

Feng, D., Nagy, J.A., Hipp, J., Dvorak, H.F. and Dovorak, A.M. (1996) Vesiculo-vacuolar organelles and the regulation of venule permeability to macromolecules by vascular permeability
factor, histamine, and serotonin. *J. Exp. Med.* 183:1981–1986.

Ferrara, N. and Henzel, W.J. (1989) Pituitary follicular cells secrete a novel heparin-binding
growth factor specific for vascular endothelial cells. *Biochem. Biophys. Res. Commun.*
161:851–858.

Ferrara, N., Winer, J., Burton, T., Rowland, A. Siegel, M., Phillips, H.S., Terrell, T., Keller, G.A.
and Levinson, A.D. (1993) Expression of vascular endothelial growth factor does not
promote transformation but confers a growth advantage in vivo to Chines hamster ovary
cells. *J. Clin. Invest.* 91:160–170.

Finkenzeller, G., Marme, D., Weich, H.A. and Hug, H. (1992) Platelet-derived growth factor-
induced transcription of the vascular endothelial growth factor gene is mediated by protein
kinase C. *Cancer Res.* 52:4821–4823.

Finkenzeller, G., Technau, A., Marme, D. (1995) Hypoxia-induced transcription of the vascular
endothelial growth factor gene is independent of functional AP-1 transcription factor. *Biochem. Biophys. Res. Commun.* 208:432–439.

Fischer, S., Sharma, H.S., Karliczek, G.F. and Schaper, W. (1995) Expression of vascular
permeability factor/vascular endothelial growth factor in pig cerebral microvascular
endothelial cells and its upregulation by adenosine. *Brain Res. Mol. Brain Res.* 28:141–
148.

Flamme, I., Breier, G. and Risau, W. (1995) Vascular endothelial growth factor (VEGF) and VEGF receptor 2 (flk-1) are expressed during vasculogenesis and vascular differentiation in the quail embryo. *Dev. Biol.* 169:699–712.

Fong, G.H., Rossant, J., Gertsenstein, M., Breitman, M.L. (1995) Role of the Flt-1 receptor tyrosine kinase in regulating the assembly of vascular endothelium. *Nature* 376:66–70.

Frank, S., Hubner, G., Breier, G., Longaker, M.T., Greenhalgh, D.G. and Werner, S. (1995) Regulation of vascular endothelial growth factor expression in cultured keratinocytes. Implications for normal and impaired wound healing. *J. Biol. Chem.* 270:12607–12613.

Garrido, C., Saule, S. and Gospodarowicz, D. (1993) Transcriptional regulation of vascular endothelial growth factor gene expression in ovarian bovine granulosa cells. *Growth Factors* 8:109–117.

Gengrinovitch, S., Greenberg, S.M., Cohen, T., Gitay, G.H., Rockwell, P., Maione, T.E., Levi, B.Z. and Neufeld, G. (1995) Platelet factor-4 inhibits the mitogenic activity of VEGF121 and VEGF165 using several concurrent mechanisms. *J. Biol. Chem.* 270:15059–15065.

Gitay-Goren, H., Soker, S., Vlodavsky, I. and Neufeld, G. (1992) The binding of vascular endothelial growth factor to its receptors is dependent on cell surface-associated heparin-like molecules. *J. Biol. Chem.* 267:6093–6098.

Gitay-Goren, H., Halaban, R. and Neufeld, G. (1993) Human melanoma cells but not normal melanocytes express vascular endothelial growth factor receptors. *Biochem. Biophys. Res. Commun.* 190:702–709.

Goldberg, M. and Schneider, T. (1994) Similarities between the oxygen sensing mechanisms regulating the expression of vascular endothelial growth factor and erythropoietin. *J. Biol. Chem.* 269:4355–4359.

Gospodarowicz, D., Abraham, J.A. and Schilling, J. (1989) Isolation and characterization of a vascular endothelial cell mitogen produced by pituitary-derived folliculo stellate cells. *Proc. Natl. Acad. Sci. USA* 86:7311–7315.

Goto, F., Goto, K., Weindel, K. and Folkman, J. (1993) Synergistic effects of vascular endothelial growth factor and basic fibroblast growth factor on the proliferation and cord formation of bovine capillary endothelial cells within collagen gels. *Lab. Invest.* 69:508–517.

Green, L., Jellinek, D., Bell, C., Beebe, L., Feistner, B., Gill, S., Jucker, F. and Janjic, N. (1995) Nuclease-resistant nucleic acid ligands to vascular permeability factor/vascular endothelial growth factor. *Chemistry & Biology* 2:683–695.

Grugel, S., Finkenzeller, G., Weindel, K., Barleon, B., Marme, D. (1995) Both v-Ha-ras and v-raf stimulate expression of the vascular endothelial growth factor in NIH 3T3 cells. *J. Biol. Chem.* 270:25915–25919.

Guerrin, M., Moukadiri, H., Chollet, P., Moro, F., Dutt, K., Malacaze, F. and Plouet, J. (1995) Vasculotropin/vascular endothelial growth factor is an autocrine growth factor for human retinal pigment epithelial cells cultured *in vitro*. *J. Cell Physiol.* 164:385–394.

Guidi, A., Abu-Jawdeh, G., Berse, B., Jackman, R., Tognazzi, K., Dvorak, H. and Brown, L. (1995) Vascular permeability factor (vascular endothelial growth factor) expression and angiogenesis in cervical neoplasia. *J. Natl. Cancer Institute* 87:1237–1245.

Guo, D., Jia, Q., Song, H.Y., Warren, R.S. and Donner, D.B. (1995) Vascular endothelial cell growth factor promotes tyrosine phosphorylation of mediators of signal transduction that contain SH2 domains. Association with endothelial cell proliferation. *J. Biol. Chem.* 270:6729–6733.

Haddow, A. (1972) Molecular repair, wound healing, and carcinogenesis: Tumor production a possible overhealing? *Adv. Cancer Res.* 16:181–234.

Harada, S., Nagy, J.A., Sullivan, K.A., Thomas, K.A., Endo, N., Rodan, G.A. and Rodan, S.B. (1994) Induction of vascular endothelial growth factor expression by prostaglandin E2 and E1 in osteoblasts. *J. Clin. Invest.* 93:2490–2496.

Harrison-Woolrych, M., Sharkey, A.M., Charnock, J.D. and Smith, S.K. (1995) Localization and quantification of vascular endothelial growth factor messenger ribonucleic acid in human myometrium and leiomyomata. *J. Clin. Endocrinol. Metab.* 80:1853–1858.

Hashimoto, E., Kage, K., Ogita, T., Nakaoka, T., Matsuoka, R. and Kira, Y. (1994a) Adenosine as an endogenous mediator of hypoxia for induction of vascular endothelial growth factor mRNA in U-937 cells. *Biochem. Biophys. Res. Commun.* 204:318–324.

Hashimoto, E., Ogita, T., Nakaoka, T., Matsuoka, R., Takao, A. and Kira, Y. (1994b) Rapid induction of vascular endothelial growth factor expression by transient ischemia in rat heart. *Am. J. Physiol.* 267:1948–1954.

Hata, Y., Nakagawa, K., Ishibashi, T., Inomata, H., Ueno, H. and Sueishi, K. (1995) Hypoxia-induced expression of vascular endothelial growth factor by retinal glial cells promotes *in vitro* angiogenesis. *Virchows Arch.* 426:479–486.

Hatva, E., Kaipainen, A., Mentula, P., Jaaskelainen, J., Paetau, A., Haltia, M. and Alitalo, K. (1995) Expression of endothelial cell-specific receptor tyrosine kinases and growth factors in human brain tumors. *Am. J. Pathol.* 146:368–378.

Heslan, J.M., Branellec, A.I., Pilatte, Y., Lang, P. and Lagrue, G. (1991) Differentiation between vascular permeability factor and IL-2 in lymphocyte supernatants from patients with minimal-change nephrotic syndrome. *Clin. Exp. Immunol.* 86:157–162.

Hlatky, L., Tsionou, C., Hahnfeldt, P. and Coleman, C.N. (1994) Mammary fibroblasts may influence breast tumor angiogenesis via hypoxia-induced vascular endothelial growth factor up-regulation and protein expression. *Cancer Res.* 54:6083–6086.

Hom, D.B. and Assefa, G. (1992) Effects of endothelial cell growth factor on vascular compromised skin flaps. *Arch. Otolaryngol. Head Neck Surg.* 118:624–628.

Houck, K.A., Ferrara, N., Winer, J., Cachianes, G., Li, B. and Leung, D.W. (1991) The vascular endothelial growth factor family: Identification of a fourth molecular species and characterization of alternative splicing of RNA. *Mol. Endocrinol* 5:1806–1814.

Houck, K.A., Leung, D.W., Rowland, A.M., Winer, J. and Ferrara, N. (1992) Dual regulation of vascular endothelial growth factor bioavailability by genetic and proteolytic mechanisms. *J. Biol. Chem.* 267:26031–26037.

Iizuka, M., Yamauchi, M., Ando, K., Hori, N., Furusawa, Y., Itsukaichi, H., Fukutsu, K. and Moriya, H. (1994) Quantitative RT-PCR assay detecting the transcriptional induction of vascular endothelial growth factor under hypoxia. *Biochem. Biophys. Res. Commun.* 205: 1474–1480.

Jackson, M.R., Carney, E.W., Lye, S.J. and Ritchie, J.W. (1994) Localization of two angiogenic growth factors (PDECGF and VEGF) in human placentae throughout gestation. *Placenta* 15:341–353.

Jakeman, L., Winer, J., Bennett, G., Altar, C. and Ferrara, N. (1992) Binding sites for vascular endothelial growth factor are localized on endothelial cells in adult rat tissues. *J. Clin. Invest.* 89:244–253.

Jakeman, L.B., Armanini, M., Phillips, H.S. and Ferrara, N. (1993) Developmental expression of binding sites and messenger ribonucleic acid for vascular endothelial growth factor suggests a role for this protein in vasculogenesis and angiogenesis. *Endocrinology* 133:848–859.

Kamat, B.R., Brown, L.F., Manseau, E.J., Senger, D.R. and Dvorak, H.F. (1995) Expression of vascular permeability factor/vascular endothelial growth factor by human granulosa and theca lutein cells: Role in corpus luteum development. *Am. J. Pathol.* 146:157–165.

Keck, P.J., Hauser, S.D., Krivi, G., Sanzo, K., Warren, T., Feder, J. and Connolly, D.T. (1989) Vascular permeability factor, and endothelial cell mitogen related to PDGF. *Science* 246:1309–1312.

Kieser, A., Weich, H., Brandner, G., Marme, D. and Kolch, W. (1994) Mutant p53 potentiates protein kinase C induction of vascular endothelial growth factor expression. *Oncogene* 9:963–969.

Kim, J., Li, B., Weiner, J., Houck, K. and Ferrara, N. (1992). The vascular endothelial cell growth factor family. Identification of biologically relevant regions by neutralizing monoclonal antibodies. *Growth Factors* 7:53–64.

Koch, A.E., Harlow, L.A., Haines, G.K., Amento, E.P., Unemori, E.N., Wong, W.L., Pope, R.M. and Ferrara, N. (1994) Vascular endothelial growth factor. A cytokine modulating endothelial function in rheumatoid arthritis. *J. Immunol.* 152:4149–4156.

Kohn, S., Nagy, J.A., Dvorak, H.F. and Dvorak, A.M. (1992) Pathways of macromolecular tracer transport across venules and small veins. Structural basis for the hyperpermeability of tumor blood vessels. *Lab. Invest.* 67:596–607.

Kondo, S., Asano, M. and Suzuki, H. (1993) Significance of vascular endothelial growth factor/vascular permeability factor for solid tumor growth, and its inhibition by the antibody. *Biochem. Biophys. Res. Commun.* 194:1234–1241.

Kondo, S., Asano, M., Matsuo, K., Ohmori, I. and Suzuki, H. (1994) Vascular endothelial growth factor/vascular permeability factor is detectable in the sera of tumor-bearing mice and cancer patients. *Biochim. Biophys. Acta* 1221:211–214.

Koos, R.D. and Olson, C.E. (1991) Hypoxia stimulates expression of the gene for vascular endothelial growth factor (VEGF), a putative angiogenic factor, by granulosa cells of the ovarian follicle, a site of angiogenesis. *J. Cell Biol.* 115:421A.

Koos, R. (1995) Increased expression of vascular endothelial growth/permeability factor in the rat ovary following an ovulatory gonadotropin stimulus: potential roles in follicle rupture. *Biology of Reproduction* 52:1426–1435.

Ku, D.D., Zaleski, J.K., Liu, S. and Brock, T.A. (1993) Vascular endothelial growth factor induces EDRF-dependent relaxation in coronary arteries. *Am. J. Physiol.* 265:h586–592.

Kuroda, M., Oka, T., Oka, Y., Yamochi, T., Ohtsubo, K., Mori, S., Watanabe, T., Machinami, R. and Ohnishi, S. (1995) Colocalization of vascular endothelial growth factor (vascular permeability factor) and insulin in pancreatic islets. *Journal of Endocrinology and Metabolism* 80:3196–3200.

Ladoux, A. and Frelin, C. (1993) Hypoxia is a strong inducer of vascular endothelial growth factor mRNA expression in the heart. *Biochem. Biophys. Res. Commun.* 195(2):1005–1010.

Ladoux, A. and Frelin, C. (1994) Cobalt stimulates the expression of vascular endothelial growth factor mRNA in rat cardiac cells. *Biochem. Biophys. Res. Commun.* 204:794–798.

Lanir, N., Ciano, P.S., Van De Water, L., McDonagh, J., Dvorak, A.M. and Dvorak, H.F. (1988) Macrophage migration in fibrin gel matrices. II. Effects of clotting factor XIII, fibronectin, and glycosaminoglycan content on cell migration. *J. Immunol.* 140:2340–2349.

Leung, D.W., Cachianes, G., Kuang, W.-J., Goeddel, D.V., Ferrara, N. (1989) Vascular endothelial growth factor is secreted angiogenic mitogen. *Science* 246:1306–1309.

Levy, A.P., Levy, N.S., Loscalzo, J., Calderone, A., Takahashi, N., Yeo, K.T., Koren, G., Colucci, W.S. and Goldberg, M.A. (1995a) Regulation of vascular endothelial growth factor in cardiac myocytes. *Circ. Res.* 76:758–766.

Levy, A.P., Levy, N.S., Wegner, S. and Goldberg, M.A. (1995b) Transcriptional regulation of the rat vascular endothelial growth factor gene by hypoxia. *J. Biol. Chem.* 270:13333–13340.

Li, X.F., Gregory, J. and Ahmed, A. (1994) Immunolocalisation of vascular endothelial growth factor in human endometrium. *Growth Factors* 11:277–282.

Li, J., Perrella, M.A., Tsai, J.C., Yet, S.F., Hsieh, C.M., Yoshizumi, M., Patterson, C., Endege, W.O., Zhou, F. and Lee, M.E. (1995) Induction of vascular endothelial growth factor gene expression by interleukin-1 beta in rat aortic smooth muscle cells. *J. Biol. Chem.* 270:308–312.

Li, J., Brown, L., Hibberd, M., Grossman, J., Morgan, J. and Simons, M. (1996). VEGF, Flk-1 and Flt-1 expression in a rat myocardial infarction model of angiogenesis. *Am. J. Physiol.* 270:H 1803–H 1811.

Liu, Y., Cox, S.R., Morita, T. and Kourembanas, S. (1995) Hypoxia regulates vascular endothelial growth factor gene expression in endothelial cells. Identification of a 5′ enhancer. *Circ. Res.* 77:638–643.

Lyttle, D.J., Fraser, K.M., Fleming, S.B., Mercer, A.A. and Robinson, A.J. (1994) Homologs of vascular endothelial growth factor are encoded by the poxvirus orf virus. *J. Virol.* 68:84–92.

Maglione, D., Guerriero, V., Viglietto, G., Delli-Bovi, P. and Persico, M.G. (1991) Isolation of a human placenta cDNA coding for a protein related to the vascular permeability factor. *Proc. Natl. Acad. Sci. USA* 88:9267–9271.

Mandriota, S.J., Seghezzi, G., Vassalli, J.D., Ferrara, N., Wasi, S., Mazzieri, R., Mignatti, P. and Pepper, M.S. (1995) Vascular endothelial growth factor increases urokinase receptor expression in vascular endothelial cells. *J. Biol. Chem.* 270:9709–9716.

Maruyama, K., Tomizawa, S., Seki, Y., Arai, H., Ogawa, T. and Kuroume, T. (1994) FK 506 for vascular permeability factor production in minimal change nephrotic syndrome [letter]. *Nephron.* 66:486–487.

Matthews, W., Jordan, C., Gavin, M., Jenkins, N., Copeland, N. and Lemischka, I. (1991) A receptor tyrosine kinase cDNA isolated from a population of enriched primitive hematopoietic cells and exhibiting close genetic linkage to *c-kit. Proc. Natl. Acad. Sci. USA* 88:9026–9030.

McClure, N., Healy, D.L., Rogers, P.A., Sullivan, J., Beaton, L., Haning, R.J., Connolly, D.T. and Robertson, D.M. (1994) Vascular endothelial growth factor as capillary permeability agent in ovarian hyperstimulation syndrome. *Lancet* 344:235–236.

Midy, V. and Plouet, J. (1994) Vasculotropin/Vascular endothelial growth factor induces differentiation in cultured osteoblasts. *Biochem. Biophys. Res. Commun.* 199:380–386.

Millauer, B., Wizigmann-Voos, S., Schnurch, H., Martinez, R., Moller, N., Risau, W. and Ullrich, A. (1993) High affinity VEGF binding and developmental expression suggest Flk-1 as a major regulator of vasculogenesis and angiogenesis. *Cell* 72:835–846.

Millauer, B., Shawver, L., Plate, K., Risau, W., Ullrich, A. (1994) Glioblastoma growth inhibited in vivo by a dominant-negative Flk-1 mutant. *Nature* 367:576–579.

Miller, J.W., Adamis, A.P., Shima, D.T., D'Amore, P.A., Moulton, R.S., O'Reilly, M.S., Folkman, J., Dvorak, H.F., Brown, L.F., Berse, B. Yeo, T.-K. and Yeo, K.-T. (1994) Vascular endothelial growth factor/vascular permeability factor is temporally and spatially correlated with ocular angiogenesis in a primate model. *Am. J. Pathol.* 145:574–584.

Minchenko, A., Bauer, T., Salceda, S. and Caro, J. (1994a) Hypoxic stimulation of vascular endothelial growth factor expression *in vitro* and *in vivo. Lab. invest.* 71:374–379.

Minchenko, A., Salceda, S., Bauer, T. and Caro, J. (1994b) Hypoxia regulatory elements of the human vascular endothelial growth factor gene. *Cell. Mol. Biol. Res.* 40:35–39.

Monacci, W.T., Merrill, M.J. and Oldfield, E.H. (1993) Expression of vascular permeability factor/vascular endothelial growth factor in normal rat tissues. *Am. J. Physiol.* 264: c995–c1002.

Morii, K., Tanaka, R., Washiyama, K., Kumanishi, T. and Kuwano, R. (1993) Expression of vascular endothelial growth factor in capillary hemangioblastoma. *Biochem. Biophys. Res. Commun.* 194:749–755.

Mukhopadhyay, D., Tsiokas, L., Zhou, X.M., Foster, D., Brugge, J.S. and Sukhatme, V.P. (1995) Hypoxic induction of human vascular endothelial growth factor expression through c-Src activation. *Nature* 375:577–581.

Murata, T., Ishibashi, T., Khalil, A., Hata, Y., Yoshikawa, H. and Inomata, H. (1995) Vascular endothelial growth factor plays a role in hyperpermeability of diabetic retinal vessels. *Ophthalmic Research* 27:48–52.

Mustonen, T. and Alitalo, K. (1995) Endothelial receptor tyrosine kinases involved in angiogenesis. *J. Cell. Biol.* 129:895–898.

Myoken, Y., Kayada, Y., Okamoto, T., Kan, M., Sato, G.H. and Sato, J.D. (1991) Vascular endothelial cell growth factor (VEGE) produced by A-431 human epidermoid carcinoma cells and identification of VEGF membrane binding sites. *Proc. Natl. Acad. Sci. USA* 88:5819–5823.

Nagy, J.A., Herzberg, K.T., Masse, E.M., Zientara, G.P. and Dvorak, H.F. (1989) Exchange of macromolecules between plasma and peritoneal cavity in ascites tumor-bearing, normal, and serotonin-injected mice. *Cancer Res.* 49:5448–5458.

Nagy, J.A., Masse, E.M., Harvey-Bliss, V.S., Meyers, M.S., Sioussat, T.M., Senger, D.R. and Dvorak, H.F. (1991) Immunochemical localization of vascular permeability factor (vascular endothelial growth factor) in ascites tumors: Distribution in peritoneal wall microvasculature. *J. Cell. Biol.* 115:264a.

Nagy, J.A., Herzberg, K.T., Dvorak J.M. and Dvorak, H.F. (1993) Pathogenesis of malignant ascites formation: Initiating events that lead to fluid accumulation. *Cancer Res.* 2631–2643.

Nagy, J., Morgan, E., Herzberg, K., Manseau, E., Dvorak, A. and Dvorak, H. (1995a) Pathogenesis of ascites tumor growth. Angiogenesis, vascular remodeling and stroma formation in the peritoneal lining. *Cancer Res.* 55:376–385.

Nagy, J.A., Masse, E.M., Herzberg, K.T., Meyers, M.S., Yeo, K.-T., Yeo, T.-K., Sioussat, T.M. and Dvorak, H.F. (1995b) Pathogenesis of ascites tumor growth. Vascular permeability factor, vascular hyperpermeability and ascites fluid accumulation. *Cancer Res.* 55:360–368.

Nagy, J.A., Meyers, M.S., Masse, E.M., Herzberg, K.T. and Dvorak, H.F. (1995c) Pathogenesis of ascites tumor growth. Fibrinogen influx and fibrin accumulation in tissues lining the peritoneal cavity. *Cancer Res.* 55:369–375.

Olander, J.V., Connolly, D.T. and DeLarco, J.E. (1991) Specific binding of vascular permeability factor to endothelial cells. *Biochem. Biophys. Res. Commun.* 175:68–76.

Olson, T.A., Mohanraj, D., Carson, L.F. and Ramakrishnan, S. (1994) Vascular permeability factor gene expression in normal and neoplastic human ovaries. *Cancer Res.* 54:276–280.

Palade, G.E. (1988) The microvascular endothelium revisited. *In:* N. Simionescu and M. Simionescu (eds): *Endothelial Cell Biology in Health and Disease.* Plenum Press, New York, pp 3–22.

Park, J., Keller, G.-A. and Ferrara, N. (1993) The vascular endothelial growth factor (VEGF) isoforms: differential deposition into the subepithelial extracellular matrix and bioactivity of extracellular matrix-bound VEGF. *Mol. Biol. Cell* 4:1317–1326.

Park, J.E., Chen, H.H., Winer, J., Houck, K.A., Ferrara, N. (1994) Placenta growth factor. Potentiation of vascular endothelial growth factor bioactivity, in vitro and in vivo, and high affinity binding to Flt-1 but not to Flk-1/KDR. *J. Biol. Chem.* 269:25646–25654.

Pe'er, J., Shweiki, D., Itin, A., Hemo, I., Gnessin, H. and Keshet, E. (1995) Hypoxia-induced expression of vascular endothelial growth factor by retinal cells is a common factor in neovascularizing ocular diseases. *Lab. Invest.* 72:638–645.

Pekala, P., Marlow, M., Heuvelman, D. and Connolly D. (1990) Regulation of hexose transport in aortic endothelial cells by vascular permeability factor and tumor necrosis factor-alpha, but not by insulin. *J. Biol. Chem.* 265:18051–18054.

Pepper, M.S., Ferrara, N., Orci, L. and Montesano, R. (1991) Vascular endothelial growth factor (VEGF) induces plasminogen activators and plasminogen activator inhibitor-1 in micro-vascular endothelial cells. *Biochem. Biophys. Res. Commun.* 181:902–906.

Pepper, M.S., Ferrara, N., Orci, L. and Montesano, R. (1992) Potent synergism between vascular endothelial growth factor and basic fibroblast growth factor in the induction of angiogenesis *in vitro*. *Biochem. Biophys. Res. Commun.* 189:824–831.

Peretz, D., Gitay-Goren, M., Safran, M., Kimmel, N., Gospodarowicz, D. and Neufeld, G. (1992) Glycosylation of vascular endothelial growth factor is not required for its mitogenic activity. *Biochem. Biophys. Res. Commun.* 1882:1340–1347.

Pertovaara, L., Kaipainen, A., Mustonen, T., Orpana, A., Ferrara, N., Saksela, O. and Alitalo, K. (1994) Vascular endothelial growth factor is induced in response to transforming growth factor-$b in fibroblastic and epithelial cells. *J. Biol. Chem.* 269:6271–6274.

Peters, K., DeVries, C., Williams, L. (1993) Vascular endothelial growth factor receptor expression during embryogenesis and tissue repair suggests a role in endothelial differentiation and blood vessel growth. *Proc. Natl. Acad. Sci.* 90:8915–8919.

Phillips, H.S., Hains, J., Leung, D.W. and Ferrara, N. (1990) Vascular endothelial growth factor is expressed in rat corpus luteum. *Endocrinology* 127:965–967.

Phillips, G.D., Stone, A.M., Jones, B.D., Schultz, J.C., Whitehead, R.A. and Knighton, D.R. (1994) Vascular endothelial growth factor (rhVEGF165) stimulates direct angiogenesis in the rabbit cornea. *In Vivo* 8:961–965.

Pierce, E.A., Avery, R.L., Foley, E.D., Aiello, L.P. and Smith, L.E. (1995) Vascular endothelial growth factor/vascular permeability factor expression in a mouse model of retinal neovascularization. *Proc. Natl. Acad. Sci. USA* 92:905–909.

Plate, K.H., Breier, G., Weich, H.A. and Risau, W. (1992) Vascular endothelial growth factor is a potential tumour angiogenesis factor in human gliomas *in vivo*. *Nature* 359:845–848.

Plate, K.H., Breier, G., Millauer, B., Ullrich, A. and Risau, W. (1993) Up-regulation of vascular endothelial growth factor and its cognate receptors in an rat glioma model of tumor angiogenesis. *Cancer Res.* 53:5822–5827.

Plate, K.H., Breier, G., Weich, H.A., Mennel, H.D. and Risau, W. (1994) Vascular endothelial growth factor and glioma angiogenesis: coordinate induction of VEGF receptors, distribution of VEGE protein and possible *in vivo* regulatory mechanisms. *Int. J. Cancer* 59:520–529.

Potgens, A.J., Lubsen, N.H., Vermeulen, R. van, Bakker, A., Schoenmakers, J.G., Ruiter, D.J. and de Waal, R. (1994) Covalent dimerization of vascular permeability factor/vascular endothelial growth factor is essential for its biological activity. Evidence from Cys to Ser mutations. *J. Biol. Chem.* 269:32879–32885.

Praloran, V., Mirshahi, S., Favard, C., Moukadiri, H. and Plouet, J. (1991) Vasculotropin is mitogenic for human peripheral lymphocytes. *C. R. Acad. Sci. Paris* 313 III:21–26.

Qu-Hong, Nagy, J.A., Senger, D.R., Dvorak, H.F. and Dvorak, A.M. (1995) Ultrastructural localization of vascular permeability factor/vascular endothelial growth factor (VPF/VEGF) to the abluminal plasma membrane and vesicuolo-vacuolar organelles of tumor micro-vascular endothelium. *J. Histochem. Cytochem.* 43:381–389.

Quinn, T., Peters, K., DeVries, C., Ferrara, N. and Williams, L. (1993) Fetal liver kinase 1 is a receptor for vascular endothelial growth factor and is selectively expressed in vascular endothelium. *Proc. Natl. Acad. Sci. USA* 90:7533–7537.

Rak, J., Filmus, J., Finkenzeller, G., Grugel, S., Marme, D. and Kerbel, R. (1995a) Oncogenes as inducers of tumor angiogenesis. *Cancer and Metastasis Research* 14:263–277.

Rak, J., Mitsuhashi, Y., Bayko, L., Filmus, J., Shirasawa, S., Sasazuki, T. and Kerbel, R. (1995b) Mutant *ras* oncogenes upregulate VEGF/VPF expression: Implications for induction and inhibition of tumor angiogenesis. *Cancer Res.* 55:4575–4580.

Ravindranath, N., Little-Ihrig, L., Phillips, H.S., Ferrara, N. and Zeleznik, A.J. (1992) Vascular endothelial growth factor messenger ribonucleic acid expression in the primate ovary. *Endocrinology* 131:254–260.

Roberts, W. and Palade, G. (1995) Increased microvasculatur permeability and endothelial fenestration induced by vascular endothelial growth factor. *J. Cell Science* 108:2369–2379.

Samoto, K., Ikezaki, K., Ono, M., Shono, T., Kohno, K., Kuwano, M. and Fukui, M. (1995) Expression of vascular endothelial growth factor and its possible relation with neo-vascularization in human brain tumors. *Cancer Res.* 55:1189–1193.

Sato, K., Terada, K., Sugiyama, T., Takahashi, S., Saito, M., Moriyama, M., Kakinuma, H., Suzuki, Y., Kato, M. and Kato, T. (1994) Frequent overexpression of vascular endothelial growth factor gene in human renal cell carcinoma. *Tohoku J. Exp. Med.* 173:355–360.

Sato, K., Yamazaki, K., Shizume, K., Kanaji, Y., Obara, T., Ohsumi, K., Demura, H., Yamaguchi, S. and Shibuya, M. (1995) Stimulation by thyroid-stimulating hormone and Graves' immuno-globulin G of vascular endothelial growth factor mRNA expression in human thyroid follicles *in vitro* and *flt* mRNA expression in the rat thyroid *in vivo. J. Clin. Invest.* 96:1295–1302.

Seetharam, L., Gotoh, N., Maru, Y., Neufeld, G., Yamaguchi, S. and Shibuya, M. (1995) A unique signal transduction from FLT tyrosine kinase, a receptor for vascular endothelial growth factor VEGF. *Oncogene* 10:135–147.

Senger, D.R. and Dvorak, H.F. (1993) Vascular permeability factor in astrocytomas and other tumors. *In:* P.M. Black and L. Lampson (eds): *Astrocytomas: Diagnosis, Treatment, and Biology.* Blackwell Scientific Publications, London, pp 250–260.

Senger, D.R., Galli, S.J., Dvorak, A.M., Perruzzi, C.A., Harvey, V.S. and Dvorak, H.F. (1983) Tumor cells secrete a vascular permeability factor that promotes accumulation of ascites fluid. *Science* 219:983–985.

Senger, D.R., Perruzzi, C.A., Feder, J. and Dvorak, H.F. (1986) A highly conserved vascular permeability factor secreted by a variety of human and rodent tumor cell lines. *Cancer Res.* 46:5629–5632.

Senger, D.R., Connolly, D.T., Van De Water, L., Feder, J. and Dvorak, H.F. (1990) Purification and NH$_2$-terminal amino acid sequence off guinea pig tumor-secreted vascular permeability factor. *Cancer Res.* 50:1774–1778.

Senger, D., Van De Water, L., Brown, L., Nagy, J., Yeo, K.-T., Yeo, T.-K., Berse, B., Jackman, R., Dvorak, A. and Dvorak, H. (1993) Vascular permeability factor (VPF, VEGF) in tumor biology. *Cancer Metastasis Rev.* 12:303–324.

Shalaby, F., Rossant, J., Yamaguchi, T.P., Gertsenstein, M., Wu, X.F., Breitman, M.L. and Schuh, A.C. (1995) Failure of blood-island formation and vasculogenesis in Flk-1-deficient mice. *Nature* 376:62–66.

Sharkey, A., Charnock-Jones, Boocock, C., Brown, K. and Smith, S. (1993) Expression of mRNA for vascular endothelial growth factor in human placenta. *J. Reproduction and Fertility* 99:609–615.

Shen, H., Clauss, M., Ryan, J., Schmidt, A.M., Tijburg, P., Borden, L., Connolly, D., Stern, D. and Kao, J. (1993) Characterization of vascular permeability factor/vascular endothelial growth factor receptors on mononuclear phagocytes. *Blood* 81:2767–2773.

Shifren, J.L., Doldi, N., Ferrara, N., Mesiano, S. and Jaffe, R.B. (1994) In the human fetus, vascular endothelial growth factor is expressed in epithelial cells and myocytes, but not vascular endothelium: implications for mode of action. *J. Clin. Endocrinol. Metab.* 79:316–322.

Shweiki, D., Itin, A., Soffer D. and Keshet, E. (1992) Vascular endothelial growth factor induced by hypoxia may mediate hypoxia-initiated angiogenesis. *Nature* 359:843–845.

Shweiki, D., Itin, A., Neufeld, G., Gitay, G.H. and Keshet, E. (1993) Patterns of expression of vascular endothelial growth factor (VEGF) and VEGF receptors in mice suggest a role in hormonally regulated angiogenesis. *J. Clin. Invest.* 91:2235–2243.

Shweiki, D., Neeman, M., Itin, A. and Keshet, E. (1995) Induction of vascular endothelial growth factor expression by hypoxia and by glucose deficiency in multicell spheroids: implications for tumor angiogenesis. *Proc. Natl. Acad. Sci. USA* 92:768–772.

Simon, M., Grone, H.J., Johren, O., Kullmer, J., Plate, K.H., Risau, W. and Fuchs, E. (1995) Expression of vascular endothelial growth factor and its receptors in human renal ontogenesis and in adult kidney. *Am. J. Physiol.* 268:f240–250.

Sioussat, T.M., Dvorak, H.F., Brock, T.A. and Senger, D.R. (1993) Inhibition of vascular permeability factor (vascular endothelial growth factor) with anti-peptide antibodies. *Arch. Biochem. Biophys.* 301:15–20.

Soker, S., Svahn, C.M. and Neufeld, G. (1993) Vascular endothelial growth factor is inactivated by binding to alpha 2-macroglobulin and the binding is inhibited by heparin. *J. Biol. Chem.* 268:7685–7691.

Stavri, G.T., Hong, Y., Zachary, I.C., Breier, G., Baskerville, P.A., Yla, H.S., Risau, W., Martin, J.F. and Erusalimsky, J.D. (1995a) Hypoxia and platelet-derived growth factor-BB synergistically upregulate the expression of vascular endothelial growth factor in vascular smooth muscle cells. *FEBS Lett* 358:311–315.

Stavri, G.T., Zachary, I.C., Baskerville, P.A., Martin, J.F. and Erusalimsky, J.D. (1995b) Basic fibroblast growth factor upregulates the expression of vascular endothelial growth factor in vascular smooth muscle cells. Synergistic interaction with hypoxia. *Circulation* 92:11–14.

Stepnick, D.W., Peterson, M.K., Bodgan, C., Davis, J., Wasman, J. and Mailer, K. (1995) Effects of tumor necrosis factor alpha and vascular permeability factor on neovascularization of the rabbit ear flap. *Arch. Otolaryngol. Head Neck Surg.* 121:667–672.

Strugar, J., Rothbart, D., Harrington, W. and Criscuolo, G.R. (1994) Vascular permeability factor in brain metastases: correlation with vasogenic brain edema and tumor angiogenesis. *J. Neurosurg.* 81:560–566.

Sugihara, T., Kaul, S.C., Mitsui, Y. and Wadhwa, R. (1994) Enhanced expression of multiple forms of VEGF is associated with spontaneous immortalization of murine fibroblasts. *Biochim. Biophys. Acta* 1224:365–370.

Takahashi, A., Sasaki, H., Kim, S.J., Tobisu, K., Kakizoe, T., Tsukamoto, T., Kumamoto, Y., Sugimura, T. and Terada, M. (1994) Markedly increased amounts of messenger RNAs for vascular endothelial growth factor and placenta growth factor in renal cell carcinoma associated with angiogenesis. *Cancer Res.* 54:4233–4237.

Takashita, S., Pu, L.Q., Stein, L.A., Sniderman, A.D., Bunting, S., Ferrara, N., Isner, J.M. and Symes, J.F. (1994a) Intramuscular administration of vascular endothelial growth factor induces dose-dependent collateral artery augmentation in a rabbit model of chronic limb ischemia. *Circulation* 90:1228–1234.

Takeshita, S., Zheng, L.P., Brogi, E., Kearney, M., Pu, L.Q., Bunting, S., Ferrara, N., Symes, J.F. and Isner, J.M. (1994b) Therapeutic angiogenesis. A single intraarterial bolus of vascular endothelial growth factor augments revascularization in a rabbit ischemic hind limb model. *J. Clin. Invest.* 93:662–670.

Terman, B.I., Carrion, M.E., Kovacs, E., Rasmussen, B.A., Eddy, R.L. and Shows, T.B. (1991) Identification of a new endothelial cell growth factor receptor tyrosine kinase. *Oncogene* 6:1677–1683.

Terman, B.I., Dougher-Vermazen, M., Carrion, M.E., Dimitrov, D., Armellino, D.C., Gospodarowicz, D. and Böhlen, P. (1992) Identification of the KDR tyrosine kinase as a receptor for vascular endothelial cell growth factor. *Biochem. Biophys. Res. Commun.* 187:1579–1586.

Thieme, H., Aiello, L.P., Takagi, H., Ferrara, N. and King, G.L. (1995) Comparative analysis of vascular endothelial growth factor receptors on retinal and aortic vascular endothelial cells. *Diabetes* 44:98–103.

Tischer, E., Mitchell, R., Hartman, T., Silva, M., Gospodarowicz, D., Fiddes, J.D. and Abraham, J.A. (1991) The human gene for vascular endothelial growth factor. Multiple protein forms are encoded through alternative exon splicing. *J. Biol. Chem.* 266:11947–11954.

Toi, M., Hoshina, S., Takayanagi, T. and Tominaga, T. (1994) Association of vascular endothelial growth factor expression with tumor angiogenesis and with early relapse in primary breast cancer. *Jpn. J. Cancer Res.* 85:1045–1049.

Tomizawa, S., Nagasawa, N., Maruyama, K., Shimabukuro, N., Arai, H. and Kuroume, T. (1990) Release of the vascular permeability factor in minimal change nephrotic syndrome is related to CD^{4+} lymphocytes [letter]. *Nephron.* 56:341–342.

Tuder, R.M., Flook, B.E. and Voelkel, N.F. (1995) Increased gene expression for VEGF and the VEGF receptors KDR/Flk and Flt in lungs exposed to acute or to chronic hypoxia. Modulation of gene expression by nitric oxide. *J. Clin. Invest.* 95:1798–1807.

Unemori, E.N., Ferrara, N., Bauer, E.A. and Amento, E.P. (1992) Vascular endothelial growth factor induces interstitial collagenase expression in human endothelial cells. *J. Cell. Physiol.* 153:557–562.

Vaisman, N., Gospodarowicz, D. and Neufeld, G. (1990) Characterization of the receptors for vascular endothelial growth factor. *J. Biol. Chem.* 265:19461–19466.

Vinores, S., Kuchle, M., Mahlow, J., Chiu, C., Green, W. and Campochiaro, P. (1995) Blood-ocular barrier breakdown in eyes with ocular melanoma. A potential role for vascular endothelial growth factor/vascular permeability factor. *Am. J. Pathol.* 147:1289–1297.

Waltenberger, J., Claesson, W.L., Siegbahn, A., Shibuya, M. and Heldin, C.H. (1994) Different signal transduction properties of KDR and Flt1, two receptors for vascular endothelial growth factor. *J. Biol. Chem.* 269:26988–26995.

Weindel, K., Marmé, D. and Weich, H.A. (1992) AIDS-associated Kaposi's sarcoma cells in culture express vascular endothelial growth factor. *Biochem. Biophys. Res. Commun.* 183:1167–1174.

Weindel, K., Moringlane, J.R., Marmé, D. and Weich, H.A. (1994) Detection and quantification of vascular endothelial growth factor/vascular permeability factor in brain tumor tissue and cyst fluid: the key to angiogenesis? *Neurosurgery* 35:439–448.

Williams, B., Baker, A.Q., Gallacher, B. and Lodwick, D. (1995a) Angiotensin II increases vascular permeability factor gene expression by human vascular smooth muscle cells. *Hypertension* 25:913–917.

Williams, B., Quinn, B.A. and Gallacher, B. (1995b) Serum and platelet-derived growth factor-induced expression of vascular permeability factor mRNA by human vascular smooth muscle cells *in vitro*. *Clin. Sci. (Colch)* 88:141–147.

Wilting, J., Christ, B. and Weich, H.A. (1992) The effects of growth factors on the day 13 chorioallentoic membrane (CAM): A study of VEGF$_{165}$ and PDGF-BB. *Anatomy and Embryology* 186:251–257.

Wizigmann-Voos, S., Breier, G., Risau, W. and Plate, K.H. (19995) Up-regulation of vascular endothelial growth factor and its receptors in von Hippel-Lindau disease-associated and sporadic hemangioblastomas. *Cancer Res.* 55:1358–1364.

Yeo, T.-K. and Dvorak, H.F. (1995) Tumor stroma. *In:* Colvin, R., Bhan, A. and McCluskey, R. (eds): *Diagnostic Immunopathology*. 2, Raven Press, New York, pp 685–697.

Yeo, T.-K., Brown, L. and Dvorak, H.F. (1991a) Alterations in proteoglycan synthesis common to healing wounds and tumors. *Am. J. Pathol.* 138:1437–1450.

Yeo, T.-K., Senger, D.R., Dvorak, H.F., Freter, L. and Yeo, K.-T. (1991b) Glycosylation is essential for efficient secretion but not for permeability-enhancing activity of vascular permeability factor (vascular endothelial growth factor). *Biochem. Biophys. Res. Commun.* 179:1568–1575.

Yeo, K.-T., Wang, H.H., Nagy, J.A., Sioussat, T.M., Ledbetter, S.R., Hoogewerf, A.J., Zhou, Y., Masse, E.M., Senger, D.R., Dvorak, H.F. and Yeo, T.-K. (1993) Vascular permeability factor (vascular endothelial growth factor) in guinea pig and human tumor and inflammatory effusions. *Cancer Res.* 53:2912–2918.

Angiogenesis inhibition

Regulation of Angiogenesis
ed. by I.D. Goldberg & E.M. Rosen
© 1997 Birkhäuser Verlag Basel/Switzerland

Angiostatin: An endogenous inhibitor of angiogenesis and of tumor growth

M. S. O'Reilly

Department of Surgery, Children's Hospital, Boston, and Department of Cellular Biology, Harvard Medical School, Boston, Massachusetts 02115, USA

Summary. Angiostatin, an internal fragment of plasminogen, is a potent inhibitor of angiogenesis, which selectively inhibits endothelial cell proliferation. When given systemically, angiostatin potently inhibits tumor growth and can maintain metastatic and primary tumors in a dormant state defined by a balance of proliferation and apoptosis of the tumor cells. We identified angiostatin while studying the phenomenon of inhibition of tumor growth by tumor mass and have elucidated one mechanism for this phenomenon. In our animal model, a primary tumor almost completely suppresses the growth of its remote metastases. However, after tumor removal, the previously dormant metastases neovascularize and grow. When the primary tumor is present, metastatic growth is suppressed by a circulating angiogenesis inhibitor. Serum and urine from tumor-bearing mice, but not from controls, specifically inhibit endothelial cell proliferation. The activity copurifies with a 38 kD plasminogen fragment which we have sequenced and named angiostatin. Human angiostatin, obtained from a limited proteolytic digest of human plasminogen, has similar activities. Systemic administration of angiostatin, but not intact plasminogen, potently blocks neovascularization and growth of metastases and primary tumors. We here show that the inhibition of metastases by a primary mouse tumor is mediated, at least in part, by the angiogenesis inhibitor angiostatin.

Introduction

In cancer patients, the removal of a malignant tumor, either alone or in combination with adjuvant therapies, may be curative. However, fifty percent of all cancer patients will have metastases (Fidler, 1994). These patients can be loosely grouped into five clinical presentations (Folkman, 1995b) based upon the pattern of growth of metastases in relation to the primary tumor. In one of these five patterns of the presentation of metastatic disease, a primary tumor can inhibit the growth of its metastases. This phenomenon of the inhibition of tumor growth by tumor mass has been referred to as concomitant immunity or concomitant tumor resistance (Prehn, 1991, 1993). For example, the removal of a large primary tumor, such as a fibrosarcoma, may in some cases lead to the rapid growth of previously undetected distant metastases (Clark et al., 1989; Sugarbaker et al., 1977; Warren et al., 1977; Woodruff, 1990; Woodruff, 1980). This inhibition of tumor growth by tumor mass is not specific for any particular tumor type. The inhibition of metastases of one tumor can be seen with a different type of primary tumor, i.e. a breast carcinoma can inhibit metastases of a melanoma. In melanoma, partial spontaneous regression of the primary tumor may be followed by rapid growth of metastases. Furthermore, if one portion of a primary tumor is removed, as is done in debulking operations, the remaining tumor may increase its growth rate (Lange et al., 1980).

A number of experimental models in animals of the inhibition of tumor growth by tumor mass have been described. For example, rodent models exist in which a primary tumor can inhibit the growth of its remote metastases (Bonfil et al., 1988, Gorelik et al., 1978, 1980; Greene and Harvey, 1960; Marie and Clunet, 1910; Milas et al., 1974; Schatten, 1958; Tyzzer, 1913). As is seen clinically, the partial removal of such a tumor may increase the growth rate in the residual tumor (Fisher et al., 1983, 1989; Simpson-Herren et al., 1976). Metastatic growth can also suppress the growth of a primary tumor (Yuhas and Pazmino, 1974). In other models of the inhibition of tumor growth by tumor mass, a primary tumor will suppress the growth of one of several secondary tumor inoculums (Bashford et al., 1907; Ehrlich and Apolant, 1905; Gershon et al., 1967; Gorelik, 1983a, 1983b; Gorelik et al., 1981; Gunduz et al., 1979; Lausch and Rapp, 1969; Ruggiero et al., 1985). The ability of the primary tumor to inhibit the secondary tumors is inversely proportional to the size of the secondary tumor and directly proportional to the size of the first tumor. In these animal models, a threshold size is necessary for the inhibitory effect to occur, and some primary tumors can inhibit a secondary tumor of a different type (Gorelik, 1983a, 1983b; Gorelik et al., 1981; Himmele et al., 1986).

Although several hypotheses have been proposed to explain how a primary tumor might inhibit the growth of its metastases (Gorelik, 1983a; Prehn, 1991, 1993), none has provided an adequate mechanism or led to an accepted theory. We recently proposed a novel hypothesis to explain the phenomenon (Holmgren et al., 1995; O'Reilly et al., 1993, 1994a, b). Briefly, we proposed that a primary tumor might produce an excess of stimulators of angiogenesis which would result in neovascularization in its own vascular bed. As a result, the primary tumor can grow. However, we further proposed that some primary tumors could also produce inhibitors of angiogenesis which would accumulate in the circulation in excess of the stimulators and would thereby inhibit angiogenesis in the vascular bed of a metastasis or secondary tumor (Fig. 1). This hypothesis states that a primary tumor can initiate its own neovascularization by generating angiogenic stimulator(s) in excess of angiogenesis inhibitor(s). However, we have shown that an angiogenesis inhibitor produced by the primary tumor, by virtue of its longer half-life in the circulation, reaches the vascular bed of a secondary tumor in excess of angiogenic stimulator escaping from the primary tumor or generated by the secondary tumor. As a result, there is an inhibition of the growth of the metastasis or secondary tumor.

The studies of Bouck and her colleagues (Good et al., 1990; Rastinejad et al., 1989) support a model in which the net balance of the stimulators and inhibitors of angiogenesis controls tumor growth. To promote angiogenesis, tumor cells produce a number of stimulatory factors, including

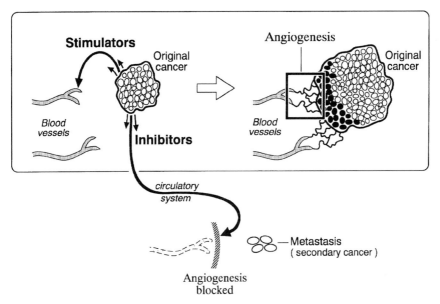

Figure 1. Mechanism of the inhibition of tumor growth by tumor mass. A primary tumor (original cancer in left upper schematic) produces angiogenesis stimulators which act to promote its own neovascularization and growth (right upper schematic). However, the original cancer also produces and releases angiogenesis inhibitors. These inhibitors accumulate in the circulation, in excess of the stimulators, and can thereby inhibit the growth of the tumor's metastases (lower portion of schematic). This model shows one mechanism to explain the inhibition of tumor growth by tumor mass.

vascular permeability factor/vascular endothelial cell growth factor (VPF/VEGF); basic and acidic fibroblast growth factors (bFGF, aFGF), interleukin-8 (IL-8), hepatocyte growth/scatter factor (HGF) and several others. However, Bouck and her colleagues have further demonstrated that tumor cells also downregulate the production of angiogenesis inhibitors during the transition to an angiogenic phenotype. They found that transformed hamster cells upregulate the production of angiogenesis stimulators but must also downregulate the production of thrombospondin, an angiogenesis inhibitor, in order to form angiogenic lesions.

Recently, an animal model was developed in mice in which a Lewis lung carcinoma could almost completely suppress the growth of its metastases (O'Reilly et al., 1994a, 1994b). In this animal model, the primary tumor generates angiostatin which inhibits the growth of the tumor's metastases. Angiostatin is a 38 kD internal fragment of plasminogen which is a specific inhibitor of endothelial cell proliferation *in vitro* and a potent inhibitor of angiogenesis *in vivo*. The experiments that led to the discovery of angiostatin and its potential applications are reviewed below.

Materials and methods

*Development of a murine model of the inhibition of tumor growth
by tumor mass*

Several murine tumors, including variants of Lewis lung carcinoma, sarcoma-180, B-16F10 melanoma, and colon-38 adenocarcinoma, were compared for the ability of a dorsal subcutaneous primary tumor to inhibit the growth of its distant metastases. These tumors were implanted into syngeneic mice and when tumors were 1500 mm^3 in size, approximately 12–16 days after implant, mice underwent a surgical removal of the tumor. Another group had a sham surgical procedure in which tumors were manipulated but left intact. When mice in either the tumor intact group or the tumor resected group were dying of tumor burden, which generally occurred within 14–28 days of the resections, all mice were autopsied. Metastases were evaluated by counting the number of visible metastases (4× magnification) and by weighing the organs involved. In this manner, primary tumors were compared for their ability to inhibit their metastases.

Collection of serum and urine from mice

Blood was obtained from anesthetized mice with 1500 mm^3 Lewis lung carcinomas or normal male mice of comparable age. The blood was obtained via sterile closed cardiac puncture, was pooled into separate glass tubes, allowed to clot, chilled to 4 °C, and centrifuged for 20 minutes at 3000 rpm. Serum was removed and immediately filtered (0.22 μm).

Urine was obtained from mice with Lewis lung carcinomas and from comparable control mice. Mice with an initial tumor volume of 600 mm^3 were placed in metabolic cages and urine was collected until tumors were 3500 mm^3. Urine was collected into a central reservoir (at 4 °C) and cages were cleaned daily. To promote diuresis, 10% sucrose was added to the drinking water. Every 24 h, collected urine was centrifuged, filtered (0.45 μm), and dialyzed (molecular weight cut off = 6–8000) against distilled H$_2$O (changed at least every 6 h) at 4 °C for 48–72 h. Urine was obtained from normal mice in a similar manner.

*Treatment of metastatic growth with serum or urine
from tumor-bearing mice*

After resection of 1500 mm^3 Lewis lung carcinomas (LLC-LM), mice received daily intraperitoneal injections with saline, 0.3 ml of serum from tumor-bearing animals, or 0.3 ml of serum from normal animals. In a separate series of experiments, mice received daily intraperitoneal injections with saline, 0.4 ml of dialyzed and concentrated (250-fold) urine from tumor-bearing animals, or 0.4 ml of dialyzed and concentrated urine from normal animals. In both sets of experiments, one group of tumor-bearing animals had only a sham procedure and were treated with saline injections. Treatments were continued until control mice were dying of metastatic tumor burden (14–21 days), at which point all mice were sacrificed and autopsied.

Mouse corneal angiogenesis assay

All of the corneal angiogenesis assays were performed by Catherine Chen. A corneal pocket was created with a cataract knife in the eyes of 8–10-week-old male C57Bl6/J mouse or immunocompromised male SCID mice. Into this pocket, a 0.34×34 mm sucrose aluminum sulfate (Bukh Meditec, Copenhagen, Denmark) pellet coated with hydron polymer type NCC (Interferon Sciences, New Brunswick, NJ) containing approximately 80 ng of bFGF was implanted (Kenyon et al., 1996; Chen et al., 1995; O'Reilly et al., 1994). Pellets were implanted into the eyes of mice with growing primary Lewis lung carcinomas of greater than 2000 mm³, or into mice without a primary tumor. The corneas of all mice were observed daily and were photographed by means of a slit-lamp stereomicroscope at $10 \times$ magnification.

Capillary endothelial cell proliferation assay

Bovine capillary endothelial cells were obtained and maintained as previously described (Folkman et al., 1979). A cell suspension was made with DMEM/10% BCS/1% antibiotics, the concentration was adjusted to 25000 cells/ml after hemocytometer count, and cells were plated onto gelatinized 24-well culture plates (0.5 ml/well). After incubation (37°C, 10% CO_2) for 24 h, the medium was replaced with 0.25 ml of DMEM/5% BCS/1% antibiotics and the test sample applied. After 20 min, medium and bFGF were added to each well to obtain a final volume of 0.5 ml of DMEM/5% BCS/1% antibiotics/1 ng/ml bFGF. After 72 h, cells were dispersed in trypsin and counted by Coulter counter.

Purification of endothelial inhibitory activity from serum and urine of tumor-bearing mice

Urine or serum from tumor-bearing animals was pooled and applied to a heparin-Sepharose column (30×1.5 cm) which was equilibrated with 50 mM NaCl in 10 mM Tris-HCl at pH 7.2. After being washed with the equilibration buffer, the column was eluted with a gradient of 50 mM–2 M NaCl in 10 mM Tris-HCl, pH 7.2. Fractions which inhibited endothelial cell proliferation were concentrated *in vacuo* and dialyzed (MWCO = 6–8000) against 0.15 M NaCl in 10 mM Tris-HCl, pH 7.2.

The inhibitory sample from heparin-Sepharose chromatography was applied to a Bio-Gel A-0.5 m column (75×1.5 cm) equilibrated with 0.15 M NaCl in 10 mM Tris-HCl, pH 7.2. The column was eluted with the equilibration buffer. Fractions which inhibited endothelial proliferation were concentrated and dialyzed as above.

The inhibitory sample from Bio-Gel A-0.5 m chromatography was applied to a SynChropak RP-4 (100×40 mm) column which was equilibrated with $H_2O/0.1\%$ trifluoroacetic acid (TFA). The column was re-equilibrated and then eluted with a gradient of acetonitrile/0.1% TFA. An aliquot of each was evaporated, resuspended in H_2O, and applied to endothelial cells. Inhibitory activity was further purified by two subsequent cycles on the same C 4 column.

Heparin-Sepharose (Pharmacia, Uppsala, Sweden), Bio-Gel A-0.5 m medium grade agarose beads (Bio-Rad Laboratories, Richmond, CA), and the SynChropak RP-4 (100×4.6 mm) C 4 reverse-phase column (Synchrom, Inc., Lafayette, IN) were prepared according to the manufacturer's recommendations.

Purification of inhibitory fragments of human plasminogen

For our initial studies, angiostatin was purified from plasminogen lysine-binding site I (Sigma Chemical Company, St. Louis, MO), which consists of fragments of plasminogen derived from a proteolytic cleavage with elastase. After resuspending the lysine-binding site I, the samples were applied to a C 4 reverse phase HPLC column and the column was eluted as described above. An aliquot of each fraction was applied to capillary endothelial cells as above. Inhibitory activity was further purified by at least two subsequent cycles on the C 4 column.

Currently, the following protocol is used to purify plasminogen from human plasma and to digest plasminogen with pancreatic elastase. This protocol was developed from previously described methods (Bok and Mangel, 1985; Deutsch and Mertz, 1970; Lerch et al., 1980; Machovich et al., 1990; Machovich and Owen, 1989; Sottrup-Jensen et al., 1978) to increase the yield and maximize the activity of any angiostatin produced from it. Recovered outdated human plasma (obtained from the blood bank at Children's Hospital, Boston) is centrifuged (7500 rpm$\times 30$ min), filtered (0.45 μM), and diluted with an equal volume of PBS. The diluted plasma is applied to a 2.6×35 cm lysine-Sepharose (Pharmacia, Uppsala, Sweden) column prepared according to the manufacturer's recommendations and equilibrated with PBS. Unless otherwise noted, all chromatography and dialyses is performed at 4 °C. The column is then reequilibrated with PBS followed by 0.3 M phosphate buffer with 3 mM EDTA, pH 7.4 (at room temperature) until the baseline (A_{280}) is stable. Plasminogen is then eluted as a single peak with 0.2 M aminocaproic acid (ACA), pH 7.4. The eluant is diluted with an equal volume of chloroform and the mixture is centrifuged for 15 min at 3000 rpm. The aqueous phase is removed and dialyzed (MWCO = 6 – 8000) for at least 48 h against several volumes of 20 mM Tris-HCl, pH 7.6. The plasminogen, as assessed by SDS-PAGE, appears as a 92 kD band.

Angiostatin is produced from human plasminogen, purified as above, by a limited proteolytic digest. Plasminogen in 20 mM Tris-HCl, pH 7.6, is diluted to 0.5 mg/ml. Porcine pancreatic elastase (from Calbiochem, CA or Elastin Products Company, Inc., Owensville, MO) is added (0.8 units elastase/mg of plasminogen) to 100–250 mg of plasminogen. The resulting solution is filtered (0.45 µM), warmed to 37 °C, and incubated on a shaker at 37 °C for 5 h. The reaction is quenched by applying the solution to a 1.5 × 35 cm lysine-Sepharose column equilibrated with 50 mM phosphate buffer, pH 7.4 at 4 °C. The column is then reequilibrated with 50 mM phosphate buffer, pH 7.4, followed by PBS until the baseline (A_{280}) is stable. Angiostatin is subsequently eluted as a single peak with 0.2 M ACA and dialyzed (MWCO = 15000) against PBS for 12 h, followed by water for an additional 36 h. The angiostatin, as assessed by SDS-PAGE, should appear as three bands (45, 52.5, and 40 kD, reduced) and there should be no residual kringle 4 (10 kD band). All purified angiostatin is tested for inhibitory activity on bovine capillary endothelial as described above.

Between purifications, the column is equilibrated with 2 M NaCl 10 mM Tris, pH 7.4, and can be stored in 20% ethanol. The digestion of plasminogen by different preparations of pancreatic elastase or by protocols other than the one described here, can result in diminished yield or reduced activity of the angiostatin.

Protein microsequencing

Microsequence analyses was performed by William S. Lane (Harvard Microchemistry Facility, Cambridge, MA). The 38 kD inhibitor of endothelial cell proliferation from the urine of tumor-bearing animals, and the 40, 42.5, and 45 kD inhibitory fragments of human plasminogen, were purified to homogeneity. Each was resolved by SDS-PAGE, electroblotted onto PVDF (Bio-Rad, Richmond, CA), detected by Ponceau S stain, and excised from the membrane. N-terminal sequence was determined by automated Edman degradation on an ABI Model 477 A protein sequencer (Foster City, CA). Separate samples of the 38 kD inhibitor from the urine of tumor-bearing mice and of the 40 kD inhibitory fragment of human plasminogen were isolated and submitted to *in situ* tryptic digest. The resulting peptide mixture was separated by narrow-bore HPLC using a Vydac C18 (2.1 × 150 mm) reverse-phase column on a Hewlett-Packard 1090 HPLC with a 1040 diode array detector. Optimum fractions were chosen based on symmetry, resolution, ultraviolet absorbance and spectra, further screened for length and homogeneity by matrix-assisted laser desorption mass spectrometry on a Finnigan Lasermat (Hemel, UK), and submitted to automated Edman degradation as above. Strategies for peak selection, reverse-phase separation, and protein microsequencing have been previously described (Lane et al., 1991).

Sequence library searches and alignments were performed against combined GenBank, Brookhaven Protein, SWISS-PROT, and PIR databases. Searches were performed at the National Center for Biotechnology Information through the use of the BLAST network service.

Treatment of mice with angiostatin

Mice implanted with Lewis lung carcinomas underwent resections as above. After surgery, mice received daily intraperitoneal injections of human angiostatin, intact human plasminogen, or saline. Mice received 24 µg (1.2 mg/kg) of angiostatin or plasminogen on the day of operation followed by a daily dose of 12 µg (0.6 mg/kg/day) via intraperitoneal injection. One group of animals, that had the primary tumor left in place after a sham procedure, was treated with saline injections. When the control mice became sick from metastatic disease (i.e. after 13 days of treatment), all mice were sacrificed and autopsied.

Chick chorioallantoic membrane (CAM) angiogenesis assay

Three-day-old fertilized white Leghorn eggs (Spafas, Norwich, CT) were cracked, and embryos with intact yolks were placed in 100 × 20 mm petri dishes. After 3 days of incubation (37 °C/3 % CO_2), a methylcellulose (Fisher Scientific, Fair Lawn, N.J.) disc containing angiostatin purified from human plasminogen was applied to the CAM of individual embryos. The discs were made by desiccation of 10 µl of 0.45 % methylcellulose (in H_2O), with or without angiostatin, on teflon rods. After 48 h of incubation, embryos and CAMs were observed by means of a stereomicroscope.

Antibody depletion of angiostatin from the urine of tumor-bearing mice

The antibody depletion experiments were carried out by Yihai Cao. A monoclonal antibody against the kringle 1 – 3 region of human plasminogen (Vap) was provided by S. G. McCance and F. J. Castellino (University of

Notre Dame, IN). The antibody (5 mg) was coupled to CNBr-activated Sepharose 4B (2 gm) (Pharmacia, Uppsala, Sweden) in an end-over-end mixer with 15 ml of coupling buffer (0.1 M NaHCO$_3$ and 0.5 M NaCl, pH 8.3) at 4°C for 20 h. The coupling buffer was removed by centrifugation and the remaining active groups of Sepharose were blocked by incubation in 1 M ethanolamine, pH 8.0, at room temperature for 2 h. Excess absorbed proteins were removed by washing with coupling buffer and 0.2 M acetate buffer with 0.5 M NaCl (pH 4.0). The antibody-Sepharose conjugate was transferred to a 15 ml column and equilibrated with PBS.

Urine from tumor-bearing mice (7.5 liters) was dialyzed, concentrated to 30 ml, and applied to the column. The sample was recirculated through the column for 15 h at 4 °C. The flow-through fraction, in the same volume as the starting material, was harvested. Aliquots of the flow-through were assayed on bovine capillary endothelial cells and compared to dialyzed and concentrated (250-fold) urine from tumor-bearing or normal mice.

Mice were implanted with Lewis lung carcinomas and underwent resections of the primary tumors as described above. The mice were treated with daily intraperitoneal injections of 0.4 ml of the flow-through obtained from the plasminogen antibody-Sepharose column, 0.4 ml of concentrated urine from tumor-bearing mice, 0.4 ml of concentrated urine from normal mice, or with saline. One group of mice had its tumors left intact and were treated with saline. After 15 days, all mice were sacrificed and autopsied.

Immunohistochemistry

All immunohistochemistry was performed by Lars Holmgren. Lung tissue was fixed in Carnoy's fixative (4 h) followed by ethanol and then embedded in paraffin. Sections (5 mm) were permeabilized with Protein-ase K (2 mg/ml, 37 °C) for 15 min and washed in PBS. Peroxidase activity was quenched by incubation with 0.3% H$_2$O$_2$ in PBS for 15 min followed by three PBS washes. Sections were incubated with rabbit antiserum against human von Willebrand Factor (Dako) in 5% goat serum in PBS. Antibody binding was detected by sequential incubation of the sections with biotinylated goat anti-rabbit serum and streptavidin-peroxidase complex (Vector). Positive staining was detected by substrate reaction with diaminobenzidine. Sections were counterstained with Gill's Hematoxylin and mounted in Permount (Fisher).

Results

Inhibition of the growth of metastases by the presence of a primary tumor

After screening several murine tumor cell lines for the ability of a primary tumor to inhibit its metastases, a variant of Lewis lung carcinoma, designated Lewis lung carcinoma-low metastatic (LLC-LM), which most potently suppressed its lung metastases, was identified. Within five days of removal of a primary Lewis lung carcinoma (800 to 2000 mm^3), the growth of previously dormant metastases was first observed. Within 13−21 days of removal of a primary Lewis lung carcinoma, the number of visible surface lung metastases had increased by greater than 10-fold (50 ± 5) compared to control mice with an intact tumor (4 ± 2) (p < 0.001). Lung weight, which correlates with total tumor burden, increased by at least 400% relative to mice in which the primary tumor was intact (p < 0.001). Comparable results were obtained in immunodeficient SCID mice lacking both T and B lymphocytes (data not shown), suggesting that inhibition of metastatic growth was not dependent on an intact immune system.

Inhibition of metastatic growth is due to an increase in apoptosis associated with the inhibition of angiogenesis

In mice bearing a primary tumor, histological and immunohistochemical studies revealed the presence of microscopic metastases, either as perivascular cuffs of 8−9 cell layers around a lung venule, or as 2 cell layer colonies of tumor cells on the pleural surface. Immunohistochemical staining with antibodies against von Willebrand factor revealed normal lung vascularization and no evidence of angiogenesis of the micrometastases.

In contrast, within 5 days of removal of the primary tumor, histological sections showed enlarging metastases with numerous neovascular capillary sprouts. The proliferative index of the metastases, as determined by bromodeoxyuridine incorporation, was nearly identical for both the metastases inhibited by an intact primary tumor and for the rapidly growing metastases after primary tumor removal. In both cases, the proliferative index was 25−30% (Holmgren et al., 1995). The apoptotic index of the tumor cells in the inhibited metastases was 7.4% ± 0.3, as determined by the TUNEL assay (Gavrieli et al., 1992), versus only 1.9% ± 0.3 (p < 0.001) in the growing metastases (Holmgren et al., 1995). These data suggest that the primary tumor inhibits angiogenesis of the metastases and that the metastases remain dormant due to a balance between apoptosis and proliferation of the tumor cells. After removal of the primary tumor angiogenesis occurs and a decrease in apoptosis of the metastatic tumor cells results in their rapid growth.

Systemic therapy with serum of urine from tumor-bearing mice inhibits the growth of metastases

After resection of a Lewis lung carcinoma, mice were treated with daily intraperitoneal injections of serum or dialyzed and concentrated urine from

Lung Metastases
(Primary Tumor Removed)

Urine
from
tumor-
bearing
mice

Normal
Urine

Saline

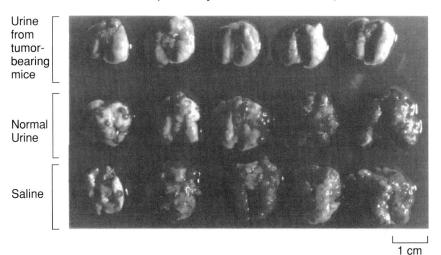

1 cm

Figure 2. Inhibition of the growth of lung metastases by systemic therapy with urine from tumor-bearing mice. Mice were treated with urine from tumor-bearing mice (upper panel), urine from normal mice (middle panel), or with saline (lower panel) after primary tumor removal. The growth of the tumor's metastases was almost completely suppressed by the treatment with the urine of tumor-bearing mice. Injections of urine from normal mice or saline had no effect on the growth of the metastases. Comparable results were seen when serum from tumor-bearing or normal mice was used in place of the urine.

tumor-bearing mice. After 13 days of treatment, the mice treated with serum or urine derived from tumor-bearing animals had fewer than 5 visible surface metastases per lung. Lung weight, which closely correlates with tumor burden, remained nearly normal. Similar results were observed in mice in which primary tumor was left intact. In contrast, in mice treated with normal urine, normal serum, or PBS, there was a 10-fold increase in visible surface metastases in the lung, and a 4-fold increase in lung weight ($p < 0.001$) (Fig. 2).

A primary tumor can inhibit angiogenesis systemically

To determine if the presence of a primary tumor inhibited angiogenesis systemically we implanted a pellet containing bFGF into a corneal pocket of mice. In normal mice, new capillary vessels grew from the corneal limbus and across the cornea in response to the sustained release of bFGF from the pellet within 3 days. Within 6 days of implantation, the new vessels had grown across the cornea and into the pellet. In contrast, in mice with a primary Lewis lung carcinoma of at least 2000 mm³ in volume, there was a virtually complete absence of corneal neovascularization in response to the bFGF pellet.

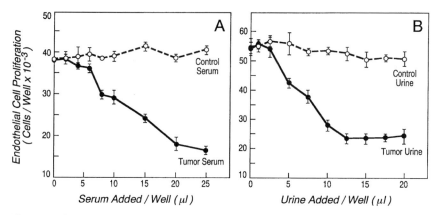

Figure 3. Inhibition of capillary endothelial cell proliferation by serum or urine of tumor-bearing mice. Serum (A) or dialyzed and concentrated urine (B) was applied to bovine capillary
endothelial cells. Serum and urine from tumor-bearing mice inhibited the proliferation of the
endothelial cells in a dose-dependent manner, as compared to controls. Each point represents
the mean and standard error (from O'Reilly et al., Cell (1994) 79:315–328).

Serum or urine from tumor-bearing mice specifically inhibits
the proliferation of endothelial cells

In an bFGF-induced proliferation assay of bovine capillary endothelial
cells *in vitro*, proliferation of the endothelial cells was inhibited to 70% of
that of untreated controls by the serum of tumor-bearing mice. The inhibition was dose-dependent and reversible. Normal serum had no effect on the
proliferation of endothelial cells (Fig. 3A).

Dialyzed and concentrated urine from tumor-bearing mice also inhibited
endothelial proliferation *in vitro* in a manner comparable to the serum.
Urine from normal mice had no effect on endothelial cell proliferation in
the same assay (Fig. 3B). After removal of the Lewis lung carcinoma, the
serum inhibitory activity gradually diminished (Fig. 4) over a 5–7 day
period. The half-life of the inhibitory activity of capillary endothelial cell
proliferation was approximately 2–3 days.

Serum and urine from tumor-bearing mice were tested for the ability to
inhibit the proliferation of a wide variety of cells of endothelial and non-
endothelial cell origin. Both serum and urine from tumor-bearing mice
inhibited bovine capillary endothelial cells, bovine aortic endothelial cells,
and an endothelial cell line from a murine hemangioendothelioma
(EOMA) in a dose-dependent manner. In contrast, a wide variety of cells
of non-endothelial cell origin were not inhibited by the serum or urine of
tumor-bearing mice. The cell types not inhibited included fibroblasts
(WI38 human fetal lung, EFN murine fetal, LM murine) epithelial
(mink lung, bovine retinal pigment, MDCK canine renal), smooth muscle

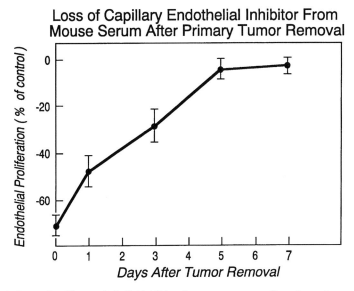

Figure 4. Loss of capillary endothelial inhibitor from mouse serum after primary tumor removal. Serum was obtained from mice 1, 3, 5, and 7 days after removal of a primary tumor and was applied to bovine capillary endothelial cells in a 72 h proliferation assay. A progressive decrease in the endothelial inhibitory activity after tumor removal was observed. The half-life of the circulating inhibitory activity was 2–3 days. Serum from normal mice was used as a control. Each point represents the mean and standard error as a percent of control.

(bovine aortic), or tumor cells (Lewis lung). Mouse Balb/c 3T3 embryo cells have shown variable inhibition with a range of 0–30% in response to the serum of tumor-bearing mice.

The endothelial cell inhibitory activity from serum and urine is independent of an intact immune system

When primary tumors were grown in athymic (*nu/nu*) mice or in SCID mice, we found antiproliferative activity against capillary endothelial cells. The inhibition was comparable to that of the serum of urine of tumor-bearing immunocompetent mice (data not shown). These date indicate that the endothelial inhibitory activity was not dependent on the presence of B or T lymphocytes. No inhibitory activity was found in the serum of urine of athymic or SCID mice that lacked a primary tumor.

A high metastatic variant of Lewis lung carcinoma does not produce detectable levels of the endothelial inhibitory activity

A subline of Lewis lung carcinoma arose after repeated transplantation in which the primary tumor lost the ability to suppress the growth of its metastases. This tumor cell line (LLC-HM, i.e. high metastatic) forms primary tumors that give rise to numerous large lung metastases, which are well vascularized, even when the primary tumor is left intact. Serum and urine from mice carrying this LLC-HM subline did not inhibit endothelial cell proliferation *in vitro* (data not shown). These results suggest that the presence of endothelial inhibitory activity in the serum and urine of tumor-bearing animals correlates directly with inhibition of the growth of metastases.

Purification of a 38 kD protein from serum and urine of tumor-bearing mice which inhibits endothelial proliferation

The capillary endothelial cell inhibitory activity, as determined in a 72 h *in vitro* assay of bFGF-stimulated bovine capillary endothelial cells, was purified from pooled serum or urine from tumor-bearing mice. For both purifications, the inhibitory activity eluted at a concentration of 30–35% acetonitrile from the C4 column and was associated with a single band on an SDS-polyacrylamide gel of Mr 38 kD. The estimated amount of protein required for half-maximal inhibition was 200 ng/ml. The inhibitory activity of the purified product was heat-($100\,°C \times 10$ minutes) and trypsin-sensitive and was associated with the aqueous phase of a chloroform extraction (data not shown). Equivalent volumes of pooled serum or urine from normal animals had no detectable inhibitory activity (data not shown) even after fractionation by heparin-Sepharose and gel filtration chromatographies.

Microsequence analysis of the 38 kD protein

After two separate purifications from pooled urine of tumor-bearing mice, the 38 kD inhibitory protein was electroblotted from an SDS-polyacrylamide gel to a polyvinylidene difluoride (PVDF) membrane. One sample (purified from 20 liters of mouse urine) was used for N-terminal analysis. The other sample (purified from 100 liters of mouse urine) was used for tryptic digest and sequence analysis of the resultant peptides.

N-terminal analysis of the first 18 amino acid residues of the first sample revealed identity to an internal fragment of plasminogen with an amino terminus at amino acid 98. Microsequence analysis of the peptides from the tryptic digest of the second sample confirmed identity (greater than 98%) of the endothelial inhibitor to an internal fragment of plasminogen which includes amino acids 98 through 306. The inhibitory fragment has a predicted (by molecular weight) carboxyl terminus at amino acid 440 and would include the first four of the five kringle regions (Lerch et al., 1980; Ponting et al., 1992) of plasminogen. We have named this fragment angiostatin (Fig. 5).

Plasminogen

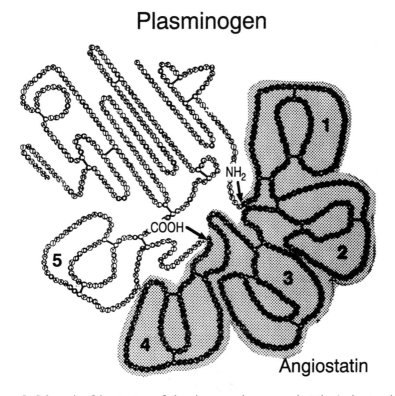

Figure 5. Schematic of the structure of plasminogen and mouse angiostatin. Amino-terminus (amino acid 98) and presumed carboxyl terminus of mouse angiostatin are indicated by arrows. Angiostatin consists of the first four (shaded area) of the triple-loop kringle structures of plasminogen. The amino-terminus of human angiostatin, derived from a proteolytic digest of plasminogen, is nearly identical (amino acid 97 or 99) to that of mouse angiostatin.

Depletion of angiostatin from the urine of tumor-bearing mice removes the anti-endothelial and anti-metastatic effects

To confirm that the angiostatin generated by a Lewis lung primary tumor is responsible for the anti-metastatic and anti-angiogenic effects, dialyzed and concentrated urine was depleted of angiostatin using antibody affinity chromatography. The flow-through from the column (urine depleted of angiostatin) had no effect on the proliferation of bovine capillary endothelial cells (O'Reilly et al., 1994b).

We next removed Lewis lung primary tumors and treated these mice with daily intraperitoneal injections of dialyzed and concentrated urine of tumor-bearing animals from which angiostatin had been depleted as above. Comparable mice were treated with urine derived from normal mice or with urine of tumor-bearing animals that was not depleted of angiostatin. Mice treated with urine of tumor-bearing mice had almost no visible

surface metastases per lung (2 ± 1). Lung weight, and therefore tumor burden, remained nearly normal (data not shown). In contrast, when mice were
treated with urine from tumor-bearing animals that was first depleted of
angiostatin, after removal of the primary tumor, there was a greater than 20-
fold increase in visible surface metastases in the lung (49 ± 4) (p < 0.001),
and a 4-fold increase in lung weight (p < 0.001) (O'Reilly et al., 1994b).
These data show that angiostatin generated by a primary tumor is the major
factor responsible for the inhibition of the growth of metastases.

Identification and purification of a fragment of human plasminogen with endothelial inhibitory activity

Human angiostatin was produced from human plasminogen by a limited
proteolysis with elastase. SDS-polyacrylamide gel electrophoresis of the
active fractions revealed three distinct bands of apparent Mr of 40, 42, and
45 kD. Each of these three peptides inhibited proliferation of bFGF-stimulated capillary endothelial cells (data not shown) equally well. Microsequence analysis revealed an amino terminal at amino acid 97 or 99 for
each of the fragments and identity to plasminogen.

Purified human angiostatin (unpublished data) was found to inhibit
endothelial cell proliferation specifically. At angiostatin concentrations
2–3 log-fold higher than those required for maximal inhibition of endothelial cells, several other cell types of non-endothelial origin, including
the Lewis lung carcinoma cell line, were not significantly inhibited. Intact
plasminogen and other fragments of plasminogen which did not contain the
kringle structures (i.e. miniplasminogen) had no effect on the growth of
capillary endothelial cells. It was of interest that fragments of plasminogen
corresponding to lysine-binding site II (Lerch et al., 1980; Sottrup-Jensen
et al., 1978), which includes kringle 4 alone, and purified kringle 4, in
some cases stimulated proliferation of capillary endothelial cells by at least
20% above control (data not shown).

Human angiostatin inhibits angiogenesis

Human angiostatin over a dose range of 0.1 µg up to 100 µg was tested in the
chick chorioallantoic membrane (CAM) assay for its ability to inhibit angiogenesis. Inhibition of angiogenesis (i.e. induction of avascular zones by 48 h
after implantation of the test substance on the 6-day-old CAM) from human
angiostatin was dose-dependent. The inhibition began at 20 µg and reached
saturation at approximately 40 µg (O'Reilly et al., 1994b). No detectable
inflammation of toxicity was observed in any of the treated eggs. Recently,
we have tested human angiostatin in the mouse corneal angiogenesis assay.
When given systemically, angiostatin inhibited corneal angiogenesis, in response to a sustained release of bFGF, by 80% or greater (data not shown).

Human angiostatin suppresses the growth of metastases

Approximately 1–2 mg of active human angiostatin was purified from the lysine-binding site I preparation. To test for the ability of human angiostatin to inhibit the growth of metastases, we removed Lewis lung primary tumors from mice. Half of these mice were treated with daily intraperitoneal injections of human angiostatin (24 µg/20 gram mouse on day 1, followed by 12 µg/day for the duration of experiment). In all of the treated mice, an almost complete suppression of the growth of metastases, with an efficacy equal to or greater than the presence of a primary tumor, was observed. An 18-fold or greater reduction in the number of visible metastases (p < 0.001) was observed compared to control mice treated with plasminogen or with saline (Fig. 6). The angiostatin-treated mice showed no significant increase in lung weight versus a 3- to 5-fold increase in lung weight in control mice treated with plasminogen or saline (p < 0.001).

The rapidly growing metastases in control mice treated with intact plasminogen or saline were highly neovascularized, as revealed by immunohistochemical staining with endothelial specific antibodies against von

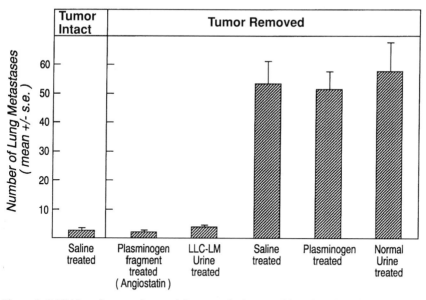

Figure 6. Inhibition of metastatic growth by systemic therapy with angiostatin. After removal of a primary tumor, mice were treated with daily intraperitoneal injections of human angiostatin, intact human plasminogen, urine from tumor-bearing mice, urine from normal mice, or saline. One group of mice had its primary tumor left intact. Human angiostatin almost completely suppressed the growth of the metastases. The inhibition seen with systemic angiostatin was comparable to that of the primary tumor and to the urine of tumor-bearing mice. Intact human plasminogen and normal mouse urine had no effect on the growth of the metastases.

Willebrand factor. In contrast, the metastases of mice treated with angiostatin remained microscopic in size (less than 100 µg radius). They contained no new vessels or were very sparsely vascularized (O'Reilly et al., 1994b).

Angiostatin suppresses the growth of primary tumors in mice

The growth of solid tumors is dependent on angiogenesis. For this reason, we speculated that angiostatin, a potent angiogenesis inhibitor, would inhibit the growth of primary tumors as well as the growth of metastases. We therefore performed dose-seeking experiments to determine if the growth of Lewis lung carcinomas, implanted subcutaneously, could be inhibited by angiostatin. Human angiostatin, when given daily via subcutaneous or intraperitoneal injection, inhibits the growth of Lewis lung carcinoma at doses of 10 mg/kg/day. Increasing dose and frequency of administration of angiostatin results in an improved efficacy in the treatment of both Lewis lung carcinoma and T241 murine fibrosarcoma primary tumors. We are currently treating mice with human angiostatin at a dose of 50 mg/kg given twice daily. We have found that a wide variety of murine tumors can be potently suppressed by angiostatin. Further, we have found that human tumors growing subcutaneously in immunocompromised SCID mice can be suppressed and even regressed by systemic therapy with human angiostatin. We have recently submitted this data as a manuscript which outlines the response seen in tumor growth.

Discussion

These experiments provide a mechanism by which a primary tumor can inhibit the growth of its metastases by producing angiostatin, a potent inhibitor of angiogenesis. The phenomenon of the inhibition of tumor growth by tumor mass can therefore be explained by differences in the balance of the inhibitors and stimulators of angiogenesis. A primary tumor grows because of the local accumulation of angiogenesis stimulators. However, its metastases are inhibited because of the relative excess of angiogenesis inhibitors in the circulation. Angiostatin is one example of such an inhibitor. It accumulates as a result of the larger mass achieved by a primary tumor before a secondary tumor begins its growth, the diffusion of the angiogenesis inhibitor(s) into the circulation, and the significantly prolonged retention of angiogenesis inhibitor(s) in the circulation relative to angiogenic stimulator(s). Although other mechanisms may exist, our model clearly demonstrates the relevance of angiogenesis inhibitors, such as angiostatin, in the regulation of tumor growth.

The presence of the primary tumor almost completely inhibits neovascularization, and therefore growth, of the metastases. These studies, and those of Holmgren et al., 1995) demonstrate a new mechanism to explain tumor dormancy. Dormancy can now be defined as a balance of apoptosis to proliferation of tumor cells which arises after the suppression of angiogenesis. Removal of a primary tumor is not only associated with increased neovascularization, but also with a 3–5-fold decrease in apoptosis. This diminished apoptosis is highly significant and appears to be a critical event for rapid growth of metastases. An apoptotic index of 2–3% in the rat liver, for example, results in a 25% loss of cells per day (Bursch et al., 1990). In the developing rat optic nerve, an apoptotic index which does not exceed 0.25% produces a 50% loss of oligodendrocytes per day (Barres et al., 1992). Thus, the observed differences in apoptosis between the inhibited and the growing metastases could explain the differences in their growth.

In recent studies, the growth of primary tumors can also be potently suppressed by systemic therapy with angiostatin. So far all of the primary tumors that have been tested, which includes three murine tumor types in syngeneic mice, and three human tumor types in immunocompromised mice, have been inhibited by angiostatin. There has been no evidence of any toxicity in any of the treated mice and no evidence of drug resistance seen, and the efficacy has been comparable to or better than that of other agents. For the human tumors, angiostatin therapy has resulted in an almost complete regression of primary tumors and a similar pattern of dormancy as that seen for the metastases. Presumably the relatively high doses of angiostatin needed to treat primary tumors is due to the accelerated clearance of the human protein once injected into mice (i.e. a circulating half-life of 4 h for the human protein versus 2–3 days for the mouse protein). Preliminary data using recombinant murine angiostatin from *E. coli* (Jie Lin et al., work in progress) and from baculovirus (Yuen Shing et al., work in progress) show improved efficacy over human angiostatin at 10–20-fold lower doses.

The ability of angiostatin to maintain metastases and primary tumors in a state of dormancy adds additional direct evidence of the dependence of tumor growth on angiogenesis, as first proposed by Judah Folkman (Folkman, 1971). Numerous lines of direct evidence have now established that the growth and expansion of tumors, and their metastases, are dependent upon angiogenesis (Brown et al., 1993; Folkman, 1995, Folkman, 1989; Hori et al., 1991; Kim et al., 1993; Millauer et al., 1994; O'Reilly et al., 1994; Rastinejad et al., 1989). Before the onset of angiogenesis, tumor growth is limited to an *in situ* stage in which tumor volume is limited to a few cubic millimeters. In order for a tumor to grow beyond this stage, neovascularization is needed to produce new vessels which are required to perfuse the tumor with nutrients and oxygen and for the removal of metabolic catabolites. However, the endothelial cell themselves also produce a number of paracrine growth factors which are stimulatory for tumor growth (Hamada et al., 1992; Nicosia et al., 1986). As a result, tumor growth is dependent upon an endothelial cell compartment and a tumor cell compartment. This two-compartment model of tumor cell growth was recently proposed by Judah Folkman (1996).

Angiogenesis inhibitors, such as angiostatin, may be used to target the endothelial cell compartment of a tumor. For angiostatin, it is now possible to induce a state of harmless dormancy which is not possible with conventional modalities. The marked efficacy of angiostatin in the treatment of malignant tumors combined with its lack of toxicity makes it unique among the other anti-neoplastic agents currently used. The field of angiogenesis has clear and far-reaching clinical implications. An angiogenesis inhibitor as potent, specific, and non-toxic as angiostatin should find wide clinical use to potentiate and improve the treatment of malignant tumors and other angiogenic diseases (Folkman, 1995a, 1995b).

Acknowledgments
I would like to thank Dr Judah Folkman for his guidance, teaching, and ideas and for his superb role as a mentor. I would also like to acknowledge my co-authors on the original manuscripts as outlined above. This work would not have been possible without their extraordinary efforts and abilities.

References

Barres, B.A., Hart, I.K., Coles, H.S., Burne, J.F., Voyvodic, J.T., Richardson, W.D. and Raff, M.C. (1992) Cell death and control of cell survival in the oligodendrocyte lineage. *Cell* 70:31–46.

Bashford, E.F., Murray, J.A. and Cramer, W. (1907) The natural and induced resistance of mice to the growth of cancer. *Proc. Royal Soc. London* 79:164–187.

Bok, R.A. and Mangel, W.F. (1985) Quantitative characterization of the lysine binding of plasminogen to intact fibrin clots, lysine-Spharose, and fibrin cleaved by plasmin. *Biochemistry* 24:3279–3286.

Bonfil, R.D., A., R.R., Bustuoabad, O.D., Meiss, R.P. and Pasqualini, C.D. (1988) Role of concomitant resistance in the development of murine lung metastases. *Int. J. Cancer* 41:415–422.

Brown, L.F., Berse, B., Jackman, R.W., Tognazzi, K., Manseau, E.J., Senger, D.R. and Dvorak, H.F. (1993) Expression of vascular permeability factor (vascular endothelial growth factor) and its receptors in adenocarcinomas of the gastrointestinal tract. *Cancer Res.* 53: 4727–4735.

Bursch, W., Paffe, S., Putz, B., Barthel, G. and Schulte-Hermann, R. (1990) Determination of the length of the histological stages of apoptosis in normal liver and in altered hepatic foci of rats. *Carcinogenesis* 11:847–853.

Chen, C., Parangi, S., Tolentino, M.J. and Folkman, J. (1995) A Strategy to Discover Circulating Angiogenesis Inhibitors Generated by Human Tumors. *Cancer Res.* 55:4230–4233.

Clark, W.H.J., Elder, D.E., Guerry, D.I.V., Braitman, L.E., Trock, B.J., Schultz, D., Synnestevdt, M. and Halpern, A.C. (1989) Model predicting survival in stage I melanoma based on tumor progression. *J. Natl. Cancer Inst.* 81:1893–1904.

Deutsch, D.G. and Mertz, E.T. (1970) Plasminogen: purification from human plasma by affinity chromatography. *Science* 170:1095–1096.

Ehrlich, P. and Apolant, H. (1905) Beobachtungen über maligne Maustumoren. *Berl. klin. Wschr.* 42:871–874.

Fidler, I.J. (1994) The implications of angiogenesis for the biology and therapy of cancer metastasis. *Cell* 79:185–188.

Fisher, B., Gunduz, N., Coyle, J., Rudock, C. and Saffer, E. (1989) Presence of a growth-stimulating factor in serum following primary tumor removal in mice. *Cancer Res.* 49:1996–2001.

Fisher, B., Gunduz, N. and Saffer, E.A. (1983) Influence of the interval between primary tumor removal and chemotherapy on kinetics and growth of metastases. *Cancer Res.* 43: 1488–1492.

Folkman, J. (1995a). Angiogenesis in cancer, vascular, rheumatoid and other disease. *Nature Medicine* 1:27–31.

Folkman, J. (1995b). Clinical applications of angiogenesis research. *N. Engl. J. Med.* 333: 1757–1763.

Folkman, J. (1996) Tumor angiogenesis and tissue factor. *Nature Medicine* 2:167–168.

Folkman, J. (1971) Tumor angiogenesis: Therapeutic implications. *N. Engl. J. Med.* 285: 1182–1186.

Folkman, J. (1989) What is the evidence that tumors are angiogenesis dependent? *J. Natl. Cancer Inst.* 82:4–6.

Folkman, J., Hausdenschild, C. and Zetter, B. (1979) Long term culture of capillary endothelial cells. *Proc. Natl. Acad. Sci. USA* 76:5217–5221.

Gavrieli, Y., Sherman, Y. and Ben-Sasson, S.A. (1992) Identification of programmed cell death *in situ* via specific labeling of nuclear DNA fragmentation. *J. Cell Biol.* 119: 493–501.

Gershon, R.K., Carter, R.L. and Kondo, K. (1967) On concomitant immunity in tumor-bearing hamsters. *Nature* 213:674–676.

Good, D.J., Polverini, P.J., Rastinejad, F., Le Beau, M.M., Lemons, R.S., Frazier, W.A. and Bouck, N.P. (1990) A tumor suppressor-dependent inhibitor of angiogenesis is immunologically and functionally indistinguishable from a fragment of thrombospondin. *Proc. Nat. Acad. Sci. USA* 87:6624–6628.

Gorelik, E. (1983a) Concomitant tumor immunity and the resistance to a second tumor challenge. *Adv. Cancer Res.* 39:71–120.

Gorelik, E. (1983b) Resistance of tumor-bearing mice to a second tumor challenge. *Cancer Res.* 43:138–145.

Gorelik, E., Segal, S. and Feldman, M. (1980) Control of lung metastasis progression in mice: role of growth kinetics of 3LL Lewis lung carcinoma and host immune reactivity. *J. Nat. Cancer Inst.* 65:1257–1264.

Gorelik, E., Segal, S. and Feldman, M. (1978) Growth of a local tumor exerts a specific inhibitory effect on progression of lung metastases. *Int. J. Cancer* 21:617–625.

Gorelik, E., Segal, S. and Feldman, M. (1981) On the mechanism of tumor "concomitant immunity". *Int. J. Cancer* 27:847–856.

Greene, H.S.N. and Harvey, E.K. (1960) The inhibitory influence of a transplanted hamster lymphoma on metastasis. *Cancer Res.* 20:1094–1103.

Gunduz, N., Fisher, B. and Saffer, E.A. (1979) Effect of surgical removal on the growth and kinetics of residual tumor. *Cancer Res.* 39:3861–3865.

Hamada, J., Cavanaugh, P.G., Lotan, O. and Nicolson, G.L. (1992) Separable growth and migration factors for large-cell lymphoma cells secreted by microvasular endothelial cells derived from target organs for metastasis. *Brit. J. Cancer* 66:349–354.

Himmele, J.C., Rabenhorst, B. and Werner, D. (1986) Inhibition of Lewis lung tumor growth and metastasis by Ehrlich ascites tumor growing in the same host. *J. Cancer Res. Clin. Oncol.* 111:160–165.

Holmgren, L., O'Reilly, M.S. and Folkman, J. (1995) Dormancy of micrometastases: balanced proliferation and apoptosis in the presence of angiogenesis suppression. *Nature Medicine* 1:149–153.

Hori, A., Sasada, R., Matsutani, E., Naito, K., Sakura, Y., Fujita, T. and Kozai, Y. (1991) Suppression of solid tumor growth by immunoneutralizing monoclonal antibody against human basic fibroblast growth factor. *Cancer Res.* 51:6180–6184.

Kim, K.J., Li, B., Winer, J., Armanini, M., Gillett, N., Phillips, H.S. and Ferrara, N. (1993) Inhibition of vascular endothelial growth factor-induced angiogenesis suppresses tumor growth *in vivo. Nature* 362:841–844.

Kenyon, B.M., Vuest, E.E., Chen, C.C., Flynn, E., Folkman, J., D'Amato, R.J. (1996) A model of angiogenesis in the mouse cornea. *Invest. Ophtha. Vis Sci.* (in press).

Lane, W.S., Galat, A., Harding, M.W. and Schreiber, S.L. (1991) Complete amino acid sequence of the FK506 and rapamycin binding protein, FKBP, isolated from calf thymus. *J. Prot. Chem.* 10:151–160.

Lange, P.H., Heknat, K., Bosl, G., Kennedy, B.J. and Fraley, E.E. (1980) Accelerated growth of testicular cancer after cytoreductive surgery. *Cancer Res.* 45:1498–1506.

Lausch, R.N. and Rapp, F. (1969) Concomitant immunity in hamsters bearing DMBA-induced tumor transplants. *Int. J. Cancer* 4:226–231.

Lerch, P.G., Rickli, E.E., Lergier, W. and Gillessen, D. (1980) Localization of individual lysine-binding regions in human plasminogen and investigations on their complex-forming properties. *Eur. J. Biochem.* 107:7–13.

Machovich, R., Himer, A. and Owen, W.G. (1990) Neutrophil proteases in plasminogen activation. *Blood Coag. Fibrinol.* 1:273–277.

Machovich, R. and Owen, W.G. (1989) An elastase-dependent pathway of plasminogen activation. *Biochemistry* 28:4517–4522.

Marie, P. and Clunet, J. (1910) Frequence des metastases viscerales chez les souris cancereuses apres ablation chirurgicale de leur tumeur. *Bulletin de l'Association Francaise pour l'Etude Cancer* 3:19–23.

Milas, M., Hunter, N., Mason, K. and Withers, H.R. (1974) Immunological resistance to pulmonary metastases in C3Hf/Bu mice bearing syngeneic fibrosarcoma of different sizes. *Cancer Res.* 34:61–71.

Millauer, B., Shawver, L.K., Plate, K.H., Risau, W. and Ullrich, A. (1994) Glioblastoma growth inhibited *in vivo* by a dominant-negative Flk-1 mutant. *Nature* 367:576–579.

Nicosia, R.F., Tchao, R. and Leighton, J. (1986) Interactions between newly formed endothelial channels and carcinoma cells in plasma clot culture. *Clin. Expl. Metastasis* 4:91–104.

O'Reilly, M., Rosenthal, R., Sage, E.H., Smith, S., Holmgren, L., Moses, M., Shing, Y. and Folkman, J. (1993) The suppression of tumor metastases by a primary tumor. *Surgical Forum.* 44:474–476.

O'Reilly, M.S., Holmgren, L., Shing, Y., Chen, C., Rosenthal, R.A., Cao, Y., Moses, M., Lane, W.S., Sage, E.H. and Folkman, J. (1994a). Angiostatin: a circulating endothelial cell inhibitor that suppresses angiogenesis and tumor growth. *Cold Spring Harbor Symp. Quant. Biol.* 59:471–482.

O'Reilly, M.S., Holmgren, L., Shing, Y., Chen, C., Rosenthal, R.A., Moses, M., Lane, W.S., Cao, Y., Sage, E.H. and Folkman, J. (1994b). Angiostatin: A novel angiogenesis inhibitor that mediates the suppression of metastases by a Lewis lung carcinoma. *Cell* 79:315–328.

Ponting, C.P., Marshall, J.M. and Cederholm-Williams, S.A. (1992) Plasminogen: a structural review. *Blood Coag. Fibrinol.* 3:605–614.

Prehn, R.T. (1991) The inhibition of tumor growth by tumor mass. *Cancer Res.* 51:2–4.

Prehn, R.T. (1993) Two competing influences that may explain concomitant tumor resistance. *Cancer Res.* 53:3266–3269.

Rastinejad, F., Polverini, P.J. and Bouck, N.P. (1989) Regulation of the activity of a new inhibitor of angiogenesis by a cancer suppressor gene. *Cell* 56:345–355.

Ruggiero, R.A., Bustuoabad, O.D., Bonfil, R.D., Meiss, R.P. and Pasqualini, C.D. (1985) "Concomitant immunity" in murine tumors of non-detectable immunogeneicity. *Br. J. Cancer* 51:37–48.

Schatten, W.E. (1985) An experimental study of postoperative tumor metastases. *Cancer* 11:455–459.

Simpson-Herren, L., Sanford, A.H. and Holmquist, J.P. (1976) Effects of surgery on the cell kinetics of residual tumor. *Cancer Treat. Rep.* 60:1749–1760.

Sottrup-Jensen, L., Claeys, H., Zajdel, M., Petersen, T.E. and Magnusson, S. (1978) The Primary structure of human plasminogen: isolation of two lysine-binding fragments and one "mini-"plasminogen (MW, 38,000) by elastase-catalyzed-specific limited proteolysis. *In*: J.F. Davidson, R.M. Rowan, M.M. Samama and P.C. Desnoyers (eds): *Progress in Chemical Fibrinolysis and Thrombolysis.* Raven Press, New York, 191–209.

Sugarbaker, E.V., Thornthwaite, J. and Ketcham, A.S. (1977) Inhibitory effect of a primary tumor on metastasis. *In*: S.B. Day, W.P.L. Myers, P. Stansly, S. Garattini and M.G. Lewis (eds): *Progress in Cancer Research and Therapy*, Raven Press, New York, pp 227–240.

Tyzzer, E.E. (1913) Factors in the production and growth of tumor metastases. *J. Med. Res.* 28:309–333.

Warren, B.A., Chauvin, W.J. and Philips, J. (1977) Blood-borne tumor emboli and their adherence to vessel walls. *In*: S.B. Day, W.P.L. Myers, P. Stansly, S. Garattini and M.G. Lewis (eds): *Progress in cancer research and therapy*. Raven Press, New York, pp 185–197.

Woodruff, M. (1990) *Cellular Variation and Adaptation in Cancer*. Oxford University Press, New York, pp 64–70.

Woodruff, M. (1980) *The Interactions of Cancer and the Host*. Grune & Stratton, New York, pp 59–66.

Yuhas, J.M. and Pazmino, N.H. (1974) Inhibition of subcutaneously growing line 1 carcinomas due to metastatic spread. *Cancer Res.* 34:2005–2010.

Regulation of Angiogenesis
ed. by I.D. Goldberg & E.M. Rosen
© 1997 Birkhäuser Verlag Basel/Switzerland

Thrombospondin as a regulator of angiogenesis

L. A. DiPietro

Burn and Shock Trauma Institute, Loyola University Medical Center, Maywood, Ilinois 60153, USA

Introduction

The process of angiogenesis involves multiple interactions between endothelial cells and the proteins of the extracellular matrix (ECM) (Furcht, 1986; Ingber and Folkman, 1989; Ingber, 1991). ECM proteins can regulate endothelial cell function, such as proliferation and chemotaxis, and may play a role in stabilizing newly formed capillaries (Madri and Williams, 1983; Madri et al., 1988; Montesano et al., 1995). Several components of the ECM, including laminin, SPARC (osteonectin), collagens, and tenascin influence endothelial cell morphology (Montesano et al., 1993; Form et al., 1986; Grant et al., 1989; Murphy-Ullrich et al., 1991; Sage et al., 1989). Another ECM molecule that has profound effects on endothelial cell function is thrombospondin (TSP). TSP is a large multi-functional ECM glycoprotein that can influence endothelial cell function *in vitro* (Tuszynski et al., 1987; Murphy-Ullrich and Mosher, 1987; Lahav, 1988; Murphy-Ullrich and Höök, 1989) and angiogenesis both *in vitro* and *in vivo* (Rastinejad et al., 1989; Good et al., 1990; Iruela-Arispe et al., 1991a; DiPietro et al., 1994).

The role of TSP in modulating endothelial cell function has been widely investigated *in vitro* since the 1970s. However, it was the novel findings of Bouck and colleagues that generated exceptional interest in the function of TSP as a regulator of angiogenesis. These studies described TSP as an inhibitor of angiogenesis whose expression was linked to a tumor suppressor gene (Rastinejad et al., 1989; Good et al., 1990). This study expanded scientific awareness of the role of extracellular matrix molecules and the angiogenic process. Although it was known that ECM molecules could modulate endothelial cell function, such large molecules had not been previously been investigated as inhibitors of angiogenesis *in vivo*. More importantly, this finding challenged previous assumptions that the acquisition of the angiogenic phenotype depended solely upon an increased production of angiogenic cytokines. The discovery that the development of the angiogenic phenotype could involve the loss of an inhibitor provided a new paradigm. Indeed, further investigations have proved this model correct in several system (Dameron et al., 1994; Weinstat-Saslow et al., 1994).

Despite the clarity of the above findings, recent investigations of the regulatory role of TSP in angiogenesis have generated much conflicting and apparently contradictory data. Depending upon the system, TSP has been shown to be either inhibitory or stimulatory to angiogenesis, leading to much debate regarding the *in vivo* function of this molecule. This chapter will summarize current evidence as to the modulation of angiogenesis by TSP, and will provide a model for understanding the apparently inconsistent information about the role of TSP as a regulator of angiogenesis.

The structure and distribution of TSP

TSP was initially identified and characterized as a protein released from the alpha granules of platelets upon activation by thrombin (Baenziger et al., 1971). TSP is now known to be produced by many cell types and is found widely distributed in numerous tissues (Lawler, 1986; Wight et al., 1985) (Table 1). TSP exhibits a disconcerting array of binding sites and capacities (Bornstein, 1995), and binds cell surfaces and ECM proteins through one of several identified receptors (Table 1). TSP also undergoes active endocytosis and lysosomal degradation through attachment to the low density lipoprotein receptor-related protein (Godyna et al., 1995). Its widespread distribution, as well as the ability of the molecule to bind several types of cells and proteins, suggests that TSP1 has many *in vivo* functions (Bornstein, 1995).

Table 1. The extensive functional interactions of TSP1

A. Cell types that produce TSP1
Endothelial cells
Fibroblasts
Smooth muscle cells
Keratinocytes
Monocytes/macrophages
Neutrophils

B. Cell surface receptors for TSP1
Integrins
CD36
Syndecans
Betaglycans
Low density lipoprotein receptor-related protein
Other proteoglycans

C. Proteins that bind to TSP1
Fibronectin
SPARC/osteonectin
Serine proteases
Collagen (I, V)
Plasminogen
Plasminogen activator
Fibrin
Fibrinogen
TGF-β

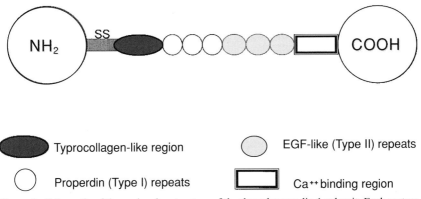

Figure 1. Schematic of the molecular structure of the thrombospondin 1 subunit. Each mature TSP1 molecule is composed of three covalently linked subunits. Homology regions within the stalk of the molecule are indicated.

The multi-functional nature of TSP is illustrated by its complex molecular structure (Fig. 1). TSP exists as a trimer of three identical subunits; the trimeric molecule has a molecular weight of approximately 450 kd. Each subunit contains two globular terminal domains joined by a central stalk (Galvin et al., 1985; Lawler et al., 1985). The amino terminal globular domains is heparin-binding while the carboxy domain is involved in cell attachment. The stalk can be further divided into five regions: (1) a region possessing the cysteines involved in interchain disulfide bonding, (2) a region homologous to procollagen type I, (3) a region containing three properdin (Type I) repeats, (4) a region containing three EGF-like (Type II) repeats, and (5) a Ca^{++} binding region.

Four additional TSP-related molecules that are encoded by four separate loci have recently been identified (Adams and Lawler, 1993; Bornstein and Sage, 1994). The TSP family of glycoproteins therefore now includes TSP1 described above) as well as TSP2, TSP3, TSP4 and TSP5. TSP1 and TSP2 are quite similar in form, as both are trimeric proteins with similar structural motifs. The related structures of TSP1 and TSP2 imply that these molecules may have some overlapping functions. However, TSP1 and TSP2 exhibit differential tissue distribution, suggesting that these molecules have distinct physiologic functions. TSP3, TSP4, and TSP5 are structurally less similar to TSP1, as they are pentameric molecules. TSP3, TSP4, and TSP5 also appear to have a more limited and specific tissue distribution. Because the members of the TSP family other than TSP1 have only recently been described, little is known about the role of these newly discovered proteins in modulating the angiogenic process. This chapter therefore focuses on TSP1 as it is the best studied member of the TSP family of molecules.

Table 2. Effects of soluble or bound TSP1 on endothelial cell function *in vitro*

Endothelial cell function

TSP1 form	Proliferation	Attachment	Chemotaxis	Sprout/tube formation
Soluble	↓	↓	↓↑*	↓
Substrate-bound	None	↑	?	↑**

↑ Promotes
↓ Inhibits
? Unknown
* Concentration dependent
** In aortic explants

TSP1 as a regulator of *in vitro* endothelial cell function

More than 100 investigations have studied the role of TSP1 in modulating endothelial cell function *in vitro*. The effect of TSP1 on several parameters of endothelial cell function that are relevant to the angiogenic process has been examined. The influence of TSP1 on endothelial proliferation, adhesion, and chemotaxis has been reported, and the regulation of *in vitro* angiogenesis by TSP1 has also been examined. Although *in vitro* investigations have employed several types of endothelial cells, including those of aortic, microvascular, and umbilical vein origin, the studies generally present a consistent set of findings regarding the ability of TSP1 to influence endothelial cell function *in vitro* (Table 2).

Most studies of the effect of TSP1 on endothelial cell function involve the addition of exogenous purified TSP1 to cultured endothelial cells. However, endothelial cells themselves produce large amounts of TSP1, and the level of production is influenced by the phenotype. Confluent endothelial cells, when plated on gelatin or collagen, and exhibiting a typical cobblestone morphology, actively produce TSP1. TSP1 synthesis diminishes if the cells differentiate or are induced to differentiate into a sprouting morphology (Canfield et al., 1990; Iruela-Arispe et al., 1991 b; DiPietro et al., 1994). These studies suggest that the effect of adding exogenous TSP1 may vary depending upon the morphologic and biosynthetic state of the culture system. Another consideration of *in vitro* studies is the molecular form of TSP1. The functional capacity of TSP1 depends upon its molecular form as the effect of matrix-bound TSP1 is often different from that of the soluble form.

TSP1 and endothelial cell proliferation

When added to cultured endothelial cells, soluble TSP1 is generally antiproliferative (Iruela-Arispe et al., 1991 a; Taraboletti et al., 1990; Bagavandoss and Wilks, 1990). Soluble TSP1 has been demonstrated to inhibit the proliferation of a wide variety of endothelial cells including those derived from rabbit corpus luteum, bovine adrenal cortex, bovine pulmonary artery, and human umbilical vein (Bagavandoss and Wilks, 1990). For example, soluble TSP1 strongly inhibits the proliferation of lung capillary endothelial cells in response to either fetal calf serum or bFGF in a dose-dependent fashion. Such results have been substantiated by work in trans-formed endothelial cell lines. These lines, which exhibit decreased levels of endogenous TSP1 production, are growth inhibited by the addition of exo-genous TSP1 (RayChaudhury et al., 1994).

In contrast to the effect of soluble TSP1, matrix-bound TSP1 appears to be a permissive substrate for endothelial cell proliferation (Nicosia and Tuszynski, 1994). When plated on a TSP1-collagen matrix, rat aortic endo-thelial cells do not proliferate. However, in response to bFGF, a three-fold increase in cell proliferation is observed in endothelial cells plated on the TSP1-collagen substrate. These results emphasize that the effect of TSP1 on endothelial cells is dramatically affected by the solubility state of the molecule. In addition, these studies suggest that, in order to prevent an autocrine inhibition of proliferation, the endogenous TSP1 that is produced by rapidly proliferating endothelial cells must either be rapidly degraded or quickly sequestered into the ECM.

TSP1 and endothelial cell adhesion

Investigations of the effect of TSP1 on endothelial cell adhesion have pro-duced some apparently conflicting results. TSP1 has been reported to modulate focal adhesions in endothelial cells (Murphy-Ullrich and Höök, 1989) and to interfere with endothelial cell attachment (Murphy-Ullrich and Mosher, 1987; Lahav, 1988). However, TSP1 has also been reported to promote endothelial cell adhesion and spreading (Tuszynski et al., 1987; Taraboletti et al., 1990). Murphy-Ullrich and Höök (1989) have shown that the effect of TSP1 on endothelial cell function is critically influenced by whether the TSP1 is in a soluble or an insoluble form. Taken together, the evidence supports the hypothesis that endothelial cells may adhere to insoluble, matrix-bound TSP1, while soluble TSP1 may saturate surface receptors and produce an anti-adhesive effect. The anti-adhesive activity of soluble TSP1 might be expected to interfere with the angiogenic process by preventing the cell-to-substrate or cell-to-cell interactions necessary for endothelial cell migration and capillary formation.

TSP1 and endothelial cell chemotaxis

Several studies have demonstrated that exogenous TSP1 has a negative effect on endothelial cell chemotaxis. When added to a modified Boyden chamber assay, purified TSP1 significantly inhibits endothelial cell chemotaxis to bFGF (Good et al., 1990; Tolsma et al., 1993). Given that soluble TSP1 is anti-adhesive for endothelial cells, this finding is expected, as TSP1 could inhibit chemotaxis by preventing the endothelial cell attachments that are necessary for cell movement. Interestingly, the effect of exogenous TSP1 on endothelial cell chemotaxis appears to be concentration-dependent. At high concentrations, TSP1 can stimulate rather than inhibit endothelial migration (Tolsma et al., 1993; Taraboletti et al., 1990). In addition, at high concentrations, TSP1 alone has been shown to be chemotactic for endothelial cells (Taraboletti et al., 1990). The inhibitory effect of TSP1 on chemotaxis has been localized to the central stalk region of the molecule, while the stimulatory activity has been localized to the amino terminal heparin binding domain of the molecule (Tolsma et al., 1993; Taraboletti et al., 1990).

The effect of endogenously produced TSP1 on endothelial cell chemotaxis has also been investigated. Endothelial cells that have been engineered to produce reduced amounts of endogenous TSP1 exhibit enhanced chemotactic activity (DiPietro et al., 1994). This description further substantiates an anti-chemotactic effect of TSP1. *In vivo*, the dominant activity of TSP1 on chemotaxis may be dictated by the local concentration and solubility, as well as the accessibility of specific binding domains.

TSP1 and in vitro *angiogenesis*

TSP1 is actively synthesized by proliferating endothelial cells, but synthesis decreases significantly as endothelial cells organize into capillary-like structures *in vitro* (Canfield et al., 1990; Iruela-Arispe et al., 1991b). This observation suggests that the production of TSP1 might interfere with the cell-to-cell or cell-to matrix interactions necessary for capillary tube formation. The effect of TSP1 on the formation of capillary cords and tubes *in vitro* has been described in several systems, including those in which endothelial cells undergo spontaneous tube formation, and another in which *in vitro* angiogenesis is induced by plating the cells on gelled basement membrane matrix (Matrigel). Spontaneous *in vitro* capillary tube formation proceeds more rapidly in the presence of anti-TSP1 antibodies (Iruela-Arispe et al., 1991a). In another study, endogenous TSP1 production by endothelial cells was specifically disrupted through the transfection of an antisense TSP1-expressing plasmid (DiPietro et al., 1994). As compared to control, cells in which TSP1 production was reduced exhibited a notably augmented *in vitro* angiogenic response. Cells in which TSP1 production was reduced displayed more rapid formation of endothelial cords when plated on gelled basement membrane matrix, forming two-fold more cord structures within 24 h. Interestingly, the decrease in TSP1 production had no observable effect upon cell morphology or growth when the cells were grown in standard conditions.

Due to the complexity of interactions between TSP1 and cells, it is not surprising that TSP1 has been shown to promote rather than to inhibit *in vitro* angiogenesis in some systems. Working in a system that investigates capillary angiogenesis as an outgrowth of aortic explants, Nicosia and Tuszynski (1994) have demonstrated that TSP1 can promote *in vitro* angiogenesis. In this system, aortic rings are placed in fibrin or collagen gels, and capillary outgrowth monitored. When TSP1 was added to these gels, the number of microvessels that formed was approximately doubled. The authors went on to demonstrate that the effect of TSP1 in this system was most likely via the myofibroblast rather than the endothelial cell itself. TSP1 was found to promote myofibroblast growth, which in turn was believed to stimulate angiogenesis. This finding remarkably emphasizes two aspects of TSP1 function. First, the data reiterate that the form of the molecule may profoundly influence the observed effects. In this study, TSP1 was incorporated into matrix, as opposed to addition in soluble form. Second, this study demonstrates that TSP1 can influence angiogenesis via interactions with cell types other than endothelial cells. For example, TSP1 may augment angiogenesis when myofibroblasts are present, yet may inhibit angiogenesis in systems containing endothelial cells alone.

The *in vitro* findings that soluble TSP1 inhibits endothelial proliferation, chemotaxis, and tube formation, suggest that the soluble TSP1 concentration would generally be low in areas of active angiogenesis. If local TSP1 production is high, an angiogenic environment might prevail if either TSP1 turnover or ECM sequestration was rapid. These factors would eliminate any substantial increase in local concentration of the soluble molecule.

While the *in vitro* evidence that TSP1 plays a role in modulating angiogenesis is impressive, a correlation between *in vitro* experiments and the *in vivo* processes of physiologic or pathologic angiogenesis has not been forthcoming. Due to the complexity of the *in vivo* system, direct experiments that assess the role of TSP1 in *in vivo* angiogenesis are difficult to perform. The amount of soluble versus matrix-bound TSP1, which is clearly critical to the role of this molecule in influencing endothelial cell function, has not been assessed *in vivo*. In addition, *in vivo*, the influence of TSP1 on cell types other than endothelial cells is difficult to determine.

Other interactions of TSP1 that may influence angiogenesis

In addition to direct effects on cells, TSP1 displays interactions with several proteins that may be critical within angiogenic environments. The most relevant of these accessory interactions of TSP1 are described here.

TSP as an inhibitor of proteases

Recent reports have produced the unexpected finding that TSP1 can function as a protease inhibitor (Hogg, 1994). These findings suggest that, apart from direct effects on cells, TSP1 may also influence angiogenesis by affecting ECM turnover and composition. TSP1 inhibits several different types of enzymes, many of which are prominent in areas of inflammation. TSP1 inhibits neutrophil elastase and cathepsin G, both serine proteases derived from the azurophilic granules of neutrophils (Hogg et al., 1993a, Hogg et al., 1993b). There is also evidence that TSP1 may inhibit the proteolytic enzymes of the fibrinolytic pathway, including plasmin and urokinase plasminogen activator (Hogg et al., 1992; Mosher et al., 1992). Each of these enzymes is prominent in sites of injury or inflammation, which are often areas of active angiogenesis. The modulation of protease activity provides a potentially crucial role for TSP1 in regulating angiogenesis, as the correct balance of protease activity is essential for angiogenesis. Capillary growth requires active remodeling of the ECM, including degradation and resynthesis (Ingber, 1991). The *in vivo* influence of TSP1 on the local protease balance remains to be examined.

The interaction of TSP1 with TGF-β

Our understanding of the role of TSP1 in the angiogenic process has been furthered by the recent finding that TSP1 can bind and activate transforming growth factor-β (Murphy-Ullrich et al., 1992; Schultz-Cherry and Murphy-Ullrich, 1993; Schultz-Cherry et al., 1994a). The finding that TSP1 can cause TGF-β activation suggests a novel function for this molecule in influencing cell growth and differentiation. The significance of the interaction of TSP1 with TGF-β to the process of angiogenesis may be great, as TGF-β has been shown to inhibit endothelial cell growth in culture, yet is angiogenic *in vivo* (Mueller et al., 1987; Roberts et al., 1986). Many cell types that can be found in areas of active angiogenesis produce both TGF-β and TSP1. For example, TGF-β and TSP1 are both found in the alpha granules of platelets, and are both produced by endothelial cells and macrophages (Assoian et al., 1987; DiPietro and Polverini, 1993; Schultz-Cherry and Murphy-Ullrich, 1993).

TGF-β, which is secreted by most cell types in an inactive complex, is activated by the dissociation of a latency-associated peptide. The activation of TGF-β is mediated by specific peptide sequences within the properdin-like region (Type I repeats) of the TSP1 molecule (Schultz-Cherry et al., 1994b; Schultz-Cherry et al., 1995). The activation of TGF-β by TSP1 can occur in the absence of cells (Schultz-Cherry and Murphy-Ullrich, 1993). Therefore, the potential exists for TSP1 to regulate TGF-β activation physiologically in areas of angiogenesis. While at least part of the biologic activity of TSP1 may therefore be attributed to the activation of TGF-β (Murphy-Ullrich et al., 1992), the extent of this influence *in vivo* is currently unclear.

TSP1 and *in vivo* angiogenesis

Several studies have demonstrated that purified TSP1 can inhibit *in vivo* angiogenesis. The first of these, as described earlier, was the report of TSP1 as an inhibitor of angiogenesis whose expression was linked to a tumor suppressor gene (Rastinejad et al., 1989). The *in vivo* assay that has been used for the study of TSP1 as a regulator of angiogenesis is the corneal assay of angiogenic activity. In this assay, inert polymer pellets containing the substance to be tested are placed into surgical pockets within the normally avascular cornea. After a period of 5 to 14 days, the amount of capillary outgrowth from the vascular limbus of the eye toward the pellet is recorded. Angiogenic stimuli generally result in a brisk growth of numerous capillary loops from the limbus toward the pellet. If angiogenesis is inhibited, few or no capillary sprouts are observed.

Utilizing the rat corneal assay of angiogenic activity, Good et al. (1990) demonstrated that either truncated TSP1 or purified platelet-derived TSP1 completely inhibited the angiogenic activity of bFGF. This inhibition could be blocked through the addition of antibodies directed against TSP1 to the pellet. Similarly, TSP1 has also been shown to inhibit the angiogenic activity of conditioned media from tumor cells (Rastinejad et al., 1989; Dameron et al., 1994). Because tumor angiogenic activity is often complex, and involves several different angiogenic mediators, these results suggest that the inhibitory activity of TSP1 extends to multiple different angiogenic factors. The majority of the *in vivo* anti-angiogenic activity of TSP1 has been localized to the central stalk region of the molecule (Tolsma et al., 1993). Further analysis, using truncated forms of recombinant TSP1, has more specifically localized the anti-angiogenic activity to the procollagen-like region of the stalk, and several peptides from this region are able to block angiogenesis *in vivo*. Anti-angiogenic activity has also been described for the peptides derived from the properdin repeat region. Thus at least two different structural domains may contribute to the anti-angiogenic activity of the TSP1 molecule.

One conflicting account regarding the anti-angiogenic activity of TSP1 *in vivo* has been reported. BenEzra et al. (1993) has shown that, in a rabbit corneal assay of angiogenic activity, human platelet-derived TSP1 augmented, rather than inhibited, the angiogenic activity of bFGF. These investigators have also shown that TSP1, while not independently angiogenic, augmented the angiogenic response to LPS. This augmentation was associated with an observed infiltration of leukocytes in response to TSP1. The direct conflict of this report with preceding reports is difficult to explain. The source and preparation of TSP1 were similar in each report, as was the source of bFGF and the weight:weight ratio of TSP1 to bFGF. One difference between the two studies involves the species used for the corneal assay. An anti-angiogenic effect of TSP1 was demonstrated in rat, while the pro-angiogenic activity was demonstrated in a rabbit model. Whether subtle variances in amount of TSP1, TSP1 : bFGF ratios, or species differences contribute to the observed biologic differences remains to be investigated. One significant technical divergence between the studies involves the method of incorporation of test substances into the inert polymers. In the studies that demonstrate an inhibitory effect of TSP1, bFGF and TSP1 were mixed prior to incorporation into a single pellet, which was then implanted. In contrast, studies describing a stimulatory effect of TSP1 on angiogenesis involved independent preparation of pellets containing either bFGF or TSP1. Two pellets were then implanted simultaneously. Given that both substances would be expected to diffuse into the cornea, this technical difference may not explain the disparity in results. However, as others have suggested, TSP1 may not be readily diffusible (Nicosia and Tuszynski, 1994). Because TSP1 has been shown to bind growth factors (Murphy-Ullrich et al., 1992), it is possible that when mixed directly with bFGF, TSP1 sequesters this growth factor. Thus, when incorporated together, a direct interaction of TSP1 with the growth factor occurs within the pellet. This hypothesis is readily testable, and could explain the results described in these particular experiments. Nonetheless, several pieces of experimental evidence suggest that sequestration of angiogenic cytokines does not account for all of the observed anti-angiogenic activity of TSP1. TSP1 inhibits the angiogenic response to multiple different angiogenic factors and substances (Tolsma et al., 1993). In addition, soluble TSP1 inhibits neovascularization in wounds, a situation in which numerous different angiogenic cytokines are likely to be produced (Tolsma et al., 1993).

TSP1 and tumor angiogenesis

A down-regulation of TSP1 synthesis, in concert with the development of a tumorigenic and angiogenic phenotype, has been documented in multiple tumor systems (Good et al., 1990; Dameron et al., 1994; Zabrenetzsky et al., 1994). This finding was first described in a chemically transformed hamster kidney cell line (Rastinejad et al., 1989). Since this initial description, several other cell lines in which a decrease in TSP1 production is associated with the acquisition of the malignant, angiogenic, phenotype have been described. These include cells derived from melanoma, breast, and epithelial tumors (Zabrenetzsky et al., 1994). In these cell lines, TSP1 is thought to serve as the major inhibitor of the angiogenic phenotype prior to malignant transformation. The ability of TSP1 to modulate not only the angiogenic, but also the *in vivo* tumorigenic phenotype of cells has recently been described (Weinstat-Saslow et al., 1994). A human breast cell carcinoma line that was metastatic and produced relatively low levels of TSP1 was used in these studies. Transfection of a TSP1 cDNA into this cell line resulted in reduced tumor growth and metastases. The artificial upregulation of TSP1 in these cells led to substantially decreased angiogenesis at the tumor site, which was quantitated as a reduction in capillary density. A similar finding has also been documented in a transformed endothelial cell line that forms hemangiomas *in vivo* and expresses low levels of TSP1 (Sheibani and Frazier, 1995). Transfection of a TSP1 expression vector into these cells resulted in cells with decreased growth rate *in vitro*. Further, transfected cells no longer formed tumors *in vivo*. These two studies indicate that the production of TSP1 can substantially inhibit tumor progression and angiogenesis *in vivo*.

Decreased TSP1 production has been linked to the loss of a tumor suppressor gene in several tumor lines. Such a linkage has been elegantly demonstrated in fibroblasts from Li-Fraumeni patients (Dameron et al., 1994). These patients, who exhibit an increased risk of developing tumors, carry one mutant and one normal allele of the p53 suppressor gene. When fibroblasts from Li-Fraumeni patients are cultured, the loss of the wild-type p53 suppressor allele correlates with a reduced expression of TSP1, and the simultaneous acquisition of the angiogenic phenotype. In further experiments, the replacement of wild type p53 by transfection restored TSP1 production and the ability of supernatant from these cell to inhibit angiogenesis.

The above studies in tumor cell systems provide compelling evidence for the negative regulation of the angiogenic phenotype in tumor cells by TSP1. As powerful as this evidence may be, other studies suggest a positive role for TSP1 in tumor growth and progression. Several investigations have described the active production of TSP1 by tumor cells both *in vitro* and *in vivo*. Both TSP1 and one of its specific receptors have been localized *in situ* in human breast carcinoma (Tuszynski and Nicosia, 1994). *In vitro*, cell lines derived from breast cancer, melanoma, squamous cell carcinoma, and osteosarcoma cells (Clezardin et al., 1989; Varani et al., 1987; Apelgren and Bumol, 1989; Pratt et al., 1989) have each been shown to exhibit marked production of TSP1. TSP1 can, at least in some cases, apparently promote tumor growth and metastasis. In at least two systems, the production of TSP1 by the tumor line appears to influence metastatic capacity positively, suggesting a positive functional role for this protein in tumor progression (Castle et al., 1991; Tuszynski et al., 1987). Some tumor cells utilize TSP1 for the attachment that is required for the development of metastases. A recent study of oral squamous cell carcinoma demonstrates that TSP1 facilitates tumor invasion, most probably through the activation of TGF-β (Wang et al., 1995). Of note is that, in the above studies, the effect of increased levels of TSP1 on tumor angiogenesis has not been specifically addressed.

It is generally accepted that all solid tumors must be angiogenic to grow beyond a few mm in diameter. How then, is it possible to reconcile the documented production of the apparent angiogenic inhibitor TSP1 by presumably angiogenic malignant cells? First, this paradox serves as a reminder of our poor state of understanding of the multiplicity of functions of TSP1. The contradictory finding leads to the conclusion that within some specific biologic systems, TSP1 does not exert an anti-angiogenic effect. There is ample evidence for this possibility, as the influence of TSP1 on angiogenesis clearly depends upon local concentration and solubility. The *in vivo* variability of these parameters within solid tumors is currently unknown, as the production of TSP1 is most often monitored in *in vitro* experimentation. If TSP1 is quickly bound to the ECM, or rapidly endocytosed and degraded, the influence on angiogenesis may be minimal. The *in vivo* effect of tumor-derived TSP1 on functions related to angiogenesis, including (1) protease activity, (2) growth factor activation, and (3) the function and proliferation of cells other than endothelial cells, remains to be rigorously examined.

TSP1 in inflammation and wound healing

Sites of inflammation and injury often exhibit robust angiogenesis. In the case of normal healing wounds, this angiogenesis is tightly regulated. Investigations of TSP1 production and function in these scenarios have contributed to our knowledge of the *in vivo* regulation of angiogenesis.

TSP and inflammatory cells

Two inflammatory cells that have been shown to produce TSP1 include neutrophils and macrophages (Jaffe et al., 1985; DiPietro and Polverini, 1993; Kreis et al., 1989). This finding is of particular interest because macrophages are known to be angiogenic, and may contribute to angiogenesis in the setting of wounds, autoimmune diseases, and tumors. Macrophages in wounds, as well as in other inflammatory settings, actively produce TSP1. The finding that angiogenic macrophages produce TSP1, a molecule that can function to inhibit angiogenesis, at first seems puzzling. However, there are several possible functions for macrophage-derived TSP1 that are unrelated to angiogenesis. Macrophages have surface receptors for TSP1 (Silverstein et al., 1989), and TSP1 has been shown to promote monocyte chemotaxis (Mansfield and Suchard, 1994) and to facilitate macrophage migration through endothelial cell layers *in vitro* (Huber et al., 1992). TSP1 is also involved in the recognition and clearance of senescent neutrophils by macrophages, and may therefore link macrophages and neutrophils in the inflammatory process (Savill et al., 1992). Finally, as described above, TSP1 may activate TGF-β from its latent form. In this context then, TSP1 production by macrophages might enhance rather than inhibit the angiogenic profile. The many possible functions of TSP1, as well as its active production by wound macrophages, provide an exciting basis for the examination of the role of this protein in inflammatory and reparative processes. Some of the current evidence regarding the role of macrophage-derived TSP1 in the context of specific inflammatory diseases and healing wounds is reviewed below.

TSP1 in inflammatory diseases

The production of TSP1 has been investigated in several inflammatory diseases, including psoriasis, atherosclerosis, glomerulonephropathies, and rheumatoid arthritis. Perhaps the best studied of these with respect to the functional significance of TSP1 is psoriasis. This skin disease is character- ized by an excessive growth of keratinocytes with associated inflammation and neovascularization. The angiogenic component of this disease is medi- ated, at least in part, by psoriatic keratinocytes (Nickoloff et al., 1994). Conditioned medium from psoriatic keratinocytes induces a vigorous angiogenic response in the rat corneal assay. In contrast, conditioned medi- um from normal keratinocytes is non-angiogenic or only weakly positive. The increase in the angiogenic capability of psoriatic keratinocytes has been linked to changes in the level of both TSP1 and interleukin-8 (IL-8) synthesis by these cells. Compared to normal keratinocytes, psoriatic kera- tinocytes exhibited a seven-fold reduction in TSP1 production. Concomitant with decreased TSP1 production, psoriatic keratinocytes exhibit a 10- to 20-fold increase in the production of IL-8, a potently angiogenic cytokine. This finding once again illustrates the concept that the acquisition of the angiogenic phenotype can involve a decrease in the production of angio- genic inhibitor and an associated increased in positive angiogenic cytokines.

In contrast to psoriasis, increased TSP1 production has been noted in several other types of inflammatory lesions that have an angiogenic component, including rheumatoid arthritis and atherosclerosis (Botney et al., 1992; Koch et al., 1993). Immunostaining of rheumatoid synovium indicates that both macrophages and endothelial cells within these lesions are positive for TSP1. Because these lesions encompass active angio- genesis as a component of their pathophysiology, TSP1 production in these tissues prevents yet another paradox in our understanding of TSP1 as a regulator of angiogenesis.

TSP1 and wound healing

Normal wound repair represents a well-defined example of inflammation that includes macrophage infiltration, physiologic angiogenesis, and fibroplasia. The *in vivo* production of TSP1 in normal dermal wound repair has been investigated by several groups (Raugi et al., 1987; Reed et al., 1993; DiPietro et al., 1996). TSP1 is released by platelets as a result of the hemostasis that occurs following injury, and is found in the thrombus at sites of injury. In incisional wounds, TSP1 production by cells within the wounds is modest. *In situ* hybridization and immunostaining studies of rat incisional wounds have documented the transient expression of TSP1 protein and mRNA within the thrombus (Reed et al., 1993). TSP1 protein is also observed at the wound edges. In human incisional wounds, TSP1 has been immunolocalized along the lateral and deep margins of the wounds, as well as in vascular channels at the wound edges and in the new granulation tissue (Raugi et al., 1987).

By comparison TSP1 production in full thickness excisional wounds is more pronounced. Full thickness excisional dermal wounds are characterized by significant inflammation and granulation tissue formation prior to resolution. In these wounds, a detectable increase in the level of TSP1 mRNA is seen during the early inflammatory phase of wound repair. *In situ* hybridization studies have shown that macrophages containing high levels of TSP1 mRNA are present within the early wound bed, but not within normal skin (DiPietro et al., 1996). The early wound environment has been documented to be remarkably angiogenic, and this observation therefore again provides evidence for alternative roles for TSP1. TSP1 within the early wound may be quickly bound and sequestered, or may undergo rapid turnover. Alternatively, macrophage-derived TSP1 within wounds may perform a novel function. As described above, TSP1 may influence macrophage migration, or the activation of TGF-β. Because activated TGF-β can promote both fibrous tissue repair and capillary regrowth, the production of TSP1 in the wound may serve to augment the repair response (Roberts et al., 1986).

The functional role of local TSP1 production within wounds has been further studied in TSP1 knockout mice (Polverini et al., 1995). Initially, wounds of knockout mice exhibit a delayed healing response. However, with time, the wounds of the knockout mice fail to resolve, manifesting as increased granulation tissue formation, prolonged neovascularization, and sustained monocyte/macrophage infiltration. The introduction of pure TSP1 into wounds in knockout mice is able to reverse these effects partially, resulting in a more rapid onset and earlier resolution of wound healing (Polverini et al., personal communication). These findings suggest a dual role for TSP1 in wound repair. In the early phase of repair, TSP1 synthesis by macrophages would facilitate the repair process, either by activating TGF-β or through local interaction with macrophages or other inflammatory cells. As repair progresses, soluble forms of TSP1 would function as anti-angiogenic agents and cause a cessation of neovascularization.

A model for the regulation of angiogenesis by TSP1

The multiple and complex roles of TSP1 as a regulator of angiogenesis may seem difficult to reconcile. Experimental evidence that TSP1 is a key inhibitor of angiogenesis in several specific systems, such as tumors and psoriasis, is strong and convincing. However, TSP1 is abundant in other tissues where angiogenesis is prominent. This conflicting evidence can be taken to indicate that the effect of TSP1 on angiogenesis in any single system depends upon the interplay of several functions (Bornstein, 1995).

One simple consideration that may dictate the particular influence of TSP1 on angiogenesis is the relative level of production of this molecule. The overall angiogenic response in a tissue clearly depends upon the relative balance of angiogenic and anti-angiogenic factors. Thus, in the face of high levels of pro-angiogenic molecules, the local production of TSP1 may not inhibit angiogenesis. Conversely, exceedingly high levels of TSP1 would generally overwhelm angiogenic factors, and result in an anti-angiogenic profile.

While this simple paradigm may provide a framework for understanding TSP1 as a modulator of angiogenesis *in vivo*, the actual *in vivo* situation is likely to be much more complex. Any model of TSP1 function in angiogenesis must encompass several compelling facts. First, soluble TSP1, and specific soluble fragments of TSP1, are anti-angiogenic. Second, bound TSP1 may promote the angiogenic response. Third, TSP1 may indirectly influence the angiogenic process via interactions with cell types other than endothelial cells. Finally, TSP1 can bind other angiogenic factors and influence their activity. By taking each of these details into account, a model of TSP1 influence on angiogenesis can be developed (Fig. 2). Simply put, whether or not TSP1 inhibits or supports angiogenesis would depend upon the local environment. The local concentration and predominant form of TSP1 would be dictated by (1) the presence of ECM molecules that bind TSP1, (2) local protease activity, and (3) the TSP1 turnover rate. TSP1 would be anti-angiogenic in its soluble form, particularly if it was cleaved into small diffusible fragments. Matrix-bound TSP1 would instead support angiogenesis, either by interactions with endothelial cells or other cell types. The scenario would be further modified by the simultaneous production of TGF-β, which could be activated by TSP1.

A model for a dual influence of TSP1 would explain many of the apparently conflicting observations regarding this protein, particularly in view of its role in tumor growth. For example, in tumors where TSP1 levels appear to be high, TSP1 might reside bound within the ECM. In this way, TSP1 could promote tumor cell adhesion, positively influence angiogenesis through its interaction with non-endothelial cells such as myofibroblasts, and even facilitate metastatic growth. In contrast, in other circumstances, large amounts of TSP1 or TSP1 fragments may remain soluble. In this context, TSP1 would inhibit endothelial cell growth, chemotaxis,

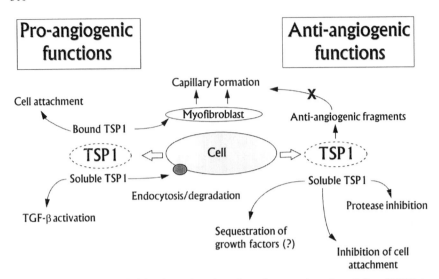

Figure 2. Theoretical model for the anti-angiogenic and pro-angiogenic activity of TSP1 *in vivo*. Cell synthesis of TSP1 may produce a pro-angiogenic or anti-angiogenic environment. The effect of TSP1 on angiogenesis depends upon the solubility of the molecule, the presence of TGF-β, the endocytic turnover rate, and the likelihood of degradation of TSP1 into smaller diffusible anti-angiogenic peptides. The angiogenic profile of the tissue may also be influenced by the balance of TSP1 production to angiogenic cytokine production (not shown).

and capillary tube formation. The loss of TSP1 production would then facilitate the transformation to a tumorigenic phenotype by allowing for angiogenesis.

Numerous excellent investigations have helped scientists to paint the current portrait of TSP1 as a regulator of angiogenesis. However, as each new brushstroke appears, it becomes more obvious that the picture is not yet complete. Additional investigations, both *in vitro* and *in vivo*, will be necessary to comprehend fully the precise influence of TSP1 on the *in vivo* angiogenic process.

Acknowledgments
I thank my colleagues Dr Peter J. Polverini and Dr Mark Lingen for helpful discussions. This work was supported by the Dr Ralph and Marion C. Falk Foundation and the National Institutes of Health.

References

Adams, J. and Lawler, J. (1993) The thrombospondin family. *Curr. Biol.* 3:188–190.

Apelgren, L.D. and Bumol, T.F. (1989) Biosynthesis and secretion of thrombospondin in human melanoma cells. *Cell Biol. Intl. Rep.* 13:189–195.

Assoian, R.K., Fleur De Lys, B.E., Stevenson, H.C., Miller, P.J., Madtes, D.K., Raines, E.W., Ross, R. and Sporn, M.B. (1987) Expression and secretion of type beta transforming growth factor by activated human macrophages. *Proc. Natl. Acad. Sci. USA* 84:6020–6024.

Bagavandoss, P. and Wilks, J.W. (1990) Specific inhibition of endothelial cell proliferation by thrombospondin. *Biochem. Biophys. Res. Comm.* 170:867–872.

Baenziger, N.L., Brodie, G.N. and Majerus, P.W. (1971) A thrombin-sensitive protein of human platelet membranes. *Proc. Natl. Acad. Sci. USA* 68:240–243.

BenEzra, D., Griffin, B.W., Maftzir, G. and Aharonov, O. (1993) Thrombospondin and *in vivo* angiogenesis induced by basic fibroblast growth factor of lipopolysaccharide. *Invest. Ophthalmol. Vis. Sci.* 34:3601–3608.

Bornstein, P. (1995) Diversity of function is inherent in matricellular proteins: an appraisal of thrombospondin 1. *J. Cell Biol.* 130:503–506.

Bornstein, P. and Sage, E.H. (1994) Thrombospondins. *Methods Enzymol.* 245:62–85.

Botney, M.D., Kaiser, L.R., Cooper, J.D., Mecham, R.P., Parghi, D., Roby, J. and Parks, W.C. (1992) Extracellular matrix protein gene expression in atherosclerotic hypertensive pulmonary arteries. *Am. J. Pathol.* 140:357–364.

Canfield, A.E., Boot-Handford, R.P. and Schor, A.M. (1990) Thrombospondin gene expression by endothelial cells in culture is modulated by cell proliferation, cell shape and the substratum. *Biochem. J.* 268:225–230.

Castle, V., Varani, J., Fligiel, S., Prochownik, E. and Dixit, V. (1991) Antisense-mediated reduction in thrombospondin reverses the malignant phenotype of a human squamous carcinoma. *J. Clin. Invest.* 87:1883–1888.

Clezardin, P., Jouishomme, H., Chavassieux, P. and Marie, P.J. (1989) Thrombospondin is synthesized and secreted by human osteoblasts and osteosarcoma cells. *Eur. J. Biochem.* 181:721–726.

Dameron, K.M., Volpert, O.V., Tainsky, M.A. and Bouck, N. (1994) Control of angiogenesis in fibroblasts by p53 regulation of thrombospondin-1. *Science* 265:1582–1584.

DiPietro, L.A. and Polverini, P.J. (1993) Angiogenic macrophages produce the angiogenic inhibitor thrombospondin 1. *Am. J. Pathol.* 143:678–684.

DiPietro, L.A., Nissen, N.N., Gamelli, R.L., Koch, A.E., Pyle, J.M. and Polverini, P.J. (1996) Thrombospondin 1 synthesis and function in wound repair. *Am. J. Pathol.* 148:1861–1869.

DiPietro, L.A., Nebgen, D.R. and Polverini, P.J. (1994) Downregulation of endothelial cell thrombospondin 1 enhances *in vitro* angiogenesis. *J. Vasc. Res.* 31:178–185.

Form, D.M., Pratt, B.M. and Madri, J.A. (1986) Endothelial cell proliferation during angiogenesis: *in vitro* modulation by basement membrane components. *Lab. Invest.* 55:521–530.

Furcht, L.T. (1986) Critical factors controlling angiogenesis: cell products, cell matrix and growth factors. *Lab. Invest.* 55:505–509.

Galvin, N.J., Dixit, V.M., O'Rourke, K.M., Santoro, S.A., Grant, G.A. and Frazier, W.A. (1985) Mapping of epitopes for monoclonal antibodies against human platelet thrombospondin with electron microscopy and high sensitive amino acid sequencing. *J. Cell Biol.* 101:1434–1441.

Godyna, S., Liau, G., Popa, I., Stefansson, S. and Argraves, W.S. (1995) Identification of the low density lipoprotein receptor-related protein (LRP) as an endocytic receptor for thrombospondin-1. *J. Cell Biol.* 129:1403–1410.

Good, D.J., Polverini, P.J., Rastinejad, F., LeBeau, M.M., Lemons, R.S., Frazier, W.A. and Bouck, N.P. (1990) A tumor suppressor-dependent inhibitor of angiogenesis is immunologically and functionally indistinguishable from a fragment of thrombospondin. *Proc. Nat. Acad. Sci. USA* 87:6624–6628.

Grant, D.S., Tashiro, K.I., Segui-Real, B., Yamada, Y., Martin, G.R. and Kleinman, H.K. (1989) Two different laminin domains mediate the differentiation of human endothelial cells into capillary-like structures *in vitro*. *Cell* 58:933–943.

Hogg, P.J. (1994) Thrombospondin 1 as an enzyme inhibitor. *Thromb. Haemostasis* 72:787–792.

Hogg, P.J., Stenflo, J. and Mosher, D.F. (1992) Thrombospondin is a slow tight-binding inhibitor of plasmin. *Biochemistry* 31:265–269.

Hogg, P.J., Owensby, D.A., Mosher, D.F., Misenheimer, T.M. and Chesterman, C.N. (1993a) Thrombospondin is a tight-binding competitive inhibitor of neutrophil elastase. *J. Biol. Chem.* 268:7139–7146.

Hogg, P.J., Owensby, D.A. and Chesterman, C.N. (1993b) Thrombospondin-1 is a tight-binding competitive inhibitor of neutrophil cathepsin-G. Determination of the kinetic mechanism of inhibition and localization of cathepsin-G binding to the thrombospondin-1 type 3 repeats. *J. Biol. Chem.* 268:21811–21818.

Huber, A.R., Ellis, S., Johnson, K.J., Dixit, V.M. and Varani, J. (1992) Monocyte diapedesis through an *in vitro* vessel wall construct: inhibition with monoclonal antibodies to thrombospondin. *J. Leukoc. Biol.* 52:524–528.

Ingber, D. (1991) Extracellular matrix and cell shape: Potential control points for inhibition of angiogenesis. *J. Cell. Biochem.* 47:236–241.

Ingber, D.E. and Folkman, J. (1989) Mechanochemical switching between growth and differentiation during fibroblast growth factor-stimulated angiogenesis *in vitro*: role of extracellular matrix. *J. Cell Biol.* 109:317–330.

Iruela-Arispe, M.L., Bornstein, P. and Sage, H. (1991a) Thrombospondin exerts an antiangiogenic effect on cord formation by endothelial cells *in vitro*. *Proc. Natl. Acad. Sci. USA* 88:5026–5030.

Iruela-Arispe. M.L., Hasselaar, P. and Sage, H. (1991b) Differential expression of extracellular proteins is correlated with angiogenesis *in vitro*. *Lab. Invest.* 64:174–186.

Jaffe, E.A., Ruggiero, J.T. and Falcone, D.J. (1985) Monocytes and macrophages synthesize and secrete thrombospondin. *Blood* 65:79–84.

Koch, A.E., Friedman, J. Burrows, J.C., Haines, G.K. and Bouck, N.P. (1993) Localization of the angiogenesis inhibitor thrombospondin in human synovial tissues. *Pathobiology* 61:1–6.

Kreis, C., La Fleur, M., Ménard, C., Paquin, R. and Beaulieu, A.D. (1989) Thrombospondin and fibronectin are synthesized by neutrophils in human inflammatory joint disease and in a rabbit model of *in vivo* neutrophil activation. *J. Immunol.* 143:1961–1968.

Lahav, J. (1988) Thrombospondin inhibits adhesion of endothelial cells. *Exp. Cell. Res.* 177:199–204.

Lawler, J. (1986) The structural and functional properties of thrombospondin. *Blood* 67:1197–1209.

Lawler, J., Derick, L.H., Connolly, J.E., Chen, J.H. and Chao, F.C. (1985) The structure of human platelet thrombospondin. *J. Biol. Chem.* 260:3762–3772.

Madri, J.A. and Williams, S.K. (1983) Capillary endothelial cell cultures: phenotypic modulation by matrix components. *J. Cell Biol.* 97:153–165.

Madri, J.A., Pratt, B.M. and Tucker, A.M. (1988) Phenotypic modulation of endothelial cells by transforming growth factor-β depends upon the composition and organization of the extracellular matrix. *J. Cell Biol.* 106:1375–1384.

Mansfield, P.J. and Suchard, S.J. (1994) Thrombospondin promotes chemotaxis and haptotaxis of human peripheral monocytes. *J. Immunol.* 153:4219–4229.

Montesano, R., Mouron, P. and Orci, L. (1985) Vascular outgrowths from tissue explants embedded in fibrin or collagen gels: a simple *in vitro* model of angiogenesis. *Cell Biol. Intl. Reports* 9:869–875.

Montesano, R., Orci, L. and Vassali, P. (1983) *In vitro* rapid organization of endothelial cells into capillary-like networks is promoted by collagen matrices. *J. Cell Biol.* 97:1648–1652.

Mosher, D.F., Misenheimer, T.M., Stenflo, J. and Hogg, P.J. (1992) Modulation of fibrinolysis by thrombospondin. *Ann. NY Acad. Sci.* 667:64–69.

Mueller, G., Behrens, J. Nussbaumer, U., Bohlen, P. and Birchmeier, W. (1987) Inhibitory action of TGF-β on endothelial cells. *Proc. Natl. Acad. Sci. USA* 84:5600–5604.

Murphy-Ullrich, J.E. and Mosher, D.F. (1987) Interactions of thrombospondin with cells in culture: rapid degradation of both soluble and matrix thrombospondin. *Semin. Thromb. Hemostasis* 13:343–351.

Murphy-Ullrich, J.E. and Höök, M. (1989) Thrombospondin modulates focal adhesions in endothelial cells. *J. Cell Biol.* 109:1309–1319.

Murphy-Ullrich, J.E., Lightner, V.A., Aukhil, I., Yan, Y.Z., Erickson, H.P. and Höök, M. (1991) Focal adhesion integrity is downregulated by the alternatively spliced domain of human tenascin. *J. Cell Biol.* 115:1127–1136.

Murphy-Ullrich, J.E., Schultz-Cherry, S. and Höök, M. (1992) Transforming growth factor-β complexes with thrombospondin. *Mol. Biol. Cell* 3:181–188.

Nickoloff, B.J., Mitra, R.S., Varani, J., Dixit, V.M. and Polverini, P.J. (1994) Aberrant production of interleukin-8 and thrombospondin-1 by psoriatic keratinocytes mediates angiogenesis. *Am. J. Pathol.* 144:820–828.

Nicosia, R.F. and Tuszynski, G.P. (1994) Matrix-bound thrombospondin promotes angiogenesis *in vitro*. *J. Cell Biol.* 124:183–193.

Polverini, P.J., DiPietro, L.A., Dixit, V.M., Hynes, R.O. and Lawler, J. (1995) Thrombospondin 1 knockout mice show delayed organization and prolonged neovascularization of skin wounds. *FASEB J.* 9:A272.

Pratt, D.A., Miller, W.R. and Dawes, J. (1989) Thrombospondin in malignant and non-malignant breast tissue. *Eur. J. Cancer Clin. Oncol.* 25:343–350.

Rastinejad, F., Polverini, P.J. and Bouck, N.P. (1989) Regulation of the activity of a new inhibitor of angiogenesis by a cancer suppressor gene. *Cell* 56:345–355.

Raugi, G.J., Olerud, J.E. and Gown, A.M. (1987) Thrombospondin in early human wound tissue. *J. Invest. Dermatol.* 89:551–554.

RayChaudhury, A., Frazier, W.A. and D'Amore, P.D. (1994) Comparison of normal and tumorigenic endothelial cells: differences in thrombospondin production and responses to transforming growth factor-β. *J. Cell Sci.* 107:39–46.

Reed, M.J., Puolakkainen, P., Lane, T.F., Dickerson, D., Bornstein, P. and Sage, E.H. (1993) Differential expression of SPARC and thrombospondin 1 in wound repair: immunolocalization and *in situ* hybridization. *J. Histochem. Cytochem.* 41:1467–1477.

Roberts, A.B., Sporn, M.B., Assoian, R.K., Smith, J.M., Roche, N.S., Wakefield, L.M., Heine, U.I., Liotta, L.A., Falanga, V., Kerhl, J.H. and Fauci, A.S. (1986) Transforming growth factor type-β: rapid induction of fibrosis and angiogenesis *in vivo* and stimulation of collagen formation *in vitro*. *Proc. Natl. Acad. Sci. USA* 83:4167–4171.

Sage, H., Vernon, R.B., Funk, S.E., Everitt, E.A. and Angello, J. (1989) SPARC, a secreted protein associated with cellular proliferation, inhibits cell spreading *in vitro* and exhibits Ca^{+2}-dependent binding to the extracellular matrix. *J. Cell Biol.* 109:341–356.

Savill, J., Hogg, N., Ren, Y. and Haslett, C. (1992) Thrombospondin cooperates with CD36 and the vitronectin receptor in macrophage recognition of neutrophils undergoing apoptosis. *J. Clin. Invest.* 90:1513–1522.

Schultz-Cherry, S. and Murphy-Ullrich, J.E. (1993) Thrombospondin causes activation of latent transforming growth factor-β secreted by endothelial cells by a novel mechanism. *J. Cell Biol.* 122:923–932.

Schultz-Cherry, S., Ribeiro, S., Gentry, L. and Murphy-Ullrich, J.E. (1994a) Thrombospondin binds and activates the small and large forms of latent transforming growth factor-β in a chemically defined system. *J. Biol. Chem.* 269:26775–26782.

Schultz-Cherry, S., Lawler, J. and Murphy-Ullrich, J.E. (1994b) The type 1 repeats of thrombospondin 1 activate latent transforming growth factor-β. *J. Biol. Chem.* 43:26783–26788.

Schultz-Cherry, S., Chen, H., Mosher, D.F., Misenheimer, T.M., Krutzsch, H.C., Roberts, D.D. and Murphy-Ullrich, J.E. (1995) Regulation of transforming growth factor-β activation by discrete sequences of thrombospondin 1. *J. Biol. Chem.* 270:7304–7310.

Sheibani, N. and Frazier, W.A. (1995) Thrombospondin 1 expression in transformed endothelial cells restores a normal phenotype and suppresses their tumorigenesis. *Proc. Natl. Acad. Sci. USA* 92:6788–6792.

Silverstein, R.L., Asch, A.S. and Nachman, R.L. (1989) Glycoprotein IV mediates thrombospondin-dependent platelet-monocyte and platelet-U937 cell adhesion. *J. Clin. Invest.* 84:546–552.

Taraboletti, G., Roberts, D., Liotta, L.A. and Giavazzi, R. (1990) Platelet thrombospondin modulates endothelial cell adhesion, motility, and growth: a potential angiogenesis regulatory factor. *J. Cell Biol.* 111:765–772.

Tolsma, S.S., Volpert, O.V., Good, D.J., Frazier, W.A., Polverini, P.J. and Bouck, N. (1993) Peptides derived from two separate domains of the matrix protein thrombospondin-1 have anti-angiogenic activity. *J. Cell Biol.* 122:497–511.

Tuszynski, G.P. and Nicosia, R.F. (1994) Localization of thrombospondin and its cysteine-serine-valine-threonine-cysteine-glycine-specific receptor in human breast carcinoma. *Lab. Invest.* 70:228–233.

Tuszynski, G.P., Rothman, V., Murphy, A., Seigler, K., Smith, L., Smith, S., Karczewski, J. and Knudsen, K.A. (1987) Thrombospondin promotes cell-substratum adhesion. *Science* 236:1570–1573.

Varani, J. Carey, T.E., Fligiel, S.E., McKeever, P.E. and Dixit, V. (1987) Tumor type-specific differences in cell-substrate adhesion among human tumor cell lines. *Int. J. Cancer* 39:397–403.

Wang, T.N., Qian, X., Granick, M.S., Solomon, M.P., Rothman, V.L. and Tuszynski, G.P. (1995) The effect of thrombospondin on oral squamous carcinoma cell invasion of collagen. *Am. J. Surg.* 170:502–505.

Weinstat-Saslow, D.L., Zabrenetzky, V.S., VanHoutte, K., Frazier, W.A., Roberts, D.D. and Steeg, P.S. (1994) Transfection of thrombospondin 1 complementary DNA into a human breast carcinoma cell line reduces primary tumor growth, metastatic potential, and angiogenesis. *Cancer Res.* 54:6504–6511.

Wight, T.N., Raugi, G.J., Mumby, S.M. and Bornstein, P. (1985) Light microscopic immunolocalisation of thrombospondin in human tissues. *J. Histochem. Cytochem.* 33:295–302.

Zabrenetzsky, V., Harris, C.C., Steeg, P.S. and Roberts, D.D. (1994) Expression of the extracellular matrix molecule thrombospondin inversely correlates with malignant progression in melanoma, lung, and breast carcinoma cell lines. *Int. J. Cancer* 59:191–195.

Regulation of angiogenesis by cell-matrix cell-cell and other interactions

Regulation of Angiogenesis
ed. by I.D. Goldberg & E.M. Rosen
© 1997 Birkhäuser Verlag Basel/Switzerland

Regulation of capillary formation by laminin and other components of the extracellular matrix

D. S. Grant[1] and H. K. Kleinman[2]

[1] *Cardeza Foundation for Hematological Research, Department of Medicine, Jefferson Medical College, Thomas Jefferson University, Philadelphia, Pennsylvania 19107, USA*
[2] *Laboratory of Developmental Biology, Building 30, National Institute of Dental Research, NIH, Bethesda, Maryland 20892, USA*

Summary. The process of angiogenesis (vessel formation) and the resulting stabilization of the mature vessel are complex events that are highly regulated and require signals from both serum and the extracellular matrix. Endothelial cells rest on a specialized thin extracellular matrix known as the basement membrane. Endothelial cells lining normal blood vessels are usually quiescent. When a proper stimulus is present, angiogenesis begins when endothelial cells degrade their basement membrane and invade the surrounding extravascular matrix. Formation of new vessels involves the migration and proliferation of cells. To assist the cells in their migration, the extravascular matrix provides an environment rich in stromal collagen fibers, fibrin, hyaluuronic acid, vitronectin and fibronectin. Once the endothelial cells assemble to form a new vessel, the cells secrete a basement membrane that helps to stabilize and maintain the vessel wall. The basement membrane adheres tightly to cells comprising the vessel wall, provides inductive signals, and plays a important role in the homeostasis of new vessels. We have demonstrated that two major components of the basement membrane, laminin and collagen IV, possess endothelial cell binding sites which regulate vessel stability. In this chapter, we will define the role of these molecules in endothelial cell behavior.

Introduction

Angiogenesis, the process of new vessel formation from preexisting vasculature, is essential in normal tissue growth, repair, and occurs in several pathological conditions such as tumor formation (Folkman, 1992; Folkman and Shing, 1992). Although the vascular system is generally stable, it has the ability to respond quickly to injury or to tumor cells. The stability of vessels is maintained by three major cell types, endothelial cells, smooth muscle cells and fibroblasts, all of which are regulated by their associated extracellular matrix and numerous cytokines/growth factors. Cytokines derived from parenchymal cells, leukocytes and interstitial cells and localized in the serum and/or extracellular matrix cause these stable vascular cells to respond rapidly, disassociate, and branch to form new vessels (Klagsbrun, 1991; Maciag, 1990; Montrucchio et al., 1994; Scott and Harris, 1994; Vlodavsky et al., 1991). The interactions of the matrix molecules and cytokines with the vascular cells are the key to initiation and regulation of angiogenic events.

New vessel formation does not always follow such a regulated and organized pattern in response to cytokines. Under some pathological conditions, such as diabetic retinopathy and psoriasis, excessive cytokine production results in too much angiogenesis and can lead to tissue dysfunction. Many of these pathological conditions may be due to improper gene expression. Therefore, a better understanding of the factors expressed during angiogenesis which regulate vessel growth is essential to understanding, diagnosing, and treating these diseases. The specific molecular events which stabilize vessels or initiate angiogenesis to form new blood vessels are still unclear. It has been observed, however, that cells in the terminal stages of angiogenesis slow migration and increase protein synthesis, especially of basement membrane components (Maragoudakis et al., 1988; Ingber, 1992; Nicosia et al., 1994; Sage and Vernon, 1994; Shima et al., 1995; Taylor et al., 1991). The exact mechanism, molecular biological changes, and components involved in angiogenesis are currently under intense investigation. In this chapter, we will present what is known regarding the role of extracellular matrix molecules in vessel formation.

Normal vessel formation

All vessels are composed of a nonthrombogenic layer of endothelial cells resting on a thin extracellular matrix which lines the intimal surface of the vessel walls (Fig. 1). The differentiated state of this cell layer is maintained by components (factors) present in both the blood and extracellular matrix. The contribution of the matrix to vascular wall homeostasis has been unclear, even though matrix comprises a significant portion of the vasculature. The thin, specialized extracellular layer beneath the endothelial cells is known as basement membrane (Deen and Ball, 1994; Grant et al., 1991; Kleinman and Schnaper, 1993). The basement membrane provides not only support and an adhesive surface for the endothelium, but also maintains the normal differentiated phenotype of the cell layer (Kleinman and Schnaper, 1993). Vessel walls are also composed of other vascular cells such smooth muscle cells, pericytes and fibroblasts (Fig. 1). The former two have their own basement membrane, and the latter is surrounded by a collagenous interstitium (the adventitia), and in some cases, elastic fibers. Studies which examine the cells comprising the vessel walls must also evaluate the role of the matrix in the maintenance of its structure. This organized matrix extends from the artery to the capillary bed and to post-capillary venules and veins. The basement membrane associated with arterial endothelial cells tends to be thicker than that of the capillary basement membrane. Additionally, the endothelial cells lining capillaries are more discontinuous than those of the arterial network, and luminal blood cells can migrate through the vessel wall into the interstitial space, when stimulated.

   New vessel formation occurs in the region of the capillary bed and post-capillary venules (Fig. 1). When stimulated, the endothelium first releases proteases that degrade its basement membrane and surrounding structural

Angiogenesis and the Vessel Wall

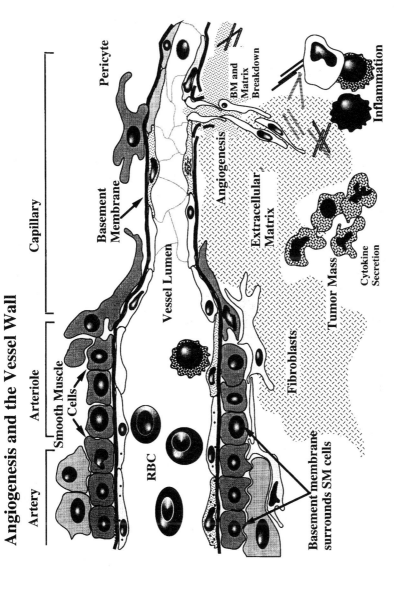

Figure 1. Diagram of blood vessel wall in artery, arteriole and capillary. A basement membrane lines the endothelial cell layer in each vessel. Angiogenesis occurs mainly in the capillaries or post-capillary venules. In response to cytokine stimulation, endothelial cells break down the basement membrane, migrate into the extravascular space, proliferate, and reorganize to form new vessel.

elements (Ruiter et al., 1993; Thorgeirsson et al., 1993). The cells then migrate into the extravascular space, and endothelial cells lining the vessel wall proliferate to replace previously migrated cells. The endothelial cells will reorganize into a new vessel and secrete new basement membrane to maintain the differentiated state of the vessel. Following these processes, the vessel is further stabilized by surrounding pericytes and smooth muscle cells (Nehls et al., 1994), both of which have their own basement membrane.

Structure of the vascular matrix and basement membrane

Most vessel walls are composed of an intimal, medial and adventitial layer. All three layers are made up of specialized cells and their associated extra-cellular matrix. Interstitial collagen fibres, fibronectin and microfibrils make up a large portion of the vessel wall but are primarily found in the adventitial layer on the outer part of the vessel wall. In contrast the intimal region of vessels (arterioles and venules) has a network of basement membrane matrix. Unlike the unorganized connective tissue of the extravascular extracellular matrix, the basement membrane is a thin matrix closely associated with endothelial and smooth muscle cells comprising the vessel wall.

Basement membranes have been shown play an important role in the behavior of several cell types, including endothelial cells (Liley, 1992). *In vitro* studies have found that basement membranes can promote epithelial cell polarity and differentiation into specialized tissues (Kleinman et al., 1987). Basement membranes are composed of a meshwork of non-fibrillar collagen (type IV), the glycoproteins laminin, entactin, and fibronectin, proteoglycans (perlecan and chondroitin sulfate proteoglycan) (Kleinman et al., 1987; Yurchenco and Schittny, 1990) and various growth factors (Table 1). Laminin (Kleinman et al., 1993) and collagen type IV are large molecules that are the most abundant components in this matrix and have been shown to be potent regulators of angiogenesis (Grant et al., 1992; Grant et al., 1989; Kaneko, 1992; Madri and Pratt, 1986; Nicosia et al., 1994; Schnaper et al., 1993). The importance *(in vivo)* of type IV collagen in angiogenesis is well documented, as shown by experiments in which an inhibitor of collagen synthesis blocked angiogenesis in the growing chorio-allantoic membrane (CAM) of the chicken embryo (Haralabopoulos et al., 1994).

Most of the information and the understanding about these basement membrane molecules has been derived from components extracted from the EHS tumor (Kleinman et al., 1987; Kleinman et al., 1986). This tumor produces a large quantity of basement membrane which can be extracted and reconstituted *in vitro*. A urea extract, termed Matrigel, contains laminin, collagen IV, heparan sulfate proteoglycan, and several growth

Table 1. Basement membrane components

Component	Size Mr	Structure	Receptor

Laminin (EHS)
  S-Laminin
  Merosin
  H-Laminin
50–70%
	800 000 –
	1 000 000 D
		A B2 B1 B	32/67 kD LBP
			Integrins $\beta 1$, $\beta 4$,
			$\alpha 1$, $\alpha 2$, $\alpha 3$, $\alpha 6$

Collagen IV Alpha Chains 30–50%	540 000 D 180 000 D	NC1 7s	Integrins $\beta 1$, $\alpha 1$, $\alpha 2$,
Entactin/Nidogen 10%	150 000 D		Integrins?

Proteoglycans
  HSPG-Lg
  HSPG-Sm
  CSPG
1–3%
	600 000 D
	800 000 D
	130 000 D
	600 000–
	50 000 D
			45 kD Protein

Others (variable) SPARC/Osteonectin Collagen VII Amyloid P Component Agrin bFGF TGFβ EGF PDGF Plasminogen Activator Gelatinase A and B	33 000 D 350 000 D 150 000		

factors, i.e., transforming growth factor-beta (TGF-β), epidermal growth factor (EGF), insulin-like growth factor 1, basic fibroblast growth factor (bFGF), and platelet-derived growth factor (PDGF) (Vukicevic et al., 1992), all of which are present in authentic basement membrane. Matrigel can form a solid gel *in vitro* at 37 °C that has biochemical and morphological features similar to authentic basement membrane. Matrigel, a liquid at 4 °C, can thus be used to coat tissue culture plastic in order to improve cell attachment. This extracellular matrix is biologically active, and has been shown *in vitro* to stimulate cell attachment and promote differentiation in both normal and transformed cells. It promotes the differentiation of a variety of cells (Kleinman et al., 1987), including endothelial cells which form tube-like structures resembling capillaries. It is now believed that several of the growth factors found in Matrigel may also be involved in angiogenesis and in endothelial cell differentiation.

Laminin has important roles in many of the steps of angiogenesis. Laminin has several biological functions including cell attachment, growth promotion, protease secretion, self-assembly (to form basement membranes) and interactions with other components of the extracellular matrix (Kleinman et al., 1993; Schnaper et al., 1993). The laminins form a large family of glycoproteins exhibiting structurally and functionally different subclasses or isoforms (Table 2) (Burgeson et al., 1994). The nomenclature for these laminins has recently been revised to allow for the characterization of additional family members which likely exist (see Table 2). The first described and most intensively studied laminin, laminin-1, is derived from the matrix of the EHS tumor. It is an 800 kD molecule that is composed of three chains (see Table 1), designated $\alpha 1 \beta 1 \gamma 1$ (Table 2). These chains bind to each other to form a cross-shaped molecule with several globular domains (Burgeson et al., 1994; Engel, 1993; Kleinman et al., 1993).

Besides having the ability to self-assemble, laminin can bind to several cell surface binding proteins, including several integrins (Davis and Camarillo, 1995; Engel, 1993; Kramer et al., 1993; Luscinskas and Lawler, 1994; Nicosia et al., 1994). There are many cell surface receptor molecules which bind laminin, with integrins being the best studied. Integrins are a large family of transmembrane heterodimeric glycoproteins that serve as receptors for extracellular matrices and for other cell surface molecules (Albelda, 1992; Kramer et al., 1993). Each integrin is composed on an alpha and beta subunit that are noncovalently associated. Normal blood vessels *in vivo* express a high density of $\alpha 3$, $\beta 1$, $\alpha 1 \beta 1$, $\alpha v \beta 3$ and $\alpha 5 \beta 1$ integrins. The relative density of each of these integrins varies depending on the angiogenic state of the endothelium (Davies and Camarillo, 1995; Enenstein et al., 1992). Several distinct sites in the EHS laminin have been identified as cell binding domains. These sites promote biological activities *in vitro* such as cell spreading, migration, and cell differentiation (Kleinman and Schnaper, 1993). It is believed that many of these activities occur through integrin-mediated transmembrane signaling (Albelda, 1992; Kramer et al., 1993).

Table 2. Forms of laminin

Isoform	Chain Composition	Molecular Weight
Mouse EHS tumor	$\alpha 1 \beta 1 \gamma 1$	440, 220, 220 kD
Mouse Heart m-Laminin	$\alpha 2 \beta 1 \gamma 1$	300, 220, 180 kD
Bovine Aortic EC S-Laminin	$\alpha 1 \beta 2 \gamma 1$	200, 180 kD
Schwannoma	$\alpha 2 \beta 1 \gamma 1$	300 kD
Human Placenta	$\alpha 1 \beta 1 \gamma 1, \alpha 2 \beta 1 \gamma 1$	350, 200, 185, 195 kD
K-Laminin	$\alpha 4 \beta 1 \gamma 1$	90 – 170 kD, ?
S-Merosin	$\alpha 2 \beta 2 \gamma 1$	300, 180, 180 kD
KS-Laminin	$\alpha 3 \beta 2 \gamma 1$	200, 160, ?, kD

Matrix models that mimic angiogenesis

Angiogenesis has been studied *in vivo* and *in vitro* (Folkman and Shing, 1992; Garrido et al., 1995; Grant et al., 1991; Kubota et al., 1988; Montesano et al., 1987). While *in vivo* methods relate directly to many events occurring during normal blood vessel formation, they have several drawbacks. First, *in vivo* angiogenic models are often difficult to control and it is hard to discern whether the action of the stimulant is direct or the result of a secondary product of inflammatory cells. It is also difficult to quantitate these assays. The most widely used model is the chicken chorioallantoic membrane (CAM) assay. The CAM assay has been used to show inhibition of capillary formation by numerous substances such as fumagillin, cortisone and a laminin-derived synthetic peptide. Drawbacks to this assay are that it takes up to four days to complete and there can be great variability between the eggs. The rabbit eye model of angiogenesis provides an avascular system that can be used to test the stimulatory effect of angiogenic compounds. This assay, however, is time-consuming and expensive since it needs many rabbits. Other *in vivo* models employ a subcutaneous implantation of a Goretex sponge or polyvinyl disk containing bFGF into mice (Maciag, 1990). There is a certain amount of variability between all of these methods and thus they require large number of test samples to obtain reliable results.

Endothelial cells from various vessels have been isolated and maintained in culture (Bauer et al., 1992; Diaz et al., 1994; Folkman and Shing, 1992; Jaffe et al., 1973). If the cells are permitted to become superconfluent without changing the culture medium, structures form above the monolayer which resemble capillary-like vessels (Fig. 2). This differentiation can be accelerated by the addition of gelatin, fibrin, or collagen to the surface of the plastic. Others have shown that if cultured endothelial cells are incubated in a fibrin clot or with collagen IV, vessel formation will occur within a week (Nicosia et al., 1993; Nicosia et al., 1984; Rohr et al., 1992). With these systems, capillary-like structures are observed. The response is generally slow, requiring several days, and in some cases the vessels are inside out, secreting basement membrane material and interstitial collagen into the lumen.

Tube formation on Matrigel

Matrigel, a basement membrane extract enriched with laminin, has the ability to promote the differentiation of epithelial and neuronal cells (Kleinman et al., 1987; Kleinman and Schnaper, 1993). Human endothelial cells were first placed on Matrigel by Kubota et al. (1988) and were observed to organize into capillary-like structures within 20 h. Under both the light and electron microscope, we showed that these capillary-like networks had lumens and the endothelial cells possessed membrane specializations similar to those observed in vessels *in vivo* (Grant et al.,

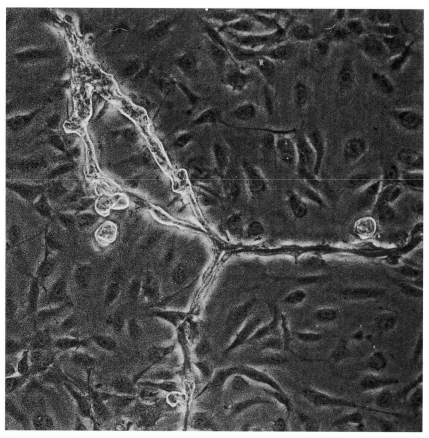

Figue 2. Tube formation of human umbilical vein endothelial cells (HUVEC) in culture. The monolayer was cultured in 20% bovine calf serum and maintained at superconfluency for 48 h. Photo taken with a Nikon phase contrast microscope at an original magnification of 10×.

1991). We developed a Matrigel assay system to determine the events and regulatory molecules important for tube formation on Matrigel. Cells undergoing morphological changes on Matrigel required an intact cyto-skeleton, protein synthesis and receptor-mediated interaction with the matrix molecules in Matrigel. We have examined and defined the role(s) of matrix molecules such as collagen IV and laminin and their specific cell-binding sites in this Matrigel model (Grant et al., 1994; Grant et al., 1990). This was accomplished by seeding cells onto Matrigel and then adding synthetic peptide antibodies to see if they could block tube formation (Fig. 3). Synthetic peptides derived from two regions in laminin, the YIGSR peptide on the $\beta 1$ chain and the ALRGDNP peptide sequence in the $\alpha 1$ chain, were also added to the tube forming assays on Matrigel. These peptides were able to block 40–45% of tube formation on Matrigel, and further confirm the role of laminin in tube formation. The addition of antiserum to laminin re-

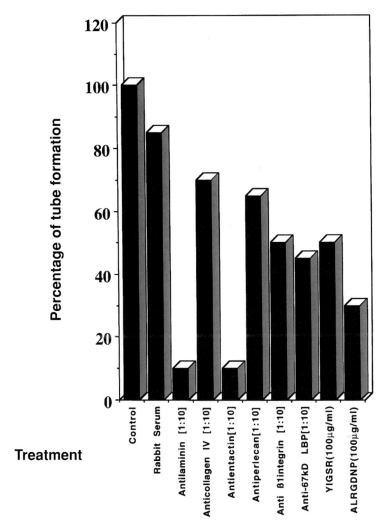

Figure 3. Graph of relative percentage of tube formation on Matrigel (for 20 h) with different antisera or peptide treatments. Forty thousand HUVECs were cultured on Matrigel in the presence of the peptides YIGSR or RGD, normal rabbit serum, or antiserum to laminin, collagen IV, entactin, perlecan, β1 integrin, or the laminin binding receptor (LBP) 32/67 at a concentration of 1:10. The resulting tube network was stained and quantified using a Nikon microscope and the NIH Image program.

duced tube formation by 90%. Tube formation was also reduced by 90% by antiserum to entactin. The addition of collagen IV antiserum reduced tube formation by only 25% (Fig. 3). This difference in blocking ability may be due to the low level of collagen IV in Matrigel. These results illustrate that endothelial cell interactions with both laminin and entactin are essential for normal tube formation on Matrigel.

We also incubated human endothelial cells with antiserum to $\beta 1$ integrin and to an antiserum to the laminin receptor (LBP 32/67 kD) and found that they could individually inhibit tube formation by approximately 50% on Matrigel. We examined light microscopic morphological changes in tube formation in the presence of antiserum to the laminin binding receptor B1 integrin and LBP-32/67. In the presence of control antiserum alone the cells were still able to form tubes (Fig. 4, top). When antiserum to $\beta 1$ integrin was added, the endothelial cells formed large clumps with bridges between them, unlike the delicate tube structures observed in normal tube assays (Fig. 4, bottom). After the addition of an antibody to LBP-32/67, the cells formed clusters in many areas and did not organize into tubes (Fig. 4, middle). These results indicate that the endothelial cell interaction(s) with laminin during tube formation are specific and that laminin induces morphological changes in the endothelial cells through receptor mediated signal transduction.

The role of laminin in vessel formation

Laminin is biologically very active and several active sites have been identified at the synthetic peptide level. These sites have previously been shown to promote cellular activities *in vitro* such as cell attachment, spreading, migration, and cell differentiation (Kleinman et al., 1987). The response of the cells is dependent on the cell type and the active site the peptide employs. For example, the YIGSR (tyr-ile-gly-ser-arg) site in laminin has been shown to inhibit tumor metastasis and growth either when co-injected with melanoma cells or injected intraperitoneally 4 days after the tumor cells (Iwamoto et al., 1987). The RGD (arg-gly-asp) site in laminin, a potent cell binding domain in the molecule, has also been shown to inhibit tumor cell invasion *in vitro* (Tashiro et al., 1991; Thompson et al., 1991). In contrast, a third site in the laminin $\alpha 1$ chain, SIKVAV (ser-ile-lys-val-ala-val), stimulates tumor growth *in vivo* and neurite outgrowth *in vitro* (Tashiro et al., 1989; Sweeney et al., 1991). This site in laminin occurs at a region in which the chains form a coiled-coil structure. Since laminin is known to bind to other components within the basement membrane network (such as collagen IV, entactin, fibronectin and heparin) in this region, the SIKVAV site may be cryptic under some conditions. Thus, the laminin molecule possesses multiple different and in some cases opposing activities and it is likely that some of these sites are not active or available at all times.

The identification of several active sites on laminin which influence endothelial cell behavior *in vitro* and *in vivo* has confirmed its role in vessel maintenance and angiogenesis. Using the tube forming assay, we have demonstrated that the YIGSR, RGD and SIKVAV sites in laminin all play specific roles in tube formation. Our studies showed that endothelial cells

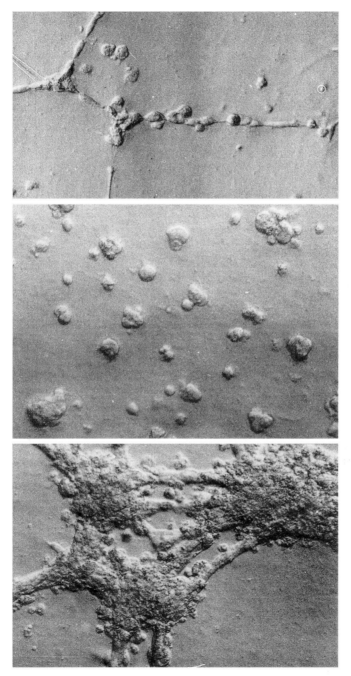

Figure 4. Morphology of HUVEC cultured on Matrigel for 20 h in the presence of normal rabbit serum (top panel), antiserum to LBP 32/67 (middle panel) or β1 integrin (lower panel). Viewed with Nomarsky optics on a Nikon microscope at an original magnification of 10× (top panel) and 40× (lower panel).

utilize the laminin RGD sequence for attachment to laminin, while the YIGSR sequence is involved in both attachment and cell-cell interactions. YIGSR also has the ability to change the shape of cells. Cells incubated on YIGSR accumulate vacuoles which fuse to form signet ring-shaped cells (Fig. 5, middle). It appears that the cell cytoskeleton is involved in this process singe normal actin stress fibers, which are usually randomly arranged in endothelial cells on plastic, realign when incubated with the peptide YIGSR to form linear arrays around and across the vacuolated cells (Fig. 5, middle).

Endothelial cells seeded on the SIKVAV peptide changed their morphology and appeared fibroblastic with many extended processes, unlike the normal cobblestone morphology observed on tissue culture plastic. In many case, cells incubated with SIKVAV on plastic aligned into structures resembling forming tubes (Fig. 5, bottom). When the SIKVAV peptide was added to the tube forming assay on Matrigel, cells formed short irregular structures, some of which penetrated the matrix, and sprouting was more apparent. Analysis of endothelial cell conditioned media of cells cultured in the presence of this peptide indicated degradation of the Matrigel, and zymograms demonstrated active collagenase IV (gelatinase) at 68 and 62 Kd. The *in vivo* murine-Matrigel implantation angiogenesis assay (Kibbey et al., 1992; Passaniti et al., 1992) and the chick yolk sac/chorio-allantoic membrane assays with the peptide, both demonstrated increased endothelial cell mobilization, capillary branching and vessel formation (Grant et al., 1992). These data suggest that the SIKVAV site plays an important role in initiating branching and in the formation of new capillaries from the parent vessels. Such behavior is observed *in vivo* in response to tumor growth or in the normal vascular response to injury.

It is remarkable that laminin could possess different activities within the same molecule; i.e. RGD site for cell attachment, YIGSR site for cell alignment, lumen and tube formation, and SIKVAV sites for initiating invasion and angiogenesis. It should be noted, however, that it is not known whether these sites are available to endothelial cells at all times *in vivo* (Grant et al., 1994; Grant et al., 1992). Temporal and spatial differences in exposure to these sites may provide a regulatory mechanism for control of endothelial cells in quiescent versus an active angiogenic state. This concept of differential sites exposure is further strengthened by the fact that not all laminin chains are found in all basement membranes, nor do all laminin α chains possess RGD and SIKVAV sequences (Kleinman et al., 1993; Yamada and Kleinman, 1992).

Figure 5. Actin staining in HUVEC on plastic (upper panel), or YIGSR-(middle panels) and SIKVAV-(lower panels) coated plastic. Cells were fixed with 4% formaldehyde in PBS, permeabilized and stained with rhodamine-phalloidon (Molecular probes Inc., Oregon). Actin filaments are observed next to the phase contrast picture. Original magnification 60×.

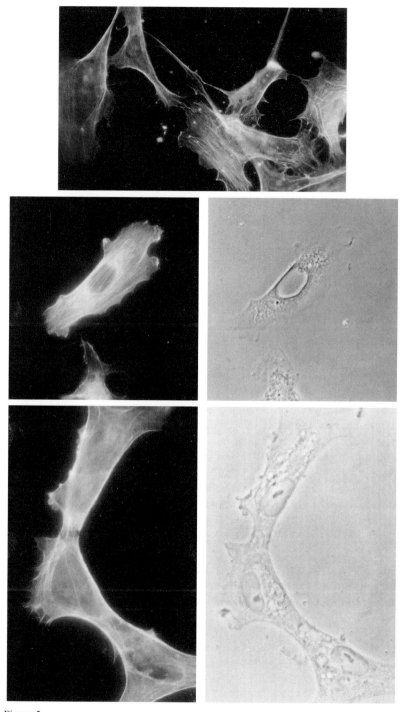

Figure 5

Basement membrane-induced molecular changes during angiogenesis

In order to understand better the process of angiogenesis, numerous *in vivo* and *in vitro* models have been developed that mimic at least some aspects of the angiogenesis, and these models provide insight into the steps and initiating factors involved in stimulating endothelial cells to form and maintain structurally intact vessels (Ingber and Folkman, 1989; Ingber et al., 1987; Kramer and Fuh, 1985; Maciag, 1990; Madri and Pratt, 1986). Many of these models have shown that protein synthesis and gene expression are involved in angiogenesis (Grant et al., 1995). For example, Maciag (1990) has shown that when confluent, cultured endothelial cells are treated with phorbol esters, approximately 50% of the cells form tube networks. Using this model of endothelial cell differentiation with subtractive hybridization, a novel protein, EDG-1, was identified and shown to increase during tube formation (Hla and Maciag, 1990a, b). The morphological differentiation of endothelial cells on Matrigel has also been successfully used to identify angiogenic and anti-angiogenic factors and induced genes (Grant et al., 1994; Grant et al., 1993; Grant et al., 1991).

Presently, we are examining the mechanisms of tube formation on Matrigel at the cellular and molecular levels. We have used a subtractive hybridization model to determine what genes are expressed early in endothelial cells plated on Matrigel (Grant et al., 1995). Differential gene expression of endothelial cells on plastic vs. Matrigel at 4 h resulted in 17 clones which were found to range in size from 500 to 3000 basepairs. Northern blot analysis verified that 10 of these cDNA clones were expressed as mRNA when endothelial cells were incubated on Matrigel for 4 h. One clone was identical in nucleotide sequence to human thymosin $\beta4$, which was not previously known to be involved in vessel differentiation. Cells cultured on plastic do not significantly change their expression of thymosin $\beta4$ over a 5 h period, whereas cells cultured on Matrigel show a 5-fold increase at 4 to 6 h. This demonstrates that on Matrigel there is an increase in the synthesis of thymosin $\beta4$ mRNA. Recent experiments now show that the addition of thymosin $\beta4$ to endothelial cells plated on Matrigel increases tube formation.

These results indicate that the presence of the matrix (in this case laminin-enriched Matrigel) can induce protein synthesis. The resulting proteins appear to be necessary for endothelial cell differentiation into capillary-like tubes. More information needs to be obtained on the surface receptors involved in this process. Finally, to understand better the regulation of endothelial cell differentiation by basement membrane molecules, transmembrane signaling events and potential nuclear binding regulatory proteins need to be studied.

References

Albelda, S.M. (1992) Differential expression of integrin cell-substratum adhesion receptors on endothelium. *In:* R. Steiner, P.B. Weisz and R. Langer (eds): Angiogenesis: Key principles–Science–Technology–Medicine (EXS 61) Birkhäuser Verlag, Basel, pp 188–192.

Bauer, J., Margolis, M., Schreiner, C., Edgell, C.J., Azizkhan, J., Lazarowski, E. and Juliano, R.L. (1992) In vitro model of angiogenesis using a human endothelium-derived permanent cell line: contributions of induced gene expression, G-proteins, and integrins. *J. Cell Physiol.* 153:437–449.

Burgeson, R.E., Chiquet, M., Deutzmann, R., Ekblom, P., Engel, J., Kleinman, H., Martin, G.R., Meneguzzi, G., Paulsson, M., Sanes, J., Timpl, R., Tryggvason, K., Yamada, Y. and Yurchenco, P.D. (1994) A new nomenclature for the laminins. *Matrix Biology* 14:209–211.

Davis, G.E. and Camarillo, C.W. (1995) Regulation of endothelial cell morphogenesis by integrins, mechanical forces, and matrix guidance pathways. *Exp. Cell Res.* 216:113–123.

Deen, S. and Ball, R.Y. (1994) Basement membrane and extracellular interstitial matrix components in bladder neoplasia-evidence of angiogenesis. *Histopathology* 25: 475–481.

Diaz, F.L., Gutierrez, R. and Varela, H. (1994) Angiogenesis: an update. *Histol. Histopathol.* 9:807–843.

Enenstein, J., Waleh, N.S. and Kramer, R.H. (1992) Basic FGF and TGF-beta differentially modulate integrin expression of human microvascular endothelial cells. *Exp. Cell Res.* 203:499–503.

Engel, J. (1993) Structure and function of laminin. *In:* D.H. Rohrbach and R. Timpl (ed): *Molecular and cellular aspects of basement membranes.* Academic Press, San Diego, pp 147–176.

Folkman, J. (1992) Angiogenesis – retrospect and outlook. *In:* R. Steiner, P.B., Weisz and R. Langer (eds): Angiogenesis: Key principles–Science–Technology–Medicine. Birkhäuser Verlag, Basel, pp 4–13.

Folkman, J. and Shing, Y. (1992) Angiogenesis. *J. Biol. Chem.* 267:10931–10934.

Garrido, T., Riese, H.H., Aracil, M. and Perez, A.A. (1995) Endothelial cell differentiation into capillary-like structures in response to tumour cell conditioned medium: a modified chemotaxis chamber assay. *Br. J. Cancer* 71:770–775.

Grant, D.S., Kibbey, M.C., Kinsella, J.L., Cid, M.C. and Kleinman, H.K.. (19994) The role of basement membrane in angiogenesis and tumor growth. *Pathol. Res. Pract.* 190: 854–863.

Grant, D.S., Kinsella, J.L., Fridman, R., Auerbach R., Piasecki, B.A., Yamada, Y., Zain, M. and Kleinman, H.K. (1992) Interaction of endothelial cells with a laminin A chain peptide (SIKVAV) *in vitro* and induction of angiogenic behavior *in vivo. J. Cell Physiol.* 153: 614–625.

Grant, D.S., Kinsella, J.L., Kibbey, M.C., Laflamme, S., Burbelo, P.D., Goldstein, A.L. and Kleinman, H.K. (1995) Matrigel induces thymosin B4 gene in differentiating endothelial cells. *J. Cell Sci.* 108:1–10.

Grant, D.S., Kleinman, H.K., Goldberg, I.D., Bhargava, M.M., Nickoloff, B.J., Kinsella, J.L., Polverini, P. and Rosen, E.M. (1993) Scatter factor induces blood vessel formation in vivo. *Proc. Natl. Acad. Sci. USA* 90:1937–1941.

Grant, D.S., Kleinman, H.K. and Martin, G.R. (1990) The role of basement membranes in vascular development. *Ann. N.Y. Acad. Sci.* 588:61–72.

Grant, D.S., Lelkes, P.I., Fukuda, K. and Kleinman, H.K. (1991) Intracellular mechanisms involved in basement membrane induced blood vessel differentiation *in vitro. In Vitro Cell Dev. Biol.* 27:327–335.

Grant, D.S., Tashiro, K.-I., Segui-Real, B., Yamada, Y., Martin, G.R. and Kleinman, H.K. (1989) Two different laminin domains mediate the differentiation of human endothelial cells into capillary-like structures *in vitro. Cell* 58:933–943.

Haralabopoulos, G.C., Grant, D.S., Kleinman, H.K., Lelkes, P.I., Papaioannou, S.P. and Maragoudakis, M.E. (1994) Inhibitors of basement membrane collagen synthesis prevent endothelial cell alignment in matrigel *in vitro* and angiogenesis *in vivo. Lab. Invest.* 71:575–582.

Hla, T. and Maciag, T. (1990a) An abundant transcript induced in differentiating human endothelial cells encodes a polypeptide with structural similarities to G-protein-coupled receptors. *J. Biol. Chem.* 265:9308–9313.

Hla, T. and Maciag, T. (1990b) Isolation of immediate-early differentiation mRNAs by enzymatic amplification of subtracted cDNA from human endothelial cells. *Biochem. Biophys. Res. Commun.* 167:637–643.

Ingber, D.E. (1992) Extracellular matrix as a solid-state regulator in angiogenesis: identification of new targets for anti-cancer therapy. *Semin. Cancer Biol.* 3:57–63.

Ingber, D.E. and Folkman, J. (1989) How does extracellular matrix control capillary morphogenesis? *Cell* 58:803–805.

Ingber, D.E., Madri, J.A. and Folkman, J. (1987) Endothelial growth factors and extracellular matrix regulate DNA synthesis through modulation of cell and nuclear expansion. *In Vitro* 23:387–394.

Iwamoto, Y., Robey, F.A., Grat, J., Sasaki, M., Kleinman, H.K., Yamada, Y. and Martin, G.R. (1987) YIGSR, a synthetic laminin pentapeptide, inhibits experimental metastasis formation. *Science* 238:1132–1134.

Jaffe, E.A., Nachman, R.L., Becker, C.G. and Minick, C.R. (1973) Culture of human endothelial cells derived from umbilical veins-identification by morphological and immunological criteria. *J. Clin. Invest.* 52:2745–2756.

Kaneko, T. (1992) [Relationship between endothelial cells and extracellular matrix: investigation using the model of angiogenesis in vitro]. *Nippon Geka Hokan* 61:134–149.

Kibbey, M.C., Grant, D.S. and Kleinman, H.K. (1992) role of the SIKVAV site of laminin in promotion of angiogenesis and tumor growth: an in vivo Matrigel model. *J. Natl. Cancer Inst.* 84:1633–1638.

Klagsbrun, M. (1991) Regulators of angiogenesis: stimulators, inhibitors, and extracellular matrix. *J. Cell Biochem.* 47:199–200.

Kleinman, H.K., Graf, J., Iwamoto, Y., Kitten, G.T., Ogle, R.C., Sasaki, M., Yamada, Y., Martin, G.R. and Luckenbill-Edds, L. (1987) Role of basement membrane in cell differentiation. *Ann. N.Y. Acad. Sci.* 513:134–145.

Kleinman, H.K., McGarvey, M.L., Liotta, L.A., Gehron-Robbey, P., Tryggvasson, K. and Martin, G.R. (1982) Isolation and characterization of type IV procollagen, laminin and heparan sulfate proteoglycan from the EHS sarcoma. *Biochemistry* 24:6188–6193.

Kleinman, H.K. and Schnaper, H.W. (1993) Basement membrane matrices in tissue development [comment]. *Am. J. Respir. Cell Mol. Biol.* 8:238–239.

Kleinman, H.K., Weeks, B.S., Schnaper, H.W., Kibbey, M.C., Yamamura, K. and Grant, D.S. (1993) The laminins: a family of basement membrane glycoproteins important in cell differentiation and tumor metastases. *Vitam. Horm.* 47:161–186.

Kramer, R.H., Enenstein, J. and Waleh, N.S. (1993) Integrin structure and ligand specificity in cell-matrix interactions. *In:* D.H. Rohrbach and R. Timpl (eds): *Molecular and cellular aspects of basement membranes.* Academic Press, San Diego, pp 2349–287.

Kramer, R.H. and Fuh, G.M. (1985) Type IV collagen synthesis by cultured human microvascular endothelial cells and its deposition in the subendothelial basement membrane. *Biochemistry* 24:7423–7430.

Kubota, Y., Kleinman, H.K., Martin, G.R. and Lawley, T.J. (1988) Role of laminin and basement membrane in the morphological differentiation of human endothelial cells into capillary-like structures. *J. Cell Biol.* 107:1589–1598.

Liley, H.G. (1992) The contributions of the extracellular matrix to vascular form and function. *Semin. Perinatol.* 16:155–160.

Luscinskas, F.W. and Lawler, J. (1994) Integrins as dynamic regulators of vascular function. *FASEB J.* 8:929–938.

Maciag, T. (1990) Molecular and cellular mechanisms of angiogenesis. *Important Adv. Oncol.* 85:85–98.

Madri, J.A. and Pratt, B.M. (1986) Endothelial cell-matrix interactions: in vitro models of angiogenesis. *J. Histochem. Cytochem.* 34:85–91.

Maragoudakis, M.E., Panoutsacopoulou, M. and Sarmonika, M. (1988) Rate of basement membrane biosynthesis as an index to angiogenesis. *Tissue Cell* 20:531–539.

Montesano, R., Pepper M.S., Vassalli, J.-D. and Orci, L. (1987) Phorbol ester induces cultured endothelial cells to invade a fibrin matrix in the presence of fibrinolytic inhibitors. *J. Cell. Physiol.* 132:509–516.

Montrucchio, G., Lupia, E., Battaglia, E., Passerini, G., Bussolino, F., Emanuelli, G. and Camussi, G. (1994) Tumor necrosis factor alpha-induced angiogenesis depends on in situ platelet-activating factor biosynthesis. *J. Exp. Med.* 180:377–382.

Nehls, V., Schuchardt, F. and Drenckhahn, D. (1994) The effect of fibroblasts, vascular smooth muscle cells, and pericytes on sprout formation of endothelial cells in a fibrin gel angiogenesis system. *Microvasc. Res.* 48:349–363.

Nicosia, R.F., Bonanno, E. and Smith, M. (1993) Fibronectin promotes the elongation of microvessels during angiogenesis in vitro. *J. Cell Physiol.* 154:654–661.

Nicosia, R.F., Bonanno, E., Smith, M. and Yurchenco, P. (1994) Modulation of angiogenesis in vitro by laminin-entactin complex. *Dev. Biol.* 164:197–206.

Nicosia, R.F., McCormick, J.F. and Bielunas, J. (1984) The formation of endothelial webs and channels in plasma clot culture. *Scan. Elect. Microsc.* 2:793–799.

Passaniti, A., Taylor, R.M., Pili, R., Guo, Y., Long, P.V., Haney, J.A., Pauly, R.R., Grant, D.S. and Martin, G.R. (1992) A simple, quantitative method for assessing angiogenesis and antiangiogenic agents using reconstituted basement membrane, heparin, and fibroblast growth factor. *Lab. Invest.* 67:519–528.

Rohr, S., Toti, F., Brisson, C., Albert, A., Freund, M., Meyer, C. and Cazenave, J.P. (1992) Quantitative image analysis of angiogenesis in rats implanted with a fibrin gel chamber. *Nouv. Rev. Fr. Hematol.* 34:287–294.

Ruiter, D.J., Schlingemann, R.O., Westphal, J.R., Denijn, M., Rietveld, F.J. and De, W.R. (1993) Angiogenesis in wound healing and tumor metastasis. *Behring Inst.* 92:258–272.

Sage, E.H. and Vernon, R.B. (1994) Regulation of angiogenesis by extracellular matrix: the growth and the glue. *J. Hypertens. Suppl.* 12:s145–152.

Schnaper, H.W., Kleinman, H.K. and Grant, D.S. (1993) Role of laminin in endothelial cell recognition and differentiation. *Kidney Int.* 43:20–25.

Scott, P.A. and Harris, A.L. (1994) Current approaches to targeting cancer using antiangiogenesis therapies. *Cancer Treat. Rev.* 20:393–412.

Shima, D.T., Sauers, B., Gougos, A. and D'Amore, P.A. (1995) Alterations in gene expression associated with changes in the state of endothelial differentiation. *Differentiation* 58:217–226.

Sweeney, T.M., Kibbey, M.C., Zain, M., Fridman, R. and Kleinman, H.K. (1991) Basement membrane and the SIKVAV laminin-derived peptide promote tumor growth and metastases. *Cancer Metas. Rev.* 10:245–254.

Tashiro, K., Sephel, G.C., Greatorex, D., Sasaki, M., Shirashi, N., Martin, G.R., Kleinman, H.K. and Yamada, Y. (1991) The RGD containing site of the mouse laminin A chain is active for cell attachment, spreading, migration and neurite outgrowth. *J. Cell Physiol.* 146:451–459.

Tashiro, K.-I., Sephel, G.C., Weeks, B., Sasaki, M., Martin, G.R., Kleinman, H.K. and Yamada, Y. (1989) A synthetic peptide containing the IKVAV sequence from the A chain of laminin mediates cell attachment, migration and neurite outgrowth. *J. Biol. Chem.* 264:16174–16182.

Taylor, C.M., Thompson, J.M. and Weiss, J.B. (1991) Matrix integrity and the control of angiogenesis. *Int. J. Radiat. Biol.* 60:61–64.

Thompson, H.L., Burbelo, P.D., Yamada, Y., Kleinman, H.K. and Metcalfe, D.D. (1991) Identification of an amino acid sequence in the laminin A chain mediating mast cell attachment and spreading. *Immunology* 72:144–149.

Thorgeirsson, U.P., Lindsay, C.K., Cottam, D.W. and Gomez, D.E. (1993) Tumor invasion, proteolysis, and angiogenesis. *J. Neurooncol.* 18:89–103.

Vlodavsky, I., Fuks, Z., Ishai, M.R., Bashkin, P., Levy, E., Korner, G., Bar, S.R. and Klagsbrun, M. (1991) Extracellular matrix-resident basic fibroblast growth factor: implication for the control of angiogenesis. *J. Cell Biochem.* 45:167–176.

Vukicevic, S., Kleinman, H.K., Luyten, F.P., Roberts, A.B., Roche, N.S. and Reddi, A.H. (1992) Identification of multiple active growth factors in basement membrane Matrigel suggests caution in interpretation of cellular activity related to extracellular matrix components. *Exp. Cell Res.* 202:1–8.

Yamada, Y. and Kleinman, H.K. (1992) Functional domains of cell adhesion molecules. *Curr. Opin. Cell Biol.* 4:819–823.

Yurchenco, P.D. and Schittny, J.C. (1990) Molecular architecture of basement membranes. *FASEB J.* 4:1577–1590.

Regulation of Angiogenesis
ed. by I.D. Goldberg & E.M. Rosen
© 1997 Birkhäuser Verlag Basel/Switzerland

Hypoxia and angiogenesis in experimental tumor models: Therapeutic implications

S. Rockwell and J.P.S. Knisely

Department of Therapeutic Radiology, Yale University School of Medicine, New Haven, Connecticut 06520-8040, USA

Introduction

W.C. Röntgen's discovery of X-rays (1895) was followed almost immediately by considerations of the possible medical uses of ionizing radiation. These included the use of "Röntgentherapie" in the treatment of human malignancies. By 1904, it had been found that the effects of X-rays were influenced by the presence or absence of oxygen, and the implications of this fact for Röntgentherapie were being discussed (Hahn, 1904). Tumor oxygenation was therefore a matter of interest and concern even at the dawn of modern cancer therapy. Interest in this problem continues today, and has even increased, because technology now becoming available allows us to explore the mechanisms by which oxygen modulates the efficacy of cancer therapy and may allow us to alter tumor oxygenation and angiogenesis and thereby improve the treatment of malignant diseases.

Tumor angiogenesis

Angiogenesis, the formation of new blood vessels by sprouting from a pre-existing endothelium, is tightly regulated: the only angiogenesis occurring under physiologic conditions in adults is that related to the female reproductive cycle. All other angiogenesis is associated with pathologic conditions, including wound healing, psoriasis, rheumatoid arthritis, diabetic retinopathy, atherosclerotic myocardial disease, and cancer. The paradigm for angiogenesis is wound healing; tumor-associated angiogenesis has been modeled as an exaggerated wound repair (Dvorak 1986; Whalen, 1990).

Like wound repair, tumor growth is a complex process, involving many cell populations. Although the ability of the malignant cells to circumvent the homeostatic mechanisms that normally control cell proliferation is necessary for tumor growth, it is not sufficient. Because of the rate at which metabolizing cells use oxygen, the oxygen tension falls rapidly with the distance from a functional blood vessel (Thomlinson and Gray, 1955). As a result, cells lying further than 100–200 μ from the nearest functional blood vessel experience very severe hypoxia. Because the diffusion distances of glucose and many other critical nutrients are similar to that of oxygen, these cells will also experience nutritional deficiencies (Kallinowski et al., 1988; Vaupel et al., 1989). Accumulation of metabolic wastes in this poorly perfused area may also lead to the development of low extracellular pH (Wike-Hooley et al., 1984; Tannock and Rotin, 1989). Cells which cannot survive in these adverse, unphysiologic microenvironments cannot form solid tumors (Rotin et al., 1989) because macroscopic foci of malignant cells cannot develop. The earliest stages of tumor growth therefore select for the emergence of a cell population which is able to survive, and even proliferate, under unphysiologically adverse conditions.

To grow beyond a microscopic size, tumors must successfully initiate the development of a new vascular bed which will support the expanding malignant cell population (Folkman, 1971, 1985, 1992b). This development requires interactions between the malignant cells, cells resident in the tissue (for example, fibroblasts, endothelial cells, macrophages), and cells recruited from the circulation (for example, platelets, neutrophils, monocytes). Interactions between these malignant and normal cell populations therefore are essential for the development of the stromal infrastructure which supports the growing neoplasm (Dvorak et al., 1995; Hlatky et al., 1996).

Tumor angiogenesis begins with the release of angiogenic factors either by the malignant cells or by macrophages resident within the tumor. There is increasing evidence that hypoxia is the factor stimulating release of these angiogenic factors (Knighton et al., 1983; Shweiki et al., 1992, 1995; Hlatky et al., 1994, 1996). Numerous biologically active compounds have been shown to play roles in angiogenesis (Klagsbrun and D'Amore, 1991). The list includes acidic and basic fibroblast growth factors (aFGF and bFGF), platelet derived growth factor (PDGF), platelet derived endothelial cell growth factor (PD-ECGF), epidermal growth factor (EGF), angiogenin, transforming growth factors alpha and beta (TGF-α, TGF-β), tumor necrosis factor alpha (TNF-α), interleukin 8 (IL-8), prostaglandin E_2, monobutyrin, and vascular endothelial growth factor (VEGF). VEGF appears to be central to the process of angiogenesis (Plate et al., 1992; Shweiki et al., 1992; Hlatky et al., 1994, 1996). In all cell types and cell lines examined, with the notable exception of endothelial cells, VEGF mRNA is upregulated in response to hypoxia. VEGF has two important properties that support tumor growth: it is a secreted specific mitogen for endothelial cells and it increases vascular permeability, thereby promoting extravasation of plasma proteins and the development of an interstitial fibrin matrix. Because of this, this molecule has been referred to both as vascular endothelial growth factor (VEGF) and as vascular permeability factor (VPF) (Senger et al., 1986).

Paracrine mechanisms appear to predominate in tumor angiogenesis. Hypoxic tumor cells, tumor-associated fibroblasts and macrophages produce VEGF, which is rapidly bound to VEGF receptors that appear to be present on the endothelial cells in tumor-associated blood vessels. This initiates a cascade of events involving increases in blood vessel permeability, extravasation of plasma proteins into tissue to form a crosslinked fibrin gel, migration, reorganization and proliferation of the endothelial cells, and the formation of capillary loops which migrate toward the region of hypoxia (Dvorak et al., 1983, 1987, 1995; Nagy et al., 1988). In wound repair and growing tissue, this process occurs under strict temporal and spatial control. However, regulation of this process is less effective in neoplastic tissue (Dvorak, 1986, 1987; Whalen, 1990). As a result, areas of rapid tumor growth are often characterized by exaggerated neovascularization. In fact, the development of this exuberant, but disorganized, neovasculature is frequently responsible for the symptoms, such as hemoptysis or other unusual bleeding, that cause a patient to visit his/her physician and thereby leads to the diagnosis of malignancy. Neovascularization is also a common diagnostic hint of the presence of a malignancy on angiograms, diagnostic X-rays, or pathologic specimens.

Despite the exuberant angiogenesis characteristic of some tumor edges, the overall tumor vasculature is poorly organized and only marginally functional (Intaglietta et al., 1977; Shubik, 1982; Jain, 1988; Vaupel et al., 1989). The vascular bed is highly heterogeneous and does not correspond to the normal vasculature bed in which there is a regular, sequential flow of blood through artery, arteriole, capillary, postcapillary venule, and vein. Rather, the tumor vascular bed is often disorganized, with arteriolar-venular shunting and other abnormal vessel interconnections. Tumor blood vessels are often tortuous in path and irregular in shape and diameter. Moreover, the structure of the vessel walls is often atypical, lacking the smooth muscle elements that are important to the regulation of luminal volume with changes in blood pressure. In addition, invasion or compression by the growing tumor cells may result in the temporary or permanent collapse of blood vessels and in the occlusion of individual blood vessels. As a result of these structural abnormalities in the vasculature, perfusion of the tumor tissue is exceeding heterogeneous both by region and with time (Intaglietta et al., 1977; Jain, 1988). Some areas of established tumors may have normal or even abnormally high rates of perfusion. Other areas will be chronically unperfused (Thomlinson and Gray, 1955). Still other areas will experience transient variations in perfusion as a result of transient fluctuations in the flow of blood through specific blood vessels (Chaplin et al., 1987). Microscopic observations of blood flow through individual tumor blood vessels show that blood flow on a microscopic level is even more complex than this model would suggest, with flow through individual microvessels varying dramatically with time, not only in the rate, but even in the direction of flow (Intaglietta et al., 1977).

These regional and temporal variations in tumor blood flow produce concomitant heterogeneity in the microenvironments within solid tumors. This microenvironmental heterogeneity has profound effects on the tumor cells, altering their metabolism, their proliferation patterns, and their responses to radiation and antineoplastic drugs.

Microenvironmental heterogeneity in solid tumors

Oxygen tensions in different areas of solid tumors have been measured using a variety of techniques. While the parameter of greatest relevance to cancer therapy is the "radiobiological hypoxic fraction", the proportion of the clonogenic cells in the tumor which are sufficiently hypoxic to be maximally radioresistant, the techniques used to measure radiobiological hypoxic fractions, described below, are not amenable to use in human tumors. Measurements of radiobiological hypoxic fractions therefore have only been made in transplanted tumors in rats and mice, in human tumors xenografted into nude mice, and in primary mouse mammary tumor virus-induced breast cancers in C3H mice. Many other techniques for measuring O_2 levels in tumors are in use or under development; a few are being tested in clinical trials (Mueller-Klieser et al., 1991; Stone, et al., 1993). Because each of the approaches being used has limitations, and because different techniques often yield different results when applied to the same tumors, it is still unclear which technique(s) will ultimately prove to be of the greatest clinical value.

The most direct approach to measuring tumor oxygenation is to use small oxygen electrodes. This approach was first used in the 1950s (Cater and Silver, 1960) and has recently entered more widespread use because of the commercial availability of microelectrode systems which can be used in human patients and which offer several advantages over earlier equipment in terms of the precision and response time of the probes and the data acquisition/storage system (Vaupel, 1990; Vaupel et al., 1991; Höckel et al., 1994; Nordsmark et al., 1994). Measurements with microelectrodes have the advantage that the polarographic signal being measured results primarily from O_2, although other chemicals can influence the measurement, resulting in spuriously high or low readings. The major problem with this approach is that the very low O_2 concentrations which are found in tumors and are of interest in radiobiology make the measurements technically difficult and stress both the technical limitations of the available equipment (e.g. the signal to noise ratio; consumption of O_2 by the electrode during the measurement; drift; alteration of probe response by motion, pressure and temperature) and the theoretical limitations of the approach as applied to biological systems (e.g. potential changes in tissue pO_2 due to injury from insertion of the probe; electrochemical signals resulting from molecules other than O_2, such as melanin or anesthetics; quenching of signals by biological molecules). Other problems include the limited spatial resolution of the measurements, the difficulty in making repeated measurements at a single location, and the difficulty in knowing the exact location of each measurement. The last problem raises special problems in small or infiltrating malignancies, where it may be difficult to assess which measurements were made in tumor and which were in normal tissues.

Despite these problems, studies using microelectrodes to measure pO_2 have been of great importance, because they have directly demonstrated that some human tumors contain large areas of hypoxia and have demonstrated that in some situations hypoxia is a strong prognostic factor, predicting higher rates of local failure and distance metastases (Gatenby et al., 1988; Höckel et al., 1994). The pO_2s measured in solid tumors have been found to vary from values typical of arterial blood (40–60 Torr), venous blood (20–40 Torr), and well-oxygenated normal tissues (which have pO_2s lying within the preceding ranges), down to values which are indistinguishable from zero, that is, complete anoxia. Measured values of pO_2 can vary dramatically over short distances and can also vary at the same position over relatively short observation periods.

Many other approaches to monitoring tissue oxygenation are being explored in experimental systems and a few have reached clinical trials (Chapman, 1984; Mueller-Klieser et al., 1991; Stone et al., 1993). In general, these do not measure pO_2 directly, but instead measure parameters which vary with oxygenation. One series of approaches involves administration of compounds which are enzymatically reduced in metabolizing cells under hypoxic conditions. Concentrations of bound drug then can be detected using an appropriate assay. Non-invasive assays include PET for ^{18}F labeled fluoromisonidazole, gamma cameras for technetium-labeled nitroimidazoles, and fluorine NMR for certain fluorinated compounds. Immunohistochemistry or autoradiography have been used to detect bound compounds in biopsy specimens. Another series of approaches probes the status of cellular molecules which reflect the oxygenation or redox state of individual cells (e.g. measurements of oxyhemoglobin saturation in individual red cells) or of small tissue voxels (e.g. measurements of cytochrome A/A3 reduction, oxyhemoglobin saturation, lactate levels, VEGF levels, or ATP levels). Electron spin resonance can be used with appropriate soluble or particulate probes. Other approaches measure the DNA damage produced by irradiation in a population of individual cells aspirated from the tumor (the "comet assay") or in a small biopsy (by alkaline elution), then use data on DNA damage in reference samples irradiated under air and hypoxia to calculate the proportion of radiobiologically hypoxic cells. (Even these measures of DNA damage differ from true radiobiological hypoxic fractions because they do not distinguish between clonogenic and non-clonogenic cells and may not distinguish malignant and non-malignant cells.) Although none of these techniques have entered routine clinical use, many have contributed to our understanding of tumor oxygenation. Histological studies mapping the binding of nitroimidazoles, for example, showed with graphic impact that the hypoxia in established solid tumors did not reflect a "central hypoxic core" as some old models assumed; instead hypoxic cells were scattered throughout the tumor in microscopic areas which sometimes were adjacent to necrosis but at other times showed no histologic evidence of cell death or tissue injury (Chapman, 1984). Measurements of hemoglobin saturation in individual red cells showed that some tumor blood vessels which appeared quite normal on frozen biopsy specimens were actually unperfused and contained only red cells having fully reduced hemoglobin (Mueller-Klieser et al., 1981).

These findings have helped to alter our models for tumor hypoxia to assume much more complex spatial and temporal variations in blood flow within tumors and a much more dynamic set of interactions between the tumor cells and their stroma. The limited studies performed on human patients have also been of great conceptual value, because they have shown that the many concepts and predictions developed using animal model systems could be confirmed in human tumors. This supports the use of animal model systems to develop new regimens for detecting hypoxia in human patients and for using hypoxia-directed approaches in cancer therapy.

Extracellular pH has also been measured in tumors in human patients, in veterinary patients, and in experimental rodents. Again, there appears to be remarkable temporal and regional variation in extracellular pH. Measurements with pH microelectrodes show that some regions appear to maintain pH in the physiologic range (7.2–7.4). However, some tumors include regions with pH as low as 5.8–6.0 (Wike-Hooley et al., 1984; Tannock and Rotin, 1989). It has been suggested that areas of overt necrosis are relatively alkaline, reflecting the extensive cell lysis in these areas, and that the very acidic areas represent viable regions, acidified by the metabolism of the living cells. The relationship between the extracellular pH and the pH within the tumor cells remains controversial. Viable cells with intact pathways of energy metabolism are capable of maintaining a pH gradient across their extracellular membranes. Such cells can therefore maintain an intracellular pH in or near the physiological range even in an unphysiologically acid environment. There are reports that incubation in hypoxia compromises the ability of cells to maintain this transmembrane pH gradient (Gillies et al., 1982). This may occur because the high energy phosphate compounds needed to pump protons outward across the cell membrane cannot be generated efficiently without O_2 available to act as a terminal electron acceptor. It is theoretically possible that inadequate levels of co-factors or energy sources would likewise compromise the ability of cells to maintain a physiologic intracellular pH. However, there are other data which suggest that hypoxic cells are quite capable of maintaining their intracellular pH, even in an adversely acidic milieu (Boyer and Tannock, 1992). *In vitro* data therefore do not provide a firm basis for predicting whether there would be significant variations in intracellular pH in the cells of solid tumors. Data on regional variations in intracellular pH within solid tumors are scant. The pH measured by magnetic resonance spectroscopy (MRS) techniques reflects a mixture of intracellular and extracellular pH and, moreover, averages these measurements over voxels and acquisition times which are much larger than the known parameters of the spatial and temporal variations of microenvironments within tumors. There are data which suggest that rodent and human soft tissue sarcomas show greater variability in pH as measured by MRS than normal tissues and that some regions of these tumors are abnormally acidic (Sostman et al., 1990, 1991).

Tumor cells will also be exposed to variations in the levels of specific nutrients. Because of its importance to energy metabolism, glucose has been perhaps the most extensively studied of these nutrients. It appears that, for many cell lines, the rates of usage of oxygen and glucose are such that the diffusion distance of glucose through packed cells is similar to that of O_2 (Tannock, 1970, 1972; Kallinowski et al., 1988). Cells in some regions of tumors therefore should be exposed to relatively low glucose concentrations. However, the lack of cytotoxicity with the glucose analog 5-thio-d-glucose in hypoxic areas of some tumors suggests that glucose concentrations remain above 5 μM in the hypoxic areas of these tumors (Rockwell and Schulz, 1984). This may reflect the relative metabolism of oxygen and glucose within these specific tumor cell populations, with the cells depleting the available oxygen more rapidly than the available glucose. Conversely, cell lines which are exceedingly sensitive to death from glucose deprivation form tumors with remarkably low radiobiological hypoxic fractions (Wallen et al., 1980; Newell et al., 1993). It may be that the diffusion of glucose, rather than the diffusion of oxygen, is limiting in these solid tumor systems. Thus, microregional variability in the availability of a critical nutrient, as well as microregional variability in oxygen, can be important in determining the characteristics and therapeutic responses to therapy of solid tumors. Although other critical nutrients, macromolecular building blocks, and energy sources are less thoroughly studied, it is reasonable to anticipate that their availability will likewise show both regional and temporal variations within solid tumors.

Implications of microenvironmental heterogeneity for radiation therapy

Molecular oxygen (O_2) is a radiosensitizer. Radiosensitization by O_2 is independent of the physiologic effects or metabolic uses of O_2, but instead reflects the participation of this molecule in the radiation chemistry which leads to the production of cytotoxic damage after radiotherapy (Quintiliani, 1986). Radiation therapy uses ionizing radiations, most commonly high energy X-rays, gamma rays, or electrons, to treat the tumor volume. As these radiations pass through tissue, they collide with and are scattered by atoms located along their paths, resulting in the release of lower-energy photons and energetic, "fast", electrons. These fast electrons interact with atoms along their paths, knocking electrons from the shells of these atoms and leaving behind a track of ion pairs, each consisting of a negatively charged, relatively low-energy electron and a positively charged atom. Chemical reactions between these exceedingly reactive ions and nearby molecules lead to the production of large numbers of ions, free radicals, excited molecules, and other chemically reactive species. Chemical reactions between these moieties and critical biological molecules lead to biological damage. Because this damage reflects the non-specific chemical reactions occurring after the initial ionizing event, radiation produces many different kinds of lesions (Ward, 1988). Molecules of many kinds are damaged. Most can be replaced without lasting effects on the cell; however, a few DNA lesions, or even a single, critical DNA lesion can damage or even kill the cell.

The radiosensitizing effects of O_2 result from the fact that this electron-affinic molecule participates in the chemical reactions that lead to the production of DNA damage, increasing both the amount and the severity of the damage produced (Quintiliani, 1986). As a result, cells which are aerobic at the time of radiation are approximately three times more sensitive to the effects of radiation than are cells irradiated in the absence of O_2 (Fig. 1) (Gray et al., 1953). Very low concentrations of O_2 are required to produce radiosensitization (Fig. 2). The radiosensitivity of cells rises rapidly as the O_2 tension rises from zero to ~10 Torr. Cellular radiosensitivity then

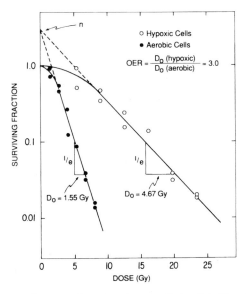

Figure 1. Survival curves for mouse breast carcinoma cells irradiated in cell culture under normal irradiation and under very severe hypoxia. The D_0s of the survival curves were used to calculate an oxygen enhancement ratio (OER), which was 3.0. (Reprinted from Rockwell, 1989.)

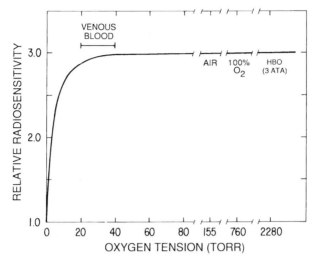

Figure 2. Variation of cellular radiosensitivity as a function of oxygen tension. This curve was derived from cell survival curves similar to those shown in Figure 1. The radiosensitizing effect of oxygen plateaus at oxygen concentrations below those of most normal tissues. (Reprinted from Rockwell, 1989.)

plateaus, and increases only marginally as the O_2 tension increases through the range found in most healthy tissues, through levels found in venous or arterial blood, and even to those O_2 tensions found in cells equilibrated with air, 100% O_2, or hyperbaric O_2. As a result, the radiobiology of most healthy normal tissues is that of fully aerobic cells. However, as described above, solid tumors contain areas in which cells are temporarily or chronically exposed to very low O_2 tensions, within the range producing maximal radioprotection.

It is clear from data on laboratory animals that cells in these hypoxic areas are viable and clonogenic, and can cause recurrence of the tumors after irradiation (Moulder and Rockwell, 1984, 1987). Moreover, it is clear that the response of solid tumors to treatment with large single doses of radiation is limited by the survival of these hypoxic cells. Figure 3 shows a dose-response curve for mouse breast carcinoma cells irradiated in solid tumors *in vivo*. The radiation dose-response curve for cells in tumors differs dramatically from the dose-response curve for cells irradiated under aerobic conditions *in vitro*. At low doses of radiation, the survival curves for cells *in vitro* and in tumors are similar, because radiation depletes the radiosensitive, aerobic, cell population which comprises 80% of the cells in these tumors. As the radiation dose increases, the slope of the survival curve changes, reflecting the radioresistance of the naturally hypoxic cells within the solid tumors. At high doses of radiation, over 10 Gy, the radiation response of the tumors in normal air-breathing animals is dominated by the response of the radioresistant hypoxic cells. Importantly, the radiation

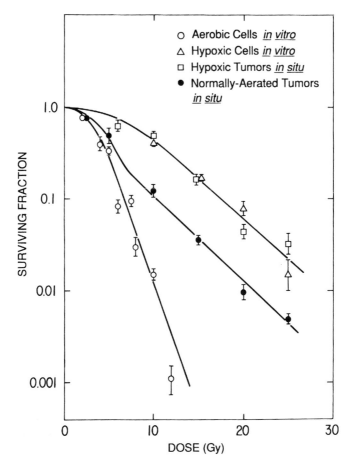

Figure 3. Comparison of the survival curves from mouse breast cancer cells irradiated in cell culture under fully aerobic and severely hypoxic conditions, in tumors acutely rendered fully hypoxic, and in the normally-aerated tumors of unanesthetized air-breathing mice. Because of their radioresistance, the naturally hypoxic cells of tumors in air-breathing mice dominate the response of these tumors to large doses of radiation. (Reprinted from Rockwell, 1989.)

response of hypoxic cells *in vitro* and of cells in tumors irradiated under maximally hypoxic conditions *in vivo* is the same, showing that the difference between aerobic cells and tumors reflects the incomplete oxygenation of the neoplasms, rather than any intrinsic physiologic difference between tumor cells growing *in vivo* and *in vitro*.

The response of solid tumors in animals to large doses of ionizing radiation is therefore limited by the radioresistance of the hypoxic tumor cells. Although the experiments shown in Figure 3 used as an endpoint *in vitro* measurements of tumor cell survival, other studies have shown that hypoxic cells also limit tumor response as measured by analyses of the post-irradiation growth of tumors subjected to subcurative radiation doses or by analyses of the doses of radiation needed to cure the tumors (Moulder and Rockwell, 1984, 1987). Mathematical analyses comparing radiation dose-response data for fully aerobic cells *in vitro*, tumors made fully hypoxic *in vivo*, and unperturbed normally-aerated tumors *in vivo* can be used to measure the proportion of the viable tumor cells which are radiobiologically hypoxic. These estimates of "the radiobiological hypoxic fractions" are in general agreement with other data on tumor hypoxia, described above.

Radiobiological hypoxic fractions in experimental tumors

Radiobiological hypoxic fractions have been measured in a large number of transplanted tumor systems in mice and rats. Our reviews provide detailed overviews of these data (Moulder and Rockwell, 1984, 1987; Rockwell and Moulder, 1990). At the sizes generally used for experimental cancer therapy studies, most tumors contain significant numbers of hypoxic cells. Most commonly, 10–20% of the viable, clonogenic tumor cells are radiobiologically hypoxic. Some tumors have higher hypoxic fractions, sometimes over 50%. Some have lower hypoxic fractions; however, virtually all macroscopic solid tumors in experimental rodents contain statistically significant numbers of hypoxic cells.

Attempts to correlate hypoxic fractions with identifiable tumor characteristics have been largely unsuccessful in experimental systems. Radiobiological hypoxic fractions do not appear to vary systematically with the degree of differentiation of the tumor or with the histology or tissue of origin of the tumor, except for the fact that leukemic infiltrates and lymphomas appear to have lower hypoxic fractions than most true solid tumors. The hypoxic fraction does not vary with the transplantation history of the tumors: primary mouse mammary carcinomas (growing in the host and site of origin) have hypoxic fractions similar to those of mammary tumors transplanted once, a few times, dozen of times, or even hundreds of times, and to those of tumors implanted from culture-adapted cell lines. No systematic differences could be identified between the hypoxic fractions measured for primary and transplanted mouse tumors, transplanted rat tumors, and human tumors xenografted into immuno-deficient mice or rats. Although some specific tumor lines have been reported to have different radiobiological hypoxic fractions when transplanted into different anatomical sites, no consistent variation of tumor oxygenation with implantation site is evident in the total data. The hypoxic fraction did not vary significantly with the tumor volume doubling time, even when data for slowly-growing human tumor xenografts and primary mouse mammary carcinomas were included in the analyses. These animal data therefore provide little foundation for predicting which tumors in human patients might have unusually high or unusually low hypoxic fractions.

At the time of their publication, these findings elicited considerable controversy, because they ran contrary to the widespread expectation that slowly-growing, well-differentiated tumors would have lower hypoxic fractions than rapidly-growing, anaplastic tumors. This expectation was based on the paradigm for tumor hypoxia existing at that time, which assumed that intrinsic differences in the cell cycle times of the malignant cells and the vascular endothelial cells caused rapidly-growing tumor cell populations to "outgrow" their slowly-growing vascular beds and thereby produce avascular areas with necrosis and hypoxia. As described in this book, recent advances in our understanding of angiogenesis have identified the central role of hypoxia in inducing angiogenesis and have revealed the complexity of the molecular signaling processes and cellular interactions involved in neovascularization. Advances in tumor pathophysiology have also shown that many factors (e.g. interstitial pressures, vessel collapse, abnormal vessel structure, and microscopic thrombosis) contribute to the regional deficiencies in tumor blood flow. These discoveries have shifted the paradigm for tumor hypoxia to a much more complex one in which the tumor cells are viewed as a cell population which adapts, evolves, and interacts with stromal cells, both responding to and modifying its environment. Under this model the hypoxic fractions of individual tumors would be expected to depend upon those physiological characteristics of the malignant cells which influence their ability to survive and proliferate in adverse microenvironments, which affect their ability to induce angiogenesis, and which influence their interactions with the established vascular bed and stromal elements within the tumor. Factors such as the tumor volume doubling time and degree of differentiation need have little effect on the hypoxic fraction.

Data on hypoxic fractions of transplanted animal tumors do provide one insight with significant clinical implications. This relates to the change in hypoxic fraction with tumor volume. While it is intuitively obvious that large tumors with extensive areas of necrosis might be expected to contain hypoxic cells, the existence and importance of hypoxic cells in small tumors have frequently been questioned. However, studies of transplanted tumors have shown conclusively that microscopic tumor deposits and tumors only 1–2 mm in diameter do contain very significant numbers of radiobiologically hypoxic, radioresistant cells (Moulder and Rockwell, 1987). This finding, which has been confirmed in several different laboratories and tumor systems, implies that hypoxia should be considered as a factor of importance even in very small primary tumors, as well as in other sites of microscopic disease, including micrometastases and surgical margins containing microscopic tumor infiltrates.

Hypoxia may present a special problem in areas subjected to recent surgery or to intensive prior radiotherapy. Prior surgery will injure blood vessels, compromise vascular function and induce transient hypoxia even in non-malignant tissues (Niinikoski et al., 1971; Knighton et al.,

1983). Intensive radiotherapy likewise will damage the vascular bed and, moreover, will destroy the ability of the vascular endothelial cells to proliferate (Song et al., 1974; Fowler, 1983). The problem of tumor hypoxia therefore may be even greater for tumors in the poorly oxygenated beds found after surgery or radiotherapy than for tumors in otherwise healthy normal tissues. This will complicate postsurgical adjuvant treatment with radiotherapy or chemotherapy and the treatment of tumors recurring in previously-treated sites.

The animal data therefore provide a firm foundation for assuming that hypoxia may be a problem in clinical cancer therapy and suggest some specific areas of concern. However, they also leave unanswered many questions of clinical importance, and therefore point to the need for investigation of the oxygenation of human tumors.

Hypoxia and radiotherapy

The response of experimental tumors to large single doses of radiation can be modified by altering the oxygenation of the tumor. As described above, making tumors uniformly hypoxic increases their radioresistance. On the other hand, improving tumor oxygenation increases the response to large doses of radiation. This has led to numerous preclinical and clinical studies examining possible approaches to improving the oxygenation of tumors during irradiation and thereby improving the outcome of radiotherapy.

Many of the treatment regimens used in radiotherapy, in fact, serve to decrease the impact of tumor hypoxia. For example, although the earliest radiotherapy treatments used one or a few very large doses of X-rays, most modern radiotherapy regimens use a large number of small daily fractions spread over several weeks. (A typical radiotherapy regimen might, for example, use thirty treatments with 2 Gy of X-rays, delivered daily over a six week period.) Alternatively, brachytherapy treatments with radioactive implants might use one or more protracted radiation treatments delivering radiation at a low dose rate over days or weeks. Protracted regimens of treatment with fractionated external-beam radiotherapy or with low dose rate brachytherapy would take advantage of the process of "reoxygenation" during the treatment regimen (Fowler, 1983; Suit, 1984). Reoxygenation has been demonstrated in animal tumors, and probably reflects several factors (Kallman, 1972). First, fluctuations in blood flow through individual tumor blood vessels, described above, lead to temporal variations in the oxygenation of specific tumor regions. Thus, cells that were hypoxic, and therefore radioresistant, during one treatment might well be aerobic, and therefore radiosensitive, during another, thus on average diminishing the impact of hypoxia on tumor cell survival. In addition, during a protracted radiotherapy regimen, tumor cells will be sterilized by the radiation and will die, resulting in decreased O_2 consumption within the tumor and in tumor shrinkage; such processes can improve the delivery of O_2 within the tumor. Dying tumor cells can continue to produce and secrete VEGF, thereby inducing angiogenesis and stimulating the development of the tumor vasculature (Hlatky, 1996). All of these processes can contribute to reoxygenation and diminish the importance of O_2 in determining the success of fractionated radiotherapy. However, studies in experimental animals and the results of clinical trials in human patients both suggest that reoxygenation during radiotherapy is not sufficient to mitigate completely the radioprotective effects of hypoxia, and that the presence of viable hypoxic cells still limits the curability of certain tumors even by modern, highly fractionated radiotherapy regimens. These data suggest that the design and implementation of regimens which circumvent the radioprotective effects of hypoxia, by oxygenating, sensitizing, or killing the hypoxic tumor cells, could improve cure rates for certain common malignancies.

Perhaps the simplest of the interventional approaches to improving tumor oxygenation is that of giving blood transfusions to patients whose hematocrits or hemoglobins fall bellow certain values. Transfusions were widely used in radiotherapy at many institutions as part of the routine supportive care for cancer patients before the recent development of concerns about transfusion-related infections. Data from several clinical trials, including some large, prospectively randomized and stratified trials, show that low hematocrits or low hemoglobin levels are prognostic factors predicting poorer local control and cure rates (Girinski et al., 1989). Some randomized trials have shown that blood transfusions can improve the responses of anemic patients to radiotherapy. A newer approach, based on the same premise, uses recombinant human erythropoietin (r-HuEPO) to increase red cell production in anemic patients. Subcutaneous r-HuEPO has been shown in randomized trials of patients receiving radiotherapy to increase hemoglobin levels by an average of 5% per week (Dusenbery et al., 1994). Definitive studies of the effect of r-HuEPO on long term local control have not yet been completed. However, as a whole, clinical data on anemia and radiotherapy remain the subject of considerable controversy, because it is unclear whether the prognostic value of anemia reflects resistance caused by tumor hypoxia, rather than an association between anemia and more aggressive or perhaps more advanced disease (Girinski et al., 1989).

Data from animals show that the response of solid tumors in animals to blood loss or to transfusion is complex (Hirst and Wood, 1987; Sevick and Jain, 1989). The acute induction of anemia, as might be expected, results in increased tumor hypoxia. However, within hours of the induction of anemia, homeostatic mechanisms altering O_2 transport and delivery within the anemic animal compensate for the acute anemia, and tumor oxygenation returns to the initial levels. Similarly, transfusion of acutely anemic hosts with appropriate regimens initially improves tumor oxygenation, but tumor radiosensitization decreases with time. The results of studies in this area confirm the fact well known in other areas of medicine: the hematologic system is tightly controlled by a variety of homeostatic mechanisms that tend to maintain oxygenation and pH in tissues within a relatively narrow range; manipulations which would shift tissues from this range will induce feedback loops which restore the physiologic values. Effective application of such approaches as blood transfusions and the use of erythropoietin as adjuncts to radiotherapy will require the development of regimens which consider the complexities of the homeostatic control of tissue oxygenation.

Alternately, many investigators have considered approaches which might alter tumor oxygenation acutely, for a short time, during the delivery of radiotherapy. As early as the 1950s, experimental animals and human patients were being treated with radiation therapy while they were breathing hyperbaric oxygen (HBO; 100% oxygen at 3–4 atmospheres) (Fowler, 1983; Dische, 1991; Overgaard and Horsman, 1996). Although theoretical analyses of changes in O_2 delivery by HBO suggest that HBO can produce only small improvements in tumor oxygenation (Fischer et al., 1986), studies in experimental animals showed that this approach could improve the oxygenation of solid tumors and increase the tumor radiation response (Suit, 1984). Because most normal tissues are already sufficiently well oxygenated to be fully radiosensitive, increases in normal tissue radiation reactions were very small. The therapeutic gain produced by HBO diminishes with fractionation because of reoxygenation, as described above, but therapeutic gain was still obtained with many fractionated regimens. Several clinical trials of hyperbaric oxygen were performed during the 1950s and 1960s. This approach fell out of favor after this time, largely because many of the trials were small and therefore inconclusive and because the use of HBO with radiotherapy was time consuming, potentially dangerous, and cumbersome, especially with the hyperbaric chambers available at that time. Interestingly, recent meta-analyses of clinical HBO data by Overgaard and Horsman (1996) suggest that, in fact, use of HBO as an adjunct to radiotherapy raised local control rates for some diseases by as much as 10%, an improvement which would be touted as a major advance if it resulted from a new chemotherapeutic agent which might represent a potential profit source.

Other approaches to improving tumor oxygenation during irradiation would include the use of perfluorochemical emulsions, or other O_2 transport vehicles (Rockwell et al., 1992; Teicher et al., 1992). Perfluorochemicals are hydrocarbons in which all of the hydrogen atoms have been replaced by fluorines; the resulting perfluorocarbons are chemically and biologically inert. Some perfluorocarbons are able to dissolve exceedingly large amounts of O_2. This high O_2-carrying capacity led to the development of perfluorochemical emulsions (PFC-E), in which very small droplets of perfluorochemicals were stabilized and suspended in salt solutions for intravenous infusion. A variety of uses were envisioned for such emulsions, ranging from emergency transfusions to surgery, to stroke, and to use in radiotherapy. There are some interesting differences between O_2 transport and delivery by PFC-E and by red blood cells, which relate to differences in the size of transporting particles (red cells vs. PFC droplets) and to differences in the nature of the association between the O_2 and the transporting molecules (Fischer et al., 1986). These could result in the delivery of O_2 to tumor regions not reached by red cells and could result in the delivery of O_2 to tissue by PFC-Es at much higher O_2 tensions than is possible with normal human hemoglobins. Moreover, because PFC-Es transport large amounts of O_2 only when patients breathe 100% O_2 or HBO, oxygen delivery to tissue could be "pulsed" during radiotherapy, eliminating the problems of adaptation described above for transfusions. Similarly, O_2 delivery systems based on hemoglobins with right-shifted O_2-dissociation curves or with large Root effects which cause them to unload O_2 selectively at high pO_2s in acidic regions (Bunn and Forget, 1986), could also offer theoretical advantages over normal human blood products for O_2 delivery to tumor tissue (Rockwell et al., 1996), as could approaches which use short-lived allosteric effectors to modulate the O_2-dissociation curve of the patient's own hemoglobin transiently during irradiation (Hirst et al., 1987). Modulation of blood flow (Jain, 1988; Jirtle, 1988) or interstitial pressure (Boucher et al., 1990) could also be effective in modulating tumor perfusion and therefore tumor oxygenation. A variety of new approaches to modulating tumor oxygenation, which offer conceptual and practical advantages over those tested previously, therefore remain to be evaluated rigorously, either in the clinic or in experimental systems.

An alternative approach to circumventing the radioprotective effects of hypoxia would be concomitant treatment with a drug that replaces O_2 in the radiochemical reactions leading to cytotoxicity and therefore selectively sensitizes hypoxic cells to the effects of radiation. Extensive laboratory studies have been performed in this area. Clinical trials of potential hypoxic cell sensitizers have been limited, because the toxicities of the available agents have precluded optimal use of these agents, which would involve administration of high doses of drug before each of 20–30 radiotherapy fractions. Importantly, however, several large, prospectively randomized and carefully stratified clinical trials have showed that this approach can have clinical benefits in certain diseases. A recent meta-analysis of the randomized trials testing radiosensitizers with radiotherapy has revealed an overall improvement of ~10% in local control rates for carcinoma of the head and neck (Overgaard, 1994, Overgaard and Horsman, 1996).

Hypoxic cell radiosensitizers are currently being examined in a clinical situation, stereotactic radiosurgery, in which a single large dose of external beam irradiation is used to treat malignant disease. In stereotactic radiosurgery, very highly collimated and precisely directed beams of ionizing radiation are targeted, from many different directions, onto a small, localized, well-defined intracranial tumor. Solitary brain metastases and primary brain tumors are being treated increasingly with radiosurgery, because the compelling need to avoid damaging critical adjacent neural tissues makes this approach preferable to either conventional radiotherapy or neurosurgery for certain patients (Phillips et al., 1994). A Phase IB study now being performed by the cooperative group RTOG is examining the effects of radiotherapy with concomitant administration of the hypoxic cell radiosensitizer etanidazole. This represents an exceptionally promising use of radiosensitizers because the single treatment allows the use of a maximal dose of drug and because, as described above, theoretical considerations and laboratory data suggest that the effects of hypoxia are most pronounced with large single radiation doses.

The outcome of therapy could also be improved by concomitantly treating solid tumors with radiotherapy and with a drug which was effective in killing the hypoxic tumor cells (Rockwell, 1992; Brown and Giaccia, 1994). Initial clinical trials of this approach were performed at our institution, using a drug with a moderate degree of selective toxicity to hypoxic cells *in vitro*, mitomycin C. This drug and the related bioreductive alkylating agents become toxic only after being enzymatically reduced within cells to active alkylating species. The selective toxicities of these drugs reflect differences in the activities and levels of several reductases occurring when cells are exposed to severe hypoxia. Our clinical trials revealed significant improvements in local/regional control when mitomycin C was administered concomitantly with radiotherapy in the treatment of carcinoma of the head and neck (Haffty et al., 1993). Ongoing studies extending this concept to the treatment of carcinoma of the cervix with concomitant mitomycin C and radiation show encouraging results in preliminary analyses. Trials testing porfiromycin, a mitomycin C analog with greater preferential toxicity towards hypoxic tumor cells, are also ongoing, again with encouraging preliminary results (Haffty et al., 1994).

The approach of using agents toxic to hypoxic cells to treat solid tumors is likely to bear increasing fruit in the future, as a result of rapidly developing knowledge of the effects of hypoxia and its associated microenvironmental inadequacies on gene expression, protein levels, and enzyme activities. It is becoming clear that hypoxia produces a large number of molecular responses which may be exploitable in cancer therapy (Coleman, 1988; Graeber et al., 1994; Dachs and Stratford, 1996). These include the induction of a number of stress proteins, some of which are also induced by exposure to low glucose and to heat (Heacock and Sutherland, 1990). Because the nuclear transcription factors HSF and NF-κB and HIF1, the proto-oncogenes *jun* and *fos*, and the tumor suppressor gene *p53* are all induced by hypoxia, a number of enzymes and metabolic pathways are also modulated in hypoxic cells. These include several oxidoreductases, e. g. DT-diaphorase, xanthine oxidase, xanthine dehydrogenase, glutathione-S-transferase, and cytochrome B_5 reductase, lactate dehydrogenase, and phosphoglycerate kinase I, which are important in the activation and detoxification of anticancer drugs. A variety of other enzymes have also been shown to be induced in specific cells, including erythropoietin, ferritin,

VEGF, PDGF, IL-6, TGF-α, TGF-β, ornithine decarboxylase and EGF receptor. This spectrum of metabolic responses suggests a variety of potential avenues for selectively targeting hypoxic tumor cells (Dachs and Stratford, 1996). Studies showing that anti-VEGF antibodies can inhibit the growth of breast cancer xenografts (Kim et al., 1993) point to the clinical potential of these specific mechanisms for attacking hypoxic tumor cells.

Further implications of hypoxia for cancer therapy

In summary, a variety of clinical trials using different approaches to circumvent the radioprotective effects of hypoxia suggest that these various approaches can improve the outcome of radiotherapy. The importance of hypoxia in limiting the success of chemotherapy in the clinic is less clear. Data from experimental systems show that hypoxic cells can be resistant to anticancer drugs, not only because oxygen is involved in the mechanism of action of some chemotherapeutic agents, such as bleomycin, but also because hypoxic cells may be relatively quiescent, and therefore resistant to cycle-active antimetabolites and antineoplastic agents (Teicher et al., 1981; Rockwell, 1992; Teicher, 1994). The deficiencies in blood flow within tumors and the high interstitial pressure within tumor tissue can result in poorer delivery of drugs within tumors than within normal tissues (Jain, 1988; Boucher et al., 1990). Moreover, many large or hydrophilic antineoplastic drugs, as well as therapeutic agents based on biological macromolecules such as monoclonal antibodies, penetrate poorly through avascular tissues; such agents remain localized in or near the patent blood vessels and do not reach hypoxic cells in effective concentrations (Boucher et al., 1990; Simpson-Herren and Noker, 1991). It has been shown in experimental animal systems that hypoxic cells can limit the effect of many chemotherapeutic agents on solid tumors (Teicher, 1994). Whether this occurs in patients being treated with standard cancer chemotherapy regimens remains uncertain, because clinical trials analogous to those performed in radiotherapy have not yet been performed. Even very small numbers of tumor cells which were resistant to drugs for any of the reasons described above would prevent chemotherapy from being curative (Fig. 4).

The implications of tumor hypoxia for the treatment of established tumors have been examined, as described above, by radiation oncologists, and more recently, by medical and surgical oncologists. The implications of hypoxia for the development and progression of malignant disease have been less thoroughly considered, but also merit attention. As described above, hypoxia develops in transplanted tumors while the developing neoplasms are still microscopic in size, and the ability of the tumor cells to survive in adverse microenvironments is a prerequisite for further tumor development. Moreover, microenvironmental heterogeneity occurs in well

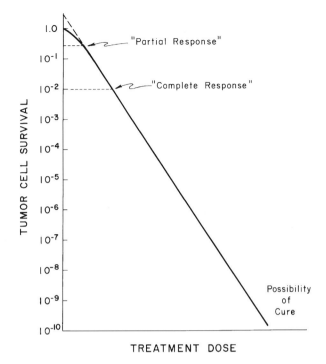

Figure 4. Schematic diagram showing the importance of considering resistant tumor cell subpopulations in the curative treatment of cancer. Because tumor cell survivals must be reduced by 9–10 logs to cure even small neoplasms, the presence of a small resistant cell population will compromise the possibility of achieving cure, even though analyses of partial and complete responses may not reveal the presence of a resistant population.

differentiated and slowly growing malignancies as well as in anaplastic, rapidly growing cancers. These facts raise the question of when hypoxia becomes a factor in the development of malignant disease. Clearly, hypoxia is an early event (Folkman et al., 1989), but it is unclear exactly when it develops in the spectrum of transition from atypia to hyperplasia to borderline malignancy to frankly malignant disease. There is reason to expect that hypoxia could be a factor contributing to the development of mutations, genomic instability, and the evolution of an increasingly aggressive and malignant phenotype (Stoler et al., 1992). Hypoxia is known to alter gene expression and to induce stress responses (Heacock and Sutherland, 1990; Graeber et al., 1994; Dachs and Stratford, 1996). Severe, prolonged hypoxia leads to perturbations in energy metabolism that ultimately result in DNA degradation and cell death (Stoler et al., 1992; Walker et al., 1994). It is therefore reasonable to assume that nonlethal exposures to hypoxia may result in significant genomic alterations. The observation that hypoxia leads to amplification of the gene for dihydrofolate reductase, resulting in the development of resistance to methotrexate

(Rice et al., 1986), offers a concrete example of a hypoxia-induced stable genetic change which worsens the prognosis for the malignancy.

An additional warning comes from work by Young et al. (1988) who examined the ability of intravenously-injected tumor cells to form lung nodules in mice. When the ability of tumor cells to form these artificial lung metastases was compared for cells cultured continuously under aerobic conditions and for cells incubated for several hours in severe hypoxia, then reoxygenated, it was found that cells which had been subjected to hypoxia and reoxygenation had higher tumor-forming efficiencies than cells which had been cultured continuously under aerobic conditions. The increased metastatic ability appeared to reflect a transient change in the cellular phenotype, rather than a permanent genetic alteration (Young et al., 1988; Young and Hill, 1990). Thus, it is possible that hypoxia occurring in solid tumors could be a factor enhancing the development of metastases. Observations by Höckel et al. (1994) and by Brizel et al. (1996) that severe hypoxia detected in human tumors by microelectrode measurements correlates with aggressive disease and with an increased incidence of metastatic disease suggest that these laboratory observations may indeed have clinical correlates, as do correlations between angiogenesis and metastases (Weidner et al., 1991). If so, hypoxia would decrease the curability of solid tumors by well executed radical surgery, as well as by radiotherapy and chemotherapy. Further study of the effects of hypoxia on the development and progression of malignant disease is clearly warranted.

Implications of angiogenesis in cancer therapy

Much of the preceding discussion has considered the therapeutic problems related to tumor hypoxia. However, the fact that tumor growth is dependent upon angiogenesis and the development of a vascular bed also points to exciting approaches for attacking solid tumors and for preventing the growth of metastases. The isolation, identification, and development of antiangiogenic substances, which could be applied exogenously to prevent angiogenesis and neovascularization, could lead to the development of drugs which would inhibit the growth of established tumors and/or prevent metastases from developing beyond microscopic sizes (Folkman, 1992a; Brem and Folkman, 1993). Because angiogenesis and neovascularization are rare events under normal conditions, treatments with such agents might have minimal toxicity. For example, thalidomide, developed as a sedative, appears to be an extremely effective inhibitor of angiogenesis *in vitro* and in animal systems (D'Amato et al., 1994). The drug was considered to be extremely non-toxic until its use for the amelioration of morning sickness was found to cause phocomelia in exposed children, probably as a result of the inhibition of angiogenesis during a critical period in the development of the limb buds. Despite this disaster, the relative lack of toxicity of the drug in adult patients supports the possibly of clinical trials testing the value of thalidomide and other antiangiogenic drugs in patients with cancer. Preliminary studies with antiangiogenic agents *in vitro* and in animal systems appear promising (Folkman et al., 1988; Folkman, 1992 a; Teicher et al., 1992; Brem and Folkman, 1993; Kusaka et al., 1994).

Alternatively, the unusual physiology and rapid growth of the developing neovasculature may provide targets which can be used to injure or kill these cells selectively, thereby damaging the developing vascular bed and indirectly leading to the death of the tumor cells (Fan et al., 1995). It has been hypothesized (Denekamp, 1992) that the high rate of cell proliferation in growing tumor blood vessels renders these cells sensitive to the cytotoxic effects of cancer chemotherapeutic agents, and is partially responsible for the effects of these drugs on solid tumors. The effects of photodynamic therapy (PDT) likewise have been hypothesized to reflect in part damage to the vascular bed by the photodynamic reaction, because the photofrins remain localized in and near the blood vessels and because O_2 is required for the cytotoxic PDT reactions (Henderson and Fingar, 1987). The delayed tumor cell death observed in tumors treated with PDT supports the concept that the death of the malignant cells is not a direct result of the PDT reaction, but instead is secondary to the obliteration of a functional vasculature. Drugs targeted specifically toward developing tumor blood vessels could have greater specificity than standard cycle-active chemotherapy and should also avoid the spatial limitations of PDT (which derive from the limited penetration of light through tissue), and therefore could have more widespread application, lesser toxicity, and greater efficacy. The finding that vascular endothelial cells involved in angiogenesis and neovascularization express receptors and secrete factors not normally found in cells of established blood vessels suggest that it should be possible to find targets specific to the developing tumor vasculature. The studies of tumor angiogenesis described elsewhere in this book therefore have the potential to lead to the development of extremely effective treatments both for eradicating solid tumors and for controlling metastatic disease.

Acknowledgements
Research in the authors' laboratories related to this chapter is supported by grants CA-35251 from the NIH, EDT-62 from the American Cancer Society and a Critical Technology Grant from the state of Connecticut. The generous support of these agencies is acknowledged and greatly appreciated.

References

Brizel, D.M., Scully, S.P., Harrelson, J.M., Layfield, L.J., Bean, J.M., Prosnitz, L.R. and Dewhirst, M.W. (1996) Tumor oxygenation predicts for the likelihood of distant metastases in human soft tissue sarcoma. *Cancer Res.* 56:941–943.
Boucher, Y., Baxter, L.T. and Jain, R.K. (1990) Interstitial pressure gradients in tissue-isolated and subcutaneous tumors: Implications for therapy. *Cancer Res.* 50:4478–4484.

Boyer, M.J. and Tannock, I.F. (1992) Regulation of intracellular pH in tumor cell lines: Influence of microenvironmental conditions. *Cancer Res.* 52:4441–4447.

Brem, H. and Folkman, J. (1993) Analysis of experimental antiangiogenic therapy. *J. Pediatr. Surg.* 28:445–450.

Brown, J.M. and Giaccia, A.J. (1994) Tumour hypoxia: the picture has changed in the 1990s. *Int.. J. Radiat. Biol.* 65:95–102.

Bunn, H.F. and Forget, B.G. (1986) *Hemoglobin: Molecular, genetic, and clinical aspects.* W.B. Saunders Co., Philadelphia, PA.

Cater, D.B. and Silver, I.A. (1960) Quantitative measurements of oxygen tension in normal tissues and in the tumours of patients before and after radiotherapy. *Acta Radiol.* 53:233–256.

Chapman, J.D. (1984) The detection and measurement of hypoxic cells in solid tumors. *Cancer* 54:2441–2449.

Chaplin, D.J., Olive, P.L. and Durand, R.E. (1987) Intermittent blood flow in a murine tumor: Radiobiological effects. *Cancer Res.* 47:597–601.

Coleman, C.N. (1988) Hypoxia in tumors: A paradigm for the approach to biochemical and physiologic heterogeneity. *J. Natl. Cancer Inst.* 80:310–317.

D'Amato, R.J., Loughnan, M.S., Flynn, E. and Folkman, J. (1994) Thalidomide is an inhibitor of angiogenesis. *Proc. Natl. Acad. Sci. USA* 91:4082–4085.

Dachs, G.U. and Stratford, I.J. (1996) The molecular response of mammalian cells to hypoxia and the potential for exploitation in cancer therapy. *Br. J. Cancer,* 74:S126–S132.

Denekamp, J. (1992) Endothelial cell proliferation as a novel approach to targeting tumour therapy. *Br. J. Cancer* 45:136–139.

Dische, S. (1991) What have we learnt from hyperbaric oxygen? *Radiother. Oncol.* 1:71–74.

Dusenbery, K.E., McGuire, W.A., Holt, P.J., Carson, L.F., Fowler, J.M., Twiggs, L.B. and Potish, R.A. (1994) Erythropoietin increases hemoglobin during radiation therapy for cervical cancer. *Int. J. Radiat. Oncol. Biol. Phys.* 29:1079–1084.

Dvorak, H.F. (1986) Tumors: Wounds that do not heal. Similarities between tumor stroma generation and wound healing. *New Engl. J. Med.* 315:1650–1659.

Dvorak, H.F. (1987) Thrombosis and cancer. *Hum. Pathol.* 18:275–284.

Dvorak, H.F., Brown, L.F., Detmar, M. and Dvorak, A.M. (1995) Vascular permeability factor/vascular endothelial growth factor, microvascular hyperpermeability, and angiogenesis. *Am. J. Pathol..* 146:1029–1039.

Dvorak, H.F., Harvey, S., Estrella, P., Brown, L.F., McDonagh, J. and Dvorak, M. (1987) Fibrin containing gels induce angiogenesis: Implications for tumor stroma generation and wound healing. *Lab. Invest.* 57:673–686.

Dvorak, H.F., Senger, D.R. and Dvorak, A.M. (1983) Fibrin as a component of the tumor stroma: Origins and biological significance. *Cancer Metast. Rev.* 2:41–73.

Fan, T.P., Jaggar, R. and Bicknell, R. (1995) Controlling the vasculature: Angiogenesis, anti-angiogenesis and vascular targeting of gene therapy. *Trends in Pharmacol. Sci.* 16:57–66.

Fischer, J.J., Rockwell, S. and Martin, D.F. (1986) Perfluorochemicals and hyperbaric oxygen in radiation therapy. *Int. J. Radiat. Oncol. Biol. Phys.* 12:95–102.

Folkman, J. (1971) Tumor angiogenesis: Therapeutic implications. *New Engl. J. Med.* 285:1182–1186.

Folkman, J. (1985) Tumor angiogenesis. *Adv. Cancer Res.* 43:175–203.

Folkman, J. (1992a) Inhibition of angiogenesis. *Semin. Cancer Biol.* 3:89–96.

Folkman, J. (1992b) The role of angiogenesis in tumor growth. *Semin. Cancer Biol.* 3:65–71.

Folkman, J., Watson, K., Ingber, D. and Hanahan, D. (1989) Induction of angiogenesis during the transition from hyperplasia to neoplasia. *Nature* 339:58–61.

Folkman, J., Weisz, P.B., Joullié, M.M., Li, W.W. and Ewing, W.R. (1988) Control of angiogenesis with synthetic heparin substitutes. *Science* 243:1490–1493.

Fowler, J.F. (1983) La Ronde – radiation sciences and medical radiology. *Radiother. Oncol.* 1:1–22.

Gatenby, R.A., Kessler, H.B., Rosenblum, J.S., Coia, L.R., Moldofsky, P.J., Hartz, W.H. and Broder, G.J. (1988) Oxygen distribution in squamous cell carcinoma metastases and its relationship to outcome of radiation therapy. *Int. J. Radiat. Oncol. Biol. Phys.* 14:831–838.

Gillies, R.J., Ogino, T., Schulman, R.G. and Ward, D.C. (1982) [31] P nuclear magnetic resonance evidence for the regulation of intracellular pH by Ehrlich ascites tumor cells. *J. Cell Biol.* 95:24–28.

Girinski, T., Pejovic-Lenfant, M.H., Bourhis, J., Campana, F., Cosset, J.M., Petit, C., Malaise, E.P., Haie, C., Gerbaulet, A. and Chassagne, D. (1989) Prognostic value of hemoglobin concentrations and blood transfusions in advanced carcinoma of the cervix treated by radiation therapy: Results of a retrospective study of 386 patients. *Int. J. Radiat. Oncol. Biol. Phys.* 16:37–42.

Graeber, T.G., Peterson, J.F., Tsai, M., Monica, K., Fornace, A.J. and Giaccia, A.J. (1994) Hypoxia induces accumulation of p53 protein, but activation of a G_1-phase checkpoint by low-oxygen conditions is independent of p53 status. *Mol. Cellular Biol.* 14:6264–6277.

Gray, L.H., Conger, A.D., Ebert, M., Hornsey, S. and Scott, O.C.A. (1953) The concentration of oxygen dissolved in tissues at the time of irradiation as a factor in radiotherapy. *Br. J. Radiol.* 26:638–648.

Haffty, B.G., Son, Y.H., Moini, M., Papac, R., Fischer, D., Rockwell, S., Sartorelli, A.C. and Fischer, J.J. (1994) Porfiromycin as an adjunct to radiation therapy in squamous cell carcinoma of the head and neck: Results of a phase I clinical trial. *Radiat. Oncol. Invest.* 1:297–304.

Haffty, B.G., Son, Y.H., Sasaki, C.T., Papac, R., Fischer, D., Rockwell, S., Sartorelli, A. and Fischer, J.J. (1993) Mitomycin C as an adjunct to postoperative radiation therapy in squamous cell carcinoma of the head and neck: Results from two randomized clinical trials. *Int. J. Radiat. Oncol. Biol. Phys.* 27:241–250.

Hahn, R. (1904) Einen Beitrag zur Röntgentherapie. *Fortschr. Geb. Röntgenstrahlen* 8:120–121.

Heacock, C.S. and Sutherland, R.M. (1990) Enhanced synthesis of stress proteins caused by hypoxia and relation to altered growth and metabolism. *Br. J. Cancer* 62:217–225.

Henderson, B.W. and Fingar, V.H. (1987) Relationship of tumor hypoxia and response to photodynamic treatment in an experimental mouse tumor. *Cancer Res.* 47:3110–3114.

Hirst, D.G. and Wood, P.J. (1987) The adaptive response of mouse tumours to anaemia and retransfusion. *Int. J. Radiat. Biol.* 51:597–609.

Hirst, D.G., Wood, P.J. and Schwartz, H.C. (1987) The modification of hemoglobin affinity for oxygen and tumor radiosensitivity by antilipidemic drugs. *Radiat. Res.* 112:164–172.

Hlatky, L., Tsionou, C., Hahnfeldt, P. and Coleman, C.N. (1994) Mammary fibroblasts may influence breast tumor angiogenesis via hypoxia-induced vascular endothelial growth factor up-regulation and protein expression. *Cancer Res.* 54:6083–6086.

Hlatky, L., Hahnfeldt, P., Tsionou, C. and Coleman, C.N. (1996) Vascular endothelial growth factor: Environmental controls and effects in angiogenesis. *Br. J. Cancer*, 74:S151–S156.

Höckel, M., Knoop, C., Schlenger, K., Vorndran, B., Knapstein, P.G. and Vaupel, P. (1994) Intra-tumor pO_2 histography as predictive assay in advanced cancer of the uterine cervix. *Adv. Exp. Med. Biol.* 345:445–450.

Intaglietta, M., Myers, R.R., Gross, J.F. and Reinhold, H.S. (1977) Dynamics of microvascular flow in implanted mouse mammary tumours. *Bibliotheca Anatomica* 15:273–276.

Jain, R.K. (1988) Determinants of tumor blood flow: A review. *Cancer Res.* 48:2641–2658.

Jirtle, R.L. (1988) Chemical modification of tumour blood flow. *Int. J. Hyperthermia* 4:355–371.

Kallinowski, F., Vaupel, P., Runkel, S., Berg, G, Fortmeyer, H.P., Baessler, K.H., Wagner, K., Mueller-Klieser, W. and Walenta, S. (1988) Glucose uptake, lactate release, ketone body turnover, metabolic micromilieu, and pH distributions in human breast cancer xenografts in nude rats. *Cancer Res.* 48:7264–7272.

Kallman, R.F. (1972) The phenomenon of reoxygenation and its implications for fractionated radiotherapy. *Radiology* 105:135–142.

Kim, K.J., Li, B., Winer, J., Armanini, M., Gillett, N., Phillips, H.S. and Ferrara, N. (1993) Inhibition of vascular endothelial growth factor-induced angiogenesis suppresses tumour growth *in vivo*. *Nature* 362:841–844.

Klagsbrun, M. and D'Amore, P.A. (1991) Regulators of angiogenesis. *Annu. Rev. Physiol.* 53:217–239.

Knighton, D.R., Hunt, T.K., Scheuenstuhl, H., Halliday, B.J., Werb, Z. and Banda, M.J. (1983) Oxygen tension regulates the expression of angiogenesis factor by macrophages. *Science* 221:1283–1285.

Kusaka, M., Sudo, K., Matsutani, E., Kozai Y, Marui, S., Fujita, T., Ingber, D. and Folkman, J. (1994) Cytostatic inhibition of endothelial cell growth by the angiogenesis inhibitor TNP-470 (AGM-1470). *Br. J. Cancer* 69:212–216.

Moulder, J.E. and Rockwell, S. (1987) Tumor hypoxia: Its impact on cancer therapy. *Cancer Metastasis Rev.* 5:313–341.

Moulder, J.E. and Rockwell, S. (1984) Hypoxic fractions of solid tumors: Experimental techniques, methods of analysis, and a survey of existing data. *Int. J. Radiat. Oncol. Biol. Phys.* 10:695–712.

Mueller-Klieser, W., Schlenger, K.-H., Walenta, S., Gross, M., Karbach, U., Hoeckel, M. and Vaupel, P. (1991) Pathophysiological approaches to identifying tumor hypoxia in patients. *Radiother. Oncol.* 20:21–28.

Mueller-Klieser, W., Vaupel, P., Manz, R. and Schmidseder, R. (1981) Intracapillary oxyhemoglobin saturation of malignant tumors in humans. *Int. J. Radiat. Oncol. Biol. Phys.* 7:1397–1404.

Nagy, J.A., Brown, L.F., Senger, D.R., Lanir, N., Van de Water, L., Dvorak A.M. and Dvorak, H.F. (1988) Pathogenesis of tumor stroma generation: A critical role for leaky blood vessels and fibrin deposition. *Biochim. Biophys. Acta* 948:305–326.

Newell, K., Franchi, A., Pouysségur, J. and Tannock, I. (1993) Studies with glycolysis-deficient cells suggest that production of lactic acid is not the only cause of tumor acidity. *Proc. Natl. Acad. Sci. USA* 90:1127–1131.

Niinikoski, J., Heughan, C. and Hunt, T.K. (1971) Oxygen and carbon dioxide tensions in experimental wounds. *Surg. Gynecol. Obstet.* 133:1003–1007.

Nordsmark, M., Bentzen, S.M. and Overgaard, J. (1994) Measurement of human tumour oxygenation status by a polarographic needle electrode: An analysis of inter- and intratumour heterogeneity. *Acta Oncologica* 33:383–389.

Overgaard, J. (1994) Clinical evaluation of nitroimidazoles as modifiers of hypoxia in solid tumors. *Oncol. Res.* 6:509–518.

Overgaard, J. and Horsman, M.R. (1996) Modification of hypoxia-induced radioresistance in tumors by the use of oxygen and sensitizers. *Seminars in Radiat. Oncol.* 6:10–21.

Phillips, M.H., Stelzer, K.J., Griffin, T.W., Mayberg, M.R. and Winn, H.R. (1994) Stereotactic radiosurgery: A review and comparison of methods. *J. Clin. Oncol.* 12:1085–1099.

Plate, K.H., Breier, G., Weich, H.A. and Risau, W. (1992) Vascular endothelial growth factor is a potential tumor angiogenesis factor in human gliomas *in vivo. Nature* 359:845–848.

Quintiliani, M. (1986) The oxygen effect in radiation inactivation of DNA and enzymes. *Int. J. Radiat. Biol.* 50:573–594.

Rice, G.C., Hoy, C. and Schimke, R.T. (1986) Transient hypoxia enhances the frequency of dihydrofolate reductase gene amplification in Chinese hamster ovary cells. *Proc. Natl. Acad. Sci. USA* 83:5978–5982.

Rockwell, S., Baserga, S.J. and Knisely, J.P.S. (1996) Artificial blood substitutes in radiotherapy. *In:* Hagen, U., Jung, H. and Streffer, C. (eds): *Radiation Research 1895–1995, Vol. 2,* in press.

Rockwell, S. (1992) Use of hypoxia-directed drugs in the therapy of solid tumors. *Semin. Oncology* 19:29–40.

Rockwell, S., Kelley, M., Irvin, C.G., Hughes, C.S., Yabuki, H., Porter, E. and Fischer, J.J. (1992) Preclinical evaluation of Oxygent™ as an adjunct to radiotherapy. *Biomater. Artif. Cells Immunobilization Biotechnol.* 20:883–893.

Rockwell, S. and Moulder, J.E. (1990) Hypoxic fractions of human tumors xenografted into mice: A review. *Int. J. Radiat. Oncol. Biol. Phys.* 19:197–202.

Rockwell, S. (1989) Principios de radiobiologia. *In:* N. Urdanetta (ed) *Manual de Radioterapia Oncologia,* Yale University, New Haven pp 9–44.

Rockwell, S. and Schulz, R.J. (1984) Failure of 5-thio-d-glucose to alter cell survival in irradiated or unirradiated EMT6 tumors. *Radiat. Res.* 100:527–535.

Röntgen, W.C. (1895) Ueber eine neue art von Strahlen. *Sitzungsberichten der Würzburger Physik-medic. Gesellschaft.*

Rotin, D., Steele-Norwood, D., Grinstein, S. and Tannock, I. (1989) Requirement of the Na^+/H^+ exchanger for tumor growth. *Cancer Res.* 49:205–211.

Senger, D.R., Perruzzi, C.A., Feder, J. and Dvorak, H.F. (1986) A highly conserved vascular permeability factor secreted by a variety of human and rodent tumor cell lines. *Cancer Res.* 46:5629–5632.

Sevick, E.M. and Jain, R.K. (1989) Viscous resistance to blood flow in solid tumors: Effect of hematocrit on intratumor blood viscosity. *Cancer Res.* 49:3513–3519.

Shubik, P. (1982) Vascularization of Tumors: A Review. *J. Cancer Res. Clin. Oncol.* 103: 211–226.

Shweiki, D., Itin, A., Soffer, D. and Keshet, E. (1992) Vascular endothelial growth factor induced by hypoxia may mediate hypoxia-initiated angiogenesis. *Nature* 359:843–845.

Shweiki, D., Neeman, M., Itin, A. and Keshet, E. (1995) Induction of vascular endothelial growth factor expression by hypoxia and by glucose deficiency in multicell spheroids: Implications for tumor angiogenesis. *Proc. Natl. Acad. Sci. USA* 92:768–772.

Simpson-Herren, L. and Noker, P.E. (1991) Diversity of penetration of anti-cancer agents into solid tumours. *Cell Prolif.* 24:355–365.

Song, C.W., Sung, J.H., Clement, J.J. and Levitt, S.H. (1974) Vascular changes in neuroblastoma of mice following x-irradiation. *Cancer Res.* 34:2344–2350.

Sostman, H.D., Charles, H.C., Rockwell, S., Leopold, K., Beam, C., Madwed, D., Dewhirst, M., Cofer, G., Moore, D., Burn, R. and Oleson, J. (1990) Soft-tissue sarcomas: Detection of metabolic heterogeneity with P-31 MR-spectroscopy. *Radiology* 176:837–843.

Sostman, H.D., Rockwell, S., Sylvia, A.L., Matwed, D., Cofer, G., Charles, H.C., Negro-Vilar, R. and Moore, D. (1991) Evaluation of BA1112 rhabdomyosarcoma oxygenation with microelectrodes, optical spectrophotometry, radiosensitivity, and magnetic resonance spectroscopy. *Mag. Res. Med.* 20:253–267.

Stoler, D.L., Anderson, G.R., Russo, C.A., Spina, A.M. and Beerman, T.A. (1992) Anoxia-inducible endonuclease activity as a potential basis of the genomic instability of cancer cells. *Cancer Res.* 52:4372–4378.

Stone, H.B., Brown, J.M., Phillips, T.L. and Sutherland, R.M. (1993) Oxygen in human tumors: Correlations between methods of measurement and response to therapy. *Radiat. Res.* 136:422–434.

Suit, H.D. (1984) Modification of radiation response. *Int. J. Radiat. Oncol. Biol. Phys.* 10:101–108.

Tannock, I.F. (1970) Population kinetics of carcinoma cells, capillary endothelial cells, and fibroblasts in a transplanted mouse mammary tumor. *Cancer Res.* 30:2470–2476.

Tannock, I.F. (1972) Oxygen diffusion and the distribution of cellular radiosensitivity in tumours. *Br. J. Radiol.* 45:515–524.

Tannock, I.F. and Rotin, D. (1989) Acid pH in tumors and its potential for therapeutic exploitation. *Cancer Res.* 49:4373–4384.

Teicher, B.A. (1995) Physiologic mechanisms of therapeutic resistance. Blood flow and hypoxia. *Hematol. Oncol. Clin. North Am.* 9:475–506.

Teicher, B.A. (1994) Hypoxia and drug resistance. *Cancer Metast. Rev.* 13:139–168.

Teicher, B.A., Herman, T.S., Frei, E., 3rd. (1992) Perfluorochemical emulsions: Oxygen breathing in radiation sensitization and chemotherapy modulation. *Important Adv. Oncol.* 39–59.

Teicher, B.A., Sotomayer, E.A. and Huang, Z. (1992) Antiangiogenic agents potentiate cytotoxic cancer therapies against primary and metastatic disease. *Cancer Res.* 52:6702–6704.

Teicher, B.A., Lazo, J.S. and Sartorelli, A.C. (1981) Classification of antineoplastic agents by their selective toxicities towards oxygenated and hypoxic tumor cells. *Cancer Res.* 41:73–81.

Thomlinson, R.H. and Gray, L.H. (1955) The histological structure of some human lung cancers and possible implications for radiotherapy. *Br. J. Cancer* 9:539–549.

Vaupel, P. (1990) Oxygenation of human tumors. *Strahlenther. Onko.* 166:377–386.

Vaupel, P., Kallinowski, F. and Okunieff, P. (1989) Blood flow, oxygen and nutrient supply, and metabolic microenvironment of human tumors: A review. *Cancer Res.* 49:6449–6465.

Vaupel, P., Schlenger, K., Knoop, C. and Höckel, M. (1991) Oxygenation of human tumors: Evaluation of tissue oxygen distribution in breast cancers by computerized O_2 tension measurements. *Cancer Res.* 51:3316–3322.

Walker, L.J., Craig, R.B., Harris, A.L. and Hickson, J.D. (1994) A role for the human DNA repair enzyme HAP1 in cellular protection against DNA damaging agents and hypoxic stress. *Nucl. Acids Res.* 22:4884–4889.

Wallen, C.A., Michaelson, S.M. and Wheeler, K.T. (1980) Evidence for an unconventional radiosensitivity of rat 9L subcutaneous tumors. *Radiat. Res.* 84:529–541.

Ward, J.F. (1988) DNA damage produced by ionizing radiation in mammalian cells: Identities, mechanisms of formation, and reparability. *Prog. Nucleic Acid Res. Mol. Biol.* 35:95–125.

Weidner, N., Semple, J.P., Welch, W.R. and Folkman, J. (1991) Tumor angiogenesis and meta-stasis: Correlation in invasive breast carcinoma. *New Engl. J. Med.* 324:1–8.

Whalen, G.F. (1990) Solid tumors and wounds: Transformed cells misunderstood as injured tissue? *Lancet* 336:1489–1492.

Wike-Hooley, J.L., Haveman, J. and Reinhold, H.S. (1984) The relevance of tumor pH to the treatment of malignant disease. *Radiother. Oncol.* 2:343–366.

Young, S.D. and Hill, R.P. (1990) Effects of reoxygenation on cells from hypoxic regions of solid tumors: Anticancer drug sensitivity and metastatic potential. *J. Natl. Cancer. Inst.* 82:371–380.

Young, S.D., Marshall, R.S. and Hill, R.P. (1988) Hypoxia induces DNA overreplication and enhances metastatic potential of murine tumor cells. *Proc. Natl. Acad. Sci. USA* 85:9533–9537.

Regulation of Angiogenesis
ed. by I.D. Goldberg & E.M. Rosen
© 1997 Birkhäuser Verlag Basel/Switzerland

The role of vascular cell integrins $\alpha_v\beta_3$ and $\alpha_v\beta_5$ in angiogenesis

J.A. Varner

Department of Medicine, University of California, San Diego, La Jolla, California 92093-0063, USA

Introduction

Six hundred thousand new cases of lung, colon, breast and prostate cancer will be diagnosed in the United States this year, accounting for 75% of new solid tumor cancers and 77% of solid tumor cancer deaths (Fraumeni et al., 1993). Although advances in therapy and in our understanding of cancer causes and risk factors have led to improved outcomes overall, many cancers will have low five year survival rates. Despite these advances in primary tumor management, 50% of patients will ultimately die of their disease, largely due to the metastatic spread of tumors to numerous or inoperable sites through the tumor-associated vasculature (Fraumeni et al., 1992; Weinstat-Saslow, 1994). It is now known that the growth and spread of solid tumor cancer *depends* on the development of a tumor-associated vasculature by a process known as angiogenesis.

Angiogenesis, or neovascularization, is the process by which new blood vessels develop from pre-existing vessels. Endothelial cells from pre-existing blood vessels become activated to migrate proliferate, and differentiate into structures with lumens, forming new blood vessels, in response to hormonal cues or hypoxic or ischemic conditions. During ischemia, the need to increase oxygenation and delivery of nutrients apparently induces the secretion of angiogenic factors by the affected tissue these factors stimulate new blood vessel formation. The growth of new blood vessels also promotes the normal physiological processes of embryonic development (Wagner and Risau, 1994; Folkman and Shing, 1992), wound healing (Clark et al., 1982, 1995), the female reproductive cycle (Phillips et al., 1990; Kamat et al., 1995) and inflammation. Except for the human reproductive cycle and the processes of wound healing or inflammation, normal adult tissues do not actively induce angiogenesis (Ausprunk et al., 1975). However, the proliferation of new blood vessels from pre-existing capillaries plays a *key* role in the pathological development of solid tumor cancers and their metastases (Folkman, 1992), as well as in childhood hemangiomas, diabetic retinopathy, age related macular degeneration, psoriasis, gingivitis, rheumatoid arthritis, and possibly osteoarthritis and inflammatory bowel disease (Auerbach and Auerbach, 1994). Because these serious diseases afflict several million people in the United States each year, considerable scientific effort has been directed toward gaining an understanding of the mechanisms regulating angiogenesis. In fact, recent advances have led to the development of anti-angiogenic compounds that may serve as therapeutic agents for these diseases.

Angiogenesis in cancer

Cancers begin as small populations of cells which have accumulated a series of mutations in growth control regulatory molecules; these transformed cells have a significant growth advantage over neighboring normal cells. Studies have demonstrated that early tumors are often no more than a few cell layers thick, or at most, about 1 mm^3 in diameter (Folkman, 1992). Folkman and colleagues demonstrated in ground-breaking studies that solid tumors establish their own blood supplies by encouraging the growth of new blood vessels into the tumor tissue in order to extend beyond the small size of approximately one cubic millimeter (Folkman, 1971; Gimbrone et al., 1972). In the absence of a blood supply, the diffusion of oxygen and nutrients across numerous cell layers limits the tumor's size. Furthermore, the development of a vasculature was shown to distinguish hyperplasias from neoplasias (Folkman et al., 1989). Analysis of neoplastic tumor architecture suggests that a given tumor cell is never far away from a capillary, thus allowing it to receive nutrients from the blood vessel (Tannock and Steel, 1969). This distance is approximately the limit of diffusability of macromolecular substances. Once the tumor vasculature is established, solid tumors acquire the nutrients which enable them to expand exponentially and ultimately to metastasize (Folkman, 1992; Gimbrone et al., 1972, 1974; Knighton et al., 1977). By establishing a vasculature they may also receive factors which alter their morphologies and growth characteristics (Brooks et al., 1995).

One of the most significant consequences of the tumor angiogenesis is the invasion of tumor cells into the vasculature. Thus, vascularization permits the survival and growth of primary tumors (Folkman, 1971; Gimbrone et al., 1972; Brem et al., 1976; Folkman and Ingber, 1992; Brooks et al., 1994b), as well as the metastatic spread of cancer (Srivastava et al., 1988; Blood and Zetter, 1990; Weinstat-Saslow and Steeg, 1994). Metastases arise from tumor cells which enter the tumor's own vasculature to be carried to local and distant sites where they create new tumors (Blood and Zetter, 1990). Tumors have typically established a vasculature and metastasized to local and distant sites by the time that primary tumors are detectable (Weinstat-Saslow and Steeg; 1994; Folkman, 1992).

The extent of angiogenesis in a tumor as determined by the density of microvessels within a tumor biopsy is prognostic of tumor stage and metastatic potential. In fact, tumor microvessel density is prognostic of a poor therapeutic outcome, particularly in carcinomas of the breast and prostate (see Hayes, 1994, for review; Weidner et al., 1991, 1992, 1993; Gasparini et al., 1992; Fregene et al., 1993) but also in lung adenocarcinoma (Yamazaki et al., 1994), in head and neck cancer (Gasparini et al., 1993) and in melanoma (Graham et al., 1994) Microvessel density is often assessed with the use of immunohistological staining of tumor biopsies with antibodies directed against the endothelial cell markers PECAM (CD31; Horak et al., 1992) or von Willebrand's Factor (vWF).

Tumor angiogenesis is responsible for some of the symptoms of solid tumor cancers. After the tumor has switched to the angiogenic phenotype, blood can sometimes be detected in bodily fluids, which is caused by the leaky angiogenic vasculature and by the high interstitial vascular pressures within the tumor which force fluids out of the blood vessels (Senger et al., 1983; Dvorak et al., 1988).

Angiogenesis in eye disease

Angiogenesis is a factor in many ophthalmic disorders which can lead to blindness. In age-related macular degeneration (ARMD), a disorder afflicting 25 % of otherwise healthy individuals over the age of 60, and in diabetic retinopathy, a condition prevalent among both juvenile and late onset diabetics, angiogenesis is induced by hypoxic conditions on the choroid or the retina, respectively (White, 1960; Patz, 1980). Hypoxia induces an increase in the secretion of growth factors including vascular endothelial growth factor (VEGF). VEGF expression in the eye may induce the migration and proliferation of endothelial cells into regions of the eye where they are not ordinarily found (Shweiki et al., 1992; Adamis et al., 1993; Miller et al., 1994; Aiello et al., 1994). Vascularization in ocular tissue has adverse effects on vision. New blood vessels on the cornea can induce corneal scarring, whereas new blood vessels on the retina can induce retinal detachment, and angiogenic vessels in the choroid may leak vision-obscuring fluids; these events often lead to blindness.

Angiogenesis in inflammatory and other diseases

Angiogenesis plays a major role in the progression and exacerbation of a number of inflammatory diseases. Psoriasis, a disease which afflicts two million Americans, is characterized by significant angiogenesis. In rheumatoid arthritis and possibly osteoarthritis, the influx of lymphocytes into joints induces blood vessels of the joint synovial lining to undergo angiogenesis; this angiogenesis appears to permit a greater influx of leukocytes and the destruction of cartilage and bone (Harris, 1990). Angiogenesis may also play a role in chronic inflammatory diseases such as ulcerative colitis and Crohn's disease. In addition, the growth of capillaries into atherosclerotic plaques is a serious problem (O'Brien et al., 1994); the rupture and hemorrhage of vascularized plaques is thought to cause coronary thrombosis (Patterson, 1938).

Mechanisms of angiogenesis

Angiogenesis is the process by which new blood vessels are formed from pre-existing capillaries or post-capillary venules (Noden, 1989). In mature organisms, vascular cells remain quiescent unless ischemic or hypoxic conditions in wounded or diseased tissues stimulate the secretion of angiogenic factors (Hobson and Denekamp, 1984; Patz, 1980; Hlatky et al., 1994; Shweiki et al., 1992). An exception to this is the cyclic angiogenesis that occurs in the follicle and corpus luteum and in the placenta in the female reproductive cycle (Phillips et al., 1990; Kamat et al., 1995). Secreted angiogenic factors include the growth factors vascular endothelial cell growth factor (VEGF), basic fibroblast growth factor (bFGF), interleukin-8 (IL-8), platelet derived growth factor (PDGF), and tumor necrosis factor-alpha (TNF-α).

Normally, angiogenesis is regulated by a balance of positive and negative regulators (Folkman, 1995; O'Reilly et al., 1994). Loss of this regulatory control is a characteristic of pathological processes in which angiogenesis plays a role. Loss of negative regulators of angiogenesis, such as thrombospondin, may permit the growth of blood vessels in response to growth factors (Dameron et al., 1994). One or more angiogenic growth factors then can stimulate both the proliferation and migration of endothelial cells in response to tumors (Klagsbrun, 1992; Plate et al., 1992; Smith et al., 1994), diseased ocular tissue (Adamis et al., 1993; Miller et al., 1994; Aiello et al., 1994; Strieter et al., 1992), psoriatic skin (Nickoloff et al., 1994) or arthritic tissue (Koch et al., 1992). Some of these factors (e.g. bFGF) are produced in such large amounts in solid tumor cancers that their elevated levels in the blood may be prognostic of a poor therapeutic outcome (Nguyen et al., 1994).

Angiogenesis is a complex biological process that can be considered in three distinct stages. First, angiogenesis is initiated by cytokine release in response to hypoxia or ischemia. In the case of a tumor, a non-vascularized nest of tumor cells secretes cytokines which diffuse in the general direction of pre-existing blood vessels. Second, these cytokines activate normally quiescent vascular cells, enabling them to proliferate and migrate toward the angiogenic source. This is potentiated by the secretion of proteolytic enzymes which degrade the extracellular matrix, thereby facilitating the vascular cell invasion process. In addition, cell adhesion receptors facilitate vascular cell migration by interacting with adhesion proteins in the extracellular matrix, including fibronectin, vitronectin, collagen, laminin and fibrinogen. Ultimately, the invasive vascular sprout reaches the tumor, where it differentiates to form a lumen and exits from the cell cycle. Lumen formation is facilitated by cell-cell adhesive contact. At this point the intact blood vessel not only serves to nourish the tumor but provides a conduit for metastatic tumor cells which leave the primary site and ultimately metastasize to local or distant locations.

A number of angiogenic growth factors are upregulated in human tumors; these include basic fibroblast growth factor (bFGF), vascular endothelial growth factor (VEGF), interleukin-8 (IL-8) and tumor necrosis factor-α (TNF-α). Experimental models of angiogenesis and the molecular dissection of these growth factors and their receptors have yielded insight into the roles they play in human solid tumor cancer and, for some, their modes of action.

Fibroblast growth factor

Basic fibroblast growth factor (bFGF or FGF-2) is one of the most important and well characterized of the endothelial cell mitogens. bFGF is a member of the fibroblast growth factor family that now includes eight structurally related proteins, including acidic FGF (Klagsbrun and Dlutz, 1993). bFGF consists of 146 amino acids with a molecular weight of 18 kDa and serves as a mitogen for both vascular endothelial cells and vascular smooth muscle cells (Klagsbrun, 1989; Gospodarowicz et al., 1987; Zhu et al., 1990; Lindner et al., 1991; Lindner and Reidy, 1991). It is also a powerful stimulant of endothelial cell and vascular smooth muscle cell motility (Biro et al., 1994; Jackson and Reidy, 1993). bFGF-induced motility *in vitro* and *in vivo* in wound repair can be blocked by anti-FGF antibodies (Biro et al., 1994; Broadley et al., 1989). bFGF stimulates angiogenesis when implanted on the chick chorioallantoic membrane, on the rabbit cornea and in rats implanted with cytokine soaked sponges (Klagsbrun, 1989). bFGF also stimulates choroidal angiogenesis when it is perfused into the choroid of minipugs (Soubrane et al., 1994). It has an established role in smooth muscle cell migration and proliferation in response to arterial injury, such as occurs during angioplasty (Lindner et al., 1991; Lindner and Reidy, 1991; Olson et al., 1992).

Both the intact 146 amino acid form of bFGF and a truncated form lacking a 15 amino acid N-terminal peptide are equally active in promoting cellular responses to the mitogen (Gospodarowicz et al., 1987). Although the gene for bFGF does not encode a hydrophobic peptide export signal sequence (Gospodarowicz et al., 1987), it is nevertheless secreted from tumor cells by an as yet undescribed mechanism. It has been postulated that some FGF release is the result of tumor cell lysis that accompanies necrosis (Folkman, 1995). bFGF has a high affinity for heparin and it is thought to bind to heparan sulfate proteoglycans in the extracellular matrix, leading to its stabilization (Vlodavsky et al., 1987; Gonzalez et al., 1990; Folkman et al., 1989; Hageman et al., 1989). bFGF can be released from this seques-tration by digestion with heaprinases and collagenases (Saksela and Rifkin, 1990). The mitogen is synthesized by tumor cells, endothelial cells (Schultze-Osthoff et al., 1990; Vlodavsky et al., 1987; Baird and Ling, 1987; Schweigerer et al., 1987) and smooth muscle cells (Weich et al., 1990). Its expression has been detected in a variety of adult tissues, includ-ing placenta, pituitary, corpus luteum, kidney, bone, prostate and retina (Esch et al., 1985; Hicks et al., 1992; Neufeld and Gospodarowicz, 1985).

bFGF has been identified in most tissues undergoing angiogenesis, including arthritic joints (Schultze-Osthoff et al., 1990), most tumors (Nguyen et al., 1994), hemangiomas (Takashita et al., 1994), the retina in diabetic retinopathy (Hanneken et al., 1991), and the choroid in macular degeneration (Soubrane et al., 1989). bFGF has been detected at high levels in serum of about 10% of all solid tumor cancer patients tested and in the urine of about 37% of them (Nguyen et al., 1994). High levels of bFGF are prognostic of poor outcome in renal cell carcinoma (Nanus et al., 1993).

Four related bFGF high affinity tyrosine kinase receptors (K_D= $2 - 15 \times 10^{-11}$ M) have been identified on endothelial cells and on vascular smooth muscle cells (Werner et al., 1992). Cell surface heparan sulfate proteoglycans are lower affinity bFGF receptors ($K_D = 2 \times 10^{-9}$ M). Apparently, binding to low affinity receptors preceeds high affinity receptor binding (Yayon et al., 1991; Rapraeger et al., 1991).

Vascular endothelial growth factor

Vascular endothelial growth factor (VEGF; for reviews, see Ferrara, 1993; Dvorak et al., 1995; Mustonen and Alitalo, 1995) was originally discovered as a tumor secreted protein that induces small veins to become hyperper-meable to macromolecules, and thus it was originally named vascular permeability factor (Senger et al., 1983, 1988). VEGF is unique among angiogenic growth factors in that its target specificity is restricted to the endothelial cell (Ferrara and Henzel, 1989). Although not produced by endothelial cells, VEGF is a potent mitogen for endothelial cells (ED_{50} $2-10$ pM) (Ferrara and Henzel, 1989; Plouet et al., 1989) in *in vitro* culture systems and in the chick chorioallantoic membrane model system (Leung et al., 1989). Besides promoting endothelial cell proliferation and chemo-taxis, VEGF induces the expression of the serine proteases urokinase-type plasminogen activator (uPA) and tissue-type plasminogen activator (tPA) as well as collagenase in cultured endothelial cells (Pepper, 1991; Unemore et al., 1992). It also induces upregulation of the uPA receptor (uPAR) on endothelial cells. The binding of uPA to uPAR leads to the activation of uPA

(Mandriota et al., 1995). These proteases are thought to contribute to the angiogenic process by degrading the matrix immediately surrounding the cells and permitting their movement through the matrix.

Four isoforms of the dimeric 23000 dalton/subunit VEGF molecule are produced by alternative splicing from a single vascular endothelial growth factor gene which has a single start transcription site near a cluster of Sp1 transcription factor binding sites (Tisher et al., 1991). The promoter also contains a hypoxia response element similar to that found in the hypoxia sensitive gene erythropoietin (Goldberg and Schneider, 1994). The four VEGF species are composed of 121, 165, 189, and 206 amino acids (Houck et al., 1991; Tisher et al., 1991). The predominant form found in normal and transformed tissues is $VEGF_{165}$. The three isoforms $VEGF_{165}$, $VEGF_{189}$, and $VEGF_{206}$ are heparin binding glycoproteins and are found associated with the extracellular matrix or the cell surface (Houck et al., 1991, 1992; Park et al., 1993). The factors can be released from the extra-cellular matrix by competition with heparin or suramin or by heparitinase digestion, indicating the growth factors are bound to extracellular matrix heparan sulfate proteoglycan (Houck et al., 1992). In contrast, only $VEGF_{121}$ and $VEGF_{165}$ are found in tissues in soluble form. $VEGF_{189}$ and $VEGF_{206}$ can also be released from the extracellular matrix and activated by plasmin digestion (Houck et al., 1992; Park et al., 1993).

VEGF expression is closely correlated with both normal and pathological angiogenic processes. During embryonic development, VEGF mRNA is expressed in trophoblast cells shortly after implantation, suggesting that it plays a role in vascular growth in the placenta (Jakeman et al., 1993). In the mouse, it is also expressed at day 7.5 in the entire endoderm and in the ventricular neuroectoderm at the capillary ingrowth stage (Breier et al., 1992). Vascular endothelial growth factor is later expressed in the heart, vertebral column, kidney, spinal cord and brain (Breier et al., 1992; Jakeman et al., 1993).

In addition to its expression in development, VEGF mRNA is also spatially and temporally associated with the proliferation of microvessels that occurs in the cyclical development of the corpus luteum (Phillips et al., 1990; Kamat et al., 1995). Approximately 60% of all the cells in the cor-pus luteum at this time are capillary endothelial cells (Phillips et al., 1990; Ravindranath et al., 1992). VEGF mRNA is also expressed in keratinocytes in healing wounds (Brown et al., 1992) and in psoriasis (Detmar et al., 1994).

VEGF is highly upregulated in human tumors and in transformed cell lines. A number of transformed cell lines express and secrete VEGF (Senger et al., 1986; Rosenthal et al., 1990; Conn et al., 1990). VEGF is also expressed at high levels in a variety of human tumors, including renal cell carcinoma, colon carcinoma, ductal breast carcinoma, stomach, bladder, pancreas and intracranial tumors, including glioblastoma (Berse et al., 1992; Shweiki et al., 1992; Plate et al., 1992; Berkman et al., 1993; Brown et al., 1992, 1993, 1995). In fact, VEGF levels can be correlated with disease severity. For example, malignant gliomas exhibit up to 50-fold higher levels of VEGF expression than glioma tumors of lower malignancy (Weindel et al., 1994; Hatva et al., 1995). These high levels of VEGF/vascular permeability factor may account for the severe edema that characterizes glioblastoma (Weindel et al., 1994). VEGF expression in highly metastatic melanoma cell lines was also significantly higher than that in melanoma lines with low metastatic potential (Pötgens et al., 1995). In contrast, little or no VEGF is detected in tumors such as lobular breast carcinoma and papillary kidney carcinoma (Brown et al., 1993, 1995), suggesting that other growth factors are responsible for angiogenesis in these tumors. A strong correlation between the degrees of vascularization and VEGF mRNA expression in direct proximity to necrotic areas of tumors has been observed (Plate et al., 1992; Shweiki et al., 1992; Berkman et al., 1993; Phillips et al., 1993). This VEGF upregulation may be a response to hypoxic conditions in the tumor environment (Minchenko et al., 1994). Even nontransformed cells in the tumor microenvironment may respond to hypoxia by upregulating VEGF. For example, mammary stromal cells cultured in hypoxic conditions have been shown to upregulate VEGF expression (Hlatky et al., 1995). In fact, a hypoxia response element similar to that present in erythropoietin has been identified in the gene for VEGF (Goldberg and Schneider, 1994).

The cellular production of VEGF is regulated by growth factors as well as hypoxia. TGF-α and EGF have been shown to stimulate an increase in VEGF RNA in keratinocytes during angiogenesis (Detmar et al., 1994). TGF-β and PDGF upregulate VEGF synthesis in fibroblasts, epithelial and smooth muscle cells (Pertovaara et al., 1994; Brogi et al., 1994). EGF also stimulates VEGF production by glioblastoma cells (Goldman et al., 1993). VEGF expression in tumor cells may be dependent on induction of p53 expression through a protein kinase C signal transduction pathway (Kieser et al., 1994).

Two high affinity receptors for VEGF have been identified on endothelial cells, VEGFR-1/flt-1 (fms-like tyrosine kinase; De Vries et al., 1992) and VEGFR-2/Flk-1 (fetal liver kinase; Millauer et al., 1993; Quinn et al., 1993). Both belong to the platelet derived growth factor family of receptor tyrosine kinases. VEGFR-1 binds VEGF with high affinity (K_d 1–20 pM) and VEGFR-2 binds VEGF with a similarly high affinity (K_d 75–770 pM). VEGFR-1 and VEGFR-2 contain 7 extracellular immunoglobulin-type domains and a tyrosine kinase domain interrupted by a kinase insert domain in the cytoplasmic tail. The expression of VEGF receptors is mainly limited to endothelial cells, whereas the activity apparently is strictly limited to endothelial cells.

VEGF receptor expression correlates strongly with ongoing angiogenesis and with disease severity. Elevated levels of both VEGFR-1 and VEGFR-2 have been observed in vascular endothelium in ductal breast carcinomas (Brown et al., 1995) and in malignant glioblastoma, whereas none was expressed in normal brain vasculature (Hatva et al., 1995) nor on less malignant gliomas (Plate et al., 1994). Further understanding of the regulation of VEGF and VEGFR expression in solid tumor cancer and in other angiogenic diseases may well serve as a starting point in developing new therapies for angiogenic disease.

A definitive role for VEGF in solid tumor growth was demonstrated when a monoclonal antibody against VEGF was able to inhibit the growth of human rhabdomyosarcomas, glioblastomas and leiomyosarcomas in nude mouse tumorigenicity assays (Kim et al., 1993). The antibody has no effect on *in vitro* tumor cell growth and hence its action is directly on the tumor vasculature. A therapeutic role for VEGF itself was demonstrated by studies in which intraarterial injections of VEGF induced the revascularization of ischemic limbs in an animal model (Takeshita et al., 1994; Bauters et al., 1995).

Interleukin-8

Another growth factor implicated in angiogenesis, interleukin-8 (IL-8) was detected in tumorigenic lung tissue as compared to normal lung tissue (Smith et al., 1994). Adenocarcinomas express IL-8, whereas in squamous cell carcinomas, IL-8 may derive from surrounding stromal cells (Smith et al., 1994). Extracts from lung tumors, but not from normal lung, were highly effective in inducing angiogenesis in rabbit from lung tumors, but not from normal lung, were highly effective in inducing angiogenesis in rabbit cornea angiogenesis assays, and this activity could be inhibited with anti-IL-8 antibodies (Smith et al., 1994). Recombinant human IL-8 has been shown potently to stimulate angiogenesis in a rabbit corneal pocket assay (Strieter et al., 1992). Monocyte-derived IL-8 was subsequently detected in arthritic tissue (Koch et al., 1992) and in psoriatic skin (Nickoloff et al., 1994) and was shown to be partially responsible for the induction of angiogenesis in inflamed tissues (Koch et al., 1992).

Tumor necrosis factor-alpha (TNF-α) is angiogenic at low concentrations (0.01–1 ng) but inhibits angiogenesis at high concentrations (1–5 ml; Fajardo et al., 1992). Since the local TNF-α concentration in tumor microenvironments is usually low, it may serve to stimulate angiogenesis (Leek et al., 1994). Its expression is induced by local hypoxia (Scannell et al., 1993), as is that of VEGF. In some tumors, such as ovarian and breast carcinomas, TNF-α is produced by tumor infiltrating macrophages (Pusztai et al., 1994; Naylor et al., 1993).

The integrin family of cell adhesion receptors

The invasion, migration, and proliferation of vascular endothelial and smooth muscle cells during angiogenesis are regulated by one or more members of the integrin family of cell adhesion proteins as demonstrated by *in vivo* and *in vitro* models of angiogenesis (Brooks et al., 1994a, 1994b; Nicosia and Bonanno, 1991; Gamble et al., 1993; Bauer et al., 1992; Davis et al., 1993). The integrin family is composed of 15 alpha and 8 beta subunits that are expressed in over twenty different $\alpha\beta$ heterodimeric combinations on cell surfaces. Integrins bind to extracellular matrix proteins or cell surface immunoglobulin family molecules through short peptide sequences present in the ligands. Several integrins recognize the tripeptide Arg-Gly-Asp (RGD; Ruoslahti and Pierschbacher, 1986; Hynes, 1992; Cheresh, 1993). While some integrins selectively recognize a single extracellular matrix protein ligand (e.g. $\alpha_5\beta_1$ recognizes only fibronectin), others bind two or more ligands (Hynes, 1992; Cheresh, 1993). Combinations of different integrins on cell surfaces allow cells to recognize and respond to a variety of different extracellular matrix proteins.

Integrins mediate cellular adhesion to and migration on the extracellular matrix proteins found in intercellular spaces and basement membranes (Hynes, 1992; Cheresh, 1993), but they also regulate cellular entry into and withdrawal from the cell cycle (Guadagno et al., 1993; Varner et al., 1995; Dike and Farmer, 1988). Ligation of integrins by their extracellular matrix protein ligands induces a cascade of intracellular signals (Juliano and Haskill, 1993) that include tyrosine phosphorylation of focal adhesion kinase (Kornberg et al., 1991, 1992; Guan and Shalloway, 1992), increases in intracellular pH (Schwartz et al., 1989, 1990, 1991; Schwartz and Lechene, 1992) and in intracellular Ca^{2+} (Pelletier et al., 1992; Schwartz 1992; Leavesley et al., 1993), inositol lipid synthesis (McNamee et al., 1993), synthesis of cyclins (Guadagno et al., 1993), and expression of immediate early genes (Varner et al., 1995). In contrast, prevention of integrin-ligand interaction suppresses cellular growth or induces apoptotic cell death (Varner et al., 1995; Meredith et al., 1993; Montgomery et al., 1994; Brooks et al., 1994b).

Useful inhibitors of integrin function include function-blocking monoclonal antibodies, peptide antagonists and peptide mimetics. While the Arg-Gly-Asp moiety can inhibit the function of several integrins, peptide antagonists that are specific for individual integrins have been developed. These have been developed by random synthesis or the use of phage display screening techniques (Pfaff et al., 1994; Koivunen et al. 1993, 1994; Healy et al., 1995). A rigid structure generated by cyclization has proven to be a key feature of peptide antagonist potency (Pfaff et al., 1994). Cyclic RGD peptides selective for α_v integrins (Pfaff et al., 1994) and for the $\alpha_v\beta_3$ integrin (Healy et al., 1995) have been described. Measurement of interatomic distances between the C_{-b} atoms of Arg and Asp for peptides selective for particular integrins indicate that distinct integrins accommodate different sizes of cyclized ring structures (Pfaff et al., 1994).

Integrins on tumor cells or vascular cells are now thought to play intricate roles in the progression of solid tumor cancer. Normal diploid cells can be induced to withdraw from the cell cycle and to become quiescent by maintenance in anchorage independent conditions (Dike and

Farmer, 1988). They are dependent on anchorage not only for growth (Dike and Farmer, 1988), but also for survival (Meredith et al., 1993; Montgomery et al., 1994; Brooks et al., 1994b). In contrast to normal cells, transformed cells are characterized by their anchorage independent growth.

Anchorage independent growth likely results from an uncoupling of the cell cycle from its dependence on signals transduced by attachment to the substratum through integrins (Guadagno et al., 1993). Loss of the cell's normal adhesive factors may be a consequence of this alteration in cell cycle regulation. The adhesion protein fibronectin was originally characterized as „Large External Transformation Sensitive" protein (LETS), since it is lost from the surface of transformed cells (Vaheri and Ruoslahti, 1975; Mautner and Hynes, 1977). Loss of binding of fibronectin to the cell surfaces of some transformed cells (Wagner et al., 1981) may result either from a transformation-associated loss of surface expression of the integrin $\alpha_5\beta_1$, the fibronectin receptor (Plantefaber and Hynes, 1989) or from an inactivation of the integrin $\alpha_5\beta_1$ via an aberrant phosphorylation event (Hirst et al., 1986).The expression and function of the integrin $\alpha_5\beta_{11}$ can actively regulate proliferation; this was demonstrated by a series of studies of tumor variants which displayed altered $\alpha_5\beta_1$ expression. Variants of MG63 osteosarcoma cells (Dedhar et al., 1987, 1989) and K562 erythroleukemia cells (Symington, 1990) selected for increased ability to attach to fibronectin exhibited a 5-fold upregulation of $\alpha_5\beta_1$ expression and displayed significantly reduced growth in assays of anchorage independent growth and tumorigenicity. Additionally, Chinese hamster ovary cells expressing 30-fold normal levels of $\alpha_5\beta_1$ by gene transfection displayed completely suppressed tumorigenicity and partially suppressed *in vitro* growth (Giancotti and Ruoslahti, 1990). In addition, total loss of integrin $\alpha_5\beta_1$ expression in Chinese hamster ovary cells enhances tumorigenicity (Schreiner et al., 1991). Thus, integrin $\alpha_5\beta_1$ expression can suppress growth in cells that overexpress it. Integrin $\alpha_5\beta_1$ expression in the absence of attachment to a substratum activates a signaling pathway that culminates in downregulation of immediate early gene expression. In contrast, attachment to a substratum induces immediate early gene expression and allows the cells to enter the cell cycle (Varner et al., 1995).

Other integrins also contribute to growth regulation, transformation and metastasis. For example, the integrin $\alpha_2\beta_1$, a collagen/laminin receptor, has been shown to impart metastatic abilities to tumor cells (Chan et al., 1992). Integrin $\alpha_v\beta_3$ has been shown to play a role in regulating the growth of melanoma cells as well as vascular endothelial cells (Felding-Habermann et al., 1992; Montgomery et al., 1994; Brooks et al., 1994b).

Role of extracellular matrix proteins and proteases in angiogenesis

Angiogenesis not only depends on the integrins and growth factors and their receptors, but is also influenced by other classes of extracellular and integral membrane molecules. For example, the migration of endothelial cells through the extracellular matrix requires the expression and action of extracellular matrix degrading enzymes, such as collagenase and plasmin, to clear a path for cellular movement (Folkman, 1992). Inhibitors of collagenase and metalloproteinases have demonstrated efficacy in inhibiting angiogenesis in animal models (Albini et al., 1994). A recently discovered endogenous inhibitor of angiogenesis, angiostatin, is a non-catalytic fragment of plasminogen (O'Reilly et al., 1994). It is interesting to speculate this fragment may somehow regulate the activity of intact plasminogen by competing for target or cell surface binding sites.

The migration of endothelial cells into tumors or regions of the eye, skin or synovium in response to growth factor secretion also requires their adhesion to proteins of the extracellular matrix, which include the proteins thrombospondin, vitronectin, fibronectin, laminin, and collagen (Good et al., 1990; Brooks et al., 1994a, 1994b). Thrombospondin in particular has been implicated in negative regulation of angiogenesis (Good et al., 1990). This trimeric extracellular matrix protein directly inhibits the proliferation of endothelial cells in culture (Bagavandoss and Wilks, 1990). It also inhibits the formation of focal contacts *in vitro* (Murphy-Ullrich and Hook, 1989) and thus may block a signaling cascade that contributes to cellular proliferation. Thrombospondin binds to one of several cell surface receptors, but its anti-angiogenic activity is mediated by a hexapeptide that is derived from a procollagen domain (Tolsma et al., 1993). The expression of thrombospondin was recently shown to be regulated by p53 expression. Loss of p53 function by the acquisition of mutations in p53, a common event in solid tumor cancers, coincides with loss of thrombospondin expression and its concomitant inhibition of angiogenesis. Restoration of p53 function by transfection restores thrombospondin expression and its inhibition of angiogenesis (Dameron et al., 1994). Although the mechanism of inhibition by thrombospondin has not yet been determined, it is possible that it directly suppresses endothelial cell cycle entry by signaling through a cell surface receptor.

The extracellular matrix protein laminin has also been implicated in angiogenesis. A synthetic peptide derived from laminin inhibits angiogenesis by preventing endothelial cell migration, but not endothelial cell proliferation (Sakamoto et al., 1991). Deposition of matrix by newly forming capillaries appears to be a critical factor in angiogenesis. Proline analogs and inhibitors of collagen synthesis and crosslinking such as b-aminoproprionitrile are able to inhibit angiogenesis (Ingber and Folkman, 1988; Haralabopoulos et al., 1992).

Adhesion molecules also play important roles in angiogenesis. For example, cell-cell adhesion molecules in the selectin family and their ligands sialyl Lewis X/A have also been implicated in the differentiation phase of angiogenesis lumen formation (Bischoff, 1995). Antibodies against sialyl Lewis X and sialyl Lewis A, as well as those against E-selectin, inhibited capillary tube formation in an *in vitro* model of angiogenesis that reproduces some but not all features of angiogenesis (Nguyen et al., 1993).

Integrin $\alpha_v\beta_3$

A number of integrin family members are expressed on any given cell type, enabling cells to respond to an array of different extracellular matrix proteins. The integrin $\alpha_v\beta_3$, in particular, is a receptor for a wide variety of extracellular matrix proteins with an exposed tripeptide Arg-Gly-Asp moiety (Cheresh, 1993) Integrin $\alpha_v\beta_3$, the most promiscuous member of the integrin family, mediates cellular adhesion to vitronectin, fibronectin, fibrinogen laminin, collagen, von Willebrand's factor, osteopontin, and adenovirus penton base, among others (Cheresh, 1987; Leavesley et al., 1992; Cheresh, 1993; Wickham et al., 1993). Thus, expression of this integrin enables are given cell to adhere to, migrate on or respond to almost any matrix protein it may encounter Despite its "promiscuous" behavior, $\alpha_v\beta_3$is not widely expressed. It is not generally expressed on epithelial cells and is expressed only at low levels on some vascular, intestinal and uterine smooth muscle cells (Brem et al., 1994; unpublished observations). This receptor is also expressed on certain activated leukocytes, on macrophages, and on osteoclasts, where it may play a role in bone resorption (Sato et al., 1990; Horton et al., 1991; Ross et al., 1993). Integrin $\alpha_v\beta_3$ is also expressed on certain invasive tumors including late-stage glioblastomas (Gladson and Cheresh, 1991) and metastatic melanoma (Albelda et al., 1990), where it may play a role in regulation of tumor proliferation (Felding-Habermann et al., 1992; Sanders et al., 1992) and in metastasis (Seftor et al., 1992; Nip et al., 1992). However, its expression appears most prominent on cytokine activated endothelial or smooth muscle cells, which may explain its intense expression on blood vessels in granulation tissue and tumors (Enenstein and Kramer, 1994; Sepp et al., 1994; Brooks et al., 1994a, 1994b). The integrin α_v subunit is widely expressed on most cell types and associates with several different beta subunits. Therefore, integrin $\alpha_v\beta_3$ expression is likely to be regulated by β_3 transcription, which is modulated by substances such as TNF-α, bFGF, granulocyte-macrophage colony-stimulating factor (GMCSF), retinoic acid and vitamin D_3 (DeNichilo and Burns, 1993; Enenstein et al., 1992; Krissanson et al., 1990; Dedhar et al., 1991; Medhora et al., 1993).

Role of vascular cell integrins $\alpha_v\beta_3$ and $\alpha_v\beta_5$ in angiogenesis

Perhaps the most significant of the physiological roles played by integrin $\alpha_v\beta_3$ is its critical role in the process of angiogenesis. Integrin $\alpha_v\beta_3$ is minimally, if at all, expressed on resting, or normal, blood vessels, but is significantly upregulated on vascular cells within human tumors (Brooks et al., 1994, 1995), within granulation tissue (Brooks et al., 194; Clark, 1995), or in response to certain growth factors *in vitro* (Sepp et al., 1994; Enenstein et al., 1992) and *in vivo* (Brooks et al., 1994; Friedlander et al., 1995). For example, basic fibroblast growth factor, but not TGF-β or interferon gamma, markedly increases β_3 mRNA and surface expression in cultured human dermal microvascular endothelial cells (Sepp et al., 1994; Enenstein et al., 1992). Basic fibroblast growth factor and tumor necrosis factor-alpha stimulate $\alpha_v\beta_3$ expression on developing blood vessels in the chick chorio-allantoic membrane (CAM) (Brooks et al., 1994; Friedlander et al., 1995) and on the rabbit cornea (Friedlander et al., 1995). Peak levels of integrin expression are observed on blood vessels 12–24 h after stimulation with basic fibroblast growth factor (unpublished observations). $\alpha_v\beta_3$ expression is also induced by human tumors cultured on the chick CAM (Brooks et al., 1994a, 1994b) and by human tumors grown in human skin explants grafted onto SCID mice (Brooks et al., 1995). Integrin $\alpha_v\beta_3$ is also significantly upregulated on vascular cells *in vivo* during wound healing (Brooks et al., 1994a; Clark et al., 1995, submitted).

The promoter for the integrin β_3 subunit has been cloned. It contains a Vitamin D response element, AP-1 and Sp1 transcription factor recognition sequences (Cao et al., 1993) as well as ets transcription factor recognition sequences. Ets is an endothelial cell specific transcription factor that regulates a number of genes required for vasculogenesis and angiogenesis including uPA, stromolysin and collagenase. Ets is not expressed, however, in adult quiescent endothelium (Wernert et al., 1992).

The highly restricted expression of $\alpha_v\beta_3$ and the upregulation of its expression during angiogenesis suggest that it may play a critical role in the angiogenic process. In fact, recent experimental evidence supports this notion. Specifically, antagonists of integrin $\alpha_v\beta_3$, but not of β_1 integrins, potential inhibit angiogenesis in a number of animal models. When angiogenesis is induced on the chick chorioallantoic membrane (CAM) with purified cytokines, $\alpha_v\beta_3$ expression is stimulated 4-fold within 72 h (Brooks et al., 1994a). Topical application of LM609, a monoclonal antibody antagonist of $\alpha_v\beta_3$, inhibited angiogenesis, but other anti-integrin antibodies were ineffective (Brooks et al., 1994a). Similarly, topical application of LM609 or cyclic RGD peptide antagonists, but not of other anti-integrin antibodies or control peptides, to tumors grown on the surface of CAMs reduced the growth of blood vessels into the tumor tissue. LM609 had no effect on pre-existing vessels (Brooks et al., 1994a). In addition, LM609 also inhibits blood vessel formation in early embryonic development. When micro-injected into quail embryos, the antibody inhibits dorsal aorta formation (Drake et al., 1995). In this case, the antibody prevents vessel maturation in the developing quail by preventing lumen formation. These finding suggest that $\alpha_v\beta_3$ plays a biological role in a late event of blood vessel formation that is common to embryonic neovascularization and angiogenesis.

Antagonists of integrin $\alpha_v\beta_3$ not only prevent the growth of new blood vessels into tumors cultured on the chick chorioallentoic membrane without affecting adjacent blood vessels (Fig. 3), but also induce tumor regression (Brooks et al., 1994b). Up to 5-fold differences in tumor sizes were observed between treated and control tumors. A single intravascular injection of 300 μg of LM609, but not of CSAT (an anti-β_1 integrin antibody) halted the growth of tumors and induced their regression as determined by tumor weight (Brooks et al., 1994b). Similarly, an injection of 300 μg of a cyclic RGD peptide antagonist of $\alpha_v\beta_3$ but not of an inactive control peptide induced tumor regression (Brooks et al., 1994b). Histological examination of the anti-$\alpha_v\beta_3$ and control treated tumors reveals that few if any viable tumor cells remain in the anti-$\alpha_v\beta_3$ treated tumors (Brooks et al., 1994b). In fact, these treated tumors contained no viable blood vessels.

It is important to note that antagonists of integrin $\alpha_v\beta_3$ also inhibit tumor growth in human skin. In exciting studies of the effect of these antagonists on human angiogenesis, Brooks and colleagues transplanted human neonatal foreskins onto SCID mice (Brooks et al., 1995). After permitting the skin to heal, they were able to demonstrate that the majority of the blood vessels within the human skin are human in origin. They then inoculated human $\alpha_v\beta_3$ negative breast cancer cells into the skin and began treating mice two weeks later with twice weekly intravenous injections of 250 μg doses of the $\alpha_v\beta_3$ antagonist LM609 and control substances. Tumor growth was either completely suppressed (8/12) or was significantly inhibited (4/12) when compared to mice treated with an irrelevant antibody control. Angiogenesis was significantly inhibited (by at least 75%) in the LM609 treated animals.

Antagonists of integrin $\alpha_v\beta_3$ also inhibit angiogenesis in the rabbit cornea (Friedlander et al., 1995). When angiogenesis is induced by TNF-α or bFGF, LM609 and a cyclic peptide inhibitor of the integrin block corneal angiogenesis by at lest 85%. In contrast, antibody inhibitors of integrin $\alpha_v\beta_5$, a related but non-identical integrin, but not inhibitors of integrin $\alpha_v\beta_3$, block angiogenesis induced by the growth factors VEGF and TGF-α. A cyclic RGD peptide that inhibits both integrins is able to inhibit angiogenesis independent of the stimulus used. Chick chorioallentoic membrane angiogenesis induced by VEGF or TGF-α is also inhibited by either the cyclic RGD peptide or the integrin $\alpha_v\beta_5$-specific antibody, but not by LM609 (Friedlander et al., 1995). Previous studies indicated that the integrin $\alpha_v\beta_5$, but not the integrin $\alpha_v\beta_3$, is activated by a protein kinase C-dependent mechanism (Klemke et al., 1994). Thus, the phorbol ester PMA activates $\alpha_v\beta_5$ in a calphostin C inhibitable manner. It is interesting to note that PMA-induced angiogenesis is $\alpha_v\beta_5$- but not $\alpha_v\beta_3$-dependent. in fact, calphostin C also inhibits VEGF, TGF-α and PMA-induced angiogenesis (Friedlander et al., 1995). Since calphostin C does not inhibit $\alpha_v\beta_3$-dependent angiogenesis, these results suggest that two different signaling pathways mediate integrin-dependent angiogenesis.

Table 1. Role of angiogenesis in disease. Angiogenesis plays a significant role in the establishment or exacerbation of a number of serious human clinical disorders, including solid tumor cancer, rheumatoid arthritis, psoriasis and inflammatory diseases.

Angiogenic Diseases

Disease	Prevalence (or Incidence)
Blindness:	
Macular degeneration	650 000
Diabetic retinopathy	300 000
Corneal transplant	100 000
Myopic degeneration	200 000
Inflammatory disease:	
Arthritis	2.1 million
Psoriasis	3.0 million
Inflammatory bowel disease and other chronic inflammatory disease	2.0 million
Solid tumor cancer:	
Lung, breast, prostate, colon, bladder, pancreas, melanoma, renal, glioblastoma, neuroblastoma and others	800 000 (new cases per year)

The expression of these receptors on cytokine stimulated blood vessels also suggests they may play roles in vascular proliferation and migration events associated with restenosis after angioplasty. Restenosis is thought go be caused by the activation of vascular smooth muscle cells to proliferate and migrate into the sites of an arterial wound, such as occurs during angioplasty (Clowes and Schwartz, 1985). This process results in the re-occlusion of blood vessels previously cleared by the angioplasty procedure. During restenosis, the healing of arterial wound, or in atherosclerotic plaques, osteopontin, an extracellular matrix protein ligand for $\alpha_v\beta_3$ and possibly $\alpha_v\beta_5$, is significantly upregulated (Giachelli et al., 1993; Liaw et al., 1993, 1995). the calcification of atherosclerotic plaques may be induced by osteopontin expression, since osteopontin is a protein with a well characterized role in bone formation and calcification (Fitzpatrick, 1994). Vascular smooth muscle cell migration on osteopontin is dependent on the integrin $\alpha_v\beta_3$ and antagonists of $\alpha_v\beta_3$ prevent both smooth muscle cell migration (Liaw et al., 1995) and restenosis in some animal models (Choi et al., 1994; Matsuno et al., 1994). In fact, 7E3, a function-blocking anti-integrin β_3 antibody that recognizes $\alpha_v\beta_3$ as well as platelet integrin $\alpha_{IIb}\beta_3$, was recently approved for use in treatment of high risk angioplasty (Topol et al., 1994).

Integrin $\alpha_v\beta_3$ and apoptosis

The mechanism of action of $\alpha_v\beta_3$ antagonists in blocking angiogenesis appears to be related to their ability to promote unscheduled programmed cell death (apoptosis) of newly sprouting blood vessels selectively. When bFGF-stimulated CAMs were isolated from chick embryos which had been treated by intravascular injection of CSAT (anti-β_1), LM609 (anti-$\alpha_v\beta_3$) or saline, a significant increase in DNA laddering was observed in LM609-treated CAMs but only after 48 h of treatment (Brooks et al., 1994b). CAMs were further analyzed by dissociation into single cell suspensions by collagenase treatment; DNA fragmentation was quantified by staining of cells with ApopTag, an immunohistological stain that detects free 3' OH groups of fragmented DNA (Gavrieli et al., 1992). ApopTag analysis revealed that a 4-fold increase in apoptotic cells was induced on the CAM by either LM609 or cyclic RGD peptide treatment but not by control antibodies or peptides (Brooks et al., 1994b). Furthermore, 25–30% of total CAM cells showed the morphological signs of apoptosis, including nuclear condensation and fragmentation, only in $\alpha_v\beta_3$ antagonist-treated CAMs. Cryostat sections from CAMs treated for 2 days prior to analysis with either CSAT or antagonists of $\alpha_v\beta_3$ were analyzed from ApopTag and LM609 immunoreactivity to identify the cell types undergoing apoptosis in these tissues. In this case, CAMs treated with antagonists of $\alpha_v\beta_3$ exhibited intense ApopTag staining that colocalized with LM609 staining, on endothelial cells lining immature blood vessels. Minimal apoptosis was detected among non-vascular cells in these tissues. In contrast, CSAT-treated CAMs exhibited minimal apoptosis of blood vessel suggesting that β_1 integrin antagonists had no effect on vascular cell survival or death (Brooks et al., 1994b).

To evaluate further the effects of these antagonists on vascular cell events, single cell suspensions were prepared from CAMs treated with bFGF and in the presence or absence of LM609. These cells were then stained with the DNA dye propidium iodide to examine the content of DNA per cell. Cells with more than one copy of DNA were presumed to have entered the cell cycle. These cells were then co-stained with ApopTag to evaluate their viability. This co-staining procedure revealed that bFGF could induce cells to enter the cell cycle and that LM609 caused ApopTag staining of these same cells. These findings demonstrate that Mab LM609 was capable of inducing apoptosis of vascular cells that had already responded to the cytokine (Brooks et al., 1994b). These results indicate that LM609 promotes apoptosis among vascular cells at a point that is late in the cell cycle. More importantly, these findings demonstrate that antagonists of $\alpha_v\beta_3$ affect a late stage of angiogenesis, perhaps during the process of vessels maturation. This is consistent with the studies by Drake et al. (1995) showing that antagonists of $\alpha_v\beta_3$ block late stage development of new blood vessels in the quail by preventing lumen formation. Together, these findings are consistent with the notion that $\alpha_v\beta_3$ provides a survival signal to proliferative vascular cells during new blood vessel growth. Presumably, after new blood vessels are fully mature, the vascular cells are refractory to antagonists of this integrin. These findings may explain why antagonists of $\alpha_v\beta_3$ selectively affect newly growing blood vessels. It is not currently known if integrin $\alpha_v\beta_5$ antagonists also induce apoptosis in angiogenic blood vessels, or if the mode of action is significantly different.

In summary, angiogenesis depends on the stimulation of quiescent vascular cells by growth factors released from tumors or other diseased tissues and also on the interaction of the integrin $\alpha_v\beta_3$ with one of its ligands (Brooks et al., 1994a, 1994b). Stimulated endothelial cells depend on $\alpha_v\beta_3$ function for survival during a critical period of the angiogenic-process, as inhibition of $\alpha_v\beta_3$/ligand interaction by antibody or peptide antagonists induces vascular cell apoptosis and inhibits angiogenesis (Brooks et al., 1994a, 1994b). Endothelial cell attachment and migration in response to extracellular matrix proteins via integrin $\alpha_v\beta_3$ generates a calcium signal that is a likely component of a signaling cascade required to maintain progression through the cell cycle (Leavesley et al., 1993). An NPXY sequence in the cytoplasmic tail of the β_3 subunit has been critically implicated in mediating these signaling events that are required for cellular migration (Filardo et al., 1995). These observations open the door to further analysis of the regulation of cellular function and cell signaling by integrins, as well as new therapeutic strategies to treat angiogenic disease.

Conclusions

The basic mechanisms regulating vascular cell proliferation, motility and differentiation *in vitro* and *in vivo* are now becoming elucidated. It should be feasible in the near future to devise therapeutic strategies to target and perturb the biological processes of angiogenic vascular cells selectively, thereby leading to effective inhibitors of angiogenesis. These strategies have led to the development of antagonists to integrin $\alpha_v\beta_3$ and integrin $\alpha_v\beta_5$ which promote unscheduled programmed cell death of newly sprouting blood vessels. These antagonists cause regression of pre-established human tumors growing in laboratory animals and thus may lead to an effective therapeutic approach for most solid tumors in humans. Studies are underway aimed at designing highly specific small organic integrin inhibitors that will serve to disrupt the signals enabling vascular cells to respond to the tumor-associated extracellular environment and to promote tumor-induced angiogenesis. Several compounds which inhibit the integrin $\alpha_v\beta_3$ and/or integrin $\alpha_v\beta_5$ function are under development in academic and industrial laboratories. These include peptide inhibitors of individual integrins as well as peptides which inhibit both integrins; non-peptidic, organic inhibitors; and chimeric or humanized antibody inhibitors of integrin $\alpha_v\beta_3$. The first of these are expected to enter initial clinical trials within the next two years. This exciting approach to a novel therapy for solid tumor cancers has great potential in the battle against cancer.

References

Abe, T., Okamura, K., Ono, M., Kohno, K., Mori, T., Mori, S. and Kuwano, M. (1993) Induction of vascular endothelial tubular morphogenesis by human glioma cells. A model system for tumor angiogenesis. *J. Clin. Invest.* 92:54–61.

Adamis, A.P., Miller, J.W. and O'Reilly, M. (1993) Vascular permeability factor (vascular endothelial growth factor) is produced in the retina and elevated levels are present in the aqueous humor of eyes with iris neovascularization. *Invest. Ophthalmol. Vis. Sci.* 34:1440.

Aiello, L.P., Avery, R.L., Arrigg, P.G., Keyt, B.A., Jampel, H.D., Shah S.T., Pasquale, L.R., Thieme, H., Iwamoto, M., Park, J.E., Nguyen, H.V., Aiello, L.M., Ferrara, N. and King, G.L. (1994) Vascular endothelial growth factor in ocular fluid of patients with diabetic retinopathy and other retinal disorders. *New Engl. J. Med.* 331:1480–1487.

Albelda, S.M., Mette, S.A., Elder, D.E., Stewart, R., Damjanovich, L., Herlyn, M. and Buck, C.A. (1990) Integrin distribution in malignant melanoma: association of the β_3 subunit with tumor progression. *Cancer Res.* 50:6757–6764.

Albini, A., Fontanini, G., Masiello, L., Tacchetti, C., Bigini, D., Luzzi, P., Noonan, D. and Stetler-Stevenson, W.G. (1994) Angiogenic potential in vivo by Kaposi's sarcoma cell-free supernatants and HIV *tat* product: Inhibition of KS-like lesions by tissue inhibitor of metalloproteinase-2. *AIDS* 8:1237–1244.

Ausprunk, D., Knighton, D. and Folkman, J. (1975) Vascularization of normal and neoplastic tissues grafted to the chick chorioallantois. *Am. J. Path.* 79:597–618.

Bagavandross, P. and Wilks, J.W. (1990) Specific inhibition of endothelial cell proliferation by thrombospondin. *Biochem. Biophys. Res. Comm.* 170:867–872.

Baird, A. and Ling, N. (1987) Fibroblast growth factors are present in the extracellular matrix produced by endothelial cells *in vitro*. Implications for a role of heparinase-like enzymes in the neovascular response. *Biochem. Biophys. Res. Commun.* 142:428–435.

Bauer, J., Margolis, M., Schreiner, C., Edgell, C.-J., Azizkhan, J., Lazarowski, E. and Juliano, R.L. (1992) In vitro model of angiogenesis using a human endothelium-derived permanent cell line: contributions of induced gene expression, G-protein, and integrins. *J. Cell Phys.* 153:437–449.

Bauters, C., Asahara, T., Zheng, P., Takeshita, S., Bunting, S., Ferrara, N, Symes, J.F. and Isner, J.M. (1995) Site-specific therapeutic angiogenesis after systemic administration of vascular endothelial growth factor. *J. Vasc. Surg.* 21:314–325.

Berkman, R.A., Merrill, M.J., Reinhold, W.C., Monacci, W.T., Saxena, A., Clark, W.C., Robertson, J.T., Ali, I.U. and Oldfield, E.H. (1993) Expression of vascular permeability/vascular endothelial growth factor gene in the central nervous system. *J. Clin. Invest.* 91:153–159.

Berse, B., Brown, L.F., Van der Water, L., Dvorak, H.F. and Senger, D.R. (1992) Vascular permeability factor (VEGF) is expressed differentially in normal tissues, macrophages and tumors. *Mol. Biol. Cell* 3:211–220.

Biro, S., Yu, Z.-X., Fu, Y.-M., Smale, G., Sasse, J., Sanchez, J., Ferrans, V. and Cassels, W. (1994) Expression and subcellular distribution of basic fibroblast growth factor are regulated during migration of endothelial cells. *Circ. Res.* 74:485–494.

Bischoff, J. (1995) Approaches to studying cell adhesion molecules in angiogenesis. *Trends in Cell Biology* 5:69–74.

Brem, H. and Folkman, J. (1993) Analysis of experimental antiangiogenesis therapy. *J. Pediatric Surg.* 28:445–450.

Brem, R.B., Robbins, S.G., Wilson, D.J., O'Rourke, L.M., Mixon, R.N., Robertson, J.E., Planck, S.R. and Rosenbaum, J.T. (1994) Immunolocalization of integrins in the human retina. *Invest. Ophthal. Vis. Sci.* 35:3466–3474.

Brem, S., Brem, H., Folkman, J., Finkelstein, D. and Patz, A. (1976) Prolonged tumor dormancy by prevention of neovascularization in the vitreous. *Cancer Res.* 36:2807–2812.

Blood, C.H. and Zetter, B.R. (1990) Tumor interactions with the vasculature: angiogenesis and tumor metastasis. *Biochem. Biophys. Acta* 1032:89–118.

Breier, G., Albrecht, U., Sterrer, S. and Risau, W. (1992) Expression of vascular endothelial growth factor during embryonic angiogenesis and endothelial cell differentiation. *Development* 114:521–534.

Broadley, K.N., Aquino, M.M., Woodward, S.C., Buckley, Sturrode, A., Sato, Y., Rifkin, D.B. and Davidson, J.M. (1989) Monospecific antibodies implicate basic fibroblast growth factor in normal wound repair. *Lab. Invest.* 61:571–575.

Brock, T.A., Dvorak, H.F. and Senger, R.R. (1991) Tumor-secreted vascular permeability factor increases cytosolic Ca^{2+} and von Willebrand factor release in human endothelial cells. *Am. J. Pathol.* 138:213–221.

Brogi, E., Wu, T, Namiki, A. and Isner, J.M. (1994) Indirect angiogenic cytokines upregulate vascular endothelial growth factor and basic fibroblast growth factor gene expression in vascular smooth muscle cells whereas hypoxia upregulates vascular endothelial growth factor expression only. *Circulation* 90:649–652.

Brooks, P.C., Clark, R.A. and Cheresh, D.A. (1994a) Requirement of vascular integrin $\alpha_v\beta_3$ for angiogenesis. *Science* 264:569–571.

Brooks, P.C., Montgomery, A.M.P., Rosenfeld, M., Reisfeld, R., Hu, T., Klier, G. and Cheresh, D. (1994b) Integrin $\alpha_v\beta_3$ antagonists promote tumor regression by inducing apoptosis of angiogenic blood vessels. *Cell* 749:1157–1164.

Brooks, P., Stromblad, S., Klemke, R., Vissher, D., Sarkar, F. and Cheresh, C. (1995) Anti-integrin $\alpha_v\beta_3$ blocks human breast cancer growth and angiogenesis in human skin. *J. Clin. Invest.* 96:1815–1822.

Brown, L.F., Yeo, K.T., Berse, B., Morgentaler, A., Dvorak, H.F. and Rosen, S. (1992) Expression of vascular permeability factor (VEGF) by epidermal keratinocytes during wound healing. *J. Exp. Med.* 176:1375–1379.

Brown, L.F., Berse, B., Jackman, R.W., Toguzzi, K., Manseau, E.T., Dvorak, H.F. and Senger, D.R. (1993) Increased expression of vascular permeability factor (vascular endothelial growth factor) and its receptors in kidney and bladder carcinomas. *Am. J. Path.* 143:1255–1262.

Brown, L.F., Berse, B., Jackman, R., Toguzzi, K., Guidd, A.D., Dvorak, H.F., Senger, D., Connolly, J. and Schitt, S. (1995) Expression of vascular permeability factor (vascular endothelial growth factor) and its receptors in breast cancer. *Human Path.* 26:86–91.

Bussolino, F., Renzo, M.F., Ziche, M., Bocchietto, E., Olivero, M., Naldinin, L., Gaudino, G., Tamagnone, L., Coffer, A. and Comoglio, P. (1992) Hepatocyte growth factor is a potent angiogenic factor which stimulates endothelial cell motility and growth. *J. Cell Biol.* 119:629–641.

Cao, X., Ross, F.P., Zhang, L., MacDonald, P.N., Chappel, J., Teitelbaum, S.L. (1993) Cloning of the promoter for the avian integrin β_3 subunit gene and its regulation by 1,25-dihydroxyvitamin D_3. *J. Biol. Chem.* 268:27371–27380.

Chan, B.M., Matsuura, N., Takada, Y., Zetter, B.R. and Hemler, M.E. (1992) *In vitro* and *in vivo* consequences of VLA-2 expression on rhabdomyosarcoma cells. *Science* 251:1600–1602.

Cheresh, D. (1987) Human endothelial cells synthesize and express an Arg-Gly-Asp directed adhesion receptor involved in attachment to fibrinogen and von Willebrand factor. *Proc. Natl. Acad. Sci. USA* 84:6471–6475.

Cheresh, D. (1993) Integrins: structure, function and biological properties. *Advances in Molecular and Cell Biology* 6:225–252.

Chiarugi, M. and Ruggerio, M. (1984) Cooperative effect of cortisone and heparin in the context of plasminogen activator. *Ital. J. Biochem.* 33:142A.

Choi, E.T., Engel, L., Callow, A.D., Sun, S., Trachtenberg, J., Santoro, S. and Ryan, U.S. (1994) Inhibition of neointimal hyperplasia by blocking $\alpha_v\beta_3$ integrin with a small peptide antagonist GpenGRGDSPCA. *J. Vasc. Surg.* 19:125–134.

Clark, R.A.F., Della Pelle, P., Manseau, E., Lanigan, J.M., Dvorak, H.F. and Colvin, R.B. (1982) Blood vessel fibronectin increases in conjunction with endothelial cell proliferation and capillary ingrowth during wound healing. *J. Invest. Dermatol.* 79:269–276.

Clark, R.A.F., Tonnesen, M.G., Gailit, J. and Cheresh, D. Transient functional expression of $\alpha_v\beta_3$ on vascular cells during wound repair (1996) *Am. J. Pathol.* 148:1407–1421.

Clowes, A.W. and Schwartz, S.M. (1985) Significance of quiescent smooth muscle migration in the injured rat carotid artery. *Circulation Research* 56:139–145.

Conn, G., Bayne, M., Soderman, D.D., Kwok, P.W., Sullivon, K.A., Palisi, T.M., Hope, D.A. and Thomas, K.A. (1990) Amino acid and cDNA sequence of a vascular endothelial growth cell mitogen homologous to platelet derived growth factor. *Proc. Natl. Acad. Sci. USA* 87:2628–2632.

Dameron, K.M., Volpert, O.V., Tainsky, M.A. and Bouck, N. (1994) Control of angiogenesis in fibroblasts by p53 regulation of thrombospondin-1. *Science* 265:1582–1584.

Davis, C.M., Danehower, S., Laurenza, A. and Molony, J.L. (1983) Identification of a role of the vitronectin receptor and proteinkinase C in the induction of endothelial cell vascular formation. *J. Cell. Biochem.* 51:206–218.

Dedhar, S., Argraves, W.S., Suzuki, S., Ruoslahti, E. and Pierschbacher, M.D. (1987) Human osteosarcoma cells resistant to detachment by an Arg-Gly-Asp-containing peptide overproduce the fibronectin receptor. *J. Cell Biol.* 105:1175–1182.

Dedhar, S., Mitchell, M.D. and Pierschbacher, M.D. (1989) The osteoblast-like differentiation of a variant of MG-63 osteosarcoma cell line correlated with altered adhesive properties. *Connect. Tiss. Res.* 20:49–61.

Dedhar, S., Robertson, K. and Gray, V. (1991) Induction of the expression of the $\alpha_v\beta_1$ and $\alpha_v\beta_3$ integrin heterodimers during retinoic acid induced neuronal differentiation of murine embryonal carcinoma cells. *J. Biol. Chem.* 266:21846–21852.

DeNichilo, M.O. and Burns, G.F. (1993) Granulocyte-macrophage and macrophage colony stimulating factors differentially regulate α_v integrin expression on cultured human macrophages. *Proc. Natl. Acad. Sci. USA* 90:2517–2521.

Detmar, M., Brown, L., Claffey, K., Yeo, K.T., Kocha, D., Jackman, R., Berse, B. and Dvorak, H. (1994) Overexpression of vascular permeability factor and its receptors in psoriasis. *J. Exp. Med.* 180:1141–1146.

DeVries, C., Escobedo, J.A., Ueno, H., Houck, K., Ferrara, N. and Williams, L.T. (1992) The fms-like tyrosine kinase, a receptor for vascular endothelial growth factor. *Science* 255:989–991.

Dike, L.E. and Farmer, S.R. (1988) Cell adhesion induces the expression of growth-associated genes in suspension-arrested fibroblasts. *Proc. Natl. Acad. Sci. USA* 85:6792–6796.

Drake, C.J., Cheresh, D.A. and Little, C.D. (1995) An antagonist of integrin $\alpha_v\beta_3$ prevents maturation of blood vessels during embryonic neovascularization. *J. Cell Sci.* 108:2655–2661.

Dvorak, H.F., Brown, L.F., Detmar, M. and Dvorak, A.M. (1995) Vascular permeability factor/vascular endothelial growth factor, microvascular hyperpermeability, and angiogenesis. *Am. J. Path.* 146:1029–1039.

Dvorak, H.F., Nagy, J.A., Dvorak, J.T. and Dvorak, A.M. (1988) Identification and characterization of the blood vessels of solid tumors that are leaky to circulating macromolecules. *Am. J. Pathol.* 133:95–109.

Enenstein, J., Waleh, N.S. and Kramer, R.H. (1992) Basic FGF and TGF$_{-\beta}$ differentially modulate integrin expression of human microvascular endothelial cells. *Exp. Cell Res.* 203:499–503.

Enenstein, J. and Kramer, R.H. (1994) Confocal microscopic analysis of integrin expression on the microvasculature and its sprouts in the neonatal foreskin. *J. Invest. Dermatol.* 103:381–386.

Ensoli, B., Markham, P., Kao, V., Barilii, G., Fiorelli, V., Gendelman, R., Raffeld, M., Zon, G. and Gallo, R.C. (1994) Block of AIDS-Kaposi's sarcoma (KS) cell growth, angiogenesis, and lesion formation in nude mice by antisense oligonucleotide targeting basic fibroblast growth factor. *J. Clin. Invest.* 94:1736–1746.

Esch, F., Baird, A., Ling, N., Ueno, N., Hill, F., Deneroy, L., Klepper, R., Gospodarowicz, D., Bohlen, P. and Guillemin, R. (1985) Primary structure of bovine pituitary basic fibroblast growth factor (FGF) and comparison with the amino terminal sequence of bovine brain acidic FGF. *Proc. Natl. Acad. Sci. USA* 82:6507–6511.

Fajardo, L.F., Kwan, H.H., Kowalski, J., Prionas, S.D. and Allison, A.C. (1992) Dual role of tumor necrosis factor-alpha in angiogenesis. *Am. J. Pathol.* 140:539–544.

Felding-Habermann, B., Mueller, B.M., Romerdahl, C.A. and Cheresh, D.A. (1992) Involvement of integrin α_v gene expression in human melanoma tumorigenicity. *J. Clin. Invest.* 89:2018–2022.

Ferrara, N. (1993) Vascular endothelial growth factor. *Trends in Cardiovascular Medicine* 3:244–250.

Ferrara, N. and Henzel, W.J. (1989) Pituitary follicular cells secrete a novel heparin-binding growth factor specific for vascular endothelial cells. *Biochem. Biophys. Res. Comm.* 161:851–855.

Filardo, E.J., Brooks, P.C., Deming, S.L. and Cheresh, D.A. (1995) Requirement of the NPXY motif in the integrin β_3 subunit cytoplasmic tail for melanoma cell migration *in vitro* and *in vivo*. *J. Cell Biol.* 130:441–450.

Fitzpatrick, L.A., Severson, A., Edwards, W.D. and Ingram, R.T. (1994) Diffuse calcification in human coronary arteries: association of osteopontin with atherosclerosis. *J. Clin. Invest.* 94:1597–1604.

Folkman, J. (1971) Tumor angiogenesis: therapeutic implications. *New Engl. J. Med.* 285: 1182–1186.

Folkman, J. (1989) Successful treatment of an angiogenic disease. *New Engl. J. Med.* 320:1211–1212.

Folkman, J. (1992) The role of angiogenesis in tumor growth. *Seminars in Cancer Biology* 3:65–71.

Folkman, J. (1995) Angiogenesis in cancer, vascular, rheumatoid and other disease. *Nature Med.* 1:27–31.

Folkman, J. and Ingber, D. (1992) Inhibition of Angiogenesis. *Seminars in Cancer Biology* 3:89–96.

Folkman, J. and Shing, Y. (1992) Angiogenesis. *J. Biol. Chem.* 267:10931–10934.

Folkman, J., Watson, K., Ingber, D. and Hanahan, D. (1989) Induction of angiogenesis during the transition from hyperplasia to neoplasia. *Nature* 339:58–61.

Fraumeni, J.F., Devesa, S.S., Hoover, R.N. and Kinlen, L.J. (1993) Epidemiology of Cancer. *In:* V. DeVita, S. Hellman and S. Rosenberg (eds): *Cancer: Principles and Practice.* J.B. Lippincott Co., Philadelphia, PA., pp 150–181.

Fregene, T., Khanuja, P.S., Noto, A., Gehani, S., Egmaont, E., Luz, D. and Pienta, K. (1993) Tumor associated angiogenesis in prostate cancer. *Anticancer Res.* 13:1377–2382.

Friedlander, M., Brooks, P.C., Shaffer, R.W., Kincaid, C.M., Varner, J.A. and Cheresh, D.A. (1995) Definition of two angiogenic pathways by distinct α_v integrins. *Science*, 270: 1500–1502.

Gamble, J., Matthias, L., Meyer, G., Kaur, P., Russ, G., Faull, R., Berndt, M. and Vadas, M.A. (1993) Regulation of in vitro capillary tube formation by anti-integrin antibodies. *J. Cell Biol.* 121:931–943.

Gasparini, G., Weidner, N., Bevilaqua, P., Maluta, S, Dalla Palma, P., Caffo, D., Barbareschi, M., Boracchi, E. and Pozza, F. (1992) Tumor microvessel density, p53 expression, tumor size and peritumoral lymphotic vessel invasion are relevant prognostic markers in node negative breast carcinoma. *J. Clin. Oncol.* 12:454–466.

Gasparini, G., Weidner, N., Maluta, S., Pozza, F., Mezzeth, M., Testolin, A. and Bevilaqua, P. (1993) Intratumoral microvessel density and p53 protein correlation with metastasis in head and neck squamous cell carcinoma. *Int. J. Cancer* 55:739–744.

Gavrieli, Y., Sherman, Y. and Ben-Sasson, S.A. (1992) Identification of programmed cell death in situ via specific labelling of nuclear DNA fragmentation. *J. Cell Biol.* 119:493–501.

Giachelli, C.M., Bae, N., Almeida, M., Denhardt, D.T., Alpers, C.E. and Schwartz, S.M. (1993) Osteopontin is elevated during neointima formation in rat arteries and is a novel component of human atherosclerotic plaques. *J. Clin. Invest.* 92:1686–1696.

Giancotti, F.G. and Ruoslahti, E. (1990) Elevated levels of the $\alpha_5\beta_1$ receptor suppresses the transformed phenotype of Chinese hamster ovary cells. *Cell* 60:849–859.

Gimbrone, M., Cotron, R., Leapman, S. and Folkman, J. (1974) Tumor growth and neovascularization: an experimental model using the rabbit cornea. *J. Natl. Canc. Inst.* 52:413–427.

Gimbrone, M., Leapman, S., Cotrans, R. and Folkman, J. (1972) Tumor dormancy in vitro by prevention of neovascularization. *J. Exp. Med.* 136:261–276.

Gladson, C. and Cheresh, D. (19991) Glioblastoma expression of vitronectin and the $\alpha v\beta 3$ integrin. *J. Clin. Invest.* 88:1924–1932.

Goldberg, M.A. and Schneider, J. (1994) Similarities between the oxygen-sensing mechanisms regulating the expression of vascular endothelial growth factor and erythropoietin. *J. Biol. Chem.* 269:4355–4359.

Goldman, C.K., Kim, J., Wong, W.L., King, V., Brock, T. and Gillespie, G.Y. (1993) Epidermal growth factor stimulates vascular endothelial growth factor production by human malignant glioma cells: a model of glioblastoma multiforme pathophysiology. *Mol. Biol. Cell* 4:121–133.

Good, D.J., Polverini, P.J., Rastinjad, F., LeBeau, M.M., Lemons, R.S., Frazier, W.A. and Bouck, N.P.(1990) A tumor suppressor dependent inhibitor of angiogenesis is immunologically and functionally indistinguishable from a fragment of thrombospondin. *Proc. Natl. Acad. Sci. USA* 87:6624–6628.

Gonzalez, A.M., Buscaglia, M., Ong, M. and Baird, A. (1990) Distribution of basic fibroblast growth factor in the 18 day rat fetus: localization in the basement membranes of diverse tissues. *J. Cell Biol.* 110:753–765.

Gospodarowicz, D., Ferrara, N., Schweigerer, L. and Neufeld, G. (1987) Structural characterization and biological functions of fibroblast growth factor. *Endocr. Rev.* 8:1–20.

Graham, C.H., Rivers, J., Kerbel, R.S., Stankiewicz, K.S. and White, W.L. (1994) Extent of vascularization as a prognostic indicator in thin (< 0.76 mm) malignant melanomas. *Am. J. Path.* 145:510–514.

Guadagno, T.M., Ohtsubo, M., Roberts, J.M. and Assoian, R.K. (1993) A link between cyclin A expression and adhesion dependent cell cycle proliferation. *Science* 262:1592–1575.

Guan, J.L. and Shalloway, D. (1992) Regulation of focal adhesion-associated protein tyrosine kinase by both cellular adhesion and oncogenic transformation. *Nature* 358:690–692.

Guo, D., Jia, Q., Song, H.-Y., Warren, R.S. and Donner, D.B. (1995) Vascular endothelial cell growth promotes tyrosine phosophorylation of mediators of signal transduction that contain SH2 domains. *J. Biol. Chem.* 270:6729–6733.

Hageman, G.S., Kirchoff-Rempe, M.A., Lewis, G.P., Fisher, S.K. and Anderson, D.M. (1991) Sequestration of basic fibroblast growth factor in the primate interphotoreceptor matrix. *Proc. Natl. Acad. Sci. USA* 88:6706–6710.

Hanneken, A., De Juan, E., Lutty, G.A., Fox, G.M., Shiffer, S. and Hjelmeland, L.M. (1991) Altered distribution of basic fibroblast growth factor in diabetic retinopathy. *Arch. Ophthalmol.* 109:1005–1011.

Haralabopoulos, G., Grant, D., Kleinman, H., Lelkes, P., Papiaonnou and Maragoudakis, M. (1992) Inhibitors of basement membrane collagen synthesis prevent endothelial cell alignment in matrigel *in vitro* and angiogenesis *in vivo*. *Lab. Invest.* 71:575–582.

Harris, E.D. (1990) Rheumatoid arthritis: Pathophysiology and implications for therapy. *New Engl. J. Med.* 322:1277–1289.

Hatva, E., Kaipainen, A., Mentula, P., Jaaskelainen, J., Paetau, A., Maltia, M. and Alitalo, K. (1995) Expression of endothelial cell specific receptor tyrosine kinases and growth factors in human brain tumors. *Am. J. Pathol.* 146:368–378.

Hayes, D.F. (1994) Angiogenesis and breast cancer. *Hematology/Oncology Clinics of North America* 8:51–71.

Healy, J., Murayama, O., Maeda, T., Yoshino, K., Sekiguchi, K. and Kikuchi, M. (1995) Peptide ligands for integrin $\alpha_v\beta_3$ selected from random phage display libraries. *Biochemistry* 34:3948–39955.

Hicks, D., Bugra, K., Faucheux, B., Jeanny, J.C., Laurent, M., Malecaze, F., Mascarelli, F., Raulais, D., Cohen, S.Y. and Courtois, A. (1992) Fibroblast growth factors in the retina. *In:* N. Osborne and G. Chader (eds): *Progress in Retinal Research*, pp 333–374.

Hirst, R., Horwitz, A.F., Buck, C. and Rohrschneider, L. (1986) Phosphorylation of the fibronectin receptor complex in cells transformed by oncogenes that encode tyrosine kinases. *Proc. Natl. Acad. Sci. USA* 83:6470–6474.

Hlatky, L., Tsionou, C., Hahnfeldt, P. and Coleman, C.N. (1994) Mammary fibroblasts may influence breast tumor angiogenesis via hypoxia-induced vascular endothelial growth factor upregulation and protein expression. *Cancer Res.* 54:6083–6086.

Hobson, B. and Denekamp, J. (1984) Endothelial proliferation in tumors and normal tissues: continous labelling studies. *Br. J. Cancer* 49:405–413.

Horak, E., Leek, R., Klenk, N., Lejeune, S., Smith, K., Stuart, N., Greenall, M., Stepniewska, K. and Harris, A. (1992) Angiogenesis, assessed by platelet/endothelial cell adhesion molecule antibodies, as indicators of node metastases and survival in breast cancer. *Lancet* 340:1120–1124.

Hori, A., Ikeyama, S. and Sudo, K. (1994) Suppression of cyclin D1 mRNA expression by the angiogenesis inhibitor TNP-470 (AGM-1470) in vascular endothelial cells. *Biochem. Biophys. Res. Comm.* 204:1067–1073.

Horton, M.A., Taylor, M.L., Arnett, T.R. and Helfrich, M.H. (1991) Arg-Gly-Asp (RGD) peptides and the anti-vitronectin receptor antibody 23C6 inhibit dentine resorption and cell spreading by osteoclasts. *Exp. Cell Res.* 195:368–375.

Houck, K.A., Ferrara, N., Winer, J., Cachianes, G., Li, B. and Leung, D.W. (1991) The vascular endothelial growth factor family: identification of a fourth molecular species and characterization of alternatively spliced RNA. *Mol. Endocrinol.* 5:1806–1814.

Houck, K.A., Leung, D.W., Rowland, S.U., Winer, J. and Ferrara, N. (1992) Dual regulator of vascular endothelial growth factor bioavailability by genetic and proteolytic mechanisms. *J. Biol. Chem.* 267:26031–26037.

Hu, D.E. and Fan, T.-P. (1995) Suppression of VEGF-induced angiogenesis by the protein tyrosine kinase inhibitor, lavendustin A. *Br. J. Pharm.* 114:262–268.

Hynes, R.O. (1992) Integrins: versatility, modulation and signaling in cell adhesion. *Cell* 69:11–25.

Ingber, D. and Folkman, J. (1988) Inhibition of angiogenesis through modulators of collagen synthesis. *Lab. Invest.* 59:44–51.

Jackson, C. and Reidy, M. (1993) Basic fibroblast growth factor: its role in the control of smooth muscle cell migration. *Am. J. Path.* 143:1024–1031.

Jakeman, J.B., Armanini, M., Phillips, H.S. and Ferrara, N. (1993) Developmental expression of binding sites and mRNA for vascular endothelial growth factor suggests a role for the protein in vasculogenesis and angiogenesis. *Endocrinology* 133:848–859.

Juliano, R.L. and Haskill, S. (1993) Signal transduction from the extracellular matrix. *J. Cell Biology* 120:577–585.

Kamat, B.R., Brown, L.F., Mansen, E.J., Sena, O.R. and Dvorak, H.F. (1995) Expression of vascular permeability factor/vascular endothelial growth factor by human granulosa and theca lutein cells: role in corpus luteum development. *Am. J. Path.* 146:157–165.

Kieser, A., Weich, H., Brandner, G. Marme, D. and Kolch, W. (1994) Mutant p53 potentiates protein kinase C induction of vascular endothelial growth factor expression. *Oncogene* 9:963–969.

Kim, K.J., Li, B., Winer, J., Armanini, M., Gillet, N., Phillips, H.S. and Ferrara, N. (1993) Inhibition of vascular endothelial growth factor-induced angiogenesis suppresses tumor growth in vivo. *Nature* 362:841–844.

Klagsbrun, M. (1989) The fibroblast growth factor family structural properties and biological properties. *Prog. Growth Factor Res.* 1:207–235.

Klagsbrun, M. (1992) Mediators of angiogenesis: the biology and significance of basic fibroblast growth factor (bFGF)-heparin and heparan sulfate interactions. *Seminars in Cell Biol.* 3:81–87.

Klagsbrun, M. and Baird, A. (1991) A dual receptor system is required for basic fibroblast growth factor activity. *Cell* 67:229–231.

Klagsbrun, M. and Dlutz, S. (1993) Smooth muscle cell and endothelial cell growth factors. *Trends in Cardiovascular Medicine* 3:213–217.

Klemke, R.L., Yebra, M., Bayna, E. and Cheresh, D.A. (1994) Receptor-tyrosine kinase signaling reguired for integrin $\alpha_v \beta_5$ – directed cell motility but not adhesion on vitronectin. *J. Cell Biol.* 127:859–866.

Knighton, D., Ausprunk, D., Tapper, D. and Folkman, J. (1977) Avascular and vascular phases of tumor growth in the chick embryo. *Br. J. Cancer* 35:347–456.

Koch, A.E., Polverini, P.J., Kunkel, S.L., Harlow, L.A., Pietra, L.A., Elner, V.M., Elner, S.G. and Strieter, R.M. (1992) Interleukin-8 as a macrophage-derived mediator of angiogenesis. *Science* 258:1798–1801.

Kohn, E., Alessandro, B., Spoonster, J., Wersto, R.P. and Liotta, L. (1995) Angiogenesis: Role of calcium mediated signal transduction. *Proc. Natl. Acad. Sci. USA* 92:1307–1311.

Koivunen, E., Gay, D.A. and Ruoslahti, E. (1993) Selection of peptides binding to the $\alpha_5 \beta_1$ integrin from a phage display library. *J. Biol. Chem.* 268:20205–20210.

Koivunen, E., Wang, B. and Ruoslahti, E. (1994) Isolation of a highly specific ligand for the $\alpha_5 \beta_1$ integrin from a phage display library. *J. Cell. Biol.* 124:373–380.

Kornberg, L.J., Earp, H.S., Turner, C.E., Prockop, C. and Juliano, R.L. (1991) Signal transduction by integrins: Increased protein tyrosine phosphorylation caused by clustering of β_1 integrins. *Proc. Natl. Acad. Sci. USA* 88:8392–8895.

Kornberg, L.J., Earp, H.S., Parsons, J.T., Schaller, M. and Juliano, R.L. (1992) Cell adhesion or integrin clustering increased phosphorylation of a focal adhesion associated kinase. *J. Biol. Chem.* 117:1101–1107.

Krissanson, G.W., Elliot, M.J., Lucas, C.M., Stomski, F.C., Berndt, M.C., Cheresh, D.A., Lopez, A.F. and Burns, G.F. (1990) Blood leukocytes bind platelet glycoprotein (IIb–IIIa) but do not express the vitronectin receptor. *J. Biol. Chem.* 265:823–830.

Leavesley, D.I., Ferguson, G.D., Wayner, E.A. and Cheresh, D.A. (1992) Requirement of integrin β_3 subunit for carcinoma cell spreading or migration on vitronectin and fibrinogen. *J. Cell Biol.* 117:1101–1107.

Leavesley, D.I., Schwartz, M.A. and Cheresh, D.A. (1993) Integrin b1- and b3-mediated endothelial cell migration is triggered through distinct signaling mechanisms. *J. Cell Biol.* 121:163–170.

Leek, R.D., Harris, A.L. and Lewis, C.E. (1994) Cytokine networks in solid human tumors: regulation of angiogenesis. *J. Leukoc. Biol.* 56:423–432.

Leung, D.W., Cachianes, G., Kuang, W.J., Goeddel, D.V. and Ferrara, N. (1989) Vascular endothelial cell growth factor is secreted angiogenic mitogen. *Science* 246:1306–1309.

Liaw, L., Almeida, M., Hart, C., Schwartz, S. and Giachelli, C. (1993) Osteopontin promotes vascular cell adhesion and spreading and is chemotactic for smooth muscle cells *in vitro. Circ. Research* 73:214–224.

Liaw, L., Skinner, M.P., Raines, E.W., Ross, R., Cheresh, D.A., Schwartz, S.M. and Giachelli, C. (1995) The adhesive and migratory effects of osteopontin are mediated via distinct cell surface integrins. *J. Clin. Invest.*, 95:713–724.

Lindner, V., Lappi, D., Baird, A., Majack, R.A. and Reidy, M.A. (1991) role of basic fibroblast growth factor in vascular lesion formation. *Circ. Res.* 68:106–113.

Lindner, V. and Reidy, M.A. (1991) Proliferation of smooth muscle after vascular injury is inhibited by an antibody against basic fibroblast growth factor. *Proc. Natl. Acad. Sci. USA* 88:3739–3743.

Majewski, S., Marczak, M., Szmurlo, A., Jablonska, S. and Bollag, W. (1994) Retinoids, interferon a, 1,25-dihydroxyvitamin D_3 and their combination inhibit angiogenesis induced by non-HPV-harboring tumor cell lines. RARa mediates the antiangiogenic effect of retinoids. *Cancer Lett.* 89:117–124.

Majewski, S., Szmurlo A., Marczak, M., Jablonska, S. and Bollag, W. (1995) Synergistic effect of retinoids and interferon a on tumor-induced angiogenesis: anti-angiogenic effect on HPV-harboring tumor cell lines. *Int. J. Cancer* 57:81–85.

Mandriota, S.J., Seghezzi, G., Vassalli, J.D., Ferrara, N., Wasi, S., Mazzieri, R., Mignatti, P. and Pepper, M.S. (1995) Vascular endothelial growth factor increases urokinase receptor expression in vascular endothelial cells. *J. Biol. Chem.* 270:9709–9716.

Matsuno, H., Stassen, J.M., Vermylen, J. and Deckman, H. (1994) Inhibition of integrin function by a cyclic RGD-containing peptide prevents neointima formation. *Circulation* 90:2203–2206.

Mautner, V. and Hynes, R.O. (1977) Surface distribution of LETS protein in relation to the cytoskeleton of normal and transformed cells. *J. Cell Biol.* 75:743–768.

McNamee, H.P., Ingber, D.E. and Schwartz, M.A. (1993) Adhesion to fibronectin stimulates inositol lipid synthesis and enhances PDGF-induced inositol lipid breakdown. *J. Cell. Biol.* 121:673–678.

Medhora, M., Teitelbaum, S., Chappel, J., Alvarez, J., Mimura, H., Ross, F.P. and Hruska, K. (1993) 1a, 25-dihydroxyvitamin D3 up-regulates expression of the osteoclast integrin $\alpha_v\beta_3$ *J. Biol. Chem.* 268:1456–1461.

Meredith, J.E., Fazeli, B. and Schwartz, M. (1993) The extracellular matrix as a cell survival factor. *Mol. Biol. Cell.* 4:953–961.

Millauer, B., Wizigmann-Voos, S., Schnurch, H., Martinez, R., Moller, N., Risau, W. and Ullrich, A. (1993) High affinity VEGF finding and developmental expression suggests Flk-1 as a major regulator of vasculogenesis and angiogenesis. *Cell* 72:835–846.

Miller, J.W., Adamis, A.P., Shima, D.T., D'Amore, P.A., Moulton, R.S., O'Reilly, M.S., Folkman, J., Dvorak, H.F., Brown, L.F., Berse, B.M. and Yeo, T.-K. (1994) Vascular endothelial growth factor/vascular permeability factor is temporally and spatially correlated with ocular angiogenesis in a primate model. *Am. J. Path.* 145:574–584.

Minchenko, A., Bauer, T., Salceda, S. and Caro, J. (1994) Hypoxic stimulation of vascular endothelial growth factor expression *in vitro* and *in vivo. Lab. Invest.* 71:374–379.

Montgomery, A.M.P., Reisfeld, R.A. and Cheresh, D.A. (1994) Integrin $\alpha_v\beta_3$ rescues melanoma cells from apoptosis in a three dimensional dermal collagen. *Proc. Natl. Acad. Sci. USA* 91:8856–8860.

Moscatelli, D. (1987) High and low affinity binding sites for basic fibroblast growth factor on cultured cells: absence of a role for low affinity binding in the stimulation of plasminogen activator production by bovine capillary endothelial cells. *J. Cell. Physiol.* 131:123–130.

Motro, B., Itin, A., Sachs, L. and Keshet, E. (1990) Pattern of interleukin 6 gene expression *in vivo* suggests a role for this cytokine in angiogenesis. *Proc. Natl. Acad. Sci. USA* 87:3092–3096.

Murphy-Ullrich, J.E. and Hook, M. (1989) Thrombospondin modulates focal adhesion in endothelial cells. *J. Cell. Biol.* 109:1309–1319.

Mustonen, T. and Alitalo, K. (1995) Endothelial receptor tyrosine kinases involved in angiogenesis. *J. Cell Biol.* 129:895–898.

Nanus, D.M., Schmitz-Dragen, B.J., Motzer, R.J., Lee, A.C., Vlamis, V., Cordon-Cardo, C., Albino, A.P. and Reuter, V.E. (1993) Expression of basic fibroblast growth factor in primary human renal tumors: correlation with poor survival. *J. Natl. Canc. Inst.* 85:1597–1599.

Naylor, M.S., Stamp, G.W., Foulkes, W.D., Eccles, D. and Balkwill, R.R. (1993) Tumor necrosis factor and its receptors in human ovarian cancer. Potential role in disease progression. *J. Clin. Invest.* 91:2194–2206.

Neufeld, G. and Gospodarowicz, D. (1995) The identification of the fibroblast growth factor receptor of baby hamster kidney cells. *J. Biol. Chem.* 260:13860–13868.

Nguyen, M., Strubel, N.A. and Bischoff, J. (1994) A role for sialyl Lewis-X/A glycoconjugates in capillary morphogenesis. *Nature* 365:267–269.

Nguyen, M., Watanabe, H., Budson, A.E., Richie, J.P., Hayes, D.F. and Folkman, J. (1994) Elevated levels of an angiogenesis peptide, basic fibroblast growth factor, in the serum of patients with a wide spectrum of cancers. *J. Natl. Canc. Inst.* 86:356–361.

Nickoloff, B.J., Mitra, R.S., Varani, J., Dixit, V.M. and Polverini, P.J. (1984) Aberrant production of IL-8 and thrombospondin-1 by psoriatic keratinocytes mediates angiogenesis. *Am. J. Pathol.* 144:820–828.

Nicosia, R.F. and Bonanno, E. (1991) Inhibition of angiogenesis in vitro by Arg-Gly-Asp-containing synthetic peptides. *Am. J. Path.* 138:829–833.

Niedbala, M.J. and Stein, M. (1991) Tumor necrosis factor induction of urokinase-type plasminogen activator in human endothelial cells. *Biochim. Biophys. Acta.* 50:327–436.

Nip, J., Shibata, H., Loskutoff, D., Cheresh, D. and Brodt, P. (1992) Human melanoma cells derived from lymphatic metastases use integrin $\alpha_v\beta_3$ to adhere to lymph node vitronectin. *J. Clin. Invest.* 90:1406–1413.

Noden, D.M. (1989) Embryonic origins and assembly of blood vessels. *Ann. Rev. Respir. Dis.* 140:1097–1103.

O'Brien, E.R., Garvin, M.R., Dev, R., Stewart, D.K., Hinohara, T.,., Simpson, J.B. and Schwartz, S.M. (1994) Angiogenesis in human coronary atherosclerotic plaques. *Am. J. Path.* 145:883–894.

O'Reilly, M.S., Holmgren, L., Shing, Y., Chen, C., Rosenthal, R.A., Moses, M., Lane, W.S., Cao, Y., Sage, F.H. and Folkman, J. (1994) Angiostatin: a novel angiogenesis inhibitor that mediates the suppression of metastases by a Lewis lung carcinoma. *Cell* 79:315–328.

Olson, N.E., Chao, S., Lindner, V., Reidy, M.A. (1992) Intimal smooth muscle cell proliferation after balloon catheter injury: the role of basic fibroblast growth factor. *Am. J. Pathol.* 140:1017–1023.

Park, J.E., Keller, G.A. and Ferrara, N. (1993) The vascular endothelial growth factor (VEGF) isoforms: differential deposition into the subepithelial extracellular matrix and bioactivity of extracellular matrix-bound VEGF. *Mol. Biol. Cell* 4:1317–1326.

Patterson, J.C. (1938) Capillary rupture with intimal hemorrhage as a causative factor in coronary thrombosis. *Arch. Pathol.* 25:474–487.

Patz, A. (1980) Studies on retinal neovascularization. *Invest. Ophthalmol. Vis. Sci.* 19:1133–1138.

Pelletier, A.J., Bodary, S.C. and A.D. Levinson (1992) Signal transduction by the platelet integrin aIIbb3: induction of calcium oscillations required for protein-tyrosine phosphorylation and ligand-induced spreading of stably transfected cells. *Mol. Biol. Cell* 3:989–998.

Pepper, M.S., Ferrara, N., Orci, L. and Montesano, R. (1991) Vascular endothelial growth factor (VEGF) induces plasminogen activators and plasminogen activator inhibitor-1 in microvascular endothelial cells. *Biochem. Biophys. Res. Comm.* 181:902–906.

Pertovaara, L., Kaipainen, A., Mustonen, T., Orpana, A., Ferrara, N., Saksela, O. and Alitalo, K. (1994) Vascular endothelial growth factor is induced in response to transforming growth factor-β in fibroblast and epithelial cells. *J. Biol. Chem.* 269:6271–6274.

Peters, K.G., DeVries, C. and Williams, C.T. (1993) Vascular endothelial growth factor receptor expression during embryogenesis and tissue repair suggests a role in endothelial differentiation and blood vessel growth. *Proc. Natl. Acad. Sci. USA* 90:8915–8918.

Pfaff, M., Tangemann, K., Muller, B, Gurrath, M., Muller, G., Kessler, H., Timpl, R. and Engel, J. (1994) Selective recognition of cyclic RGD peptides of NMR defined conformation by $\alpha_{IIb}\beta_3$, $\alpha_v\beta_3$ and $\alpha_5\beta_1$ integrins. *J. Biol. Chem.* 269:20233–23238.

Phillips, H.S., Armanini, M., Stavrou, D., Ferrara, N. and Westphal, M. (1993) Intense focal expression of vascular endothelial growth factor mRNA in human intracranial neoplasms associated with regions of necrosis. *Int. J. Oncol.* 2:913–919.

Phillips, H.S., Hains, J., Leung, D.W. and Ferrara, N. (1990) Vascular endothelial growth factor is expressed in rat corpus luteum. *Endocrinology* 127:965–967.

Plantefaber, L. and Hynes, R.O. (1989) Changes in integrin receptors on oncogenically transformed cells. *Cell* 56:281–290.

Plate, K., Breier, G., Weich, H., Mennel, H. and Risau, W. (1994) Vascular endothelial growth factor and glioma angiogenesis: coordinate induction of VEGF receptors, distribution of VEGF protein and possible *in vivo* regulatory mechanisms. *Int. J. Cancer* 59: 520–529.

Plate, K.H., Breier, G., Weich, H.A. and Risau, W. (1992) Vascular endothelial growth factor is a potential tumour angiogenesis factor in human gliomas *in vivo. Nature* 359:845–848.

Plouet, J., Schilling, J. and Gospodarowicz, D. (1989) Isolation and characterization of a newly identified endothelial cell mitogen produced by AtT-20 cells. *EMBO J.* 8:3801–3806.

Pötgens, A., Lubsen, N., Altena, M., Schoenmakers, J., Ruiter, D. and Waal, R. (1995) Vascular permeability factor expression influences tumor angiogenesis in human melanoma lines xenografted to nude mice. *Am. J. Path.* 146:197–209.

Pusztai, L., Clover, L.M., Cooper, K., Starkey, P.M., Lewis, C.E. and McGee, J.O.D. (1994) Expression of tumor necrosis factor alpha and its receptors in carcinoma of the breast. *Br. J. Cancer* 70:289–292.

Quinn, T., Peters, K., DeVaide, C., Ferrara, N. and Wilimas, L.T. (1993) Fetal liver kinase 1 is a receptor for vascular endothelial growth factor and is selectively expressed in vascular endothelium. *Proc. Natl. Acad. Sci. USA* 90:77533–7537.

Rapraeger, A.C., Krufka, A. and Olwin, B.B. (1991) Requirement of heparan sulfate for bFGF-mediated fibroblast growth and myoblast differentiation. *Science* 252:1075–1078.

Ravindranath, N., Little-Ihrig, L., Phillips, H.S., Ferrara, N. and Zeleznik, A. (1992) Vascular endothelial growth factor messenger ribonucleic acid expression in the primate ovary. *Endocrinology* 131:254–260.

Reilly, T., Taylor, D.S., Herblin, W.F., Thoolen, M.J., Chiu, A.T., Watson, D.W. and Timmermans, P.B. (1989) Monoclonal antibody directed against basic fibroblast growth factor which inhibits its biological activity in vitro and in vivo. *Biochem. Biophys. Res. Comm.* 164:736–743.

Rosen, E., Liu, D., Setter, E., Bhargava, M. and Goldberg, I. (1991) Interleukin-6 stimulates motility of vascular endothelium. *In:* I.D. Goldberg (ed): *Cell Motility Factors.* Birkhäuser Verlag, Basel, Switzerland, pp 194–205.

Rosenthal, R.A., Megyesi, J.F., Henzel, W.J., Ferrara, N. and Folkman, J. (1990) Conditioned medium from mouse sarcoma 180 cells contains vascular endothelial growth factor. *Growth Factor* 4:53–59.

Ross, F.P., Chappel, J., Alvarez, J.I., Sander, D., Butler, W.T., Farach-Carson, M.C., Mintz, K.A., Robey, P.G., Teitelbaum, S.I. and Cheresh, D. (1993) Interactions between the bone matrix proteins osteopontin and bone sialoprotein and the osteoclast integrin $\alpha_v\beta_3$ potentiate bone resorption. *J. Biol. Chem.* 268:9901–9907.

Ruoslahti, E. and Pierschbacher, M. (1986) Arg-Gly-Asp: A versatile cell recognition sequence. *Cell* 44:517–518.

Sakaguchi, M., Kajio, T., Kawahara, K. and Kato, K. (1988) Antibodies against basic fibroblast growth factor inhibit the autocrine growth of pulmonary artery endothelial cells. *FEBS Lett.* 233:163–166.

Sakamoto, N., Iwahana, I., Tanaka, N. and Osada, Y. (1991) Inhibition of angiogenesis and tumor growth by a synthetic laminin peptide CDPGYISGR-NH$_2$. *Cancer Res.* 51:903–906.

Sakamoto, N. and Tanaka, N. (1988) Mechanism of synergistic effects of heparin and cortisone against angiogenesis and tumor growth. *Cancer* 2:9–13.

Saksela, O. and Rifkin, D. (1990) Release of basic fibroblast growth factor-heparin sulfate complexes from endothelial cells by plasminogen activator mediated proteolytic activity. *J. Cell Biol.* 110:767–775.

Sanders, L.C., Felding-Habermann, B., Mueller, B.M. and Chersh, D.A. (1992) Role of av integrins and vitronectin in human melanoma cell growth. *Cold Spring Harbor Symposia on Quantitative Biology* 58:233–240.

Sato, M., Sardana, M.K., Grasser, W.A., Garsky, V.M., Murray, J.M. and Gould, R.J. (1990) Echistatin is a potent inhibitor of bone resorption in culture. *J. Cell Biol.* 111:1713–1723.

Scannell, G., Waxman, K., Kami, G.J., Ioloi, G., Gatanaga, T., Yamamoto, R. and Granger, G.A. (1993) Hypoxia induces a human macrophage cell line to release tumor necrosis factor-alpha and its soluble receptors *in vitro. J. Surg. Res.* 54:281–285.

Schreiner, C., Fisher, M., Hussein, S. and Juliano, R.L. (1991) Increased tumorigenicity of fibronectin receptor deficient Chinese hamster ovary cell variants. *Cancer Res.* 51: 1738–1740.

Schultze-Osthoff, K., Risau, W., Vollmer, E. and Sorg, C. (1990) *In situ* detection of basic fibroblast growth factor by highly specific antibodies. *Am. J. Pathol.* 137:85–92.

Schwartz, M.A., Both, G. and Lechene, C. (1989) Effect of cell spreading on cytoplasmic pH in normal and transformed fibroblasts. *Proc. Natl. Acad. Sci. USA* 86:4525–4529.

Schwartz, M.A., Cragoe, E.J. Jr., and Lechene, C.P. (1990) pH regulation in spread and round cells. *J. Biol. Chem.* 265:1327–1332.

Schwartz, M.A., Lechene, C. and Ingber, D.E. (1991) Insoluble fibronectin activates the Na/H antiporter by clustering and immobilizing integrin $\alpha_5\beta_1$, independent of call shape. *Proc. Natl. Acad. Sci. USA* 88:7849–7853.

Schwartz, M. and Lechene, C. (1992) Adhesion is required for protein kinase C dependent activation of the Na⁺/H⁺ antiporter by platelet-derived growth factor. *Proc. Natl. Acad. Sci. USA* 89:6138–6141.

Schwartz, M.A. (1992) Spreading of human endothelial cells on fibronectin or vitronectin triggers elevation of intracellular free calcium. *J. Cell Biol.* 120:1003–1010.

Schwartz, M. (1993) Signalling by integrins: implications for tumorigenesis. *Cancer Res.* 51:1503–1505.

Schweigerer, L., Neufeld, G., Friedman, J., Abraham, J.A., Fiddes, J.C. and Gospodarowicz, D. (1987) Capillary endothelia cells express basic fibroblast growth factor, a mitogen that promotes their own growth. *Nature* 325:257–259.

Seftor, R., Seftor, E., Gehlsen, K., Stetler-Stevenson, W., Brown, P., Ruoslahti, E. and Hendrix, M. (1992) role of the $\alpha_v\beta_3$ integrin in human melanoma cell invasion. *Proc. Natl. Acad. Sci. USA* 89:1557–1561.

Senger, D.R. Galli, S.J., Dvorak, A.M., Perruzzi, C.A., Harrey, V.S. and Dvorak, H. (1983) Tumor cells secrete a vascular permeability factor that promotes accumulation of ascites fluid. *Science* 219:983–985.

Senger, D., Perruzzi, C.A., Fedor, J. and Dvorak, H.F. (1988) A highly conserved vascular permeability factor secreted by a variety of human and rodent tumor cell lines. *Cancer Res.* 46:5269–5275.

Sepp, N.T., Li, L.-J., Lee, K.H., Brown, E.J., Caughman, S.W.W., Lawley, T.J. and Swerlick, R.A. (1994) Basic fibroblast growth factor increases expression of the $\alpha_v\beta_3$ complex on human microvessel endothelial cells. *J. Invest. Dermatol.* 103:295–299.

Shweiki, D., Itin, A., Soffer, D. and Keshtet, E. (1992) Vascular endothelial growth factor induced by hypoxia may mediate hypoxia-induced angiogenesis. *Nature* 359: 843–845.

Sioussat, T.M., Dvorak, H.F., Brock, T.A. and Senger, D.R. (1993) Inhibitors of vascular permeability factor (vascular endothelial growth factor) with anti-peptide antibodies. *Arch. Biochem. Biophys.* 30:15–20.

Smith, D.R., Polverini, P.J., Kunkel, S.L., Orringer, M.B., Whyte, R.I., Burdick, M.D., Wilke, C.A. and Strieter, R.M. (1994) Inhibition of interleukin-8 attenuates angiogenesis in bronchogenic carcinoma. *J. Exp. Med.* 179:1409–1415.

Soubrane, G., Cohen, S.-Y., Delayre, T., Tassin, T., Hartmann, M.-P., Coscas, G., Courtois, Y. and Jeanny, J.-C. (1994) Basic fibroblast growth factor experimentally induced choroidal angiogenesis in the minipig. *Current Eye Res.* 13:183–195.

Soubrane, G., Courtois, Y., Assouline, M., Coscas, G. and Jeanny, J.C. (1989) Iodinated bFGF specific binding to normal and newly formed vessels in different mammalian ocular tissues. *Invest. Ophthalmol. Vis. Sci.* 30(Suppl.):90.

Srivastava, A., Laidler, P., Davies, R.P., Horgan, K. and Hughes, L.E. (1988) The prognostic significance of tumor vascularity in intermediate-thickness (0.76–4 mm) skin melanoma. *Am. J. Pathol.* 133:419–422.

Strieter, R.M., Kunkel, S.L., Elner, V.M. and Martoryi, C.L. (1992) Interleukin-8: a corneal factor that induces neovascularization. *Am. J. Pathol.* 141:1279–1284.

Symington, B.E. (1990) Fibronectin receptor overexpression and loss of transformed phenotype in a stable variant of the K562 cell line. *Cell. Regul.* 1:637–648.

Symington, B.E. (1993) Fibronectin receptor modulates cyclin dependent kinase activity. *J. Biol. Chem.* 267:25744–25747.

Takahashi, K., Mulliken, J., Kozakewich, H., Rogers, A.A., Folkman, J. and Ezekowitz, A.A. (1994) Cellular markers that distinguish the phases of hemangioma during infancy and childhood. *J. Clin. Invest.* 93:2357–2364.

Takashita, S., Zheng, L.P., Brogi, E., Kearney, M., Bunting, S., Ferrara, N., Symes, J.F. and Isner, J.M. (1994) Therapeutic angiogenesis: a single intraarterial bolus of vascular endothelial growth factor augments revascularization in a rabbit ischemic hind limb model. *J. Clin. Invest.* 93:662–670.

Tannock, I.F. and Steel, G.G. (1969) Quantitative technique for study of the anatomy and function of small blood vessels in tumors. *J. Natl. Canc. Inst.* 42:771–787.

Tisher, E., Mitchell, R., Hartman, T., Silva, M., Gospodarowicz, Fiddes, J.C. and Abraham, J.A. (1991) The human gene for vascular endothelial growth factor. Multiple forms are encoded through alternative exon splicing. *J. Biol. Chem.* 266:11947–11954.

Tolsma, S., Volpert, O., Good, D., Frazier, W., Polverini, P. and Bouck, N. (1993) Peptides derived from two separate domains of the matrix protein thrombospondin-1 have anti-angiogenic activity. *J. Cell Biol.* 122:497–511.

Topol, E.J., Califf, R.M., Weisman, H.F., Ellis, S.G., Tcheng, J.E., Worley, S., Ivanhoe, R., George, B.S., Fintel, D., Weston, M., Sigmon, K., Anderson, K.M., Lee, K.L. and Willerson, J.T. (1994) Randomised trial of coronary intervention with antibody against platelet IIb/IIIa integrin for reduction of clinical restenosis: results at six months. *Lancet* 343: 881–886.

Unemore, E., Ferrara, N., Bauer, E.A. and Amato, E.P. (19992) Vascular endothelial growth factor induces interstitial collagenase expression in human endothelial cells. *J. Cell. Physiol.* 153:557–562.

Vaheri, A. and Ruoslahti, E. (1975) Fibroblast surface antigen produced but not retained by virus-transformed human cells. *J. Exp. Med.* 142:530–535.

Varner, J.A., Emerson, D. and Juliano, R. (1995) Integrin $\alpha_5\beta_1$ expression negatively regulates cell growth: reversal by attachment to fibronectin. *Mol. Biol. Cell* 6:725–740.

Visscher, D.W., Lawrence, W.D. and Boman, S. (1994) Angiogenesis in breast carcinoma – Clinicopathologic relevance and potential use as a quantifiable surrogate endpoint biomarker. *J. Cell Biochem., Supp.* 19:146–152.

Vlodavsky, I., Folkman, J., Sullivan, R., Fridman, R., Ishai-Michaeli, R., Sasse, J. and Klagbrun, M. (1987) Endothelial cell-derived basic fibroblast growth factor synthesis and deposition into subendothelial extracellular matrix. *Proc. Natl. Acad. Sci. USA* 84:2292–2296.

Wagner, D., Ivatt, R., Destree, A. and Hynes, R. (1981) Similarities and differences between fibronectins of normal and transformed hamster cells. *J. Biol. Chem.* 256:11708–11715.

Wagner, E.F. and Risau, W. (1994) Oncogenes in the study of endothelial cell growth and differentiation. *Semin. Cancer Biol.* 5:137–145.

Weich, H.E., Iberg, N., Klagsbrun, M. and Folkman, J. (1990) Expression of acidic and basic fibroblast growth factors in human and bovine vascular smooth muscle cells. *Growth Factors* 2:2313–320.

Weidner, N., Carroll, P.R., Flax, J., Blumenfeld, W. and Folkman, J. (1939) Tumor angiogenesis correlates with metastasis in invasive prostate carcinoma. *Am. J. Pathol.* 143:401–409.

Weidner, N., Folkman, J., Pozza, F., Bevilaqua, P., Allred, E.N., Moore, D.H., Meli, S. and Gasparini, G. (1992) Tumor angiogenesis: a new significant and independent prognostic indicator in early stage breast carcinoma. *J. Natl. Cancer Inst.* 84:1875–1887.

Weidner, N., Semple, J.P., Welch, W.R. and Folkman, J. (1991) Tumor angiogenesis – correlation in invasive breast carcinoma. *New Engl. J. Med.* 324:2–8.

Weindel, K., Moringlane, J., Marme, D., Weich, H.A. (1994) Detection and quantification of vascular endothelial growth factor/vascular permeability factor in brain tumor tissue and cyst fluid: the key to angiogenesis? *Neurosurgery* 35:439–449.

Weinstat-Saslow, D. and Steeg, P. (1994) Angiogenesis and colonizing in the tumor metastatic process: basic and applied advances. *FASEB J.* 8:401–407.

Werner, S., Duan, D.-S.R., DeVries, C., Peters, K.G., Johnson, D.E. and Williams, L.T. (1992) Differential splicing in the extracellular region of fibroblast growth factor receptor 1 generates receptor variants with different ligand binding specificities. *Mol. Cell Biol.* 12:82–88.

Wernert, N., Raes, M.B., Lassalle, P., Dehouck, M.-P., Gosselin, B., Vandenbunder, B. and Stehelin, D. (1992) c-ets1 protooncogene is a transcription factor expressed in endothelial cells during tumor vascularization and other forms of angiogenesis in humans. *Am. J. Pathol.* 140:119–127.

White, P. (1960) Childhood diabetes: its course and influence on the second and third generations. *Diabetes* 9:345–355.

Wickham, T.J., Mathias, P., Cheresh, D.A. and Nemerow, G.R. (1993) Integrins $\alpha_v\beta_3$ and $\alpha_v\beta_5$ promote adenovirus internalization but not virus attachment. *Cell* 73:309–319.

Yamazaki, K., Abe, S., Takekawa, H., Sukoh, N., Watanabe, N., Ogura, S., Nakajima, I., Isobe, H., Inoue, K. and Kawakami, Y. (1994) Tumor angiogenesis in human lung adenocarcinoma. *Cancer* 74:2245–2250.

Yayon, Y., Klagsbrun, M., Esko, J.D., Leder, P. and Orniz, D.M. (1991) Cell surface heparin-like molecules are required for binding of basic fibroblast growth facto to its high affinity receptor. *Cell* 64:841–848.

Zhu, X., Komiya, H. and Chirino, A. (1990) Three dimensional structures of acidic and basic fibroblast growth factor. *Science* 251:90–93.

Regulation of Angiogenesis
ed. by I.D. Goldberg & E.M. Rosen
© 1997 Birkhäuser Verlag Basel/Switzerland

Role of fibrin and plasminogen activators in repair-associated angiogenesis: *In vitro* studies with human endothelial cells

V. W. M. van Hinsbergh, P. Koolwijk and R. Hanemaaijer

Gaubius Laboratory TNO-PG, Zernikedreef 9, 2333 CK Leiden, The Netherlands

Summary. Angiogenesis, the formation of new blood vessels from existing ones, plays a central role in development and in a number of pathological conditions. Tissue repair-associated angiogenesis usually involves cell invasion into a fibrin structure and the presence of inflammatory cells. In this chapter the role of plasminogen activators in the dissolution of fibrin and the invasion of endothelial cells into a fibrin matrix is described. Tissue-type plasminogen activator is stored in endothelial cells and can be released acutely into the vessel lumen upon stimulation of the endothelium to activate fibrinolysis and to prevent fibrin deposition. At the basolateral side of the cell, urokinase-type plasminogen activator (uPA) bound to a specific cellular receptor is involved in the proteolytic modulation of matrix proteins and cell-matrix interaction. The cytokine tumor necrosis factor-α (TNF-α) cooperates with the angiogenic factors basic fibroblast growth factor (bFGF) and vascular endothelial growth factor (VEGF) in inducing human microvascular endothelial cells *in vitro* to invade a three dimensional fibrin matrix and to form capillary-like tubular structures. The formation of these capillary-like tubules requires cell-bound uPA activity.

Introduction

Angiogenesis, the outgrowth of new blood vessels from existing ones, is an essential process during development, but normally stops when the body becomes adult. With the exception of the female reproductive system, angiogenesis in the adult is associated with tissue repair after wounding or inflammation. The difference in rapidly growing and adult tissues is also reflected in the proliferation rate of the vascular endothelial cells. While the half life of endothelial cells is several days in placenta and tumors, it increases to 100 to 10000 days in normal tissues of the adult body (Hobson and Denekamp, 1984).

Angiogenesis in embryonic development

Embryonic development requires the formation of new blood vessels. After fusion of blood islands into vascular structures (vasculogenesis) (Risau, 1990), new blood vessels bud off from existing ones (angiogenesis) (Ausprunk and Folkman, 1977; Folkman and Shing, 1992; Risau, 1990). The recent development of technologies which enable the deletion of genes in mice and other animals ("knock-out" technology) has provided a treasure of information regarding genes that are required for proper blood vessel formation. These genes comprise angiogenic growth factors, growth factor receptors, and cell receptors for cell-cell and cell-matrix interaction including integrins (Shalaby et al., 1995; Fong et al., 1995; Sato et al., 1995). It has been postulated on the basis of the co-existence of urokinase-type plasminogen activator (uPA), uPA receptor and plasminogen activator inhibitor type-1 (PAI-1) expression in the development of the vascular system in embryos that the uPA system is involved in developmental angiogenesis (Valinsky et al., 1981; Sappino et al., 1989). However, the fact that uPA-deficient, combined uPA- and tissue-type plasminogen activator (tPA)-deficient (Carmeliet et al., 1994), PAI-1-deficient (Carmeliet et al., 1993 a, b) and uPA receptor-deficient mice (Bugge et al., 1995) develop with a normal vascular system has shown that in these animals the development of the vascular tree does not depend necessarily on plasminogen activator activity.

Angiogenesis in the adult

In the adult organism the vascular tree is fully developed. In the absence of injury, overt angiogenesis is limited to the reproductive system of females (formation of corpus luteum and placenta) (Bacharach et al., 1992). However, the formation of new blood vessels is an essential factor in tissue repair (formation and regression of granulation tissue), which is necessary to restore healthy tissue after wounding and/or inflammation. Furthermore, angiogenesis is associated with many pathological conditions, such as chronic inflammation including rheumatoid arthritis (Weinberg et al., 1991), malignancies (Folkman and Shing, 1992), and retinopathy caused by metabolic dysregulation in particular diabetes (Aiello et al., 1994). In these conditions angiogenesis is commonly accompanied by vascular leakage (Dvorak et al., 1995), the occurrence of inflammatory cells (Polverini, 1989), and the presence of fibrin (Dvorak, 1986; Colvin, 1986). These latter factors are absent in angiogenesis during embryonic development. Therefore, the possibility exists that "developmental angiogenesis" and adult "repair-associated or pathologic angiogenesis" are two processes with many identical features, but with different properties with respect to their regulation.

Fibrin, a temporary repair matrix

Fibrin is a temporary matrix which is formed after wounding of a blood vessel and when plasma leaks from blood vessels forming a fibrous exudate, often seen in areas of inflammation and in tumors (Dvorak et al.,

1992). The fibrin matrix not only acts as a barrier preventing further blood loss, but also provides a structure in which new microvessels can infiltrate during wound healing. Proper timing of the outgrowth of microvessels as well as the subsequent (partial) disappearance of these vessels is essential to ensure adequate wound healing and to prevent the formation of scar tissue. An important role in the invasion of a fibrin matrix by endothelial cells is played by the plasminogen activator/plasmin system, which lyses fibrin at the basolateral side of the cell. However, the endothelium must respond differently to fibrin depending on whether the fibrin is present at its luminal or abluminal side. If fibrin is generated at the luminal side, the vessel may be occluded, which will cause serious damage to the distal tissues. To prevent this the endothelium is able to increase instantaneously the fibrinolytic activity in the blood compartment. If, however, this would lead to strong systemic fibrinolytic activity, recurrent bleeding might occur. Therefore, fibrinolysis must be restricted to a limited distance of the endothelium. To prevent local intravascular fibrin accumulation, while permitting a differently timed contribution of fibrinolysis in neovascularisation and tissue repair, endothelial cells are equipped with a complex regulatory system, which involves inhibitors, a tPA storage pool, cell polarity and cellular receptors. After discussing the components of the plasminogen activator/plasmin system, we will discuss the contribution of plasminogen activators in fibrinolysis and pericellular proteolytic events associated with cell migration and the invasion of capillary-like tubular structures into fibrin matrices.

Components of the plasmin/plasminogen activator system

Figure 1 summarizes the proteases and the inhibitors involved in fibrinolysis. Fibrin degradation and probably also activation of several matrix metalloproteinases is accomplished by the serine protease plasmin.

Proteases

Plasmin is formed from its zymogen plasminogen by proteolytic activation by plasminogen activators (PAs). Two types of physiological PAs are presently known: tissue-type plasminogen activator (tPA) and urokinase-type plasminogen activator (uPA) (Bachmann, 1987; Wallén, 1987). The three serine proteases, plasminogen, tPA and uPA, are synthesized as single polypeptide chains, and each of them is converted by specific proteolytic cleavage to a molecule with two polypeptide chains connected by a disulphide bond. The carboxy-terminal part of the molecule (the so-called B-chain) contains the proteolytically active site, whereas the

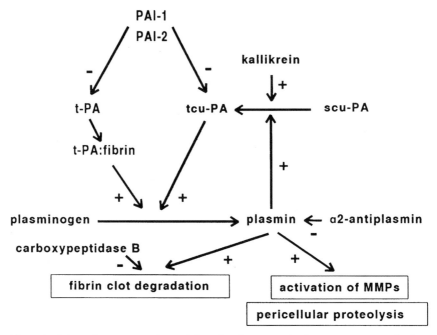

Figure 1. Schematic representation of the plasminogen activation system. +: activation; −: inhibition. PA: plasminogen activator; tPA: tissue type PA; uPA: urokinase type PA; scuPA: single-chain uPA; tcuPA: two-chain uPA; PAI: PA inhibitor; MMPs: matrix-degrading metalloproteinases.

amino-terminal part of the molecule (the A-chain) is built up of domains that determine the interaction of the proteases with matrix proteins and cellular receptors. The proteolytic cleavage of plasminogen and single-chain uPA to their respective two-chain forms is necessary to disclose the proteolytically active site and to activate the molecule. In contrast, tPA activity markedly increases by the interaction of tPA with fibrin. Once bound to this substrate, both the single-chain and two-chain form of tPA are active. The interaction of plasminogen with fibrin or the cell surface occurs predominantly via binding sites in the kringle structures, which recognize lysine residues of proteins, in particular carboxy-terminal lysines. Because B-type carboxypeptidases remove carboxy-terminal lysine residues from potential binding sites for plasminogen in fibrin or on the cell surface, they can act as negative regulators of the fibrinolytic system (Redlitz et al., 1995).

    The actual activity of the PAs is regulated not only by their concentration and activation, but also by their interaction with PA inhibitors (PAIs) (Sprengers and Kluft, 1987; Bachmann, 1995; Loskutoff, 1991), cellular receptors (Miles and Plow, 1988; Barnathan, 1992; Hajjar, 1995) and, as indicated above, matrix proteins.

Inhibitors

The activities of the proteases of the fibrinolytic system are controlled by potent inhibitors, which are members of the serine protease inhibitor (serpin) superfamily. Plasmin, if not bound to fibrin, is instantaneously inhibited by α2-antiplasmin (Holmes et al., 1987), but this reaction is attenuated when plasmin is bound to fibrin. The predominant regulators of tPA and uPA activities are PAI-1 and PAI-2. PAI-1 is a 50 kD glycoprotein present in blood platelets and synthesized by endothelial cells, smooth muscle cells and many other cell types in culture (Sprengers and Kluft, 1987; Loskutoff, 1991). PAI activity in human plasma is normally exclusively PAI-1. PAI-1 is also the main if not the sole inhibitor of PAs synthesized by endothelial cells and vascular smooth muscle cells. PAI-2 is produced by monocytes/macrophages and placental trophoblasts. It can be found as a glycosylated secreted molecule and as non-glycosylated intracellular molecule (Wohlwend et al., 1987).

Receptors

Regulation of fibrinolytic activity also occurs by cellular receptors. These receptors direct the action of PAs and plasmin to focal areas on the cell surface, or are involved in the clearance of the PAs. High affinity binding sites for plasminogen (Miles et al., 1988; Plow et al., 1991; Nachman, 1992; Hajjar, 1995), tPA (Hajjar, 1991, 1995) and uPA (Vassalli, 1994; Blasi et al., 1994; Danø et al., 1994) are found on various types of cells including endothelial cells.

Endothelial cells *in vitro* bind plasminogen with a moderate affinity (120 to 340 nM depending on whether the Lys or Glu form of plasminogen is used) but with a high capacity (3.9 to 14×10^5 molecules per cell) (Hajjar and Nachman, 1988; Plow et al., 1991). This binding, which is also observed with many other cell types, is mediated by the lysine binding sites of kringles 1–3 of the plasmin(ogen) molecule. Because these lysine binding sites are also involved in the interaction of plasmin with α_2-antiplasmin, occupation of lysine binding sites protects plasmin from inhibition by α_2-antiplasmin not only when plasmin is bound to fibrin (see above) but also when it is bound to the cellular receptors. The nature of the plasminogen receptors is not fully resolved. In addition to gangliosides, which directly or indirectly contribute to the plasminogen binding (Miles et al., 1989 a), at least eight proteins have been reported to be involved in plasminogen binding. Among them are members of the low density lipoprotein receptor family, such as gp330 and LRP; and annexin II, a not yet identified 45 kD protein, GbIIb/IIIa and α-enolase (see Hajjar, 1995 for review). In neural cells plasminogen binding to amphoterin was found. Lipoprotein Lp(a), which has strong structural homology with a large part of the plasminogen molecule, can compete for plasminogen binding to endothelial cells (Miles et al., 1989b; Nachman, 1992).

Specific binding of tPA to human endothelial cells has been reported (Hajjar et al., 1987; Beebe, 1987; Barnathan et al., 1988). tPA binds via its growth factor domain with high affinity to annexin II on human endothelial cells in culture (Hajjar et al., 1994). Annexin II expression on cultured endothelial cells is under metabolic control and is decreased after exposure of the cells to retinoic acid (Hajjar, 1995). Different epitopes of the annexin II molecule are involved in plasminogen and tPA binding. It is conceivable that annexin II, like fibrin, forms a ternary complex with tPA and plasminogen on the endothelial cells surface (Hajjar, 1995). In addition, tPA interacts with matrix-bound PAI-1 (Barnathan et al., 1988).

Binding of uPA to the cell surface limits plasminogen activation to focal areas such as the cellular protrusions involved in cell migration and invasion. Furthermore, uPA interaction with the cell can evoke signal transduction and specific phosphorylation of proteins (Dumler et al., 1993; Rao et al., 1995). A specific uPA receptor has been identified and cloned. It is present on many cell types, including endothelial cells (Barnathan, 1992). It is a glycosyl phosphatidylinositol(GPI)-anchored glycoprotein (Danø et al., 1994), which binds both single-chain uPA and two-chain uPA via their growth factor domain (Appella et al., 1989). The uPA receptor is heavily glycosylated. It belongs to the cysteine-rich cell surface proteins. After synthesis it is proteolytically processed at its carboxy-terminus and subsequently anchored in the plasma membrane by a GPI group (Ploug et al., 1991). It comprises three domains, which are structural homologous to snake venom α-toxins (Danø et al., 1994). The uPA receptor has been found in focal attachment sites, where integrin-matrix interactions occur, and in cell-cell contact areas (Pöllänen et al., 1988; Conforti et al., 1994). Human endothelial cells *in vitro* contain about 140000 uPA receptors per cell (Haddock et al., 1991).

The uPA receptor acts both as a site for focal pericellular proteolysis by uPA and as a clearance receptor for the uPA:PAI-1 complex. Upon secretion, single-chain uPA binds to the uPA receptor and is subsequently converted to the proteolytically active two-chain uPA. Since the endothelial cell also contains plasmin(ogen) receptors, an interplay between receptor-bound uPA and receptor-bound plasmin(ogen), and plasmin formation is likely to happen. The generated plasmin can degrade a number of matrix proteins. In addition, a direct plasmin-independent proteolytic action of uPA on matrix proteins may also occur (Quigley et al., 1987). Like free uPA activity, receptor-bound two-chain uPA is subject to inhibition by PAI-1. As a consequence uPA is only active over a short period of time. In contrast to receptor-bound single-chain or non-inhibited two-chain uPA, the uPA:PAI-1 complex is rapidly internalised together with the uPA receptor (Olson et al., 1992), followed by degradation of the uPA:PAI-1 complex and return of the empty uPA receptor to the plasma membrane. Internalisation of the GPI-linked uPA receptor probably occurs after interaction with (an)other receptor(s), such as the α_2-macroglobulin/low density lipoprotein receptor-related protein (LRP) (Nykjær et al., 1992). Recently it has been found that the uPA receptor may have an additional role. The uPA receptor occupied by uPA interacts avidly with vitronectin (Wei et al., 1994). Hence, cell adhesion may represent an additional function of the uPA receptor.

The LRP and LRP-like proteins on liver hepatocytes are involved in the clearance of plasmin-α2-antiplasmin, PA:PAI-1 complexes (Orth et al., 1992; Bu et al., 1992, 1994) and probably free PAs from the circulation. In addition, tPA can also be cleared by mannose receptors present on macrophages and liver endothelial cells (Otter et al., 1991) and by α-fucose receptors on hepatocytes (Hajjar and Reynolds, 1994).

Role of tPA in the prevention of intravascular fibrin deposition

The fibrinolytic activity in blood is largely determined by the concentration of tPA, which is synthesized by the endothelium (Rijken et al., 1980; Wun and Capuano, 1985). After release into the blood stream, tPA resides only for a short period in the blood, unless it encounters a specific binding site, in particular fibrin. This is due to the short half life of tPA in the circulation, which is 5 to 10 minutes in humans. This rapid clearance, which occurs in the liver, and the ability of endothelial cells to release a relatively large amount of tPA immediately after exposure to vasoactive agents make that the plasma concentration of tPA can change rapidly (Emeis, 1992).

Endothelial cells contain a storage pool of tPA, which can be rapidly released after exposure of the cells to vasoactive substances, such as bradykinin, platelet activating factor and thrombin (Emeis, 1992; van den Eijnden-Schrauwen et al., 1995, 1996). Intracellularly stored tPA has been demonstrated in small vesicles (van den Eijnden-Schrauwen et al., 1996), which are different from the Weibel-Palade bodies, the storage organelles for von Willebrand factor and P-selectin (Bonfanti et al., 1989). The acute tPA release mechanism produces an enhanced tPA concentration, exclusively at those sites of the vascular system where fibrin deposition is pending or occurs. Hence, it contributes to the local protection against an emerging unwanted thrombus. If a generalized stimulation of the endothelium occurs, for example by catecholamines or after dDAVP infusion, the acute release mechanism causes a rapid temporary increase in the systemic blood tPA concentration. Recent studies have demonstrated that under the proper experimental conditions, acute tPA release can also be demonstrated in cultured endothelial cells. This offers the possibility of unravelling the regulation of this process (Schrauwen et al., 1994; van den Eijnden-Schrauwen et al., 1995, 1996).

The size of the intracellular tPA pool depends on the rate of tPA synthesis. Hence, influencing tPA synthesis can influence both constitutive tPA production and the amount of acutely released tPA. Factors that influence the synthesis of tPA include activators of protein kinase C, retinoids and certain triazolobenzodiazepines. The data are summarized in a recent review by Kooistra et al. (1994). The regulation of tPA gene expression is independent of that of its inhibitor PAI-1, although certain mediators, such as thrombin, can induce both tPA and PAI-1 synthesis. Hence, the endothelial cell is able both by specifically releasing tPA from a storage pool and by differential regulation of tPA and PAI-1 expression to modulate fibrinolysis in the blood compartment.

Regulation of plasminogen activation and matrix-degrading metalloproteinase (MMP) production by inflammatory mediators

In a number of diseases PAI-1 is elevated in the blood (sepsis, mature onset diabetes, postoperative thrombosis) and/or in tissues (arteriosclerosis, sepsis). Several factors involved in inflammatory and vascular diseases, such as the cytokines tumor necrosis factor-α (TNF-α) and interleukin-1 (IL-1), endotoxin (LPS), transforming growth factor-β (TGF-β), oxidized lipoproteins and thrombin, can stimulate PAI-1 production in endothelial cells *in vitro*. Also *in vivo*, administration of TNF-α, IL-1, LPS or thrombin causes an increase in PAI-1 concentration in the circulation. After infusion of LPS in animals, PAI-1 mRNA increased in vascularized tissues and was elevated in the endothelium of various organs (Quax et al., 1990; Keeton et al., 1993). Administration of LPS or TNF-α to patients or healthy volunteers caused a large increase in blood PAI-1 levels after about 2 h, preceded by a rapid and sustained increase in tPA concentration (Suffredini et al., 1989; van Hinsbergh et al., 1990b; van Deventer et al., 1990). The mechanism underlying the stimulation of tPA synthesis *in vivo* by LPS or TNF-α is still unresolved. The rapid increase is probably due to indirect stimulation of the acute release of tPA. The continuous increased release of tPA is likely caused by the induction of tPA gene transcription. It is not yet certain whether this occurs directly through TNF-α, as has been observed in human microvascular but not in macrovascular endothelial cells (van Hinsbergh et al., 1988; Koolwijk et al., 1996), or may be the indirect result of the LPS or TNF-α infusion via the generation of another mediator *in vivo*. The large increase in PAI-1 production observed 2 h after TNF-α or LPS administration far exceeds that of tPA (Suffredini et al., 1989; van Hinsbergh et al., 1990b; van Deventer et al., 1990). This may result – after an initial rise in fibrinolytic activity – in a prolonged attenuation of the fibrinolysis process. It is generally believed that induction of PAI-1 by inflammatory mediators may contribute to the thrombotic complications in endotoxinemia and sepsis.

The inflammatory mediators TNF-α, IL-1 and LPS also elicit another effect on the regulation of plasminogen activator production in endothelial cells. At the same time as the increase in PAI-1 and tPA, these inflammatory mediators induce the synthesis of uPA in human endothelial cells (van Hinsbergh et al., 1990 a). Induction of uPA by TNF-α is associated with an increased degradation of matrix proteins (Niedbala and Stein Picarella, 1992). The enhanced secretion of uPA occurs entirely towards the basolateral side of the cell, whereas the secretion of tPA and PAI-1 proceeds equally to the luminal and basolateral sides of the cell (van Hinsbergh et al., 1990 a). The polar secretion of uPA suggests that uPA may be involved in local remodeling of the basal matrix of the cell. The increase of PAI-1 induced by inflammatory mediators may represent, in addition to a role in the modulation of fibrinolysis, a protective mechanism of the cell against uncontrolled uPA activity.

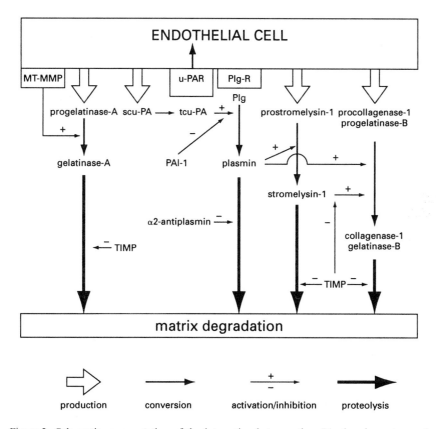

Figure 2. Schematic representation of the interaction between the uPA-plasmin system and the presumed activation of matrix-degrading metalloproteinases (MMPs). Abbreviations: PA: plasminogen activator; uPA: urokinase type PA; scuPA: single-chain uPA; tcuPA: two-chain uPA; uPAR: uPA receptor; Plg: plasminogen: Plg-R: Plg receptor: PAI-1: PA inhibitor-1; MT-MMP: membrane-type MMP; TIMP: tissue inhibitor of MMP. +: stimulation; –: inhibition.

In line with a putative role for TNF-α and IL-1 inflammation-induced local pericellular proteolysis is the observation that TNF-α also increases the production of matrix-degrading metalloproteinases (MMPs) by endothelial cells. In human microvascular and vein endothelial cells, TNF-α increases the mRNA levels and the synthesis of interstitial collagenase (MMP-1), stromelysin-1 (MMP-3) and – if protein kinase C is also activated – gelatinase-B (MMP-9), whereas the mRNA levels and synthesis of their physiological inhibitors TIMP-1 and TIMP-2 are not changed significantly (Hanemaaijer et al., 1993; Cornelius et al., 1995). Furthermore, activation of gelatinase-A (MMP-2) was observed after exposure of the cells to TNF-α (Hanemaaijer et al., 1993). This suggests an increase of membrane-type MMP (MT-MMP) activity. Interestingly, secretion of gelatinases by bovine endothelial cells occurs predominantly towards the basolateral side of the cells (Unemori et al., 1990) similar to the TNF-α-induced production of uPA (van Hinsbergh et al., 1990a).

The plasmin-plasminogen activator system and the matrix metalloproteinases cooperate in the degradation of extracellular matrix proteins (Liotta et al., 1991). Figure 2 depicts the interaction between the two systems. It should be noted, however, that this schematic picture is based on *in vitro* data, and that it is still uncertain whether all the depicted steps also act *in vivo*. Nevertheless, it will be clear from the foregoing that activation of endothelial cells by TNF-α affects multiple sites in the proteolytic cascades involved in the degradation of matrix proteins, markedly enhancing the breakdown and remodeling of the endothelial cell basal membrane.

Role of plasminogen activators in tissue remodeling and cell migration

Limitation of the action radius of plasminogen activators

The induction and activation of plasmin and matrix-degrading metalloproteinases require the cell to protect itself against excessive proteolytic activity. In addition to a tightly controlled activation of the proteases, spreading of proteolytic activity into the cell environment is restricted by two mechanisms. First, once the protease is activated, the proteolytic activity is rapidly neutralized by specific inhibitors. Second, specific cellular receptors and specific interaction sites on the substrate bind the zymogens and active forms of the various proteases. Bound proteases are often protected against their inhibitors. In this manner binding both limits the proteolytic activity to defined sites, and delays the interaction of the protease with its inhibitor(s). Figure 3 depicts schematically the release, interaction and clearance of tPA and uPA. In particular uPA acts in the immediate vicinity of the cell that has synthesized it, i.e. the cell itself or an adjacent cell.

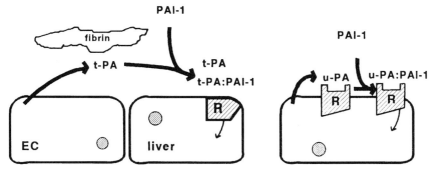

Figure 3. Schematic presentation of the synthesis and cellular uptake of plasminogen activators (PAs). Left: fibrinolysis; right: pericellular proteolysis. After secretion the PA binds to a matrix or cellular receptor (R) where it becomes active and performs its proteolytic activity until it is inhibited by its inhibitor PAI-1. The PA:PAI-1 complex is subsequently internalized by the same or another cell. PAs that do not interact with a suitable matrix or receptor are cleared by the liver. EC: endothelial cell.

Involvement of uPA and uPA receptor in cell invasion and migration

Concentration of uPA activity at the cellular protrusions of invading cells has been frequently observed (Danø et al., 1985; Estreicher et al., 1990). Blasi (1993) hypothesized that a continuous activation and removal of uPA bound to the receptor could contribute to the formation and detachment of focal attachment sites and hence to locomotion of the cell. Indeed, migrating and invading cells, such as monocytes and tumor cells, express uPA activity bound to uPA receptors on their cellular protrusions (Estreicher et al., 1990) and on focal attachment sites (Pöllänen et al., 1988; Hébert and Baker, 1988). Receptor-bound uPA activity is also thought to be involved in smooth muscle and endothelial cell migration and in the formation of new blood vessels (angiogenesis). Inhibition of plasminogen activation interferes with smooth muscle cell migration *in vitro* (Schleef and Birdwell, 1982) and affects smooth muscle cell migration and proliferation *in vivo* (Clowes et al., 1990). In mice lacking uPA, intimal hyperplasia of injured arteries is less pronounced than in wild-type or tPA-deficient mice (Carmeliet and Collen, 1994). Moreover, PAI-1-deficient mice show an exacerbated intimal proliferation (Carmeliet and Collen, 1994). These data suggest a role for uPA in cell recruitment.

An involvement of uPA and the uPA receptor in endothelial cell migration has been demonstrated by Pepper et al. (1987, 1993) and Sato et al. (1988) using bovine adrenal microvascular endothelial cells. After wounding of a monolayer of these endothelial cells, the surrounding cells migrate into the wounded area. These migrating cells display uPA activity (Pepper et al., 1987), which is bound to uPA receptor (Pepper et al., 1993). It was shown by Sato et al. (1988) that the migration and expression of uPA depends on the release of bFGF from the wounded area. bFGF is a potent inducer of plasminogen activator activity, in particular uPA, in bovine endothelial cells (Saksela et al., 1987; Gualandris and Presta, 1995). bFGF also increases the number of uPA receptors on endothelial cells (Mignatti et al., 1991; Pepper et al., 1993). It should be noted that in human aorta endothelial cells the effect of bFGF on cell migration was considerably less. IL-8 and TNF-α were better inducers of cell migration in the latter cells (Szekanecz et al., 1994). The relatively poor response to the sole addition of bFGF may be due to the fact that bFGF, while it induces uPA receptor (Mignatti et al., 1991; Koolwijk et al., 1996), is unable to enhance uPA production in human endothelial cells (Koolwijk et al., 1996).

Receptor-bound uPA was found in focal adhesion sites (Hébert and Baker, 1988; Conforti et al., 1994). It may act proteolytically on these structures and hence influence cell-matrix interactions and cell migration. Alternatively, it has been suggested that uPA can act on cell migration without involvement of its proteolytic activity (Odekon et al., 1992), either by uPA receptor-dependent signal transduction (Dumler et al., 1993; Rao et al., 1995) or by interaction of the occupied uPA receptor with vitronectin (Wei et al., 1994).

Regulation of the uPA receptor by angiogenic growth factors

In addition to bFGF, two other angiogenic factors – acidic FGF (aFGF) and vascular endothelial cell growth factor (VEGF) – induce an increase in the number of uPA receptors (Mandriota et al., 1995; Koolwijk et al., 1996). The number of uPA receptors on human and bovine endothelial cells is also enhanced by activation of protein kinase C and by elevation of the cellular cAMP concentration (Langer et al., 1993; van Hinsbergh, 1992). Preliminary experiments in our laboratory have demonstrated that the induction of uPA receptor by VEGF in human endothelial cells is inhibited by protein kinase C inhibition. The effects of bFGF and VEGF on the induction of uPA receptor in human endothelial cells are regulated independently of their effects on cell proliferation. Similarly, Presta et al. (1989) have shown that the induction of uPA by bFGF in bovine endothelial cells proceeds independent of the stimulation of mitogenesis by this growth factor.

In addition to FGFs and VEGF, which induce mitogenesis, TNF-α can also induce angiogenesis, but this occurs without stimulation of cell proliferation (Leibovich et al., 1987; Fratèr-Schröder et al., 1987). TNF-α increases uPA receptor in human microvascular endothelial cells (Koolwijk et al., 1996) and in monocytes (Kirchheimer et al., 1989), but non in endothelial cells from human umbilical vein or aorta (Koolwijk et al., 1996). However, simultaneous exposure of the latter cells to TNF-α (which induces uPA synthesis), to bFGF and to VEGF (which enhance the expression of uPA receptors), potently increases cell-bound uPA activity.

Role of the plasmin/plasminogen activator system in the formation of endothelial tubes in a fibrin matrix

Proteolysis of the basement membrane of endothelial cells and invasion of endothelial cells into the underlying matrix are prerequisites for angiogenesis (Ausprunk and Folkman, 1977). The most simple form of angiogenesis is the recanalisation of a fibrin clot by invading endothelial cells. This invasion is usually preceded by infiltration of inflammatory cells, which interact with the ingrowing capillaries (Kwaan, 1966). A three dimensional fibrin matrix model was used by Pepper and Montesano to demonstrate a direct correlation between the expression of PA activity and the formation of capillary sprouts by bovine microvascular endothelial cells *in vitro* (Pepper et al., 1990; Montesano, 1992). The outgrowth of tubular structures was increased by bFGF, which increases both uPA activity and uPA receptor in bovine endothelial cells. Interestingly, the extent of tube formation and the diameter of the tubes formed were reduced by simultaneous presence of TGF-β (Pepper et al., 1990). The latter is a growth factor, which strongly enhances PAI-1 synthesis in cultured endothelial cells, and thus inhibits PA activity (Saksela et al., 1987).

Studies in human endothelial cells showed that no or only a limited number of tubular structures are formed when a quiescent monolayer of human microvascular endothelial cells grown on a fibrin matrix is exposed to bFGF. However, when bFGF and TNF-α are added simultaneously, a large number of capillary-like tubular structures are formed after 10 to 14 days (Fig. 4). These data agree with the data on bovine endothelial cells, except that in human endothelial cells a second mediator is required to induce uPA synthesis. In animals bFGF and aFGF have been shown to induce neovascularisation (Thompson et al., 1988; Broadley et al., 1989). Because it is difficult to rule out the involvement of a limited number of leukocytes in these experiments *in vivo*, it remains to be elucidated whether these growth factors act *in vivo* in concert with inflammatory mediators or independently of the latter mediators.

In addition to bFGF, VEGF can also stimulate bovine adrenal microvascular endothelial cells to form tubular structures. It acts cooperatively with bFGF in this induction (Pepper et al., 1992). Similar results were obtained in human endothelial cells, if TNF-α was also present or when uPA was supplied to the cells exogenously (Koolwijk et al., 1996). The outgrowth of tubular structures requires uPA activity and is markedly reduced by anti-uPA antibodies or by inhibiting the interaction of uPA with its receptor. Furthermore, proteolytic activation of plasminogen appears to be involved, because the plasmin inhibitor aprotinin largely inhibits the formation of tubular structures.

A recent paper of Nehls and Herrmann (1996) points to the importance of the fibrin structure in endothelial cell migration and the formation of tubular structures. These authors show that the rigidity of the fibrin gel has a strong

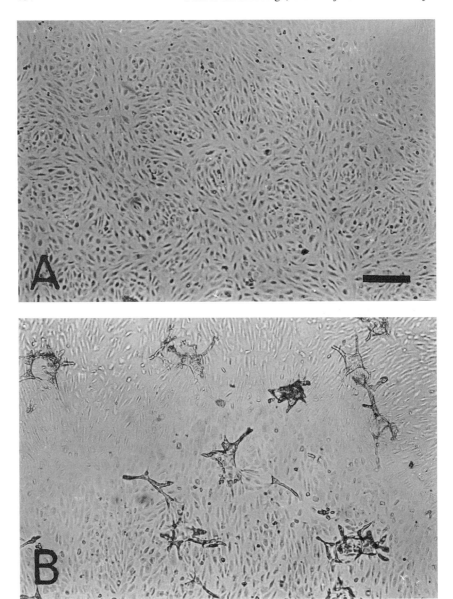

Figure 4. Formation of capillary-like tubular structures in a three dimensional fibrin matrix by human endothelial cells. A. Human microvascular endothelial cells grown under control conditions on top of a three dimensional fibrin matrix. B. Formation of tubular structures is induced by the simultaneous addition of the growth factor bFGF (20 ng/mL) and the cytokine TNF-α (20 ng/mL). The bar is 0.3 mm.

impact on tube formation by bovine endothelial cells in response to bFGF and VEGF. It remains to be determined whether this effect of the fibrin structure reflects a mechanical barrier to movement of cells by a dense fibrin network, or is due to an inadequate spacing of cell-binding epitopes in the fibrin network. It should be noted that fibrin contains cell binding domains for endothelial cells: a RGD sequence in its α-chains which binds to the vitronectin receptor, i.e. the $\alpha_v\beta_3$ integrin; and another site in the β-chain (residues 5–42) (Bunce et al., 1992), which binds to an 130 kD receptor (Erban and Wagner, 1992). Brooks et al. (1994a, b) have shown that inhibition of the $\alpha_v\beta_3$ integrin reduces angiogenesis in several *in vivo* models. These authors (Friedlander et al., 1995) recently reported that both $\alpha_v\beta_3$ and $\alpha_v\beta_5$ integrin are involved in growth factor-stimulated angiogenesis in the rabbit cornea, but via distinct mechanisms. It is of interest to note that bFGF-and TNF-α-induced angiogenesis is inhibited by an antibody against the $\alpha_v\beta_3$ integrin, whereas angiogenesis induced by VEGF or by a protein kinase C activating phorbol ester required $\alpha_v\beta_5$ integrin. It remains to be established whether both $\alpha_v\beta_3$- and $\alpha_v\beta_5$-dependent mechanisms are active in the invasion of endothelial cells into a fibrin matrix. The involvement of $\alpha_v\beta_3$ integrin interaction with the fibrin matrix is likely (Chang et al., 1995), while the role of $\alpha_v\beta_5$ integrin, which more selectively interacts with vitronectin, has to be evaluated. Irrespective of the exact role of fibrin in stimulating angiogenesis, the fibrin structure has important consequences for wound healing, and proteolytic modification of fibrin, e.g. by leukocytic elastase, or interaction of fibrin with other matrix proteins, such as vitronectin and fibronectin, may affect cell invasion and angiogenesis and the success of wound healing.

Perspective

The invasion of endothelial cells into a fibrin matrix is an attractive model to study the angiogenesis process associated with tissue repair. It is versatile because of its simplicity. However, it is also relatively simple in comparison with many pathological conditions. Probably it reflects most closely the recanalisation of a fibrin clot. More complex matrices containing collagen or mixtures of collagen and fibrin can be made and are used (Montesano, 1992; Pepper et al., 1994). These matrix models can be further adapted by inclusion of other cell types, such as pericytes, mast cells and leukocytes. Such models will be helpful in unraveling the complex cell-matrix interactions and proteolytic events that occur during the formation of new capillary structures. Together with *in vivo* data, they will contribute to the delineation of the various cellular pathways that are involved in angiogenesis. Knowledge of these pathways is essential to allow us to interfere selectively with unwanted angiogenesis, such as in rheumatoid arthritis and tumors, and to stimulate neovascularisation at sites where it is needed, such as in normal wound healing and collateral formation.

Acknowledgements
We would like to thank Dr Willem Nieuwenhuizen for critical reading of the manuscript and his valuable suggestions.

References

Aiello, L.P., Avery, R.L., Arrigg, P.G., Keyt, B.A., Jampel, H.D., Shah, S.T., Pasquale, L.R., Thieme, H., Iwamato, M.A., Park, J.E., Nguyen, H.V. and Aiello, L.M. (1994) Vascular endothelial cell growth factor in ocular fluid of patients with diabetic retinopathy and other retinal disorders. *New Engl. J. Med.* 331: 1480–1487.

Appella, E., Robinson, E.A., Ullrich, S.J., Stoppelli, M.P., Corti, A., Cassani, G. and Blasi, F. (1987) The receptor-binding sequence of urokinase: A biological function for the growth-factor module of proteases. *J. Biol. Chem.* 262: 4437–4440.

Ausprunk, D. and Folkman, J. (1977) Migration and proliferation of endothelial cells in performed and newly formed blood vessels during tumor angiogenesis. *Microvasc. Res.* 14: 52–65.

Bacharach, E., Itin, A. and Keshet, E. (1992) *In vivo* patterns of expression of urokinase and its inhibitor PAL-1 suggest a concerted role in regulating physiological angiogenesis. *Proc. Natl. Acad. Sci. USA* 89: 10686–10690.

Bachmann, F. (1995) The enigma PAI-2. Gene expression, evolutionary and functional aspects. *Thromb. Haemostas.* 74: 172–179.

Bachmann, F. (1987) Fibrinolysis. *In:* M. Verstraete, J. Vermylen, R. Lijnen and J. Arnout (eds): Thrombosis and Haemostasis 1987. Leuven University Press, Leuven, pp 277–265.

Barnathan, E.S., Kuo, A., Van der Keyl, H., McCrae, K.R., Larsen, G.R. and Cines, D.B. (1988) Tissue-type plasminogen activator binding to human endothelial cells. Evidence for two distinct sites. *J. Biol. Chem.* 263: 7792–7799.

Barnathan, E.S. (1992) Characterization and regulation of the urokinase receptor of human endothelial cells. *Fibrinolysis* 6: 1–9.

Beebe, D.P. (1987) Binding of tissue plasminogen activators to human umbilical vein endothelial cells. *Thromb. Res.* 46: 241–254.

Blasi, F. (1993) Urokinase and urokinase receptor – A paracrine/autocrine system regulating cell migration and invasiveness. *Bioessays* 15: 105–111.

Blasi, F., Conese, M., Møller, L.B., Pedersen, N., Cavallaro, U., Cubellis, M., Fazioli, F., Hernandez-Marrero, L., Limongi, P., Munoz-Canoves, P., Resnati, M., Riittinen, L., Sidenius, N., Soravia, E., Soravia, M., Stoppelli, M., Talarico, D., Teesalu, T. and Valcamonica, S. (1994) The urokinase receptor: Structure, regulation and inhibitor-mediated internalization. *Fibrinolysis* 8: 182–188.

Bonfanti, R., Furie, B.C., Furie, B. and Wagner, D.D. (1989) PADGEM (GMP140) is a component of Weibel-Palade bodies of human endothelial cells. *Blood* 73: 1109–1112.

Broadley, K.N., Aquino, A.M., Woodward, S.C., Buckley-Sturrock, A., Sato, Y., Rifkin, D. and Davidson, J.M. (1989) Monospecific antibodies implicate basic fibroblast growth factor in normal wound repair. *Lab. Invest.* 61: 571–575.

Brooks, P.C. Clark, R.A. and Cheresh, D.A. (1994a) Requirement of vascular integrin $\alpha_v \beta_3$ for angiogenesis. *Science* 264: 569–571.

Brooks, P.C., Montgomery, A.M., Rosenfeld, M., Reisfeld, R.A., Hu, T., Klier, G. and Cheresh, D.A. (1994b) Integrin $\alpha_v \beta_3$ antagonists promote tumor regression by inducing apoptosis of angiogenic blood vessels. *Cell* 79: 1157–1164.

Bu, G., Williams, S., Strickland, D.K. and Schwartz, A.L. (1992) Low density lipoprotein receptor-related protein/α_2-macroglobulin receptor is hepatic receptor for tissue-type plasminogen activator. *Proc. Natl.. Acad. Sci. USA* 89: 7427–7431.

Bu, G., Warshawsky, I. and Schwartz, A.L. (1994) Cellular receptors for the plasminogen activators. *Blood* 83: 3427–3436.

Bugge, T.H., Suh, T.T., Flick, M.J., Daugherty, C.C., Rømer, J., Solberg, H., Ellis, V., Danø, K. and Degen, J.J. (1995) The receptor for urokinase-type plasminogen activator is not essential for mouse development or fertility. *J. Biol. Chem.* 270: 16886–16894.

Bunce, L.A., Sporn, L.A. and Francis, C.W. (1992) Endothelial cell spreading on fibrin requires fibrinopeptide B cleavage and amino acid residues 15–42 of the β-chain. *J. Clin. Invest.* 89: 842–850.

Carmeliet, P. and Collen, D. (1994) Evaluation of the plasminogen/plasmin system in transgenic mice. *Fibrinolysis* 8: 269–276.

Carmeliet, P., Kieckens, L., Schoonjans, L., Ream, B., Van Nuffelen, A., Prendergast, G., Cole, M., Bronson, R., Collen, D. and Mulligan, R.C. (1993a) Plasminogen activator inhibitor-1 gene-deficient mice. I. Generation by homologous recombination and characterization. *J. Clin. Invest.* 92: 2746–2755.

Carmeliet, P., Stassen, J.M., Schoonjans, L., Ream, B., Van den Oord, J.J., De Mol, M., Mulligan, R.C., and Collen, D. (1993b) Plasminogen activator inhibitor-1 gene-deficient mice. II. Effects on hemostasis, thrombosis, and thrombolysis. *J. Clin. Invest.* 92: 2756–2760.

Carmeliet, P., Schoonjans, L., Kieckens, L., Ream, B., Degen, J., Bronson, R., De Vos, R., Van den Oord, J.J., Collen, D. and Mulligan R.E. (1994) Physiological consequences of loss of plasminogen activator gene function in mice. *Nature* 368: 419–424.

Chang, M.-C., Wang, B.-R. and Huang, T.-F. (1995) Characterization of endothelial cell differential attachment to fibrin and fibrinogen and its inhibition by Arg-Gly-Asp-containing peptides. *Thromb. Haemostas.* 74: 764–769.

Clowes, A.W., Clowes, M.M., Au, Y.P.T., Reidy, M.A. and Belin, D. (1990) Smooth muscle cells express urokinase during mitogenesis and tissue-type plasminogen activator during migration in injured rat carotid artery. *Circ. Res.* 67: 61–67.

Colvin, R.B. (1986) Wound healing processes in hemostasis and thrombosis. *In:* M.A. Gimbrone Jr. (ed): *Vascular Endothelium in Hemostasis and Thrombosis*. Churchill Livingstone, Edinburgh, pp 220–241.

Conforti, G., Dominguez-Jimenez, C., Rønne, E., Høyer-Hansen, G. and Dejana, E. (1994) Cell-surface plasminogen activation causes a retraction of *in vitro* cultured human umbilical vein endothelial cell monolayer. *Blood* 83: 994–1005.

Cornelius, L.A., Nehring, L.C., Roby, J.D., Parks, W.C. and Welgus, H.G. (1995) Human dermal microvascular endothelial cells produce matrix metalloproteinases in response to angiogenic factors and migration. *J. Invest. Dermatol.* 105: 170–176.

Danø, K., Andreasen, P.A., Grøndahl-Hansen, J., Kristensen, P., Nielsen, L.S. and Skriver, L. (1985) Plasminogen activators in tissue degradation and cancer. *Adv. Cancer Res.* 44: 139–264.

Danø, K., Behrendt, N., Brünner, N., Ellis, V., Ploug, M. and Pyke, C. (1994) The urokinase receptor: Protein structure and role in plasminogen activation and cancer invasion. *Fibrinolysis* 8: 189–203.

Dumler, I., Petri, T. and Schleuning, W.-D. (1993) Interaction of urokinase-type plasminogen activator (uPA) with its cellular receptor (uPAR) induces phosphorylation on tyrosine of a 38 kDa protein. *FEBS Lett.* 322: 37–40.

Dvorak, H.F. (1986) Tumors: wounds that do not heal: similarities between tumor stroma generation and wound healing. *New Engl. J. Med.* 315: 1650–1659.

Dvorak, H.F., Nagy, J.A., Berse, B., Brown, L.F., Yeo, K.-T., Yeo, T.-K., Dvorak, A.M., Van De Water, L., Sioussat, T.M. and Senger, D.R. (1992) Vascular permeability factor, fibrin, and the pathogenesis of tumor stroma formation. *Ann. N.Y. Acad. Sci.* 667: 101–111.

Dvorak, H.F., Brown, L.F., Detmar, M. and Dvorak, A.M. (1995) Vascular permeability factor/vascular endothelial growth factor, microvascular hyperpermeability, and angiogenesis. *Am. J. Pathol.* 146: 1029–1039.

Emeis, J.J. (1992) Regulation of the acute release of tissue-type plasminogen activator from the endothelium by coagulation activation products. *Ann. N.Y. Acad Sci.* 667: 249–258.

Erban, J.K. and Wagner, D.D. (1992) A 130-kDa protein on endothelial cells binds to amino acids 15–42 of the Bβ chain of fibrinogen. *J. Biol. Chem.* 267: 2451–2458.

Estreicher, A., Mühlhauser, J., Carpentier, J.-L., Orci, L. and Vassalli, J.-D. (1990) The receptor for urokinase type plasminogen polarizes expression of the protease to the leading edge of migrating monocytes and promotes degradation of enzyme inhibitor complexes. *J. Cell. Biol.* 111: 783–792.

Folkman, J. and Shing, Y. (1992) Angiogenesis. *J. Biol. Chem.* 267:10931–10934.

Fong, G.-H., Rossant, J., Gertsenstein, M. and Breitman, M.L. (1995) Role of the flt-1 receptor tyrosine kinase in regulating the assembly of vascular endothelium. *Nature* 376: 66–70.

Fratèr-Schröder, M., Risau, W., Hallman, R., Gautschi, P. and Böhlen, P. (1987) Tumor necrosis factor type α, a potent inhibitor of endothelial cell growth *in vitro*, is angiogenic *in vivo*. *Proc. Natl. Acad. Sci. USA* 84:5277–5281.

Friedlander, M., Brooks, P.C., Shaffer, R.W., Kincais, C.M., Varner J.A. and Cheresh, D.A. (1995) Definition of two angiogenic pathways by distinct α_v integrins. *Science* 270: 1500–1502.

Gualandris, A. and Presta, M. (1995) Transcriptional and posttranscriptional regulation of uro-kinase-type plasminogen activator expression in endothelial cells by basic fibroblast growth factor. *J. Cell. Physiol.* 162:400–409.

Haddock, R.C., Spell, M.L., Baker III, C.D., Grammer, J.R., Parks, J.M., Speidel, M. and Booyse, F.M. (1991) Urokinase binding and receptor identification in cultured endothelial cells. *J. Biol. Chem.* 266:21466–21473.

Hajjar, K.A. (1991) The endothelial cell tissue plasminogen activator receptor. Specific inter-action with plasminogen. *J. Biol. Chem.* 266:21962–21970.

Hajjar, K.A. (1995) Cellular receptors in the regulation of plasmin generation. *Thromb. Haemostas.* 74:294–301.

Hajjar, K.A. and Nachman, R.L. (1988) Endothelial cell-mediated conversion of Glu-plasmino-gen to Lys-plasminogen. Further evidence for assembly of the fibrinolytic system on the endothelial cell surface. *J. Clin. Invest.* 82:1769–1778.

Hajjar, K.A. and Reynolds, C. (1994) α-Fucose-mediated binding and degradation of tissue-type plasminogen activator by HepG2 cells. *J. Clin. Invest.* 93:703–710.

Hajjar, K.A., Hamel, N.M., Harpel, P.C. and Nachman, R.L. (1987) Binding of tissue plasmino-gen activator to cultured human endothelial cells. *J. Clin. Invest.* 80:1712–1719.

Hajjar, K., Jacovina, A. and Chacko, J. (1994) An endothelial cell receptor for plasminogen tissue plasminogen activator 1 identity with annexin II. *J. Biol. Chem.* 269:21191–21197.

Hanemaaijer, R., Koolwijk, P., Leclercq, L., De Vree, W.J.A. and Van Hinsbergh, V.W.M. (1993) Regulation of matrix metalloproteinase expression in human vein and microvascular endo-thelial cells – Effects of tumour necrosis factor-α, interleukin-1 and phorbol ester. *Biochem. J.* 296:803–809.

Hébert, C.A. and Baker, J.B. (1988) Linkage of extracellular plasminogen activator to fibroblast cytoskeleton: Colocalization of cell surface urokinase with vinculin. *J. Cell Biol.* 105: 1241–1247.

Hobson, B. and Denekamp, J. (1984) Endothelial proliferation in tumours and normal tissues: Continuous labelling studies. *Br. J. Cancer* 49:405–413.

Holmes, W.E., Nelles, L., Lijnen, H.R. and Collen, D. (1987) Primary structure of human α_2-antiplasmin, a serine protease inhibitor (Serpin). *J. Biol. Chem.* 262:1659–1664.

Keeton, M., Eguchi, Y., Swadey, M., Ahn, C. and Loskutoff, D. (1993) Cellular localization of type 1 plasminogen activator inhibitor messenger RNA and protein in murine renal tissue. *Am. J. Pathol.* 142:59–70.

Kirchheimer, J.C., Nong, Y. and Remold, H.G. (1989) IFN-γ, tumor necrosis factor-α, and uro-kinase regulate the expression of urokinase receptors on human monocytes. *J. Immunol.* 141:4229–4234.

Kooistra, T., Schrauwen, Y. and Emeis, J.J. (1994) Regulation of endothelial cell t-PA synthesis and release. *Int. J. Hematol.* 59:233–255.

Koolwijk, P., Van Erck, M.G.M., De Vree, W.J.A., Vermeer, M.A., Weich, H.A., Hanemaaijer, R. and Van Hinsbergh, V.W.M. (1996) Cooperative effect of TNF-α, bFGF and VEGF on the formation of tubular structures of human microvascular endothelial cells in a fibrin matrix. Role of urokinase activity. *J. Cell Biol.* 132:1177–1188.

Kwaan, H.C. (1966) Tissue fibrinolytic activity studied by a histochemical method. *Fed. Proc.* 25:52–56.

Langer, D.J., Kuo, A., Kariko, K., Ahuja, M., Klugherz, B.D., Ivanics, K.M., Hoxie, J.A., Williams, W.V., Liang, B.T., Cines, D.B. and Barnathan, E.S. (1993) Regulation of the endothelial cell urokinase-type plasminogen activator receptor – Evidence for cyclic AMP-dependent and protein kinase-C dependent pathways. *Circ. Res.* 72:330–340.

Leibovich, S.J., Polverini, P.J., Shepard, H.M., Wiseman, D.M., Shively, V. and Nuseir, N. (1987) Macrophage-induced angiogenesis is mediated by tumour necrosis factor-*α*. *Nature* 329: 630–632.

Liotta, L.A., Steeg, P.S. and Stetler-Stevenson, W.G. (1991) Cancer metastasis and angiogenesis – An imbalance of positive and negative regulation. *Cell* 64:327–336.

Loskutoff, D.J. (1991) Regulation of PAI-1 gene expression. *Fibrinolysis* 5:197–206.

Mandriota, S., Seghezzi, G., Vassalli, J.-D., Ferrara, N., Wasi, S., Mazzieri, R., Mignatti, P. and Pepper, M. (1995) Vascular endothelial growth factor increases urokinase receptor expression in vascular endothelial cells. *J. Biol. Chem.* 270:9709–9716.

Mignatti, P., Mazzieri, R. and Rifkin, D.B. (1991) Expression of the urokinase receptor in vascular endothelial cells is stimulated by basic fibroblast growth factor. *J. Cell Biol.* 113:1193–1201.

Miles, L.A. and Plow, E.F. (1988) Plasminogen receptors: ubiquitous sites for cellular regulation of fibrinolysis. *Fibrinolysis* 2:61–71.

Miles, L.A., Levin, E.G., Plescia, J., Collen, D. and Plow, E.F. (1988) Plasminogen receptors, urokinase receptors, and their modulation on human endothelial cells. *Blood* 72:628–635.

Miles, L.A., Dahlberg, C.M., Levin, E.G. and Plow, E.F. (1989a) Gangliosides interact directly with plasminogen and urokinase and may mediate binding of these fibrinolytic components to cells. *Biochemistry* 28:9337–9343.

Miles, L.A., Fless, G.M., Levin, E.G. Scanu, A.M. and Plow, E.F. (1989b) A potential basis for the thrombotic risks associated with lipoprotein(a). *Nature* 399:301–303.

Montesano, R. (1992) Regulation of angiogenesis in vitro. *European J. Clin. Invest.* 22:504–515.

Nachman, R.L. (1992) Thrombosis and atherogenesis: molecular connections. *Blood* 79:1897–1906.

Nehls, V. and Herrmann, R. (1996) The configuration of fibrin clots determines capillary morphogenesis and endothelial cell migration. *Microvasc. Res.* 51:347–364.

Niedbala, M.J. and Stein-Picarella, M. (1992) Tumor necrosis factor induction of endothelial cell urokinase-type plasminogen activator mediated proteolysis of extracellular matrix and its antagonism by *γ*-interferon. *Blood* 779:679–687.

Nykjær, A., Petersen, C.M., Møller, B., Jensen, P.H., Moestrup, S.K., Holtet, T.L., Etzerodt, M., Thogersen, H.C., Munch, M., Andreasen, P.A. and Gliemann, J. (1992) Purified *α₂*-macroglobulin receptor/LDL receptor-related protein binds urokinase activator inhibitor type-1 complex – Evidence that the *α₂*-macroglobulin receptor mediates cellular degradation of urokinase receptor-bound complexes. *J. Biol. Chem.* 267:14543–14546.

Odekon, L.E., Sato, Y. and Rifkin, D.B. (1992) Urokinase-type plasminogen activator mediates basic fibroblast growth factor-induced bovine endothelial cell migration independent of its proteolytic activity. *J. Cell. Physiol.* 150:258–263.

Olson, D., Pöllänen, J., Høyer-Hansen, G., Rønne, E., Sakaguchi, K., Wun, T.-C., Appella, E., Danø, K. and Blasi, F. (1992) Internalization of the urokinase-plasminogen activator inhibitor type-1 complex is mediated by the urokinase receptor. *J. Biol. Chem.* 267:9129–9133.

Orth, K., Madison, E.L., Gething, M.-J., Sambrook, J.F. and Herz, J. (1992) Complexes of tissue-type plasminogen activator and its serpin inhibitor plasminogen-activator inhibitor type-1 are internalized by means of the low density lipoprotein receptor-related protein/*α₂*-macroglobulin receptor. *Proc. Natl. Acad. Sci. USA* 89:7422–7426.

Otter, M., Barrett-Bergshoef, M.M. and Rijken, D.C. (1991) Binding of tissue-type plasminogen activator by the mannose receptor. *J. Biol. Chem.* 266:13931–13935.

Pepper, M.S., Vassalli, J.-D., Montesano, R. and Orci, L. (1987) Urokinase-type plasminogen activator is induced in migrating capillary endothelial cells. *J. Cell Biol.* 105:2535–2541.

Pepper, M.S., Belin, D., Montesano, R., Orci, L. and Vassalli, J. (1990) Transforming growth factor-*β*-1 modulates basic fibroblast growth factor induced proteolytic and angiogenic properties of endothelial cells *in vitro*. *J. Cell Biol.* 111:743–755.

Pepper, M.S., Ferrara, N., Orci, L. and Montesano, R. (1992) Potent synergism between vascular endothelial growth factor and basic fibroblast growth factor in the induction of angiogenesis *in vitro*. *Biochem. Biophys. Res. Comm.* 189:824–831.

Pepper, M.S., Sappino, A.-P., Stocklin, R., Montesano, R., Orci, L. and Vassalli, J.-D. (1993) Upregulation of urokinase receptor expression on migrating endothelial cells. *J. Cell Biol.* 122:673–684.

Pepper, M.S., Vassalli, J.-D., Wilks, J.W., Schweigerer, L., Orci, L. and Montesano, R. (1994) Modulation of bovine microvascular endothelial cell proteolytic properties of inhibitors or angiogenesis. *J. Cell. Biochem.* 55:419–434.

Ploug, M., Behrendt, N., Lober, D. and Danø, K. (1991) Protein structure and membrane anchorage of the cellular receptor for urokinase-type plasminogen activator. *Seminars in Thromb. Haemostas.* 17:183–193.

Plow, E.F., Felez, J. and Miles, L.A. (1991) Cellular regulation of fibrinolysis. *Thromb. Haemostas.* 66:32–36.

Pöllänen, J., Hedman, K., Nielsen, L.S., Danø, K. and Vaheri, A. (1988) Ultrastructural localization of plasma membrane-associated urokinase-type plasminogen activator at focal contacts. *J. Cell Biol.* 106:87–95.

Polverini, P. (1989) Macrophage-induced angiogenesis – A review. *Macrophage-Derived Cell Regulatory Factors* 1:54–73.

Presta, M., Maier, J.A.M. and Ragnotti, G. (1989) The mitogenic signalling pathway but not the plasminogen activator-inducing pathway of basic fibroblast growth factor is mediated through protein kinase C in fetal bovine aortic endothelial cells. *J. Cell Biol.* 109:1877–1884.

Quax, P.H.A., Van den Hoogen, C.R., Verheijen, J.H., Padro, T., Zeheb, R., Gelehrter, T.D., Van Berkel, T.J.C., Kuiper, J. and Emeis, J.J. (1990) Endotoxin induction of plasminogen activator and plasminogen activator inhibitor type 1 mRNA in rat tissues *in vivo*. *J. Biol. Chem.* 265:15560–15563.

Quigley, J.P., Gold, L.I., Schwimmer, R. and Sullivan, L.M. (1987) Limited cleavage of cellular fibronectin by plasmin activator purified from transformed cells. *Proc. Natl. Acad. Sci. USA* 84:2776–2780.

Rao, N.K., Shi, G.-P. and Chapman, H.A. (1995) Urokinase receptor is a multifunctional protein: Influence of receptor occupancy on macrophage gene expression. *J. Clin. Invest.* 96:465–474.

Redlitz, A., Tan, A.K., Eaton, D.L. and Plow, E.F. (1995) Plasma carboxypeptidases as regulators of plasminogen system. *J. Clin. Invest.* 96:2534–2538.

Rijken, D.D., Wijngaards, G. and Welbergen, J. (1980) Relationship between tissue plasminogen activator and the activators in blood and vascular wall. *Thromb. Res.* 18:815–830.

Risau, W. (1990) Angiogenic growth factor. *Progress in Growth Factor Research* 2:71–79.

Saksela, O.D., Moscatelli, D. and Rifkin, D. (1987) The opposing effects of basic fibroblast growth factor and transforming growth factor beta on the regulation of plasminogen activator activity in capillary endothelial cells. *J. Cell. Biol.* 105:957–963.

Sappino, A.P., Huarte, J., Belin, D. and Vassali, J.-D. (1989) Plasminogen activators in tissue remodeling and invasion: mRNA localization in mouse ovaries and implanting embryos. *J. Cell Biol.* 109:2471–2479.

Sato, T.N., Tozawa, Y., Deutsch, U., Wolburg-Buchholz, K., Fujiwara, Y., Gendron-Maguire, M., Gridley, T., Wolburg, H., Risau, W. and Qin, Y. (1995) Distinct roles of the receptor tyrosine kinases tie-1 and tie-2 in blood vessel formation. *Nature* 376:70–74.

Sato, Y. and Rifkin, D.B. (1988) Autocrine activities of basic fibroblast growth factor: Regulation of endothelial cell movement, plasminogen activator synthesis, and DNA synthesis. *J. Cell Biol.* 107:1199–1205.

Schleef, R.R. and Birdwell, C.R. (1982) The effect of proteases on endothelial cell migration in vitro. *Exp. Cell Res.* 141:503–508.

Schrauwen, Y., De Vries, R.E.M., Kooistra, T. and Emeis, J.J. (1994) Acute release of tissue-type plasminogen activator (t-PA) from the endothelium; regulatory mechanisms and therapeutic target. *Fibrinolysis* 8(Suppl. 2):8–12.

Shalaby, F., Rossant, J., Yamaguchi, T.P., Gertsenstein, M., Wu, X.-F., Breitman, M.L. and Schuh, A.C. (1995) Failure of blood-island formation and vasculogenesis in flk-1-deficient mice. *Nature* 376:62–66.

Sprengers, E.D. and Kluft, C. (1987) Plasminogen activator inhibitors. *Blood* 69:381–387.

Suffredini, A.F., Harpel, P.C. and Parrillo, J.E. (1989) Promotion and subsequent inhibition of plasminogen activation after administration of intravenous endotoxin to normal subjects. *New Engl. J. Med.* 320:1165–1172.

Szekanecz, Z., Shah, M.R., Harlow, L.A., Pearce, W.H. and Koch, A.E. (1994) Interleukin-8 and tumor necrosis factor-alpha are involved in human aortic endothelial cell migration. *Pathobiology* 62:134–139.

Thompson, J.A., Anderson, K.D., DiPietro, J.M., Zwiebel, J.A., Zametta, M., Anderson, W.F. and Maciag, T. (1988) Site-directed neovessel formation in vivo. *Science* 241:1349–1352.

Unemori, E.N., Bouhana, K.S. and Werb, Z. (1990) Vectorial secretion of extracellular matrix proteins, matrix-degrading proteinases, and tissue inhibitor of metalloproteinases by endothelial cells. *J. Biol. Chem.* 265:445–451.

Valinsky, J.E., Reich, E. and Le Douarin, N.M. (1981) Plasminogen activator in the bursa of Fabricius: correlations with morphogenic remodeling and cell migrations. *Cell* 25:471–476.

Vassalli, J.-D. (1994) The urokinase receptor. *Fibrinolysis* 8:172–181.

Van den Eijnden-Schrauwen, Y., Atsma, D.E., Lupu, F., De Vries, R.E.M., Kooistra, T. and Emeis, J.J. (1996) Intracellular signalling pathways involved in the differential release of von Willebrand factor (vWF) and tissue-type plasminogen activator (tPA) from human endothelial cells. *Blood*, in press.

Van den Eijnden-Schrauwen, Y., Kooistra, T., De Vries, R.E.M. and Emeis, J.J. (1995) Studies on the acute release of tissue-type plasminogen activator from human endothelial cells *in vitro* and in rats *in vivo*: Evidence for a dynamic storage pool. *Blood* 85:3510–3517.

Van Deventer, S.J.H., Büller, H.R., Ten Cate, J.W., Aarden, L.A., Hack, E. and Sturk, A. (1990) Experimental endotoxemia in humans: analysis of cytokine release and coagulation, fibrinolytic, and complement pathways. *Blood* 76:2520–2526.

Van Hinsbergh, V.W.M. (1992) Impact of endothelial activation on fibrinolysis and local proteolysis in tissue repair. *Ann. N.Y. Acad. Sci.* 667:151–162.

Van Hinsbergh, V.W.M., Kooistra, T., Van den Berg, E.A., Princen, H.M. G., Fiers, W. and Emeis, J.J. (1988) Tumor necrosis factor increases the production of plasminogen activator inhibitor in human endothelial cells *in vitro* and in rats *in vivo*. *Blood* 72:1467–1473.

Van Hinsbergh, V.W.M., Van den Berg, E.A., Fiers, W. and Dooijewaard, G. (1990a) Tumor necrosis factor induces the production of urokinase-type plasminogen activator by human endothelial cells. *Blood* 75:1991–1998.

Van Hinsbergh, V.W.M., Bauer, K.A., Kooistra, T., Kluft, C., Dooijewaard, G., Sherman, M.L. and Nieuwenhuizen, W. (1990b) Progress of fibrinolysis during tumor necrosis factor infusions in humans. Concomitant increase in tissue-type plasminogen activator, plasminogen activator inhibitor type-I, and fibrin(ogen) degradation products. *Blood* 76:2284–2289.

Wallén, P. (1987) Structure and function of tissue plasminogen activator and urokinase. *In:* P.J. Castellino, P.J. Gaffney M.M. Samama, and A. Takada (eds): *Fundamental and Clinical Fibrinolysis*. Elsevier, Amsterdam, pp 1–18.

Wei, Y., Waltz, D., Rao, N., Drummond, R., Rosenberg, S. and Chapman, H. (1994) Identification of the urokinase receptor as cell adhesion receptor for vitronectin. *J. Biol. Chem.* 269:32380–32388.

Weinberg, J.B., Pippen, A.M.M. and Greenberg, C.S. (1991) Extravascular fibrin formation and dissolution in synovial tissue of patients with osteoarthritis and rheumatoid arthritis. *Arthritis Rheum.* 34:996–1005.

Wohlwend, A., Belin, D. and Vassalli, J.-D. (1987) Plasminogen activator-specific inhibitors produced by human monocytes/macrophages. *J. Exp. Med.* 165:320–339.

Wun, T.-C. and Capuano, A. (1985) Spontaneous fibrinolysis in whole human plasma. Identification of tissue activator-related protein as the major plasminogen activator causing spontaneous activity *in vitro*. *J. Biol. Chem.* 260:5061–5066.

Regulation of Angiogenesis
ed. by I.D. Goldberg & E.M. Rosen
© 1997 Birkhäuser Verlag Basel/Switzerland

Tumor angiogenesis: Functional similarities with tumor invasion

W. G. Stetler-Stevenson and M. L. Corcoran

Extracellular Matrix Pathology Section, Laboratory of Pathology, Division of Clinical Sciences, National Cancer Institute, National Institutes of Health, Bethesda, Maryland 20892, USA

Summary. In this review the functional relationship between tumor cell invasion and tumor-induced angiogenesis is discussed. Emphasis is placed on the similarities of the invasive phenotype in both processes and the common themes of $\alpha_V\beta_3$ expression and gelatinase A activity.

Introduction

It is now recognized that tumor-induced angiogenesis supports the growth, expansion and eventual metastasis of many human cancers. Angiogenesis and tumor dissemination are pathologic conditions that are similar to many physiologic conditions, such as trophoblast implantation or wound healing, in that all of these processes involve cellular invasion. Studies reveal that cellular invasion in all of these processes share functional similarities. Cell invasion depends on the coordination of cellular adhesion, matrix proteolysis and migration. We will briefly review these events and highlight similarities in tumor cell and endothelial cell invasion. These similar control points in angiogenesis and tumor cell invasion could serve as targets for future therapeutic intervention in the growth and spread of human cancer.

The expanded three step hypothesis

Cell invasion is dependent on a series of coordinated cell-cell and cell-matrix interactions (Stetler-Stevenson et al., 1993). Liotta and colleagues proposed the original three step hypothesis to describe tumor cell invasion (Liotta, 1986). This original proposal was very detailed and specific in the molecular events leading to tumor cell invasion. However, as an overall scheme for cell invasion the three steps of attachment, dissolution and locomotion remain valid today. We now recognize that the three step hypothesis can be generalized to all invasive cell types, although the specific molecular events may be quite different to those originally proposed for tumor cell invasion of the basement membrane (Stetler-Stevenson et al., 1993). It is not fair to hold the original three step hypothesis to the constraints of knowledge at that time, especially since we have learned a great deal about all of these steps since then (Furcht et al., 1994). This expanded three step hypothesis recognizes that directed cellular invasion is the result of a highly coordinated series of cell-matrix interactions that has three distinct phases: (1) modification of cell-cell contacts and establishing new cell-matrix contacts; (2) proteolytic modification of the extracellular matrix; (3) migration of the invasive cell through the proteolysed matrix to establish new matrix contacts. This begins a new cycle and these events are repeated over and over again. These events must be coordinated and integrated such that the leading edge of the invasive cell is forming new contacts with the extracellular matrix while the trailing edge is breaking previously formed cell-matrix contacts. Proteolysis of the extracellular matrix must be balanced and regulated in order to preserve the critical cell-matrix contacts that allow traction to occur. Furthermore, these events are not independent of one another. We now know that integrin mediated cell-matrix interactions can influence protease production and that protease activity can alter cell attachment and spreading (Ray and Stetler-Stevenson, 1994; Seftor et al., 1992). Understanding how these molecular events of cell invasion during tumor invasion and angiogenesis are coordinated may allow identification of common mechanisms that could be targeted by novel therapeutic interventions.

Endothelial cell adhesion during angiogenesis

Recent studies have implicated the vitronectin receptor in both angiogenesis and tumor cell invasion. This receptor is probably the most promiscuous of the integrin receptors and is commonly referred to by its subunit composition $\alpha_V\beta_3$ (Felding-Habermann and Cheresh, 1993). The $\alpha_V\beta_3$ receptor is elevated in malignant melanoma cells and expression has been correlated with invasive potential in both melanoma and glioblastoma (Albelda et al., 1990; Gehlsen et al., 1992; Gladson and Cheresh, 1991). Recent studies have demonstrated that expression of functional $\alpha_V\beta_3$ on the surface of CS-1 melanoma cells promotes spontaneous metastasis of these cells *in vivo* (Filardo et al., 1995). In addition, tumor cells may adhere to exposed collagen matrix using $\alpha_2\beta_1$ type integrin receptors (Klemke et al., 1994). Following adhesion there is collagenolysis that reveals cryptic sites for $\alpha_V\beta_3$ and $\alpha_V\beta_5$ binding (Davis, 1992; Montgomery et al., 1994b). This binding acts as a survival stimulus preventing the initiation of programmed cell death and allows cellular invasion to continue, but in the case of $\alpha_V\beta_5$ this requires the intervention of additional paracrine or autocrine factors. For example, human pancreatic carcinoma cells adhere and migrate on vitronectin via $\alpha_V\beta_5$ but these events require stimulation of the cells with EGF and activation of protein kinase C (Klemke et al., 1994).

The process of angiogenesis has also been shown to depend on $\alpha_V\beta_3$ mediated cell-matrix interactions (Brooks et al., 1994a, 1994b, 1995).

Integrin $\alpha_V\beta_3$ is highly expressed on the surface of endothelial cells involved in active angiogenesis in response to bFGF stimulation in the chick chorioallantoic membrane assay (Brooks et al., 1994a). When this interaction is disrupted through the use of antagonists of the $\alpha_V\beta_3$ receptor, the endothelial cells undergo apoptosis and angiogenesis is halted. This can result in tumor regression due to disruption of the angiogenic blood vessels (Brooks et al., 1994b, 1995). Recent work has shown that the type of angiogenic stimulus may influence the integrin involved in mediating the endothelial cell invasion and the subsequent autonomy of this process from other growth factors (Friedlander et al., 1995).

Thus, expression of $\alpha_V\beta_3$ on tumor cells or endothelial cells promotes their attachment to partially proteolysed collagen via cryptic RGD binding sites, which in turn promotes survival of these cells and continuation of the invasive process. The work of Seftor and colleagues (Seftor et al., 1992) demonstrates that ligation of the $\alpha_V\beta_3$ receptor may influence protease expression in some tumor cell types. Experimental treatment of human A375M melanoma cells with antibodies against $\alpha_V\beta_3$ resulted in enhanced *in vitro* invasive ability of these cells. This was at least in part due to enhanced expression of matrix metalloproteinase activity, specifically gelatinase A (MMP-2). These findings demonstrate that the first two steps of cell invasion, that is cell-matrix attachment and extracellular matrix proteolysis, are linked with integrin binding influencing protease expression. This may function to coordinate cell attachment with matrix proteolysis.

Extracellular matrix proteolysis

Another functional link between angiogenesis and tumor cell invasion is the demonstrated requirement for extracellular matrix proteolysis in both processes. Specifically, both urokinase type plasminogen activator (uPA) and the matrix metalloproteinase-2 (MMP-2), also known as gelatinase A (formerly 72 kDa type IV collagenase), have been functionally linked to both processes (Ito et al., 1995; Mignatti and Rifkin, 1993; Stetler-Stevenson et al., 1993). The most compelling findings are the close correlation between expression of these enzymes and cellular invasion in both processes, and the use of selective protease inhibitors to block tumor cell invasion as well as angiogenesis (Mignatti and Rifkin, 1993; Stetler-Stevenson et al., 1993). Agents which induce angiogenesis, such as bFGF, will induce endothelial expression of both urokinase and plasminogen activator inhibitor-1 (PAI-1), with a slight excess in favor of the protease (Pepper et al., 1990). In these assays altering the balance to favor protease inhibitor activity slightly, by using synthetic serine protease inhibitors, changed the morphology of the endothelial cell cultures to favor solid cords of endothelial cells rather than tubes. Similar findings have been observed using endothelioma cell lines. In tumor cell experiments using either Bowes' melanoma cells, producing excess tissue type plasminogen activator (tPA) or HT1080 fibrosarcoma cells, producing excess uPA, it has been demonstrated that addition of low concentrations of serine-type protease inhibitors or anti-plasmin antibodies can enhance cell invasion (Tsuboi and Rifkin, 1990). This is subsequently blocked by higher concentrations of serine protease inhibitors or antibodies.

Tissue inhibitors of metalloproteinases, or TIMPs, are endogenous inhibitors of matrix metalloproteinases (Stetler-Stevenson et al., 1993). Three human TIMPs have been identified and all have core proteins of about 21 kDa. TIMP-1 and TIMP-3 are glycosylated and run as proteins of 28 and 24 kDa, respectively, on gel electrophoresis (Hewitt et al., 1996). All the TIMPs form high affinity, non-covalent complexes with activated MMPs. In addition, specific complexes are formed between progelatinase A and TIMP-2, as well as progelatinase B and TIMP-1. The biological function of these proenzyme-inhibitor complexes remains elusive.

TIMPs have demonstrated a functional role for MMPs in both tumor cell and endothelial cell invasion during angiogenesis (Hewitt et al., 1996; Stetler-Stevenson et al., 1993). Low TIMP expression correlated with enhanced invasive and metastatic properties in murine and human tumor cell lines. TIMP-1 and TIMP-2 have been shown to inhibit metastasis during *in vivo* animal models of tumor progression (DeClerk et al., 1992; Khokha, 1994; Khokha et al., 1992). TIMP-2 expression in *ras*-transfected rat embryo fibroblasts resulted in loss of lung colonizing ability following intravenous injection of the cells, but did not completely block the formation of pulmonary metastases form primary tumors following subcutaneous inoculation (DeClerk et al., 1992). TIMP-2 overexpression in human melanoma cells inhibited the growth of primary tumors following subcutaneous inoculation into *scid* mice or in three dimensional collagen gels, but did not prevent metastasis formation (Montgomery et al., 1994a).

Genetic alteration of the MMP-2 to TIMP-2 ratio in human melanoma cells was shown to alter cell attachment, morphology and spreading (Ray and Stetler-Stevenson, 1995). Excess inhibitor activity resulted in large, flattened cells that were more adherent to a variety of substrates including fibronectin and vitronectin. Downregulation of TIMP-2 relative to MMP-2 resulted in elongated spindle-shaped cells that were less adherent than the parental cells. These findings suggest that protease activity and cell attachment are also linked, in that excess protease activity can result in disruption of cell adhesion. Studies by Schnaper and colleagues have shown that MMP-2 and TIMP-2 balance is also critical to the vascular morphogenesis during endothelial cell invasion (Schnaper et al., 1993). Addition of exogenous gelatinase A to endothelial cells grown on matrigel could either promote or inhibit tube formation depending on the levels of TIMP-2. At low concentrations, gelatinase A increased tube formation, but at higher concentrations it was inhibitory. This inhibitory effect could be reversed by addition of TIMP-2 at low concentration. Again, these results emphasize that there is a critical balance between matrix attachment and matrix proteolysis that is required for cell migration and successful completion of the invasive phenotype.

Conclusion

We have attempted to demonstrate the functional similarities between the process of tumor cell invasion and angiogenesis. Both of these complex biological processes involve multiple interactions of the invasive cell with the extracellular matrix. Despite the very different nature of these two processes, malignant versus non-transformed host cell response, they share similar molecular effectors of cellular invasion. These are the $\alpha_V\beta_3$ integrin receptor and gelatinase A (MMR-2). These two effectors appear essential to the invasive phenotype of endothelial cells during angiogenesis and many human cancer cells as well. This raises an interesting question regarding the relationship between this integrin receptor and the protease. As pointed out above, binding of antibodies to $\alpha_V\beta_3$ can induce gelatinase A expression, suggesting a possible link through integrin control of protease expression. However, it should be pointed out that not all cells that express $\alpha_V\beta_3$ necessarily express gelatinase A. Understanding the relationship between $\alpha_V\beta_3$ expression and the role of gelatinase A in the invasive phenotype should bring greater understanding of the coordination of cellular events required for successful completion of these complex biological processes. Hopefully, this understanding will provide new therapeutic targets for disrupting disease progression.

Acknowledgements
The authors thank Dr Lance A. Liotta for his continued support and encouragement.

References

Albelda, S.M., Mette, S.A., Elder, D.E., Stewart, R., Damhanovich, L. and Herlyn, M. (1990) Integrin distribution in malignant melanoma: Association of the β_3 subunit with tumor progression. *Cancer Res.* 50:6757–6764.

Brooks, P.C., Clark, R.A., Cheresh, D.A. (1994a) Requirement of vascular integrin $\alpha_V\beta_3$ for angiogenesis. *Science* 264:569–571.

Brooks, P.C., Montgomery, A.M., Rosenfeld, M., Reisfeld, R.A., Hu, T., Klier, G. and Cheresh, D.A. (1994b) Integrin $\alpha_V\beta_3$ antagonists promote tumor regression by inducing apoptosis of angiogenic blood vessels. *Cell* 79:1157–1164.

Brooks, P.C., Stromblad, S., Klemke, R., Visscher, D., Sarkar, F.H., Cheresh, D.A. (1995) Anti-integrin $\alpha_V\beta_3$ blocks human breast cancer growth and angiogenesis in human skin. *J. Clin. Invest.* 96:1815–1822.

Davis, G.E. (1992) Affinity of integrins for damaged extracellular matrix: $\alpha_V\beta_3$ binds to denatured collagen type I through RGD sites. *Biophys. Biochem. Res. Comm.* 182:1025–1031.

DeClerk, A.A., Perez, N., Shimada, H., Bone, T.C., Langley, K.E. and Taylor, S.M. (1992) Inhibition of invasion and metastasis in cells transfected with an inhibitor of metalloproteinase. *Cancer Res.* 52:701–708.

Felding-Habermann, B. and Cheresh, D.A. (1993) Vitronectin and its receptors. *Curr. Opin. Cell Biol.* 5:864–868.

Filardo, E.J., Brooks, P.C., Deming, S.L., Damsky, C. and Cheresh, D.A. (1995) Requirement of the NPXY motif in the integrin β_3 subunit cytoplasmic tail for melanoma cell migration in vitro and in vivo. *J. Cell Biol.* 130:441–450.

Friedlander, M., Brooks, P.C., Shaffer, R.W., Kincaid, C.M., Varner, J.A. and Cheresh, D.A. (1995) Definition of two angiogeneic patways by distinct α_v integrins. *Science* 270: 1500–1502.

Furcht, L.T., Skubitz, A.P.N. and Fields, G.B. (1994) Tumor cell invasion, matrix metallo-proteinases, and the dogma. *Lab. Invest.* 70:781–783.

Gehlsen, K.R., Davis, G.E. and Sriramarao (1992) Integrin expression in human melanoma cells with differing invasive and metastatic properties. *Clin. Exp. Metastasis* 10:111–120.

Gladson, C.L. and Cheresh, D.A. (1991) Glioblastoma expression of vitronectin and the α_v/β_3 integrin. *J. Clin. Invest.* 88:1924–1932.

Hewitt, R.E., Corcoran, M.L. and Stetler-Stevenson, W.G. (1996) The activation, expression and function of gelatinase A (MMP-2). *Trends in Glycoscience Glycotech.* 8:23–36.

Ito, K.-I., Ryuto, M., Ushiro, S., Ono, M., Sugenoya, A., Kuraoka, A., Shibata, Y. and Kuwano, M. (1995) Expression of tissue type plasminogen activator and its inhibitor couples with the development of capillary network by human microvascular endothelial cells on matrigel. *J. Cell Physiol.* 162:213–224.

Khokha, R. (1994) Suppression of the tumorogenic and metastatic abilities of murine B16-F10 melanoma cells. *J. Natl. Cancer Inst.* 86:299–304.

Khokha, R., Zimmer, J., Graham, C.H., Lala, P.K. and Waterhouse, P. (1992) Suppression of invasion by inducible expression of tissue inhibitor of metalloproteinase (TIMP-1) in B16-F10 melanoma cells. *J. Natl. Cancer Inst.* 84:1017–1022.

Klemke, R.L., Yebra, Bayna, E.M. and Cheresh, D.A. (1994) Receptor tyrosine kinase signaling required for integrin $\alpha_v\beta_5$-directed cell motility but not adhesion on vitronectin. *J. Cell Biol.* 127:859–866.

Liotta, L.A. (1986) Tumor invasion and metastases-role of the extracellular matrix: Rhoads Memorial Award Lecture. *Cancer Res.* 46:1–7.

Mignatti, P. and Rifkin, D.B. (1993) Biology and biochemistry of proteinases in tumor invasion. *Physiol. Rev.* 73:161–195.

Montgomery, A.M., Mueller, B.M., Reisfeld, R.A., Taylor, S.M. and DeClerk, Y.A. (1994a) Effect of tissue inhibitor of metalloproteinase-2 on the growth and spontaneous metastasis of human melanoma cell line. *Cancer Res.* 54:5467–5473.

Montgomery, A.M., Reisfeld, R.A. and Cheresh, D.A. (1994b) Integrin $\alpha_v\beta_3$ rescues melanoma cells from apoptosis in three-dimensional dermal collagen. *Proc. Natl. Acad. Sci. USA* 91: 8856–8860.

Pepper, M.S., Belin, D., Montesano, R., Orci, L. and Vassali, J.-D. (1990) Transforming growth factor-β_1 modulated basic fibroblast growth factor-induced proteolytic and angiogenic pro-perties of endothelial cells *in vitro. J. Cell Biol.* 111:743–755.

Ray, J.M. and Stetler-Stevenson, W.G. (1995) Gelatinase A activity directly modulates melanoma cell adhesion and spreading. *EMBO J.* 14:908–917.

Schnaper, H.W., Grant, D.S., Stetler-Stevenson, G., Fridman, R., D'Orazi, G., Murphy, A.N., Bird, R.E., Hoythya, M., Fuerst, T.R., French, D.L., Quigley, J.P. and Kleinman, H.K. (1993) Type IV collagenase(s) and TIMPs modulate endothelial cell morphogenesis *in vitro. J. Cell Physiol.* 156:235–246.

Seftor, R.E., Seftor, E.A., Gehlsen, K.R., Stetler-Stevenson, W.G., Brown, P.D., Ruoslahti, E. and Hendrix, M.J. (1992) Role of the $\alpha_v\beta_3$ integrin in human melanoma cell invasion. *Proc. Natl. Acad. Sci. USA* 89:1557–1561.

Stetler-Stevenson, W.G., Aznavoorian, S. and Liotta, A. (1993) Tumor cell interactions with the extracellular matrix during invasion and metastasis. *Annu. Rev. Cell. Biol.* 9:541–573.

Tsuboi, R. and Rifkin, D.B. (1990) Bimodal relationship between invasion of the amniotic membrane and plasminogen activator activity. *Int. J. Cancer* 46:56–60.

Regulation of Angiogenesis
ed. by I.D. Goldberg & E.M. Rosen
© 1997 Birkhäuser Verlag Basel/Switzerland

Control of angiogenesis by the pericyte: Molecular mechanisms and significance

K. K. Hirschi[1] and P. A. D'Amore[2]

[1,2] *Laboratory for Surgical Research, Children's Hospital, and Departments of Surgery and*
[2] *Pathology, Harvard Medical School, Boston, Massachusetts 02115, USA*

Summary. The microvasculature consists of endothelial cells (EC) with albuminally located pericytes. A number of clinical and experimental observations suggest that pericytes contribute to the regulation of microvascular growth and function. EC and pericytes appear to have a variety of means whereby they may influence one another, including soluble growth factors, gap junctions and adhesion molecules, to name a few. Co-culture systems have provided a good deal of evidence to support the concept that these two cells interact and that these communications are central to vessel assembly, growth control and normal function.

Vascular development

Blood vessel formation occurs by one of two processes – vasculogenesis or angiogenesis. Vasculogenesis occurs early in embryogenesis and refers to the development of an initial vasculature framework, derived entirely from mesoderm (Coffin and Poole, 1988; Pardanaud et al., 1989). Vasculogenesis begins with clustering of primitive vascular cells or hemangioblasts into blood islands (Doetschman et al., 1987). These blood islands consist of endothelium and primitive blood cells and give rise to tube-like structures, which define the pattern of the vasculature (Noden, 1989). Angiogenesis is the subsequent branching and remodeling of such structures, leading to the formation of the vascular network.

Although the process of vasculogenesis is confined to embryonic development, normal angiogenesis continues to occur, in a limited and tightly-controlled manner, throughout life (Folkman, 1992; Hudlicka et al., 1992; Folkman, 1995). Angiogenesis is necessary not only for post-embryonic tissue growth and development, but is critical for the normal healing of wounds and fractures. Furthermore, in the female reproductive system, angiogenesis is required for follicle development, as well as the maintenance of the corpus luteum during ovulation and the placenta during pregnancy. Uncontrolled angiogenesis, in contrast, is known to contribute to and exacerbate such pathological conditions as solid tumor growth, diabetic retinopathy and atherosclerosis.

Vessel assembly

Vessel assembly, whether occurring via vasculogenesis or angiogenesis, is likely to involve similar processes and cell-cell interactions. Embryonic data suggest that, in either case, endothelial tube structures form first and play a role in the subsequent recruitment of the surrounding vessel layers, composed of mural cells, pericytes in small vessels and smooth muscle cells (SMC) in large vessels. Endothelial tubes are thought to invade organ primordia, thereby becoming surrounded by locally-derived mesenchymal cells (i.e. mural cell-precursors, adventitial fibroblasts) which form the mural cell layers (Nakamura, 1988). Hence, the primordia itself contributes the pericyte or SMC layer(s) to the developing vessels, and may allow for tissue-specific functional and regulatory properties of vascular mural cells. Information is sparse regarding the mechanisms of endothelial tube formation and their branching during angiogenesis and the mechanism(s) by which endothelial cells (EC) recruit mural cell-precursors during vessel formation, and subsequently induce their differentiation. Similarly, we know little about the role that mural cells play in controlling or directing vessel formation.

With regard to the recruitment of mural cells to forming vessels, we and others (Zerwes and Risau, 1987; Holmgren et al., 1991; Hirschi and D'Amore, in press) have hypothesized that EC direct the recruitment of mural cell precursors via the synthesis and secretion of a diffusible soluble factor, platelet derived growth factor (PDGF), which would function in an paracrine fashion to promote mural cell proliferation and migration. PDGF is a dimeric molecule, consisting of homo- and hetero-dimers of two subunits, PDGF-A and PDGF-B (PDGF-AA, PDGF-BB and PDGF-AB). The PDGF receptor consists of two subunits, alpha and beta. The response of any particular cell type to PDGF is dependent on its complement of receptors, as the alpha subunit binds both the A and B chains, whereas the beta subunit binds only the B chain (Williams, 1989). PDGF is a mitogen and chemoattractant for a variety of mesenchymal cells and is thus well suited to this role.

The spatial distribution of active PDGF ligand and receptor genes in human placental development provides support for a role of PDGF in vessel formation (Holmgren et al., 1991). Holmgren and coworkers found that the EC of developing vessels exhibit high levels of PDGF-B expression, but have no detectable PDGF-β receptor mRNA. Interestingly, the PDGF-β receptor is expressed at high levels in SMC and fibroblast-like cells in the surrounding mesenchyme. Although other soluble factors may play a role in the EC recruitment of mural cells, supportive evidence is limited.

We believe that subsequent differentiation of the mural cell precursors into pericytes and SMC is a contact-dependent process involving the activation of another soluble factor, transforming growth factor-beta (TGF-β) (Hirschi, Rohovsky and D'Amore, unpublished observation). We and others have shown that both EC and mural cells, grown separately, produce a latent form of TGF-β which is activated in a plasmin-mediated process upon EC-mural cell contact (Antonelli-Orlidge et al., 1989; Sato and Rifkin, 1989). This phenomenon will be discussed later in the context of *in vitro* evidence of pericyte control of angiogenesis. Activated TGF-β_1 is known to induce α-smooth muscle actin expression in myofibroblasts (Desmouliere et al., 1993) and pericytes (Verbeek et al., 1994). Furthermore, TGF-β is thought to induce other changes in both cell types which may contribute to the formation of a quiescent vessel and maintenance of growth control.

Pericyte control of angiogenesis

To date there is little direct evidence to support a role for pericytes or SMC in physically *directing* angiogenesis. One report suggests that pericytes participate in angiogenesis from the earliest stages of capillary sprouting (Nehls et al., 1992). Histological examination of rat mesentery revealed pericytes (identified by immunostaining for desmin) "at and in front of the advancing tips of endothelial sprouts." Furthermore, pericytes were found to "bridge the gap between the leading edges of opposing endothelial sprouts". These observations provide circumstantial evidence that pericytes participate in and/or direct microvessel outgrowth. Recent studies provide convincing evidence of a role of astrocytes in retinal vascularization (Stone et al., 1995). However, since astrocytes are confined to the brain and retina, some other cell type must serve an equivalent function in other tissues.

Evidence for pericyte control of angiogenesis comes from both *in vivo* observations as well as from *in vitro* studies. One piece of evidence relates to the location and number of pericytes in various tissues. Pericyte number has been shown to vary in different microvascular beds, ranging from a ratio of one pericyte per endothelial cell in brain and retina (Speiser et al., 1968) to as few as one pericyte per hundred capillary endothelial cells in skeletal muscle (Sims et al., 1994). Interestingly, there is some correlation between pericyte distribution and the turnover of microvascular cells in various tissues. Brain and retina have the lowest labeling index of all tissues examined (Engerman et al., 1967; Denekamp, 1982). In fact, Engerman and his coworkers calculate, based on labeling index of vascular cells in flat mounts of rodent retinas, that the turnover time of retinal microvascular cells in the range of three years.

Observations of neovascular processes in human retina are consistent with these findings. It has been well documented that, prior to the neovascularization that characterizes proliferative diabetic retinopathy, there is a selective loss of pericytes associated with the retinal capillaries (Speiser et al., 1968). Though it seems unlikely that these growth control mechanisms are unique to the retina, there are not data to directly support their function in other peripheral tissues. Although there is documented pericyte degeneration in other tissues of diabetics (Tilton et al., 1987), there is no evidence to indicate that neovascularization also occurs in these tissues. Ease in visualizing the retina, as well as the dramatic clinical consequences that result from the neovascular process, have focused attention on the retina. It is unclear whether one could expect clinically relevant pathology to ensue from local vaso-occlusive events in tissues other than the retina. Tedious quantitative and morphometric studies of other tissues would be necessary to document an increase in vascularity in other tissues.

The correlation between pericyte absence and vasoproliferation also holds true in other retinal neovascular diseases. Retinopathy of prematurity (ROP) is a pathologic process of premature newborns characterized by neovascularization in the retina. In this case, though there is not a loss of pericytes, as many of the growing vessels of the incompletely vascularized retina have yet to be invested by mural cells or pericytes. As a result, the neovascularization that occurs does so from an immature or incompletely formed vascular bed (Patz, 1982). These observations led to the suggestion that the absence of pericytes is permissive for the angiogenic process (Kuwabara and Cogan, 1963). Similar conclusions were drawn from examination of developing vessels during the wound healing process (Crocker et al., 1970). In an ultrastructural study the investigators noted that the investment of growing capillaries with pericytes was coincident with the deposition of basement membrane and the cessation of vessel growth. These observations led the investigators to postulate that the pericyte has an inhibitory or suppressive effect on vessel growth.

Other support for a role of pericyte-endothelial interactions in the maintenance of normal vascular functions comes from the pathology associated with diabetic retinopathy. In addition to the selective degradation of pericytes in the microvasculature of the retina, there is also a well-documented increase in the thickness of the microvascular basement membrane (de Venecia et al., 1976). The basis for this complication is not understood nor has its biological significance been demonstrated. There are suggestions that the thickened basement membrane might alter cell function and/or oxygen diffusion, thus contributing to the impaired wound healing that is often seen in diabetics. One recently described consequence of the thickening of the basement membrane is a reduction in the number of contacts between endothelial cells and pericytes. Using a galactosemic rabbit model, Robison and his coworkers demonstrated a 70 percent reduction in the number of endothelial cell-pericyte contacts in the retinal microvascular bed (Robison et al., 1989). Interestingly, the thickened basement membrane, and therefore the reduced contacts, were alleviated when animals were fed an aldose reductase inhibitor, suggesting that glucose metabolism into its impermeable alcohol derivatives, such as sorbitol, might contribute to the process. Without clear evidence about the role of the endothelial cell-pericyte contacts, the biological relevance of their reduction secondary to basement membrane thickening is not clear.

However, considering the high incidence of endothelial cell-mural cell contacts throughout the vascular systems and results from *in vitro* studies (see below), it seems reasonable to suspect that at least some of the vascular dysfunction observed in diabetic vascular disease might be the result of a disruption of the normal intercellular communication.

Endothelial cell-pericyte interactions *in vitro*

Though clinical observations suggest a role for pericytes in the control of angiogenesis, no *in vivo* experimental systems have been developed to test this hypothesis directly. The best data to date supporting a role for pericytes in regulating microvascular growth and function have come from *in vitro* studies. Co-cultures of EC and pericytes were shown to suppress endothelial proliferation (Orlidge and D'Amore, 1987) and migration (Sato and Rifkin, 1989). The mechanism for inhibition in co-cultures was shown to be the activation of transforming growth factor beta-1 (TGF-β_1). Cells in solo culture produce TGF-β in a latent form, that is, the mature protein is associated with a latency peptide. Contact between EC and pericytes leads to the activation of TGF-β (Antonelli-Orlidge et al., 1889) via a mechanism that involves the action of plasmin (Sato et al., 1990). Why the activation requires contacts between these two different cell types is not entirely clear. One report suggests that the latency peptide of TGF-β binds to the surface of SMC via the mannose-6-phosphate receptor (Sato et al., 1993). The sequestration of the latent TGF-β would then provide an efficient "solid state cell surface reaction".

Endothelial-pericyte interaction and vascular function

Though the endpoints of proliferation and migration have been the primary focus of co-culture studies to date, we speculate that a variety of other functions are moderated by pericyte-endothelial interactions. TGF-β influences a range of cell functions including differentiation and extracellular matrix production. Thus, it is reasonable to suspect that other microvascular behaviors will be influenced by the TGF-β which is activated as a result of the cell-cell interaction in the microvasculature. For example, preliminary data from our laboratory suggest that EC-mural cell interactions lead to the alteration of certain extracellular matrix components, including the differential expression of specific fibronectin splice variants (Beck, L. and D'Amore, P.A., unpublished results). Further we have observed that in EC-mural cell co-culture (SMC in this case) the expression of particular cell surface adhesion molecules, including E-selectin, are altered (Dodge et al., manuscript in preparation). Interestingly, the alteration of E-selectin expression in the co-cultures is not mediated by TGF-β.

Modes of EC-pericyte communication

The fact that we have observed changes in the co-cultures that are not due to TGF-β emphasizes the fact that endothelial cells and pericytes (or smooth muscle cells) communicate via a variety of means, including the synthesis and release of soluble mediators, synthesis of extracellular matrix molecules that in turn can signal to other cells, the formation of intercellular gap junction channels, and heterotypic binding via cell surface adhesion molecules (Figure 1).

Gap junctions

Though the nature and relevance of the gap junction communication in the vasculature is yet to be entirely elucidated, observations from other systems indicate that the passage of small molecules via gap junctions may be important in regulating both cell growth and differentiation (reviewed in Guthrie and Gilula, 1989). Although gap junctions between EC and mural cells have been detected *in vitro* (Larson et al., 1987; Sweet et al., 1988), data regarding their functionality *in vivo* are mixed. Segal and Bény (1992) injected Lucifer Yellow into EC and SMC of hamster cheek pouch arterioles and showed dye transfer efficiently among EC, but not among SMC. Furthermore, there was no heterologous transfer between EC and SMC. Little et al. (1995), however, using the same model system showed weak gap junction communication among SMC and polar dye transfer from EC to SMC but not in the reverse direction. The significance of selective or compartmentalized gap junction communication is not well understood but may reflect differences in gap junction protein composition in EC and SMC.

Gap junctions can be composed of any of a dozen different connexin (Cx) proteins and some controversy exists as to the exact protein composition of gap junction channels of EC and SMC. EC have been shown to express Cx40 (Bruzzone et al., 1993; Traub et al., 1994; Little et al., 1995) and Cx37 (Reed et al., 1993) *in vivo*. Cx43 (Little et al., 1995) was also detected in EC in one study of arteriole gap junctions. SMC are thought to express Cx43 (Bruzzone et al., 1993), although one study (Little et al., 1995) demonstrates Cx40 expression as well. Determining the exact Cx composition of vascular cells of various vessels would lead to the understanding of selective communication. It is known that each connexin protein can form functional channels with like proteins (i.e. Cx43 = Cx41) and that Cx43 = Cx37 heterochannels are functional as well, but Cx43 = Cx40 heterochannels are not. Therefore, if EC express Cx40 and Cx37, while SMC express only Cx43, heterotypic channels between Cx43 and Cx37 would be their only means of intercellular communication. This type of heterochannel communication between EC and SMC may have differing functional characteristics, and thus may serve a different purpose from heterotypic channel communication among EC and among SMC.

Soluble factors

In the intact vessel, EC and mural cells are likely to continue to influence each other's behavior and growth state via the bi-directional synthesis and secretion of regulatory factors. Though the levels of these factors are likely to change as a function of the cell growth and differentiation states, limited studies have compared growth factor levels in proliferating *(in vitro)* and quiescent *(in vivo)* cells. Whereas growing EC synthesize PDGF, quiescent EC express low levels of this factor (Barrett et al., 1984; Holmgren et al., 1991; Liaw and Schwartz, 1993). Similarly, proliferating EC express significant levels of mRNA for thrombospondin, a molecule that inhibits EC proliferation (Bagavandoss and Wilks, 1990) and migration (Dameron et al., 1994) *in vitro* and angiogenesis *in vivo* (Good et al., 1990), whereas

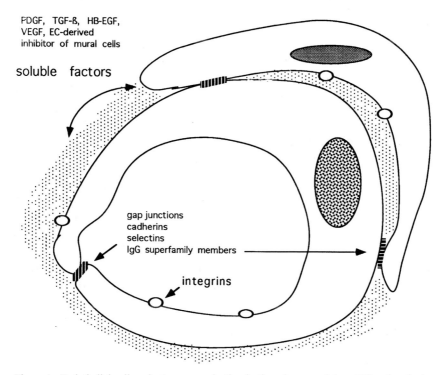

Figure 1. Endothelial cell-pericyte communication in the microvasculature. EC and pericytes have a variety of mechanisms that they may use to communicate. These include the local release of growth factors that act in a paracrine fashion to influence the other cell type. In addition, the two cell types interact via gap junctions and possibly by intercellular signalling via cell surface adhesion molecules.

freshly isolated cells have negligible levels. We and others have shown that conditioned media from confluent (but not proliferating) EC in culture inhibit the proliferation of SMC and pericytes (Willems et al., 1982; Dodge et al., 1992; Fillinger et al., 1993). Thus, EC and pericytes appear to communicate via the release of growth factors that act in a paracrine fashion to influence vascular cell growth and behavior.

Cell surface adhesion molecules

Similarly, signaling is known to occur via other cell surface adhesion molecules. The selectins, members of the IgG superfamily, and the integrins are all possible candidates for modes of endothelial cell-pericyte communication. We have documented specific binding between the basal surface of endothelial cells and a monolayer culture of pericytes and are currently in the process of elucidating the molecules involved in this specific adhesion interaction.

References

Antonelli-Orlidge, A., Saunders, K.B., Smith, S.R. and D'Amore, P.A. (1989) An activated form of TGF-β is produced by cocultures of endothelial cells and pericytes. *Proc. Natl. Acad. Sci. USA* 86:4544–4548.

Bagavandoss, P. and Wilks, J.W. (1990) Specific inhibition of endothelial cell proliferation by thrombospondin. *Biochem. Biophys. Res. Comm.* 170:867–872.

Barrett, T.B., Gajdusek, C.M., Schwartz, S.M., McDougall, J.K. and Benditt, E.P. (1984) Expression of the *sis* gene by endothelial cells in culture *in vivo. Proc. Natl. Acad. Sci. USA* 81:6772–6774.

Bruzzone, R., Haefliger, J.-A., Gimlich, R.L. and Paul, D.L. (1993) Connexin 40, a component or gap junctions in vascular endothelium, is restricted in its ability to interact with other connexins. *Mol. Biol.* 4:7–20.

Coffin, J.D. and Poole, T.J. (1988) Embryonic vascular development: immunohistochemical identification of the origin and subsequent morphogenesis of the major vessel primordia in quail embryo. *Development* 102:1–14.

Crocker, D.J., Murad, T.M. and Greer, J.C. (1970) Role of the pericyte in wound healing. An ultrastructural study. *Exp. Mol. Pathol.* 13:51–65.

Dameron, K.M., Volpert, O.V., Tainsky, M.A. and Bouck, N. (1994) Control of angiogenesis in fibroblasts by p53 regulation of thrombospondin-1. *Science* 265:1582–1584.

de Venecia, G., Davis, M. and Engerman, R. (1976) Clinicopathologic correlations in diabetic retinopathy. *Arch. Ophthalmol.* 94:1766–1778.

Denekamp, J. (1982) Endothelial cell proliferation as a novel approach to targeting tumour therapy. *Br. J. Cancer* 45:136–139.

Desmouliere, A, Genioz, A., Gabbiani, F. and Gabbiani, G. (1993) Transforming growth factor-β_1 induces α-smooth muscle cell actin expression in granulation tissue myofibroblasts and in quiescent and growing cultured fibroblasts. *J. Cell Biol.* 122:103–111.

Dodge, A.B., Gabriels, J.E. and D'Amore, P.A. (1992) Endothelial cells modulate mural cell proliferation and migration *in vitro. J. Cell. Biochem.* 16A:49.

Doetschman, T.A., Gossler, A. and Kemler, R. (1987) Blastocyst-derived embryonic stem cells as a model for embryogenesis. *In:* W. Feichtingen and P. Kemeter (eds): *Future Aspects in Human In Vitro Fertilization.* Springer-Verlag, Berlin, pp 187–195.

Engerman, R.L., Pfaffenbach, D. and Davis, M.D. (1967) Cell turnover of capillaries. *Lab. Invest.* 17:738–743.

Fillinger, M.F., O'Connor, S.E., Wagner, R.J. and Cronenwett, J.L. (1993) The effect of endothelial cell coculture on smooth muscle cell proliferation. *J. Vasc. Surg.* 17: 1058–1068.

Folkman, J. (1992) Angiogenesis – Retrospect and outlook. *In:* R. Steiner, P.B. Weisz and R. Langer (eds): *Angiogenesis: Key Principles.* Birkhäuser Verlag, Basel, Switzerland, pp 4–13.

Folkman, J. (1995) Angiogenesis in cancer, vascular, rheumatoid and other diseases. *Nature Med.* 1:27–31.

Good, D.J., Polverini, P.J., Rastinejad, F., Le Beau, M.M., Lemons, R.S., Frazier, W.A. and Bouck, N.P. (1990) A tumor suppressor dependent inhibitor of angiogenesis is immunologically and functionally indistinguishable from a fragment of thrombospondin. *Proc. Natl. Acad. Sci. USA* 87:6624–6628.

Guthrie, S.C. and Gilula, N.B. (1989) Gap junction communication and development. *Trends in Neurosciences* 12:12–16.

Holmgren, L., Glaser, A., Pfeifer-Ohlsson, S. and Ohlsson, R. (1991) Angiogenesis during human extra-embryonic development involves the spatiotemporal control of PDGF ligand and receptor gene expression. *Development* 113:749–754.

Hudlicka, O., Brown, M. and Egginton, S. (1992) Angiogenesis in skeletal and cardiac muscle. *Physiol. Rev.* 72:369–417.

Kuwabara, T. and Cogan, D.G. (1963) Retinal vascular patterns. VI. Mural cells of the retinal capillaries. *Arch. Ophthalmol.* 69:492–502.

Larson, D.M., Carson, M.P. and Haudenschild, C.C. (1987) Junction transfer of small molecules in cultured bovine brain microvascular endothelial cells and pericytes. *Microvasc. Res.* 34:184–199.

Liaw, L. and Schwartz, S.M. (1993) Comparison of gene expression in bovine aortic endothelium in vivo versus in vitro. *Arteriosclerosis and Thrombosis* 13:985–993.

Little, T.L., Beyer, E.C. and Duling, B.R. (1995) Connexin 43 and connexin 40 gap junction proteins are present in arteriolar smooth muscle and endothelium *in vivo. Am. J. Physiol.* 268:H729–739.

Nakamura, H. (1988) Electron microscopic study of the prenatal development of the thoracic aortia in the rat. *Am. J. Anat.* 181:406–418.

Nehls, V., Denzer, K. and Drenckhahn, D. (1992) Pericyte involvement in capillary sprouting during angiogenesis *in situ. Cell Tissue Res.* 270:469–474.

Noden, D.M. (1989) Embryonic origins and assembly of blood vessels. *Am. Res. Respir. Dis.* 140:1097–1103.

Orlidge, A. and D'Amore, P.A. (1987) Inhibition of capillary endothelial cell growth by pericytes and smooth muscle cells. *J. Cell Biol.* 105:1455–1462.

Pardanaud, L., Yassine, F. and Dieterlen-Lièvre, F. (1989) Relationship between vasculogenesis, angiogenesis and haemopoiesis during avian ontogeny. *Development* 105:473–485.

Patz, A. (1982) Clinical and experimental studies on retinal neovascularization. *Am. J. Ophthalmol.* 94:715–743.

Reed, K.E., Westphale, E.M., Larson, D.M., Wang, H.-Z., Veenstra, R.D. and Beyer, E.C. (1993) Molecular cloning and functional expression of human connexin 37, and endothelial gap junction protein. *J. Clin. Invest.* 91:997–1004.

Robison, W.G., Jr., Magata, M., Tillis, T.N., Laver, N. and Kinoshita, J.H. (1989) Aldose reductase and pericyte-endothelial cells contacts in retina and optic nerve. *Invest. Ophthalmol. Vis. Sci.* 30:2293–2299.

Sato, Y., Okada, F., Abe, M., Seguchi, T., Kuwano, M., Sato, S., Furuya, A., Hanai, N. and Tamaoki, T. (1993) The mechanism for the activation of latent TGF-β during co-culture of endothelial cells and smooth muscle cells. Cell-type specific targeting of latent TGF-β to smooth muscle cells. *J. Cell Biol.* 123:1249–1254.

Sato, Y. and Rifkin, D.B. (1989) Inhibition of endothelial cell movement by pericytes and smooth muscle cells: activation of a latent transforming growth factor-β-1-like molecule by plasmin during co-culture. *J. Cell Biol.* 109:309–315.

Sato, Y., Tsuboi, R., Lyons, R., Moses, H. and Rifkin, D.B. (1990) Characterization of the activation of latent TGF-β by co-cultures of endothelial cells and pericytes or smooth muscle cells: a self-regulating system. *J. Cell Biol.* 111:757–763.

Segal, S.S. and Bény, J.-L. (1992) Intracellular recording and dye transfer in arterioles during blood flow control. *Am. J. Physiol.* 263:H1–7.

Sims, D., Horne, M.M., Creighan, M. and Donald, A. (1994) Heterogeneity of pericyte populations in equine skeletal muscle and dermal microvessels: A quantitative study. *Anat. Histol. Embryol.* 23:232–238.

Speiser, P., Gittelsohn, A.M. and Patz, A. (1968) Studies on diabetic retinopathy. III. Influence of diabetes on intramural pericytes. *Arch. Ophthalmol.* 80:332–337.

Stone, J., Itin, A., Alon, T., Pe'er, J., Gnessin, H., Chan-Ling, T. and Keshet, E. (1995) Development of retinal vasculature is mediated by hypoxia-induced vascular endothelial growth factor (VEGF) expression by neuroglia. *J. Neurosci.* 15:4738–4747.

Sweet, E., Abraham, E.H. and D'Amore, P.A. (1988) Functional evidence of gay junctions between capillary endothelial cells and periutes *in vitro. Invest. Ophthal mol. Vis. Sci.* 29:109a.

Tilton, R.G., Faller, A.M., Hoffman, P.L., Kilo, C. and Williamson, J.R. (1987) Acellular capillaries and increased pericyte degeneration in the diabetic extremity. *Front Diabetes* 8:186–189.

Traub, O., Eckert, R., Lichtenberg-Frate, H., Elgfang, C., Bastide, B., Scheidtmann, K.H., Hulser, D.F. and Willecke, K. (1994) Immunochemical and electrophysiological characterization of murine connexin 40 and -43 in mouse tissues and transfected human cells. *Eur. J. Cell Biol.* 53:101–112.

Verbeek, M.M., Otte-Höller, I., Wesseling, P., Ruiter, D.J. and de Waal, R.M.W. (1994) Induction of α-smooth muscle actin expression in cultured human brain pericytes by transforming growth factor-β. *Am. J. Pathol.* 144:372–382.

Willems, C.H., Astaldi, G.C.B., De Groot, P.G., Janssen, M.C., Gonsalvez, M.D., Zeulemaker, W.P., Van Mourik, J.A. and Van Aken, W.G. (1982) Media conditioned by cultured human vascular endothelial cells inhibit the growth of vascular smooth muscle cells. *Exp. Cell Res.* 139:191–197.

Williams, L.T. (1989) Signal transduction by the platelet-derived growth factor receptor. *Science* 243:1564–1570.

Zerwes, H.-G. and Risau, W. (1987) Polarized secretion of a platelet-derived growth factor-like chemotactic factor by endothelial cells *in vitro. J. Cell Biol.* 105:2037–2041.

Subject Index

I.D. Goldberg / E.M. Rosen
Long Island Jewish Medical Center, NY, USA

Epithelial-Mesenchymal Interactions in Cancer

1995. 300 pages. Hardcover
ISBN 3-7643-5117-9
(EXS 74)

The contribution of epithelia-mesenchyme interaction to normal development (eg., tissue formation) and to neoplasia has become a subject of increasing interest to research scientists because of recent progress in deciphering the molecular signals that mediate this interaction.

Clearly, some of the same types of molecules (eg., growth factors and their receptors, proteolytic enzymes, cell adhesion molecules, and structural proteins of the extracellular matrix) mediate exchange of information between epithelia and mesenchyme during normal development and malignant growth. However, defects in the regulation of this exchange appear to contribute to malignancy by allowing growth-promoting, invasogenic, and angiogenic factors to accumulate within the microenvironment of the tumor. For example, recent studies

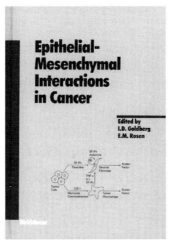

suggest that abnormal interactions between tumor epithelial cells and stromal mesenchymal cells contribute to the overproduction and accumulation of scatter factor hepatocyte growth factor), an invasogenic and angiogenic cytokine, in certain types of tumor. The production and and activation of type IV collagenase, a matrix-degrading enzyme required for tumor cell invasion, appears to require intimate cooperation between tumor and stromal cells.

The material contained in this volume highlights the state-of-the-art of knowledge of the molecular mechanisms by which epithelia and mesenchyme collaborate, and the abnormalities in these mechanisms that may lead to the development of cancer.

Birkhäuser Verlag • Basel • Boston • Berlin

U. Feige, Amgen Inc., Thousand Oaks, CA, USA / **R.I. Morimoto,** Northwestern University, Evanston, IL, USA / **I. Yahara,** Tokyo Metropolitan Institute for Medical Science, Cell Biology, Tokyo, Japan / **B.S. Polla,** UFR Cochin Port Royal, Paris, France (Eds)

Stress-Inducible
Cellular Responses

1996. 512 pages. Hardcover
ISBN 3-7643-5205-1
(EXS 77)

This book will deal with heat shock proteins and more generally with stress-related inducible gene expression as a pleiotropic adaptive response to stress. It presents a textbook-like overview of the field not only to heat shock experts, but to physiologists, pharmacologists, physicians, neuropsychologists and others as well. It is intended to be a state-of-the-art and perspective book rather than an up-to-date presentation of recent data. It should provide a basis for new experimetal approaches to fields at the edge of the classical heat shock field.

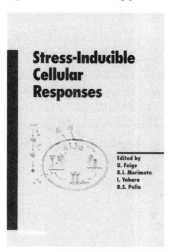

Drugs, UV irradiation and environmental toxics will be considered as important modulators of the stress response. Radical scavengers such as superoxide dismutases and inducible regulatory proteins of metallic ion status such as ferritin as well as immunophilins and protein disulfide isomerases will be considered within the frame of stress proteins.

The potential practical applications of heat shock proteins in toxicology and medicine for the diagnosis, prognosis and eventually therapy of clinical conditions associated with an increased oxidative burden will be outlined. The role of heat shock proteins in the modulation of immune responses will also be included.

The book considers heat shock from a broad perspective including fields for which heat-shock may become of importance in the very near future such as cellular responses to environmental stresses and complex stress responses under specific conditions. It was also felt timely to incorporate a whole section on medical and technological applications of stress proteins.

The book will be invaluable for all those working on stress and is intended for every „stress laboratory" as a source of knowledge and perspectives.

Birkhäuser Verlag • Basel • Boston • Berlin

M. Karmazyn, Univ. of Western Ontario, London Ont., Canada (Ed.)

Myocardial Ischemia: Mechanisms, Reperfusion, Protection

1996. 528 pages. Hardcover
ISBN: 3-7643-5269-8
(EXS 76)

Myocardial ischemia and subsequent reperfusion of the ischemic myocardium represent complex phenomena encompassing numerous physiological processes. This book aims at enhancing our understanding of these processes and stresses recent important developments in this very active area of research.

The concise, state-of-the-art reviews cover recent advances in many fields important to the area of myocardial ischemia and reperfusion including physiology, pathology, pharmacology, biochemistry and molecular biology with reference to clinical relevance and applicability of these findings. Major areas which are highlighted include vascular mechanisms resulting in myocardial ischemia, cellular events in the ischemic, postinfarcted and reperfused myocardium as well as new exciting developments in cardiac protection that involve both novel pharmacological approaches as well as endogenous cardioprotective mechanisms such as preconditioning.

Aimed at both the basic and clinical cardiovascular investigator, the book comprehensively reviews the rapid progress made in recent years in understanding the etiology of myocardial ischemia and reperfusion. It will further serve as an authoritative reference for all those interested in learning of the important developments which have evolved in the treatment of myocardial ischemic and reperfusion disorders.

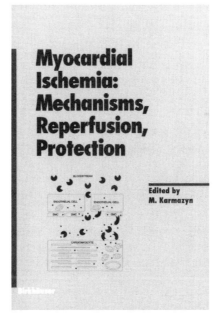

Myocardial Ischemia: Mechanisms, Reperfusion, Protection

Edited by
M. Karmazyn

Birkhäuser Verlag • Basel • Boston • Berlin